INTERNAL COMBUSTION ENGIENS AND GAS TURBINES

DR. SADHU SINGH
Formerly, Professor & Head, Mechanical Engineering Department
&
Dean, Faculty of Engineering & Technology
Govind Ballabh Pant University of Agriculture & Technology,
Pantnagar - 263 145 (Uttarakhand)
Formerly, Director (Colleges), Punjab Technical University, Jalandhar

S.K. KATARIA & SONS®
Publisher of Engineering Books
4885/109, Prakash Mahal, Dr. Subhash Bhargav Lane,
Opposite Delhi Medical Association, Daryaganj, New Delhi–110002 (INDIA)
Phone: +91-11-43551243; *Mobile:* +91-9871775858, 7678425209
e-mail: katariabook@gmail.com, katariabooks@yahoo.com
Website: www.skkatariaandsons.com
(e-books, subscriptions and bulk buying also available on our website)

Published By:

S.K. Kataria & Sons®
4885/109, Prakash Mahal, Dr. Subhash Bhargav Lane,
Opposite Delhi Medical Association, Daryaganj, New Delhi-110002 (INDIA)
Phone: +91-11-43551243; *Mobile:* +91-9871775858, 7678425209
e-mail: katariabook@gmail.com, katariabooks@yahoo.com
Website: www.skkatariaandsons.com

First Edition: 2011; Reprint: 2025

Laser Typesetting by: Stride Computers (Delhi)

Printed at : A.S. Offset, Delhi.

PREFACE

Internal Combustion Engines and Gas Turbines (ME–307F) has been introduced from the session 2011–12 by the Maharishi Dayanand University, Rohtak. It is offered in the 5th Semester to B.Tech./B.E. Mechanical Engineering students. The present book has been written strictly according to the new syllabus.

The book contains twelve chapters on : Introduction to I.C. Engines; Air standard cycles; Carburetion, Fuel Injection and Ignition Systems; Combustion in I.C. Engines; Lubrication and Cooling Systems; Engine Testing and Performance; Air Pollution from I.C. Engines; Reciprocating Air Compressors; Rotary Compressors; Centrifugal Compressors; Axial Compressors; and Gas Turbines. A large number of examples have been solved to illustrate the solution methods. An added feature of the book is that previous questions from MDU and KU examinations have been solved in each chapter. A large number of problems have been given at the end of each chapter. The book also covers the syllabus of Kurukshetra University.

The author is highly thankful to Mr. Sanjeev Kataria and Mr. Sharda Singh for their help and cooperation during the preparation of the manuscript of the book and otherwise.

The author expresses his appreciations for his family members for their tolerance and cooperation. My grand children Kanupreet and Amitoej have been a source of great encouragement during the writing of manuscript and proof reading.

Suggestions for the improvement of the book shall be welcome. The author shall appreciate receiving information from readers regarding any errors ro misprints left inadvertently.

— Dr. SADHU SINGH

CONTENTS

Chapter

1

Introduction to Internal Combustion Engines

Inside This Chapter

1.1 HEAT ENGINES

An engine may be defined as a device which converts one form of energy into mechanical energy. Mechanical energy can be further easily converted into electrical energy which is easier to transport and control. Heat engine is a device which transforms heat energy into mechanical energy.

The heat engines are classified as :

(i) External combustion engines, and

(ii) Internal combustion engines.

In an external combustion engine, the fuel is burned outside the engine and its heat is given to the working fluid of the engine for power generation. At example of external combustion engine is the steam engine in which the fuel is burned in the boiler to generate steam which is used in the engine for power generation. The working fluid is not mixed with fuel, therefore, the same working fluid (water in the form of steam) is used again and again in the system.

In an internal combustion engine, the fuel is mixed with air and burned inside the engine to generate power. In this case, the same working fluid (air-fuel mixture) cannot be used again in the cycle. Some examples of internal combustion engine are : petrol engine, diesel engine, etc.

The internal combustion engines offer special advantages over external combustion engines.

1. Greater mechanical simplicity.
2. Lower weight to output ratio.
3. Higher overall thermal efficiency.
4. Lesser water requirement.
5. Easy and quick starting.

1.2 CLASSIFICATION OF I.C. ENGINES

The I.C. engines can be classified on the basis of cycle of operation in cylinder, type of fuel, method of supply of fuel, type of ignition, etc. as shown in Fig. 1.1.

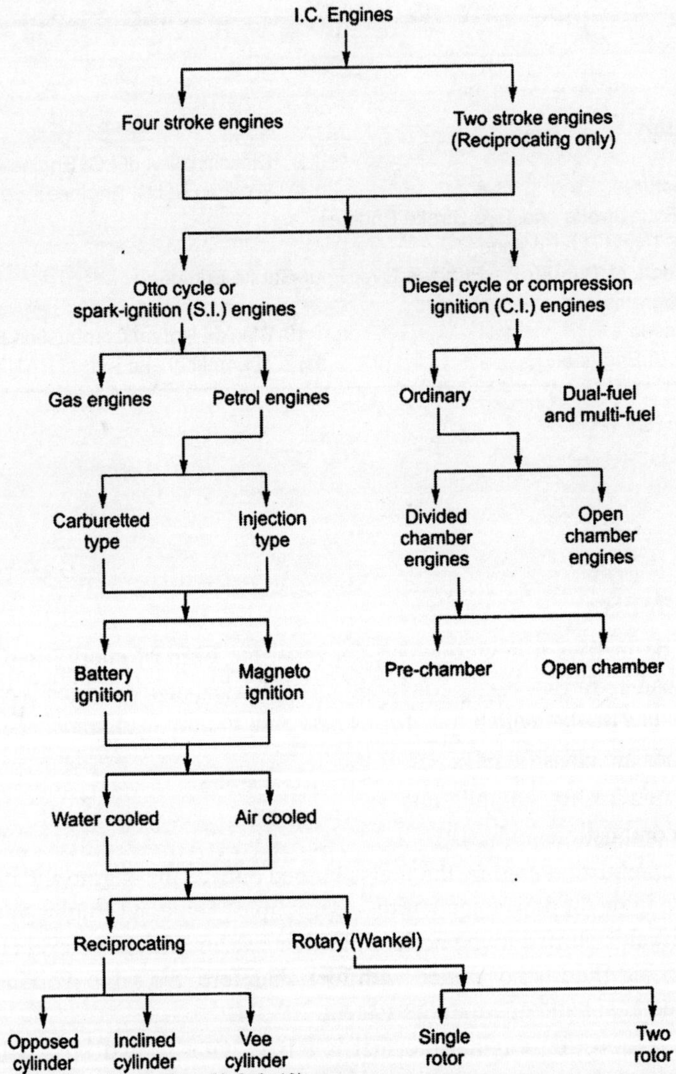

Fig. 1.1 I.C. engine classification.

1. **Basic engine design :** Reciprocating engines, rotary (Wankel) engines.
2. **Working cycle :** Otto cycle and Diesel cycle.
3. **Number of strokes :** Four-stroke and two-stroke engines.
4. **Fuel :** Gasoline (or petrol), compressed natural gas (CNG), liquefied petroleum gas (LPG), diesel oil (light diesel oil, LDO and high speed diesel oil, HSD), fuel oil, alcohols (methanol, ethanol).
5. **Fuel supply and mixture preparation :** Carburetted type and injection type.
6. **Method of ignition :** C.I. (compression-ignition), S.I. (spark-ignition) engines : battery ignition or magneto ignition.
7. **Method of cooling :** Water cooled or air cooled.
8. **Cylinder arrangement (Fig. 1.2) :** Inline, V or Vee, radial, opposed piston and opposed cylinder, delta.

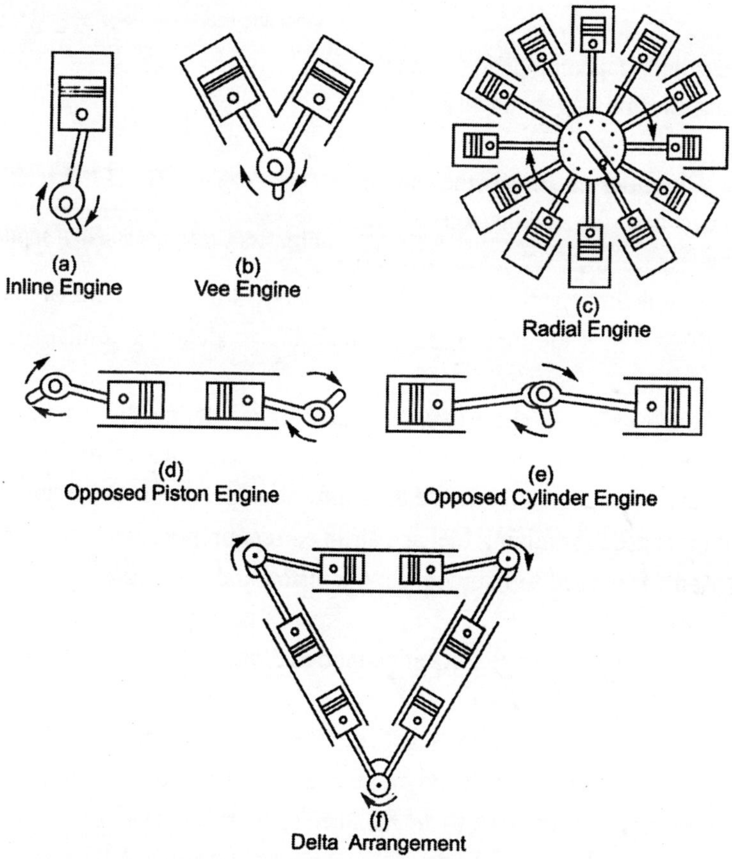

Fig. 1.2 Classification of I.C. engines as per cylinder arrangement.

9. **Valve or port design and location :** Overhead valves (I-head), side valve (L-head); in two-stroke engines : cross scavenging, loop scavenging, uniform scavenging.
10. **Application :** Automotive engines for land transport, marine engines for propulsion of ships, aircraft engines for aircraft propulsion, industrial engines, prime movers for electrical generators.

1.3 CONSTRUCTION FEATURES

(a) Four-Stroke Spark-Ignition Engine

The cross section of a four-stroke spark-ignition engine is shown in Fig. 1.3. The major components of the engine are :

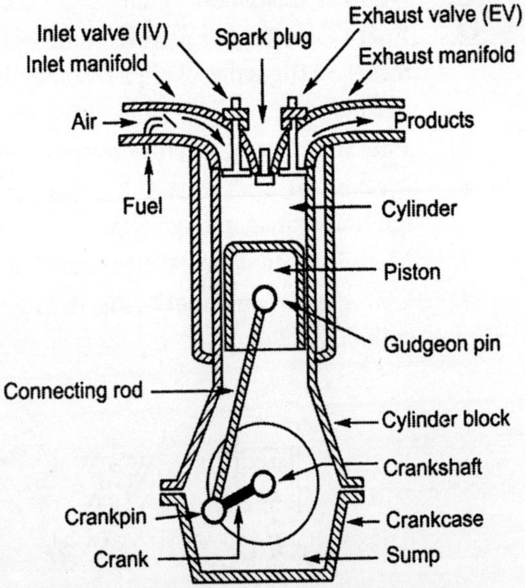

Fig. 1.3 Cross-section of a spark-ignition engine.

Cylinder Block : It is the main supporting structure for the various components. The cylinder head is mounted on the cylinder block. The bottom of the cylinder block is called the crank case. The lubricating oil is kept in the crank case sump.

Cylinder : It is a cylindrical vessel in which the piston reciprocates.

Piston : It is a cylindrical component fitted into the cylinder forming the moving boundary of the combustion system.

Combustion Chamber : It is the space between the cylinder and the piston top where combustion takes place.

Inlet Manifold : it is the pipe through which air or air-fuel mixture is drawn in the cylinder.

Exhaust Manifold : It is the pipe through which the products of combustion escape into the atmosphere.

Inlet and Exhaust Valves : The valves are used for regulating the incoming charge into the cyinder (inlet valve) or discharging the products of combustion (exhaust valve) from the cylinder.

Spark Plug : It is a component to initiate the combustion process in S.I. engine.

Carburettor : It is used for mixing fuel and air in correct proportion.

Connecting Rod : It is used to interconnect the piston and the crank to transmit force from the piston to the crankshaft.

Crankshaft : It is used to covert reciprocating motion of the piston into rotary motion of the output shaft.

Piston Rings : They provide a tight seal between the piston and the cylinder wall.

Gudgeon Pin : It connects the small end of the connecting rod to the piston.

Cam Shaft : It is a shaft on which cams are mounted to operate the valves. It is driven from the cam shaft by gears.

Fly wheel : It absorbs surplus energy during working stroke and delivers during idle stroke.

(b) Four-Stroke Compression-Ignition Engine

Except for the spark plug and carburettor, all other components are the same as for the S.I. engine. It also requires a fuel pump and fuel nozzle. Fig. 1.4 shows outline of a C.I. engine.

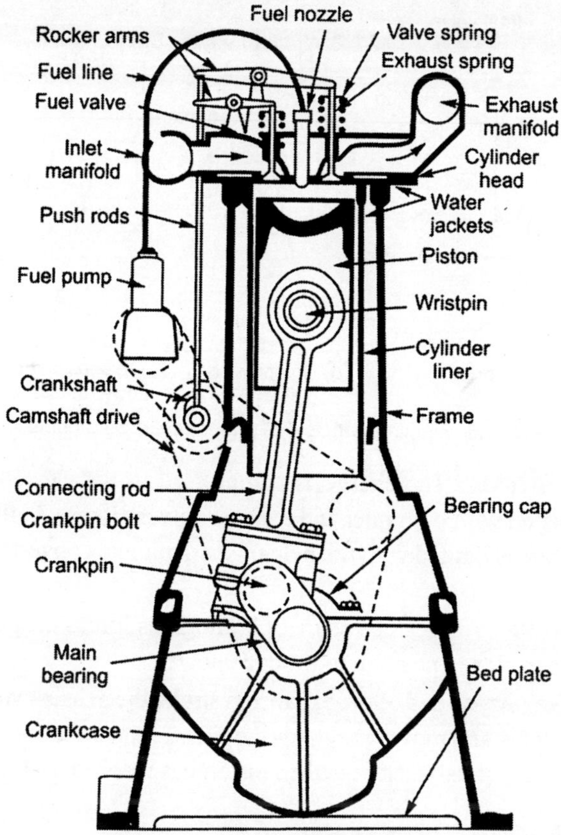

Fig. 1.4 Outline of diesel engine.

(c) Two-Stroke S.I. Engine

In the case of two-stroke engine, the valves are replaced by the ports : exhaust port and transfer port. The charging of the cylinder with air-fuel mixture is through the carburettor and ignition by a spark plug. Other components are same as for a four-stroke engine.

(d) Two-Stroke C.I. Engine

Here the carburettor and spark plug are replaced by a fuel pump and fuel nozzle. All other components are same as for two-stroke S.I. engine.

1.4 WORKING OF I.C. ENGINES

1.4.1 Four-Stroke Spark-Ignition Engine

The details of various process of a four-stroke spart-ignition engine with over head valves are shown in Fig. 1.5. During the four strokes, there are five events to be completed, viz., suction, compression, combustion, expansion and exhaust.

1. **Suction Stroke :** It starts when the piston is at T.D.C. and about to move downwards. The inlet valve is open and exhaust valve closed at this time. The charge consisting of fuel-air-mixture is drawn into the cylinder. When the piston reaches the B.D.C. the suction stroke ends and inlet valve closes.

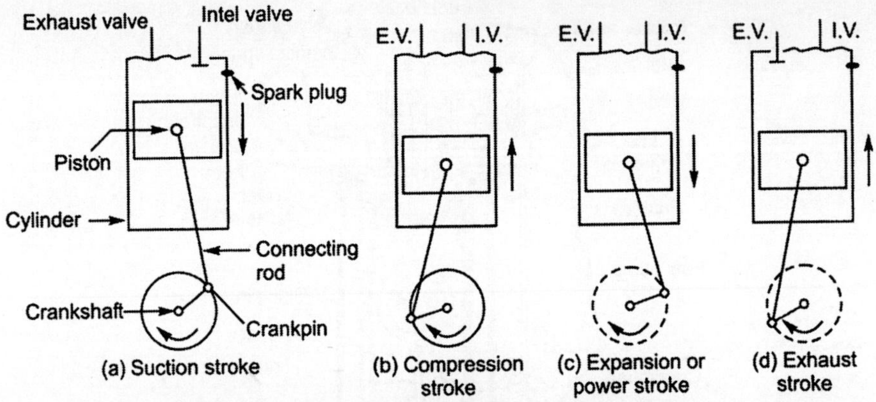

Fig. 1.5 Working principle of a four stroke S.I. engine.

2. **Compression Stroke :** The charge is compressed to the clearance volume by the return stroke of the piston with both inlet and exhaust valves closed. At the end of the compression stroke the mixture is ignited with the help of spark plug to convert chemical energy of fuel to heat energy.
3. **Expansion Stroke :** The high pressure of burnt gases forces the piston towards B.D.C. with both the valves in closed position to produce power.
4. **Exhaust Stroke :** At the end of the expansion stroke the exhaust valve opens with inlet valve closed. The piston starts moving towards the T.D.C. and sweeps the burnt gases out from the cylinder. The exhaust valve closes when piston has reached T.D.C.

1.4.2 Four-Stroke Compression-Ignition Engine

The four cycles of operation of C.I. engine are shown in Fig. 1.6.

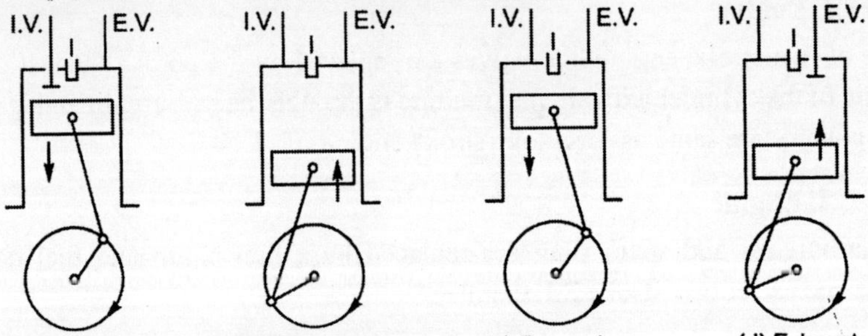

Fig. 1.6 Cycle of operation of a four-stroke C.I. engine.

(a) **Suction Stroke :** Air alone is induced during the suction stroke with inlet valve open and exhaust valve closed.

(b) **Compression Stroke :** The sucked air is compressed into the clearance volume with both valves closed.

(c) **Expansion Stroke :** Fuel injection starts nearly at the end of the compression stroke resulting into combustion. The products of combustion expand and both the valves remain closed.

(d) **Exhaust Stroke :** The piston moves from B.D.C. to T.D.C. and pushes out the products of combustion with the exhaust valve open and intake valve closed.

1.4.3 Two-Stroke Spark-Ignition Engine

In a two-stroke engine, the cycle is completed in one revolution of the crankshaft. The filling process is accomplished by the charge compressed in crankcase or by a blower. The induction of the compressed

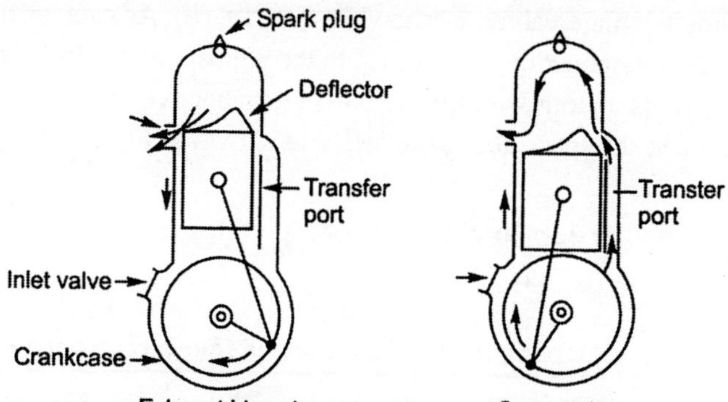

Fig. 1.7 Crankcase scavenged two-stroke S.I. engine.

charge moves out the products of combustion through exhaust ports. Therefore, two piston strokes are required for these two operations. Two strokes are sufficient to complete the cycle, one for compressing the fresh charge and the other for expansion or power stroke. Fig. 1.7 shows the crankcase scavenged two-stroke engine.

1.4.4 Two-Stroke Compression-Ignition Engine

The two-stroke compression-ignition engine is shown in Fig. 1.8. It is a crankcase scavenged type engine. S is the plate valve for admission of air in the crank case, E is the exhaust port and A is the port in the cylinder communicating to the crankcase through the cylinder block casting, Cams and valves are not required. The four operations : air induction, air compression and fuel injection, expansion and exhaust are completed in two strokes, i.e., in one revolution of crankshaft.

In the upward motion of the piston, suction is created in the crankcase and air enters through the plate valves for full 180° of crank rotation as shown in Fig. 1.8 (a). Above the piston the compression

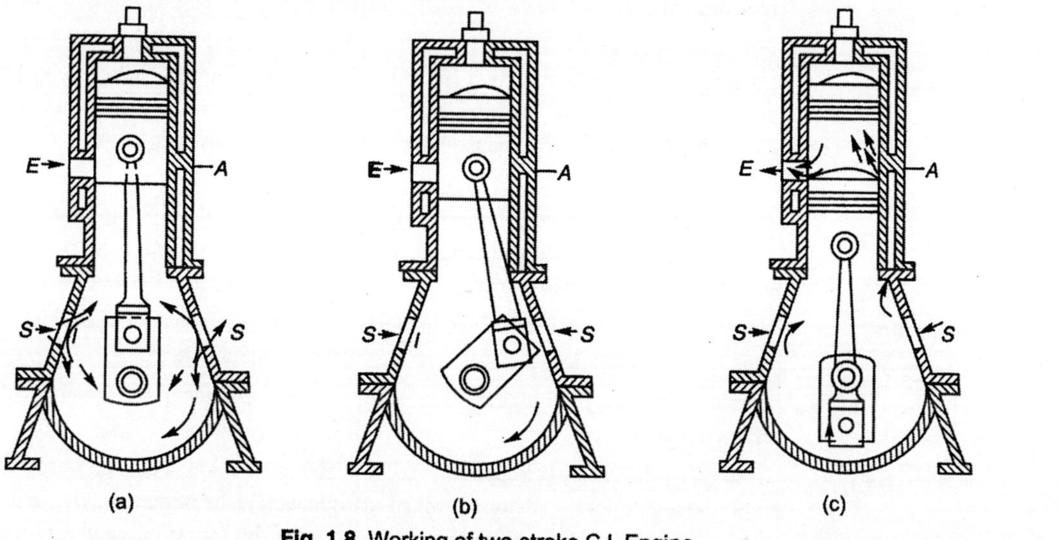

Fig. 1.8 Working of two-stroke C.I. Engine.

starts after both the points have been covered by the piston. At the end of compression fuel is injected and ignites at T.D.C. giving products of combustion at high pressure.

On the downward stroke the high prssure products of combustion expand and the air below the piston comprsses, closing the plate valves, as shown in Fig. 1.8 (b). As soon as the piston uncovers the exhaust port the products of combustion are released to the atmosphere. A little latter, on the downward stroke, the other port communicating with the crankcase gets uncovered and the air compressed in the crankcase gets transferred to the space above the piston as shown in Fig. 1.8 (c).

1.5 COMPARISON OF FOUR-STROKE AND TWO-STROKE ENGINES

The comparison is given in Table 1.1.

Table 1.1 Comparison of two-stroke and four-stroke engines.

Feature	*Four-Stroke Engine*	*Two-Stroke Engine*
1. Power strokes	The thermodynamic cycle is completed in four strokes of the piston or in two revolutions of the crankshaft. Thus, one power stroke is obtained in every two revolutions of the crankshaft.	The thermodynamic cycle is completed in two strokes of the piston or in one revolution of the crankshaft. Thus one power stroke is obtained in each revolution of the crankshaft.
2. Size of flywheel	Because of the above, turning moment is not so uniform and hence a heavier flywheel is needed.	Because of the above, turning moment is more uniform and hence a lighter flywheel is needed.
3. Power developed	Because of one power stroke for two revolutions, power produced for same size of engine is less, or for the same power the engine is heavier and bulkier.	Because of one power stroke for every revolution, power produced for same size of engine is twice, or for the same power the engine is lighter and more compact.
4. Wear and tear	Because of one power stroke in two revolutions lesser cooling and lubrication requirements. Lower rates of wear and tear.	Because of one power stroke in one revolution greater cooling and lubrication requirements. Higher rates of wear and tear.
5. Valves	Four-stroke engines have valves and valve actuating mechanisms for opening and closing of the intake and exhaust valves.	Two-stroke engines have no valves but only ports (some two-stroke engines are fitted with conventional exhaust valve or reed valve).
6. Initial cost	Because of comparatively higher weight and complicated valve mechanism, the initial cost of the engine is more.	Because of light weight and simplicity due to the absence of valve actuating mechanism, initial cost of the engine is less.
7. Volumetric efficiency	Volumetric efficiency is more due to more time for induction.	Volumetric efficiency is low due to lesser time for induction.
8. Thermal efficiency	Thermal efficiency is higher; part load efficiency is better.	Thermal efficiency is lower; part load efficiency is poor.
9. Applications	Used where efficiency is important, viz., in cars, buses, trucks, tractors, industrial engine, aeroplanes, power generation, etc.	Used where low cost, compactness and light weight are important, viz., in mopeds, scooters, motorcycles, hand sprayers, etc.

1.6 COMPARISON OF S.I. AND C.I. ENGINES

The comparison of S.I. and C.I. engines is given in Table 1.2.

Table 1.2 Comparison of S.I. and C.I. Engines

Feature	*S.I. Engine*	*C.I. Engine*
1. Basic cycle	Otto cycle	Diesel cycle
2. Fuel	Petrol (Gasoline)	Diesel oil
3. Introduction of fuel	Fuel and air mixed in proper proportion in the carburettor and introduced during the suction stroke. Throttle controls the quantity of mixture introduced.	Fuel injected directly at the end of compression stroke carburettor is not required but a high pressure fuel pump and injector are necessary.
4. Ignition	Requires ignition system with spark plug in combustion chamber.	Self ignition due to high temperature.
5. Compression ratio range	6 to 10.5	14 to 22
6. Speed	Higher maximum revolution per minute due to lighter weight.	Maximum rpm lower.
7. Efficiency	Maximum efficiency lower due to low compression ratio.	Higher maximum efficiency due to higher compression ratio.
8. Weight	Lighter	Heavier due to higher pressures.

1.7 MERITS AND DEMERITS OF TWO-STROKE ENGINES OVER FOUR-STROKE ENGINES

Merits :

1. A two-stroke engine gives twice as many power strokes as a four-stroke cycle engine at the same engine speed, therefore, the two-stroke engine of the same size should develop twice the power of a four-stroke engine. In practice, the actual power developed by two-stroke engine is about 1.7 to 1.8 times the power developed by four-stroke engine of the same dimensions and speed. This is because, some of the power is used for compressing the charge in the crank case and effective stroke is less than the actual stroke.
2. For the same power developed, the two-stroke engine is much lighter, less bulky and occupies less floor area and so is more suitable for uses in marine engines and for transport purposes.
3. It provides mechanical simplicity as valves, rocker arms, push-rods, cam and cam shafts are not required. The friction loss is also less therefore it gives higher mechanical efficiency.
4. The two-stroke engines are much easier to start.
5. A crankcase compression and valveless type two-stroke engine can run in either direction, which is useful in marine applications.
6. The initial cost of the engine is considerable less.
7. The weight/kW ratio is considerably less.

Demerits :

1. The thermodynamic efficiency of an engine is only dependent on the compression ratio. The effective compression ratio in the case of two-stroke engine is less than that for four-stroke

engine for the same stroke (actual) and clearance volume. So the thermodynamic efficiency of two-stroke cycle is always less than four-stroke cycle engine.

2. The actual efficiency of the two-stroke cycle is less than four-stroke cycle engine because greater overlapping of the ports is necessary in two-stroke engine for effective scavenging. A portion of fresh charge in the case of S.I. engine always escapes unused through the exhaust ports, therefore, the specific fuel consumption is usually higher.
3. The power-strokes per minute are twice the power stroke of four-stroke cycle engines, the capacity of the cooling system used must be higher. The cooling of the engine also presents difficulty as quantity of heat removed per minute is large. Due to the firing in each revolution, the piston is likely to get overheated and oil cooling of the piston is necessary.
4. The consumption of lubricating oil is sufficiently large because of high operating temperatures.
5. Sudden release of the gases makes the exhaust more noisy.
6. A two-stroke petrol engine with crankcase compression 50 to 60% of the swept volume filled with fresh charge while four-stroke petrol engine contains 80 to 95%. This is because, the space occupied by rotating parts in the crankcase prevents a full charge being sucked in.
7. The scavenging is not complete particularly in high speed engines as very short time is available for exhaust and hence the fresh charge is highly polluted. This polluting can be reduced using opposed piston two-stroke diesel engine which provides undirectional scavenging.
8. The turning movement of two-stroke engine is more non-uniform compared with four-stroke engine, so it requires heavier flywheel and strong foundation.

1.8 VALVE TIMING DIAGRAMS

1.8.1 Four-Stroke S.I. Engine

The valve timing diagrams for low and high speed four-stroke S.I. engine are shown in Fig. 1.9. Valve timing diagram is the regulation of the positions in the cycle at which the valves are set to open and close. The typical valve timings are given in Table 1.3.

b : before TDC : Top dead centre.

a : after BDC : Bottom dead centre

Table 1.3 Typical Valve timings for Four-Stroke S.I.Engines.

Position	*Theoretical*	*Actual*	
		Low speed engine	*High speed engine*
Inlet valve opens	TDC	10° b* TDC	10° b TDC
Inlet valve closes	BDC	10° a* BDC	60° a BDC
Inlet valve is open for	180°	200°	250°
Exhaust valve opens	BDC	25° b BDC	55° b BDC
Exhaust valve closes	TDC	5° a TDC	20° a TDC
Exhaust valve is open for	180°	210°	255°
Valve overlap spark	NIL	15°	30°
Spark	TDC	15° b TDC	30° b TDC

*a = after, b = before

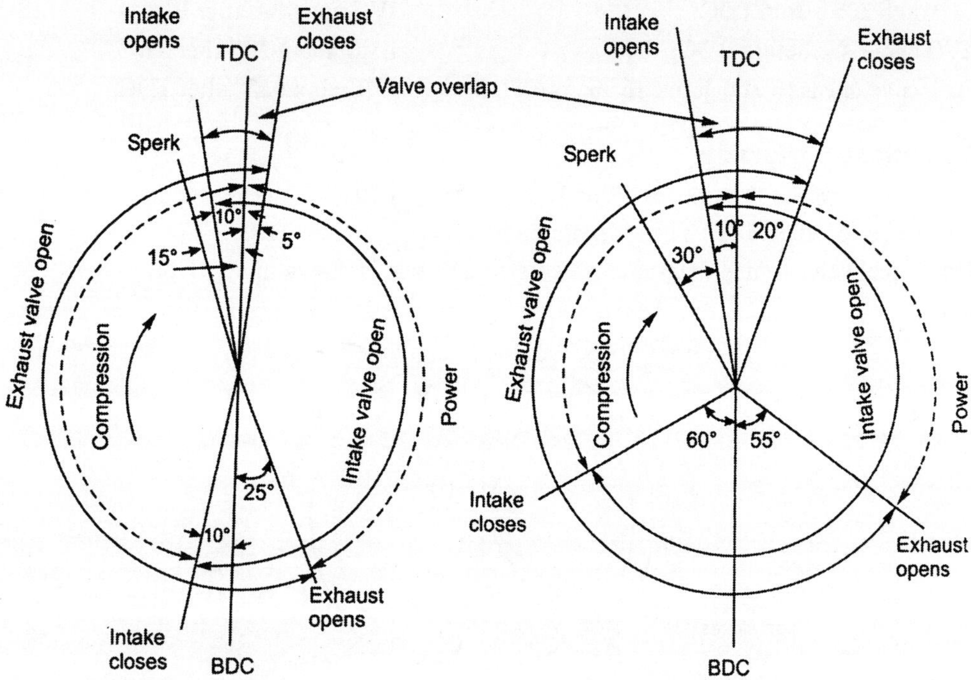

(a) Low speed engines (b) High speed engines

Fig. 1.9 Valve timing for low and high speed four-stroke S.I. engine.

1.8.2 Four-Stroke C.I. Engine

A typical valve timing diagram for a four-stroke C.I. engine is shown in Fig. 1.10. The typical timing valves are :

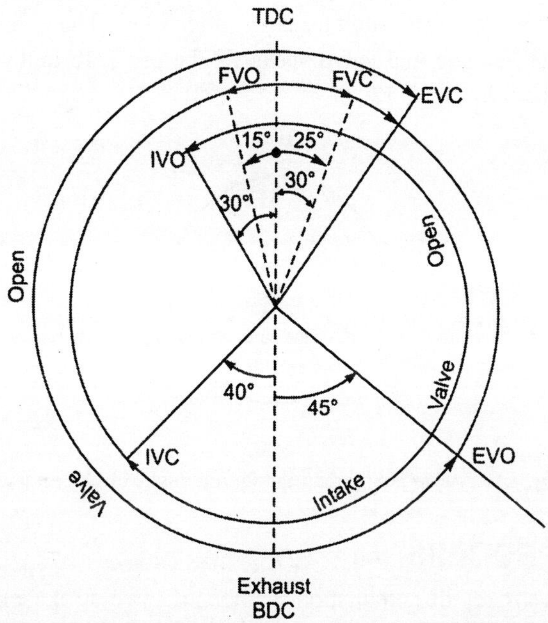

Fig. 1.10 Valve timing diagram for four-stroke C.I. engine.

IVO upto 30° before TDC; IVC upto 40° after BDC

EVO about 45° before BDC; EVC about 30° after TDC

Fuel valve opens (FVO) about 15° before TDC; FVC about 25° after TDC.

1.8.3 Two-Stroke S.I. Engine

The transfer port opens 35° before BDC and closes 35° after BDC. The exhaust port opens 45° before BDC and closes 45° after BDC. The spark ignition occurs 20° before TDC.

The typical valve timing diagram for two-stroke engine is shown in Fig. 1.11.

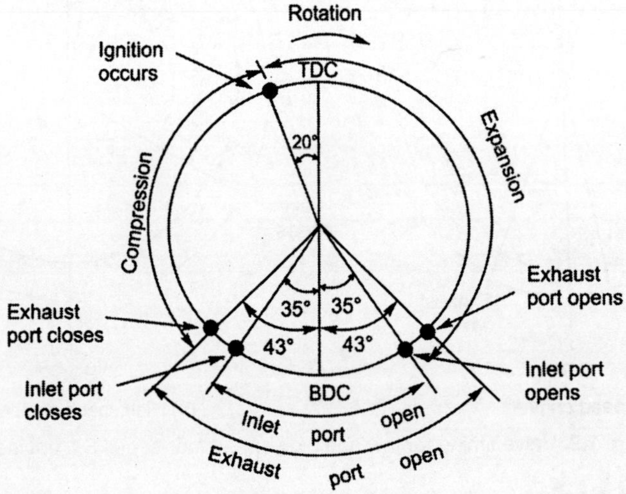

Fig. 1.11 Typical valve timing diagram of a two-stroke S.I. engine.

1.8.4 Two-Stroke C.I. Engine

The transfer port opens 45° before BDC and closes 45° after BDC. The exhaust port opens 60° before BDC and closes 60° after BDC. The fuel valve opens 15° before TDC and closes 20° after TDC. The valve timing diagram is shown in Fig. 1.12.

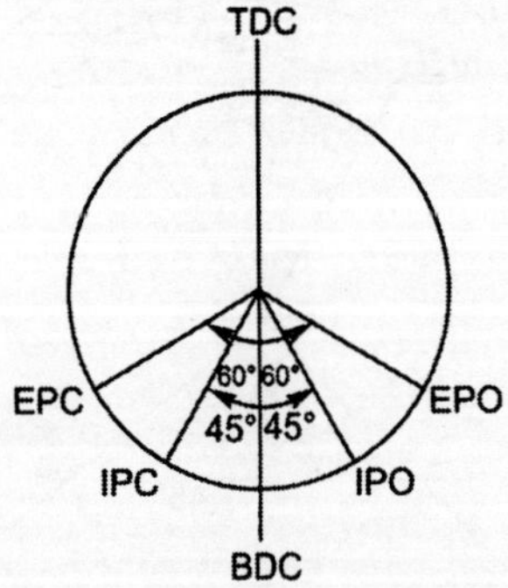

Fig. 1.12 Valve timing diagram for a two-stroke C.I. engine.

1.9. SCAVENGING PROCESS

At the end of the expansion stroke, the combustion chamber of a two-stroke engine is left full of products of combustion as there is no exhaust stroke available to clear the cylinder of burnt gases. The process of clearing the cylinder, after the expansion stroke, is called scavenging process.The scavenging process is the replacement of the products of combustion in the cylinder from the previous power stroke with fresh-air charge to be burned in the next cycle.

There are three types of scavenging systems as explained below :

1. **Uniflow Scavenging System :** In this system, shown in Fig. 1.13, the air enters the cylinder from one end and leaves it from the other end. Air acts like an ideal piston and pushes out the residual gas in the cylinder and replaces it with fresh charge. Due to absence of any eddies or turbulence it is easier in a uniflow scavenging system to push the products of combustion out of the cylinder without mixing with it and short circuiting. Thus, this system has highest scavenging efficiency.

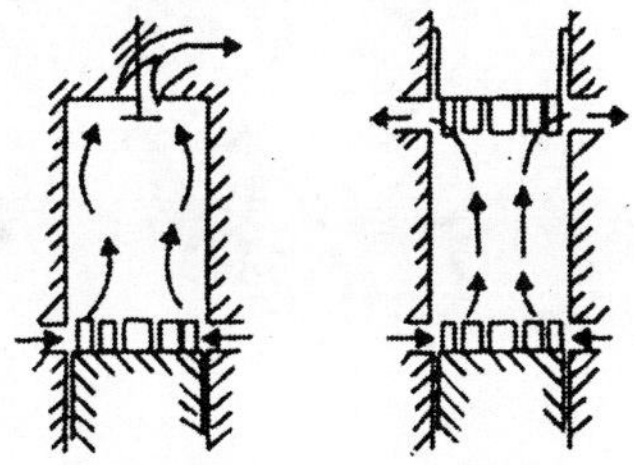
(a) (b)

Fig. 1.13 Uni-flow scavenging system.

2. **Cross Scavenging :** This process shown in Fig. 1.14, employs inlet and exhaust ports placed in oposite sides of the cylinder wall. The air moves up to combustion chamber on one side of the cylinder and then down on the other side to flow out of the exhaust ports. This requires that air should be guided by use of either a suitably shaped detector formed on piston top or by use of inclined ports. The main disadvantage of this system is that the scavenging air is not able to get rid of the layer of exhaust gas near the wall resulting in poor scavenging. Some of the fresh charge also goes directly in the exhaust port.

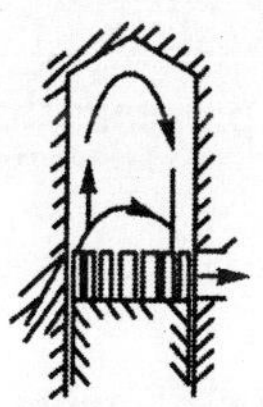

Fig. 1.14 Cross scavenging.

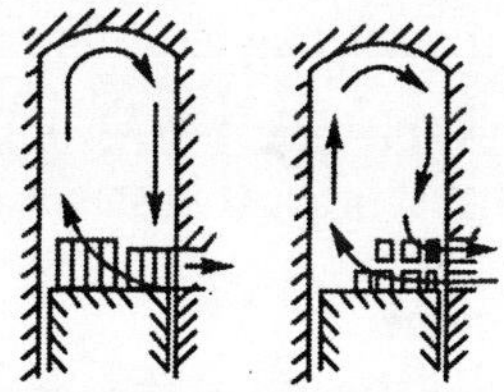
(a) (b)

Fig. 1.15 Loop or reverse scavenging.

3. **Loop or Reverse Scavenging :** Fig 1.15 shows the loop or reverse scavenging system. This avoids the short-circuiting of the cross-scavenged engine and improves upon the scavenging efficiency. The inlet and exhaust ports are placed on the same side of the cylinder wall. The major mechanical problem with this system is that of obtaining an adequate oil supply to the cylinder wall consistent with reasonable lubricating oil consumption and cylinder wear.

1.10 WANKEL ROTARY COMBUSTION ENGINE

This engine works on the ordinary Otto cycle but the piston undergoes rotary motion instead of reciprocating motion. The piston is replaced by a three lobe rotor which rotates eccentrically in an oval shape chamber. The shape of the chamber is such that it ensures proper contact of rotor throughout its rotation.

The working of the engine is illustrated in Fig 1.16 which shows the various positions of the rotor in sequence. The rotor lobes A, B and C ensure a tight sealing with the chamber. The rotor has oval shaped depressions on its three faces between the lobes; shown by dotted lines. In Fig. 1.16 (a), in side

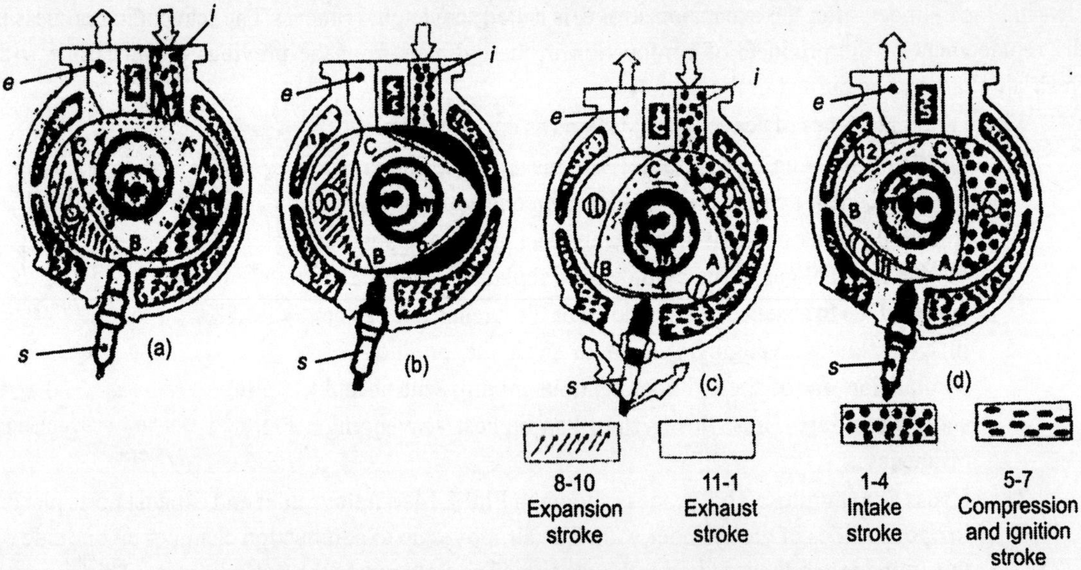

Fig. 1.16 Four phases of operation of a Wankel engine.

CA of the rotor in position 1, has passed the intake port and the induction of air-fuel mixture is going to start. The induction of charge continues through stages 2 and 3 but reaches its maximum value at stage 4, as shown in Fig. 1.16 (b). Next consider side AB of the rotor in Fig. 1.16 (a) where the compression process just starts and it gets completed through stages 5, 6 and 7. At stage 8 the spark plug fires and ignition of the charge takes place. Now the power cycle starts, which proceeds through stages 9 and 10. This is followed by exhaust process in stages 11 and 12.

Features of Rotary Engine

1. There are no reciprocating parts. Therefore, the problems of unbalance due to inertia of reciprocating parts and engine vibration is eliminated.
2. There is no separate intake-exhaust valve mechanism. Hence the operation is noise free.
3. The time for one stroke is 270° in terms of the rotating angle of output shaft, and there is one explosion for one rotation of the output shaft.

Applications :

1. High performance motor cycle.
2. Snow mobile.
3. Outboard engine.
4. Helicopter engine.
5. General purpose engine.
6. Industrial engines.
7. Chain saw engine.
8. Model plane engine.
9. Passenger cars.

1.11 APPLICATIONS OF I.C. ENGINES

The detail applications with capacity and type of engine used are listed in Table 1.4.

Table 1.4 : Applications of I.C. Engines

	Applications	*Power (kW)*	*SI or CI*	*2-Stroke or 4-Stroke*	*Cooling Method*
1.	Road vehicles				
	(i) Mopeds	0.75-1.5	SI	2	A*
	(ii) Scooters and Motor Cycles	3-7	SI	2, 4	A
	(iii) Small Car	15-75	SI	4	W*
	(iv) Large Car	75-200	SI	4	W
	(v) Lawn movers	0.75-3	SI	2, 4	A
2.	Locomotives	400-4000	CI	2, 4	W
3.	Small Air crafts				
	(i) Helicopters	40-1500	SI	4	A
	(ii) Airplanes	40-3000	SI	4	A
4.	Marine applications				
	(i) Motor boats	0.5-75	SI, CI	4	W
	(ii) Ships	3000-20000	CI	2, 4	W
5.	Industrial applications				
	(i) Electric power	40-25,000	CI	2, 4	W
	(ii) Gas pipe lines	700-5000	SI	2, 4	W
6.	Off-road vehicles				
	(i) Light vehicles (factory, airport)	1.5-15	SI	2, 4	A
	(ii) Agriculture	3-150	SI, CI	2, 4	A, W
	(iii) Earthmoving machinery	40-800	CI	2, 4	W
	(iv) Military purpose	40-2000	CI	2, 4	A, W

Type of Engine	*Applications*
1. Petrol Engines	Automobile, marine, aircraft
2. Diesel Engines	Automobile, locomotive, marine and power generation
3. Gas Engines	Industrial purposes and power generation
4. Gas Turbines	Aircraft, marine, industrial and power generation

* A = Air, W = Water

1.12 THEORETICAL AND ACTUAL P-V DIAGRAMS

1.12.1 Four-Stroke Petrol Engine

The theoretical and actual p-V diagrams for a 4-stroke petrol engine are shown in Fig. 1.17 (a) and (b) respectively.

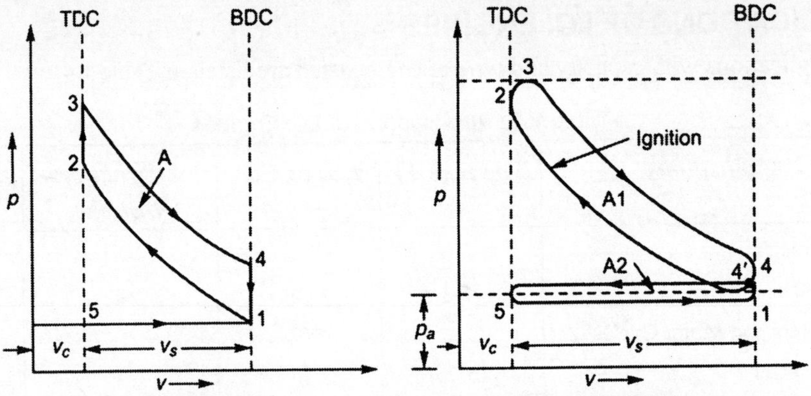

(a) Theoretical (b) Actual

Fig. 1.17 Theoretical and actual *p-v* diagrams for 4-stroke petrol engine.

The theoretical p-V diagram is drawn with the following assumptions :

1. Suction and exhaust take place at atmospheric pressure through 180° rotation of crank.
2. Compression and expansion take place through 180° rotation of crank.
3. Compression and expansion processes are isentropic.
4. The combustion takes place instantaneously at constant volume at the end of compression stroke.
5. Pressure suddenly falls to the atmospheric pressure at the end of expansion stroke.

The various processes of theoretical p-v diagram are :

5-1 : Suction stroke (p_a = const)

1-2 : Compression stroke (pv^γ = const)

2-3 : Instantaneous combustion (v = const)

3-4 : Expansion stroke (pv^γ = const)

4-1 : Sudden fall in pressure (v = const)

1-5 : Exhaust stroke (p_a = const)

In practice, the actual condition differ from the ideal as described below :

1. The suction of mixture in the cylinder is possible only if the pressure inside the cylinder is below atmospheric pressure.
2. The burnt gases can be pushed out into the atmosphere only if the pressure of the exhaust gases is above atmospheric pressure.
3. The compression and expansion do not follow the isentropic law, as there will be heat exchange during these processes.
4. Sudden pressure rise is not possible after the ignition as combustion takes some time for completion and actual pressure rise is less than theoretical considered. The pressure increase takes place through some crank rotation, or increase in volume.
5. Sudden pressure release after the opening of expansion valve is not possible and it also takes place through some crank rotation.

If all these modifications are taken into account, then the cycle can be represented on $p-v$ diagrams as shown in Fig. 1.17 (b).

The area 4′-5-1-4′ representing negative work is called negative loop or pumping loop. This work is required for admitting the fresh charge and for exhausting the burnt gases. This loss of work is known as pumping loss and power consumed for this is known as pumping horse-power.

The net work per cycle of the engine is given by the area $(A_1 - A_2)$. This area $(A_1 - A_2)$ is always less than the area A as shown in Fig. 1.17 (a) due to the actual deviations of operations from the theoretical ones.

If both the areas are represented in the form of rectangles taking v_s as base, then the ordinates give the mean effective pressures as shown in Fig. 1.18.

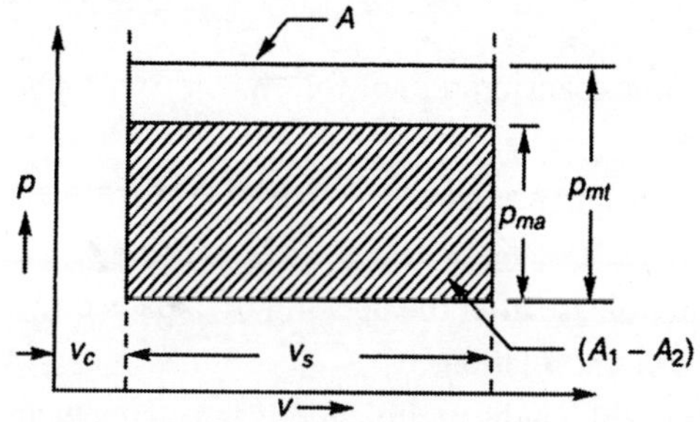

Fig. 1.18

p_{ma} (Actual mean effective pressure) = p_{mt} (theoretical mean effective pressure) × DF (diagram factor)

1.12.2 Four-Stroke Diesel Engine

The theoretical and actual p-V diagrams for a 4-stroke diesel engine are shown in Fig. 1.19 (a) and (b) respectively.

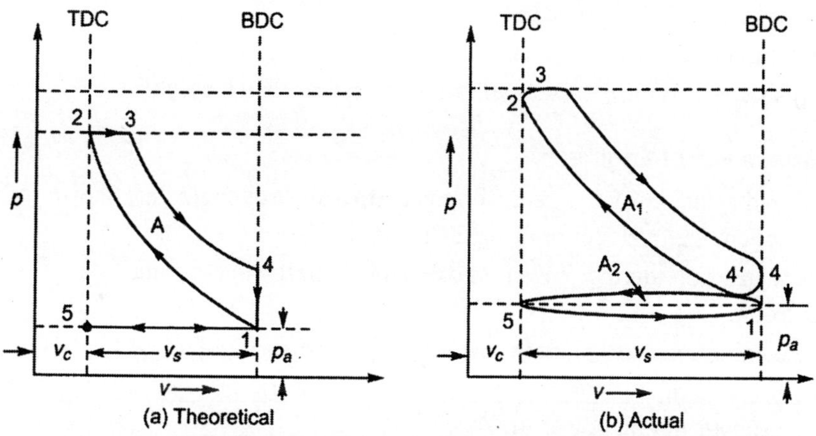

Fig. 1.19 Theoretical and actual *p-v* diagrams for 4-stroke diesel engine.

The theoretical $p - v$ diagram for the 4-stroke diesel engine is drawn with the following assumptions :

1. Suction and exhaust take place at atmospheric pressure through 180° of crank rotation.
2. Compression and expansion take place during 180°of crank rotation.
3. Compression and expansion are isentropic.

4. The combustion takes place at constant pressure during a small part of expansion stroke.
5. Pressure suddenly falls to atmospheric pressure at the end of expansion stroke.

With the above assumptions, the working cycle can be represented on $p-v$ diagram as shown in Fig. 1.19 (a), and it is similar to the theoretical Diesel cycle.

But in practice, the actual conditions differ from the ideal as described below :

1. The suction of the air inside the cylinder is possible only if the pressure inside the cylinder is below atmospheric.
2. Exhausting of gasses is possible only if the pressure of the exhaust gases is above atmospheric pressure.
3. The compression and expansion do not follow the isentropic process, as there are heat and pressure losses.
4. The combustion at constant pressure is not possible as the fuel will not burn as it is introduced into the cylinder.
5. The sudden pressure release after the opening of expansion valve is not possible and it takes place through some crank rotation.

The operations of the cycle, taking the modifications into account are represented on $p-v$ diagram in Fig. 1.19 (b).

The various processes of theoretical $p-v$ diagram are :

5-1 : Suction stroke (p_a = const)

1-2 : Compression stroke (pv^{γ} = const)

2-3 : Constant pressure combustion (p = const)

3-4 : Expansion stroke (pv^{γ} = const)

4-1 : Sudden fall in pressure (v = const)

1-5 : Exhaust stroke (p_a = const)

Actual area $(A_1 - A_2)$ on $p-v$ diagram per cycle is less than theoretical.

1.12.3 Two-Stroke Petrol Engine

The theoretical and actual $p-v$ diagrams for a two-stroke petrol engine are shown in Fig. 1.20. The following assumptions are made for drawing the theoretical $p-v$ diagram :

1. The expansion during power stroke and compression during compression stroke are isentropic.
2. The combustion takes place instantaneously at constant volume at the end of compression.
3. The pressure falls instantaneously to the atmospheric pressure as the piston uncovers the exhaust ports during power stroke.
4. The scavenging takes place at atmospheric pressure.

It may be observed from Fig. 1.20 (a) that the compression of the charge starts from '1' instead of point 6.

Effective compression ratio,

$$r_{ce} = \frac{v_{se} + v_c}{v_c}$$

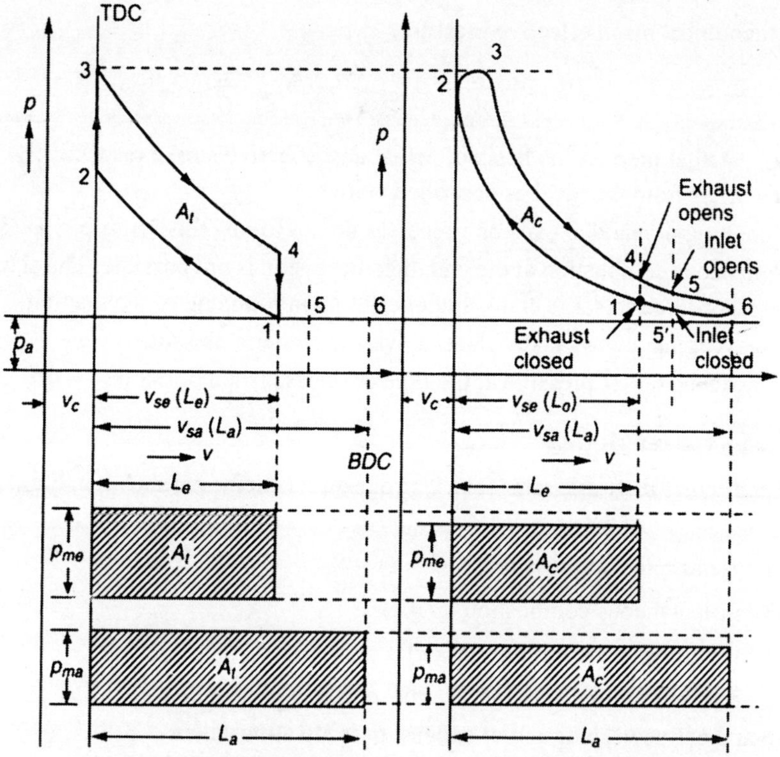

(a) Theoretical (b) Actual

Fig. 1.20 Theoretical and actual *p-v* diagrams for 2-stroke petrol engine.

where v_{se} = effective stroke volume $= \frac{\pi}{4} d^2 L_e$

v_{sa} = actual stroke volume $= \frac{\pi}{4} d^2 L_a$

L_e, L_a = effective and actual stroke lengths respectively

d = cylinder diameter.

The various processes are :

1-2 : isentropic compression ($pv^\gamma = c$)

2-3 : instananeous combustion ($v = c$)

3-4 : isentropic expansion ($pv^\gamma = c$)

4-1 : release of burned charge to atmosphere ($v = c$)

1-6 and 6-1 : sweeping out of exhaust gases to atmosphere

Point 5 : inlet port opens

5-6 and 6-5 : Charging the cylinder with fresh charge and scavenging action.

Actual theoretical mean effective pressure,

$$p_{ma} = \frac{A_t}{v_{sa}}$$

Effective theoretical mean effective pressure,

$$p_{me} = \frac{A_t}{v_{se}} = p_{ma} \cdot \frac{v_{sa}}{v_{se}} = p_{ma} \cdot \frac{L_a}{L_e}$$

p_{ma}', p_{me}' = actual mep on the basis of actual and effective stroke respectively. In practice, the actual conditions differ from the ideal as described below :

1. The compression and expansion processes do not follow the isentropic law strictly.
2. Instantaneous combustion at the end of compression is not possible. The actual pressure rise takes place through some crank angle resulting in rounding of the diagram.
3. The scavenging always takes place above atmospheric pressure.
4. Instanteneous fall of pressure at the time of release is not possible.

1.12.4 Two-Stroke Diesel Engine

The theoretical and actual $p - v$ diagrams for a 2-stroke diesel engine are shown in Fig. 1.21. The various processes are :

1-2 : isentropic compression ($pv^\gamma = c$)

2-3 : instantaneous combustion ($p = c$)

3-4 : isentropic expansion ($pv^\gamma = c$)

4-1 : release of burned charge to atmosphere ($v = c$)

1-6 and 6-1 : sweeping out of exhaust gases to atmosphere

5-6 and 6-5 : charging the cylinder with fresh charge and scavenging action.

Point 5 : inlet port opens.

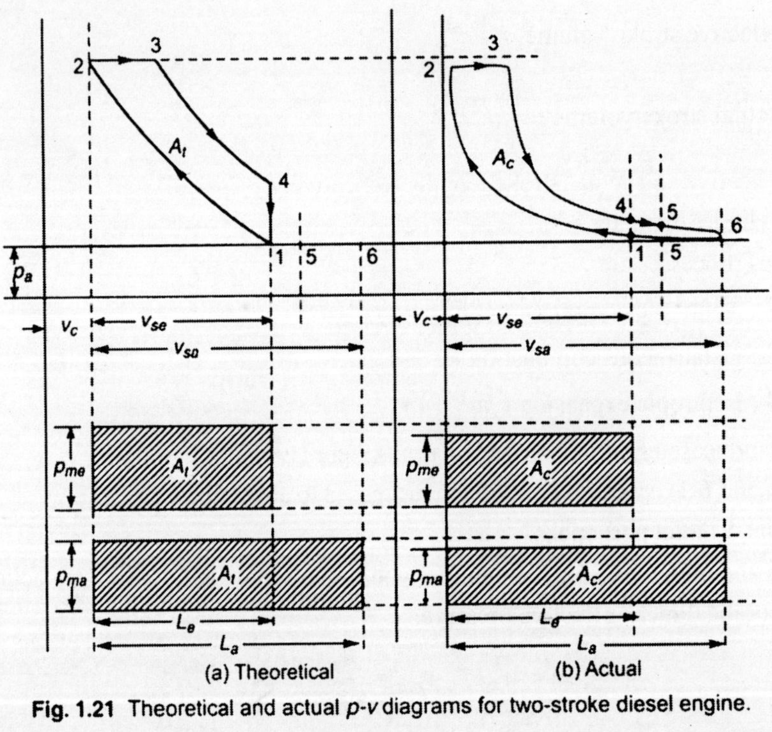

Fig. 1.21 Theoretical and actual *p-v* diagrams for two-stroke diesel engine.

SOLVED QUESTIONS FROM UNIVERSITY EXAMINATIONS

■ **Example 1.1**

(a) Explain the difference between two-stroke and four-stroke engines. With the help of neat sketch, explain the working of four-stroke engine.

(b) With the help of neat sketch, explain the working of a WANKEL-Engine. [KU, Dec., 2005]

Solution : (a) Refer to art. 1.5.

(b) Refer to art. 1.10.

■ **Example 1.2** What are the main advantages and disadvantages of the C.I. engine as compared to S.I. engine. In this light discuss the field of application of the two engines. [KU, Dec., 2006]

Solution : Refer to art. 1.6.

■ **Example 1.3**

(a) What is the difference between I.C. engines and External combustion engines ? Give examples of each type.

(b) With the help of neat sketches, explain the actual sequence of events in the cylinder of a petrol engine working on four stroke cycle. [KU, Dec., 2007]

Solution : (a) Refer to art. 1.1.

(b) Refer to art. 1.12.1.

■ **Example 1.4** What is the difference between Actual cycle and Air standard cycle. [KU, Dec., 2008]

Solution : Refer to art. 1.12.1.

■ **Example 1.5**

(a) What are different criteria for the classification of I.C. Engine ? Explain classification of I.C. Engines on the basis of fuels used.

(b) Explain the cycle of operation of four stroke petrol engine. [KU, Sept. 2009]

Solution : (a) Refer to art. 1.2.

(b) Refer to art. 1.4.1.

SHORT ANSWER TYPE QUESTIONS

1. Define a heat engine.
2. Define an internal combustion engine.
3. What is an external combustion engine.
4. Define an engine.
5. List four advantages of internal combustion engine over external combustion engine.
6. Give four important basis for the classification of internal combustion engines.
7. What is the function of flywheel in an internal combustion engine?
8. What is the necessity of valve-timing diagram?
9. What is scavenging process in I.C. engines?
10. List four applications of Wankel engine.

11. Why the actual *p-v* diagrams differ from the theoretical diagrams of I.C. engines?
12. Define diagram factor.
13. Write four assumptions generally made while drawing the theoretical *p-v* diagram for 2-stroke petrol engine.
14. Suggest the I.C. engines used for the following :
    (i) Bus (ii) Motor cycle (iii) Scooter (iv) Moped.

EXERCISES

1. What are the advantages of internal combustion engines over external combustion engines ?
2. How I.C. engines are classified ?
3. Describe the salient features of S.I. engine.
4. Differentiate between S.I. and C.I. engines.
5. What are the main differences between 4-stroke and 2-stroke engines ?
6. List the applications of I.C. engines.
7. Explain the working of :
   (a) Four-stroke petrol engine
   (b) Four-stroke diesel engine.
8. Expalin the working of 2-stroke petrol engine.
9. Compare 4-stroke and 2-stroke engines.
10. Compare S.I. and C.I. engines.
11. List the merits of two-stroke engines over 4-stroke engines.
12. Draw the $p-v$ diagrams for (a) 4-stroke petrol engine (b) 2-stroke petrol engine.
13. Why the actual $p-v$ diagram for a 4-stroke diesel engine differ from the theoretical diagram ?
14. Explain the working of 2-stroke diesel engine.

Chapter 2

Air Standard Cycles

Inside This Chapter

2.1 INTRODUCTION

The gas as the working fluid does not undergo any phase change. The engines working on the gas cycles may be either cyclic or non-cyclic. The working fluid is assumed to obey all laws of a perfect gas.

In most of th gas power cycles, the working fluid consists mainly of air. Therefore, it is convenient to analyses the gas power cycles by devising an idealized cycle known as Air Standard Cycle, in which the working fluid is pure air.

The following assumptions are made in the air-standard cycles

1. The air as the working fluid follows the perfect gas law : $pV = mRT$
2. The working fluid is homogeneous throughout and no chemical reaction takes place.
3. The specific heats of air do not vary with temperature
4. The mass of air in the cycle remains fixed.
5. The combustion process is replaced by an equivalent heat addition process.
6. The exhaust process is replaced by an equivalent heat rejection process.
7. All processes are internally reversible.

The thermal efficiency, η_{th} of air standard cycle is given by

$$\eta_{th} \text{ or } \eta_a = \frac{\text{Work Done}}{\text{Heat Supplied}} = \frac{\text{Heat Supplied} - \text{Heat Rejected}}{\text{Heat Supplied}}$$

$$= \frac{Q_s - Q_r}{Q_s} = \frac{q_s - q_r}{q_s}$$

The thermal efficiency is also called air-standard efficiency (η_a)

Relative efficiency (η_r) = Actual thermal efficiency/Air standard efficiency.

2.2 PISTON-CYLINDER ARRANGEMENT

The piston-cylinder arrangement is shown in Fig. 2.1. The various definitions are :

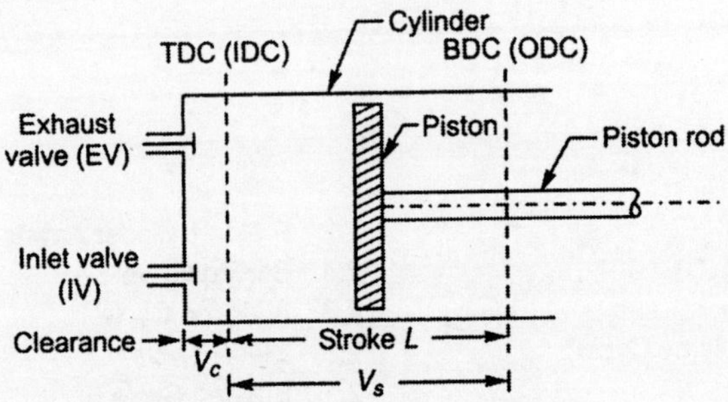

Fig. 2.1 Piston-cylinder arrangement.

Bore (d) : is the cylinder or piston diameter.

Stroke (L) : is the distance moved by the piston in one direction. It is equal to twice the crank radius.

Top Dead centre (T.D.C.) : is the extreme position of the piston near to the head of cylinder. It is also called inner dead centre (I.D.C.) position.

Bottom Dead centre (B.D.C.) : is the extreme position of the piston opposite to the head of cylinder. It is also called outer dead centre (O.D.C.) position.

$$L = \text{T.D.C.} - \text{B.D.C.}$$

Clearance Volume (V_c) : is the minimum volume of clearance between the cylinder head and the piston at T.D.C. position.

Swept Volume (V_s) : is the maximum volume swept by the piston in moving from TDC to BDC position or vice versa

$$V_s = \frac{\pi}{4} d^2 L$$

Cylinder Volume (V) : is the sum of clearance volume and swept volume.

$$V = V_c + V_s$$

Compression ratio (r) : is the ratio of cylinder volume to the clearance volume. Thus

$$r = \frac{V}{V_c} = \frac{V_c + V_s}{V_c} = 1 + \frac{V_s}{V_c} \qquad \ldots(2.3)$$

Clearance ratio (c) : is th ratio of clearance volume to stroke volume. Thus

$$c = \frac{V_c}{V_s} \text{ and } r = 1 + \frac{1}{c} = \frac{1+c}{c} \quad ...(2.4)$$

Mean effective pressure (p_m) : is the average pressure which, if acted on the piston during the entire power or outward stroke, would produce the same work output as the net work output for the actual cyclic process. Thus

Work done per cycle, $W = \frac{\pi}{4} d^2 L \cdot p_m$ or $p_m = \frac{4W}{\pi d^2 L}$

Work done per minute = $p_m LAn$

where $A = \frac{\pi}{4} d^2$

n = number o strokes per minute

= N, for two stroke engines.

= $N/2$ for four stroke engines

$N = rpm$

Indicated power, $$IP = \frac{p_m LAn}{60} \text{ kW} \quad ...(2.5)$$

where p_m is in k Pa.

2.3 CARNOT CYCLE

The Carnot cycle consists of two reversible isothermal processes (1–2 and 3–4) and two reversible adiabatic processes (2–3 and 4–1). The cycle is represented on p-v and T-s diagrams in Fig. 2.2

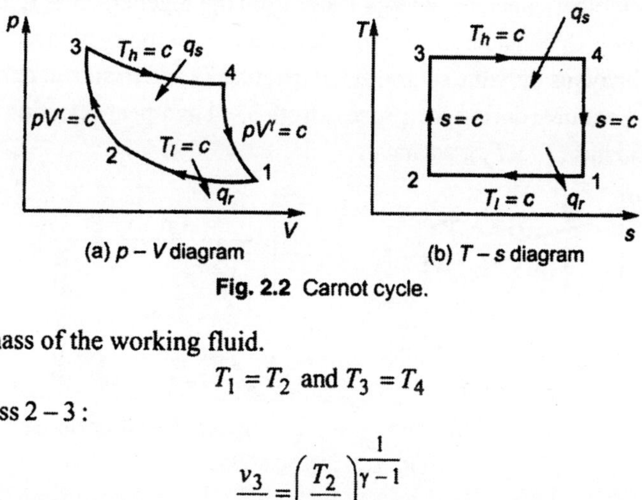

Fig. 2.2 Carnot cycle.

Consider unit mass of the working fluid.

$$T_1 = T_2 \text{ and } T_3 = T_4$$

Adiabatic process 2 – 3 :

$$\frac{v_3}{v_2} = \left(\frac{T_2}{T_3}\right)^{\frac{1}{\gamma-1}} \quad ...(1)$$

Adiabatic process 4 – 1 :

$$\frac{v_4}{v_1} = \left(\frac{T_1}{T_4}\right)^{\frac{1}{\gamma-1}} = \left(\frac{T_2}{T_3}\right)^{\frac{1}{\gamma-1}} \quad ...(2)$$

From eqs. (1) and (2), we get

$$\frac{v_3}{v_2} = \frac{v_4}{v_1} \text{ or } \frac{v_4}{v_3} = \frac{v_1}{v_2} \qquad ...(3)$$

Heat supplied during isothermal expansion process 3-4,

$$q_s = q_{3-4} = RT_h \ln\left(\frac{v_4}{v_3}\right)$$

Heat rejected during isothermal compression process 1-2

$$q_r = q_{1-2} = RT_l \ln\left(\frac{v_1}{v_2}\right) = RT_l \ln\left(\frac{v_4}{v_3}\right)$$

Net work, $$w_{net} = q_s - q_r = R(T_h - T_l) \ln\left(\frac{v_4}{v_3}\right)$$

Cycle thermal efficiency, $$\eta_{th} = \frac{w_{net}}{q_s} = \frac{T_h - T_l}{T_h} = 1 - \frac{T_l}{T_h} \qquad ..(2.6)$$

where l and h subscripts denote lower and higher respectively.

2.4 STIRLING CYCLE

This cycle works on the principle of regeneration by using a regenerator within the engine itself which would store the rejected heat energy during heat rejection process and supply the same during heat addition process. The Stirling cycle consists of two reversible isotherms and two reversible isochors. The $p-v$ and $T-s$ diagrams are shown in Fig. 2.3. We note that heat addition process 2 – 3 and heat rejection process 4 – 1 cancel each other, since the energy taken from the regenerator is returned in the latter. An alternative interpretation is to consider the regenerator as a part of the system so that the heat exchange with the surroundings involves only the source at temperature T_1 and th sink at temperature T_2. The heat supplied per cycle, is therefore, during the process from 3 – 4 and heat rejection is during the process 1 – 2. For one kg of ideal gas ($T_1 > T_2$), we have

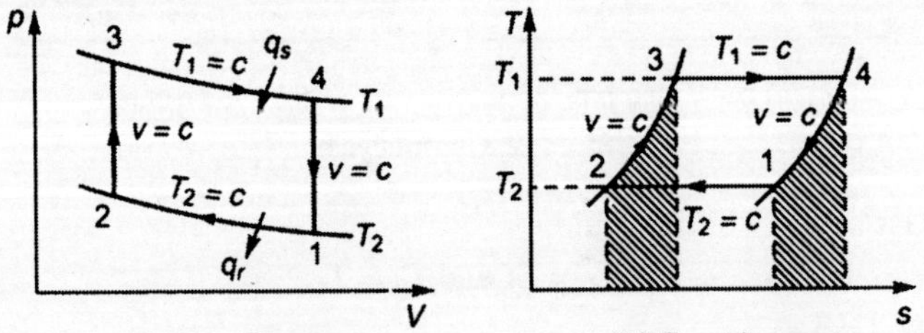

(a) $p-v$ diagram (b) $T-s$ diagram

Fig. 2.3 Stirling cycle.

Heat supplied, $$q_s = q_{3-4} = w_{3-4} = RT_1 \ln\left(\frac{v_4}{v_3}\right)$$

Heat rejected, $$q_r = q_{1-2} = w_{1-2} = RT_2 \ln\left(\frac{v_1}{v_2}\right)$$

$$\text{Thermal efficiency, } \eta_{th} = \left(\frac{q_s - q_r}{q_s}\right) = \frac{\left[RT_1 \ln\left(\frac{v_4}{v_3}\right) - RT_2 \ln\left(\frac{v_1}{v_2}\right)\right]}{RT_1 \ln\left(\frac{v_4}{v_3}\right)}$$

But $v_4 = v_1,\ v_3 = v_2$ and $r = \left(\frac{v_1}{v_2}\right)$, hence

$$\eta_{th} = \left(\frac{T_1 - T_2}{T_1}\right) = 1 - \left(\frac{T_2}{T_1}\right) \quad \text{...(2.7)}$$

Generally there is loss of heat due to radiation and poor conductivity of gas. If η_r is the efficiency of the regenerator, then

Heat supplied $= RT_1 \ln r + (1 - \eta_r) \cdot c_v (T_1 - T_2)$

Heat rejected $= RT_2 \ln r + (1 - \eta_r) \cdot c_v (T_1 - T_2)$

$$\text{Thermal efficiency, } \eta_{th} = \frac{R(T_1 - T_2) \ln r}{RT_1 \ln r + (1 - \eta_r) \cdot c_v (T_1 - T_2)} \quad \text{...(2.8)}$$

2.5 ERICSSON CYCLE

This cycle also works on the principle of regeneration. It consists of two reversible isothermals and two reversible isobars. The *p-v* and *T-s* diagram are shown in Fig. 2.4. The heat rejected during the heat rejection process 1 – 2 at constant pressure is stored in the regenerator and the same is supplied during heat addition process 3 – 4. Assuming 100% efficiency of the regenerator, we have

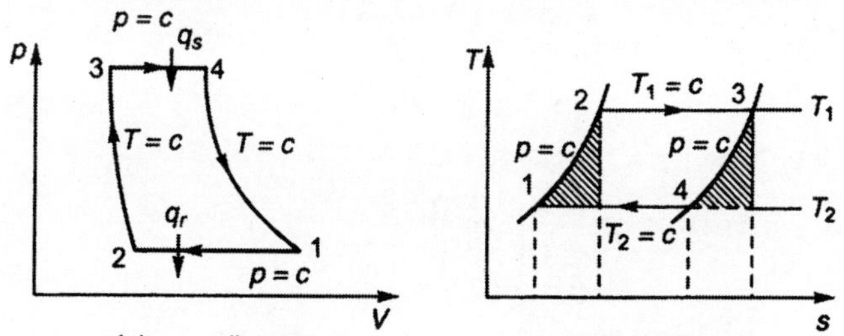

(a) *p – v* diagram (b) *T – s* diagram

Fig. 2.4 Ericsson cycle.

Heat supplied $q_s = RT_1 \ln\left(\frac{v_4}{v_3}\right) = RT_1 \ln r$

Heat rejected, $q_r = RT_2 \ln\left(\frac{v_1}{v_2}\right) = RT_2 \ln r$

where $r = \left(\frac{v_1}{v_2}\right) = \left(\frac{v_4}{v_3}\right)$

$$\text{Thermal efficiency, } \eta_{th} = \left(\frac{q_s - q_r}{q_s}\right) = \left(\frac{T_1 - T_2}{T_1}\right) = 1 - \frac{T_2}{T_1} \quad \text{...(2.9)}$$

2.6 ATKINSON CYCLE

This cycle has two reversible adiabatics (isentropics), one isobar and one isochore. The *p-v* and *T-s* diagrams are shown in Fig. 2.5. For one kg of the working fluid, we have

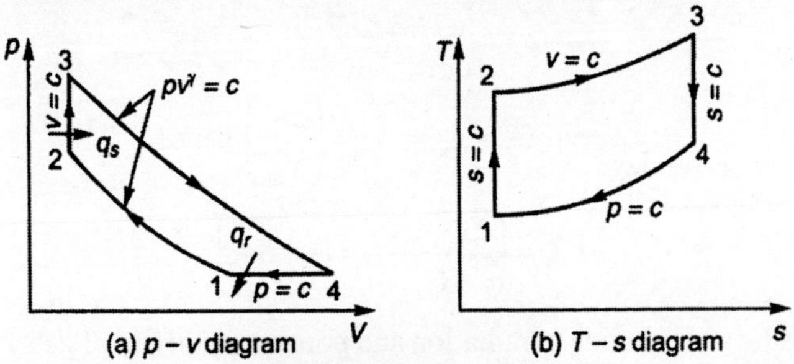

Fig. 2.5 Atkinson cycle.

Heat supplied, $q_s = c_v(T_3 - T_2)$

Heat rejected, $q_r = c_p(T_4 - T_1)$

Net work done, $w_{net} = q_s - q_r = c_v(T_3 - T_2) - c_p(T_4 - T_1)$

Thermal efficiency, $$\eta_{th} = \frac{w_{net}}{q_s} = \frac{c_v(T_3 - T_2) - c_p(T_4 - T_1)}{c_v(T_3 - T_2)} = 1 - \frac{\gamma(T_4 - T_1)}{T_3 - T_2}$$

where $$\gamma = \frac{c_p}{c_v}$$

Let compression ratio, $$r = \frac{v_1}{v_2}$$

Explosion ratio, $$\alpha = \frac{p_3}{p_2}$$

From compression process 1 – 2, we have

$$T_2 = T_1\left(\frac{v_1}{v_2}\right)^{\gamma - 1} = T_1 r^{\gamma - 1}$$

For constant volume heat addition process 2 – 3, we have

$$\frac{p_2}{T_2} = \frac{p_3}{T_3}$$

$$T_3 = T_2\frac{p_3}{p_2} = T_2 .\alpha = \alpha T_1 r^{\gamma - 1}$$

From expansion process 3 – 4, we have

$$\frac{T_3}{T_4} = \left(\frac{p_3}{p_4}\right)^{(\gamma - 1)/\gamma} = \frac{\dfrac{p_3}{p_2}}{\left(\dfrac{p_2}{p_1}\right)^{\frac{\gamma - 1}{\gamma}}} = \left(\frac{\alpha\, p_2}{p_1}\right)^{\frac{\gamma - 1}{\gamma}}$$

$$= \alpha^{(\gamma-1)/\gamma} \cdot \left(\frac{p_2}{p_1}\right)^{(\gamma-1)/\gamma}$$

Also
$$\frac{p_2}{p_1} = \left(\frac{v_1}{v_2}\right)^{\gamma} = r^{\gamma}$$

$\therefore$
$$\frac{T_3}{T_4} = \alpha^{(\gamma-1)/\gamma} \cdot r^{(\gamma-1)}$$

or
$$T_4 = \frac{T_3}{\alpha^{\frac{\gamma-1}{\gamma}} \times r^{\gamma-1}} = \frac{\alpha T_1 r^{\gamma-1}}{\alpha^{\frac{\gamma-1}{\gamma}} \cdot r^{\gamma-1}} = T_1 \alpha^{\frac{1}{\gamma}}$$

Substituting the values of T_2, T_3 and T_4, we get

$$\eta_{th} = \frac{1-\gamma\,(T_1\alpha^{\gamma-1} - T_1)}{\alpha T_1 r^{\gamma-1} - T_1 r^{\gamma-1}}$$

$$= 1 - \frac{1}{r^{\gamma-1}}\left[\frac{\gamma\left(\alpha^{\frac{1}{\gamma}} - 1\right)}{\alpha - 1}\right] \qquad \text{...(2.10)}$$

2.7 OTTO CYCLE (CONSTANT VOLUME CYCLE)

This cycle is composed of four internally reversible processes, two adiabatic and two constant volume processes. The *p-v* and *T-s* diagrams are shown in Fig. 2.6. The various processes are :

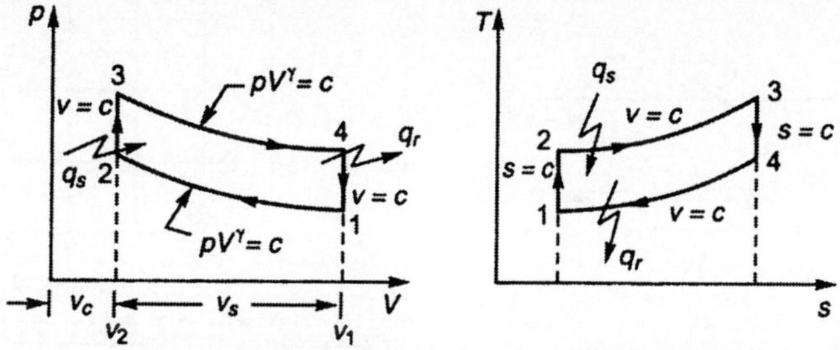

(a) $p - v$ diagram (b) $T - s$ diagram

Fig. 2.6 Otto cycle.

Process 1-2 : Isentropic compression

Process 2-3 : Constant volume heat addition

Process 3-4 : Isentropic expansion

Process 4-1 : Constant volume heat rejection

This cycle is used for spark ignition (petrol) engines.

Consider 1 kg of air flowing through the cycle. Since the air in the cylinder acts as a closed system, from first law of thermodynamics for isentropic compression and expansion, we have

$$q - w = \Delta u$$

For constant volume heat supplied and rejection processes, since $w = 0$

$$q = \Delta u - c_v \Delta T$$

Heat supplied $\quad = q_s = q_{2-3} = c_v(T_3 - T_2)$

Heat rejected $\quad q_r = q_{4-1} = c_v(T_4 - T_1)$

Work done per cycle, $w_{net} = q_s - q_r = c_v[(T_3 - T_2) - (T_4 - T_1)]$

Air standard (or thermal) efficiency $= \dfrac{\text{Work done}}{\text{Heat supplied}}$

$$\eta_a \text{ or } \eta_{th} = \frac{c_v(T_3 - T_2) - c_v(T_4 - T_1)}{c_v(T_3 - T_2)}$$

$$= 1 - \frac{T_4 - T_1}{T_3 - T_2}$$

Now $\quad T_2 = T_1 \dfrac{v_1}{v_2} = T_1 r^{\gamma - 1}$

where $\quad r = \dfrac{v_1}{v_2} = \dfrac{v_1}{v_c} = \text{compression ratio}$

$$T_3 = \frac{T_2 p_3}{p_2} = \alpha_p T_2 = \alpha_p T_1 r^{\gamma - 1}$$

where $\alpha_p = \dfrac{p_3}{p_2} = \dfrac{T_3}{T_2} = \text{pressure ratio}$

$$T_4 = T_3\left(\frac{1}{r}\right)^{\gamma - 1} = \alpha_p T_1$$

$$\therefore \eta_a = \eta_{th} = 1 - \frac{\alpha_p T_1 - T_1}{\alpha_p T_1 r^{\gamma-1} - T_1 r^{\gamma-1}}$$

$$= 1 - \frac{\alpha_p - 1}{(\alpha_p - 1) r^{\gamma - 1}}$$

$$= 1 - \frac{1}{r^{\gamma - 1}} \qquad \ldots(2.6)$$

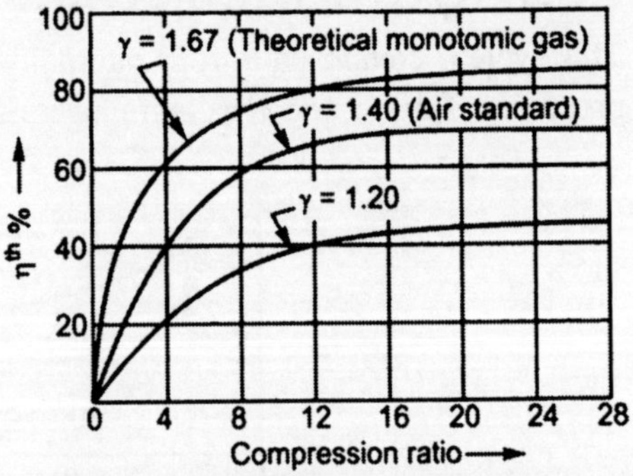

Fig. 2.7 Otto cycle thermal efficiency v's compression ratio.

Th air standard efficiency of Otto cycle depends on compression ratio only and increases as compression ratio increases (Fig. 2.7). In actual engine working on Otto cycle, the compression ratio varies from 5 to 8. This engine is used for spark ignition engines working on petrol.

Mean Effective Pressure (m.e.p.) : It may be defined as the ratio of work done to the displacement volume of piston.

Work done, $\quad w = c_v[(T_3 - T_2) - (T_4 - T_1)]$

Displacement volume, $v_s = v_1 - v_2$

$$\text{M.E.P.}, \; p_m = \frac{w}{v_s} = \frac{c_v[(T_3 - T_2) - (T_4 - T_3)]}{v_1 - v_2}$$

Now

$$v_1 - v_2 = v_1\left(1 - \frac{1}{r}\right) = \frac{RT_1}{p_1}\left(\frac{r-1}{r}\right)$$

$$p_m = \frac{p_1 r[(r^{\gamma-1}(\alpha_p - 1) - (\alpha_p - 1)]}{(\gamma - 1)(r - 1)}$$

$$= \frac{p_1 r(\alpha_p - 1) - (r^{\gamma-1} - 1)}{(\gamma - 1)(r - 1)} = \frac{(p_3 v_3 - p_4 v_4) - (p_2 v_2 - p_1 v_1)}{(\gamma - 1)(v_1 - v_2)} \quad \text{...(2.11)}$$

2.8 DIESEL CYCLE

This cycle is used for compression ignition internal combustion engines working on diesel oil. The *p-v* and *T-s* diagrams are shown in Fig. 2.8. It consists of four internally reversible processes two adiabatic, one constant pressure and one constant volume. The various processes are :

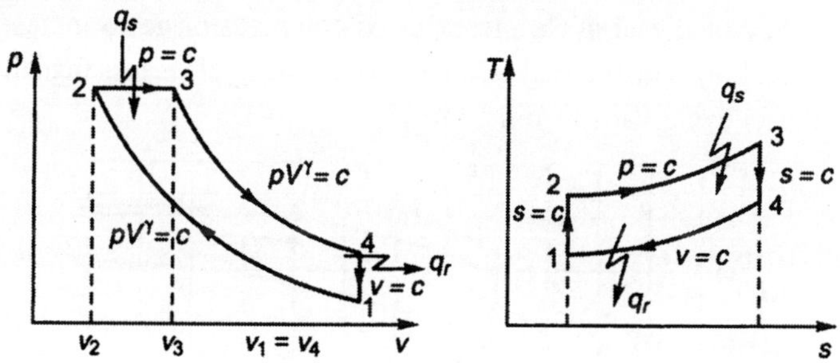

(a) $p - v$ diagram (b) $T - s$ diagram

Fig. 2.8 Diesel cycle.

Process 1-2 : Isentropic compression of air.

Process 2-3 : Heat addition at constant pressure

Process 3-4 : Isentropic expansion of air.

Process 4-1 : Heat rejection at constant volume.

Considering 1 kg of air

Heat supplied, $q_s = q_{2-3} = c_p(T_3 - T_2)$

Heat rejected, $q_r = q_{4-1} = c_v(T_4 - T_1)$

Work done per cycle, $w = q_s - q_r = c_p(T_3 - T_2) - c_v(T_4 - T_1)$

Thermal efficiency,

$$\eta_{th} = \frac{\text{Work done}}{\text{Heat supplied}} = \frac{w}{q_s}$$

$$= \frac{c_p(T_3 - T_2) - c_v(T_4 - T_1)}{c_p(T_3 - T_2)} = 1 - \frac{1}{\gamma}\left[\frac{T_4 - T_1}{T_3 - T_2}\right]$$

Let $r = \frac{v_1}{v_2} = \frac{v_4}{v_2}$ = compression ratio

Now $$T_2 = T_1\left(\frac{v_1}{v_2}\right)^{\gamma-1} = T_1 r^{\gamma-1}$$

Let $$\rho = \frac{v_3}{v_2} = \text{cut off ratio}$$

$$\rho = \frac{v_3}{v_2} = \frac{T_3}{T_2} \text{ or } T_3 = T_2\,\rho = T_1 r^{\gamma-1}\rho$$

$$T_4 = T_3\left(\frac{v_3}{v_4}\right)^{\gamma-1} = T_3\left(\frac{v_3}{v_2}\times\frac{v_2}{v_4}\right)^{\gamma-1} = T_3\left(\frac{\rho}{r}\right)^{\gamma-1}$$

$$= T_1 r^{\gamma-1}\rho\left(\frac{\rho}{r}\right)^{\gamma-1} = T_1\rho^{\gamma}$$

$$\eta_{th} = 1 - \frac{1}{\gamma}\left[\frac{T_1\rho^{\gamma} - T_1}{T_1 r^{r-1}\rho - T_1 r^{\gamma-1}}\right] = 1 - \frac{1}{r^{\gamma-1}}\left[\frac{\rho^{\gamma}-1}{\gamma(\rho-1)}\right] \qquad \text{...(2.12)}$$

Th thermal efficiency of diesel cycle increases as compression ratio increases but decreases as cut-off ratio increases (Fig. 2.9). The thermal efficiency of diesel cycle is less than that of Otto cycle. The compression ratio for diesel cycle varies from 14 to 18.

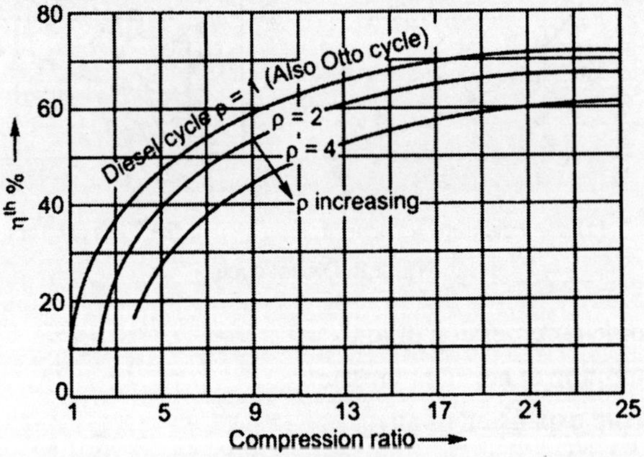

Fig. 2.9 Diesel cycle thermal efficiency v's compression ratio.

Mean effective pressure, $$p_m = \frac{\text{Work done}}{\text{Swept volume}}$$

$$= \frac{c_p(T_3 - T_2) - c_v(T_4 - T_1)}{v_1 - v_2}$$

Now $$v_1 - v_2 = v_1\left(1 - \frac{v_2}{v_1}\right) = v_1\left(1 - \frac{1}{r}\right)$$

$$= \frac{RT_1}{p_1}\left(\frac{r-1}{r}\right) - c_v(\gamma-1)\frac{T_1}{p_1}\times\left(\frac{r-1}{r}\right)$$

$$p_m = \frac{c_p(T_3 - T_2) - c_v(T_4 - T_1)}{c_v(\gamma - 1)\dfrac{T_1}{p_1} \times \left(\dfrac{r-1}{r}\right)}$$

$$= \frac{p_1 r[r(T_3 - T_2) - (T_4 - T_1)]}{(\gamma - 1)\, T_1 (r - 1)}$$

$$= \frac{p_1 r[\gamma T_1 (r^{\gamma - 1}\rho - r^{\gamma - 1}) - T_1(\rho^{\gamma} - 1)]}{(\gamma - 1) T_1 (r - 1)}$$

$$= \frac{p_1 r[\gamma r^{\gamma - 1}(\rho - 1) - (\rho^{\gamma - 1})]}{(r - 1)(\gamma - 1)} \qquad \text{...(2.13)}$$

2.9 DUAL CYCLE

In the dual cycle, part of the heat is supplied at constant volume and the rest at constant pressure. This is also called mixed or limited pressure cycle. The *p-v* and *T-s* diagrams are shown in Fig. 2.10. The various processes are :

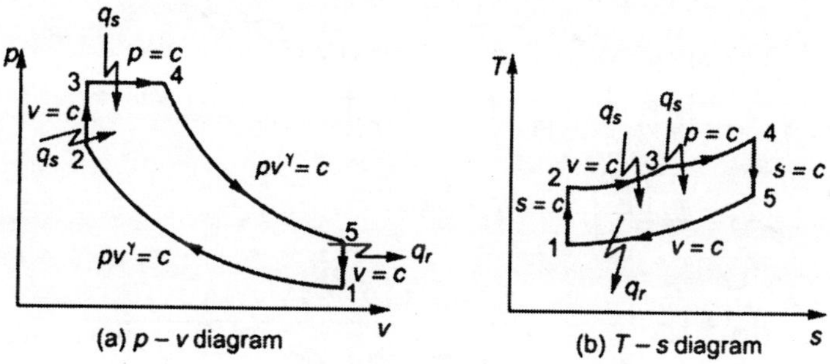

Fig. 2.10 Dual cycle.

Process 1-2 : Isentropic compression of air

Process 2-3 : Constant volume heat addition

Process 3-4 : Constant pressure heat addition

Process 4-5 : Isentropic expansion of air

Process 5-1 : Constant volume heat rejection.

Considering 1 kg of working fluid,

Heat supplied $\quad q_s = c_v(T_3 - T_2) + c_p(T_4 - T_3)$

Hat rejected, $\quad q_r = c_v(T_5 - T_1)$

Work done per cycle, $\quad w = q_s - q_r = c_v(T_3 - T_2) + c_p(T_4 - T_3) - c_v(T_5 - T_1)$

Thermal efficiency, $\quad \eta_{th} = \dfrac{w}{q_s}$

$$= \frac{c_v(T_3 - T_2) + c_p(T_4 - T_3) - c_v(T_5 - T_1)}{c_v(T_3 - T_2) + c_p(T_4 - T_3)}$$

$$= 1 - \frac{T_5 - T_1}{(T_3 - T_2) + r(T_4 - T_3)}$$

Let $r = \frac{v_1}{v_2}$ = compression ratio

$\alpha_p = \frac{p_3}{p_2}$ = pressure ratio

$\rho = \frac{v_4}{v_3} = \frac{v_4}{v_2}$ = cut-off ratio

Then from process 1-2, we have

$$T_2 = T_1 r^{\gamma-1}$$

From process 2-3, $\frac{T_3}{T_2} = \frac{p_3}{p_2} = \alpha_p$ or $T_3 = \alpha_p T_2 = \alpha_p r^{\gamma-1} T_1$

From process 3-4, $\frac{T_4}{T_3} = \frac{v_4}{v_3} = \frac{v_4}{v_2} = \rho$ or $T_4 = \rho T_3 = \rho\alpha_p r^{\gamma-1} T_1$

From process 4-5, $\frac{T_5}{T_4} = \left(\frac{v_4}{v_5}\right)^{\gamma-1} = \left[\left(\frac{v_4}{v_2}\right)\times\left(\frac{v_2}{v_5}\right)\right]^{\gamma-1}$

$$= \left[\left(\frac{v_4}{v_2}\right)\times\left(\frac{v_2}{v_1}\right)\right]^{\gamma-1} = \left(\frac{\rho}{r}\right)^{\gamma-1}$$

$$T_5 = T_4\left(\frac{\rho}{r}\right)^{\gamma-1} = T_1\alpha_p\rho^{\gamma}$$

$$\eta_{th} = 1 - \frac{(\alpha_p\rho^{\gamma} - 1)}{(\alpha_p r^{\gamma-1} - r^{\gamma-1}) + (\alpha_p\rho r^{\gamma-1} - \alpha_p r^{\gamma-1})}$$

$$= 1 - \frac{1}{r^{\gamma-1}}\left[\frac{(\alpha_p\rho^{\gamma} - 1)}{(\alpha_p - 1) + \gamma\alpha_p(\rho - 1)}\right] \quad \text{...(2.14)}$$

The variation of thermal efficiency of dual cycle is shown in Fig. 2.11.

Mean effective pressure,

$$p_m = \frac{\text{Work done}}{\text{Swept volume}} = \frac{w}{v_1 - v_2}$$

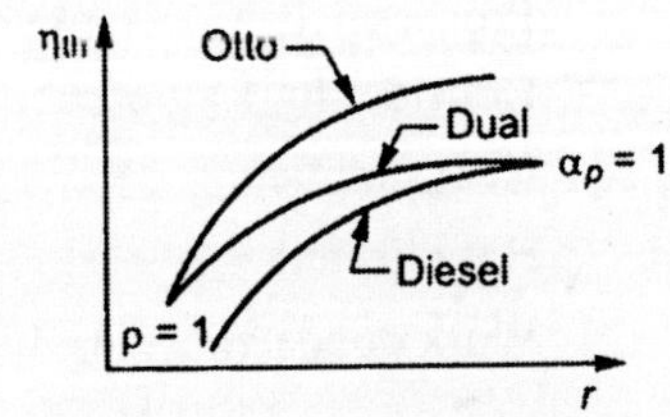

Fig. 2.11 Dual cycle efficiency variation.

Now $v_1 - v_2 = v_1\left(1 - \frac{1}{r}\right)$

$$= v_1(r-1)/r = \frac{RT_1}{p_1}\left(\frac{r-1}{r}\right)$$

$$= \frac{c_v(\gamma-1)T_1}{p_1}\left(\frac{r-1}{r}\right)$$

$$= \frac{c_v(T_3 - T_2) + c_p(T_4 - T_3) - c_v(T_5 - T_1)}{\frac{c_v(\gamma-1)T_1}{p_1}\left(\frac{r-1}{r}\right)}$$

$$= \frac{p_1 r}{(\gamma - 1)(r-1)} \left[\frac{T_3 - T_2}{T_1} + \frac{\gamma (T_4 - T_3)}{T_1} - \frac{T_5 - T_1}{T_1} \right]$$

$$= \frac{p_1 r}{(\gamma - 1)(r-1)} [(\alpha_p r^{\gamma - 1} - r^{\gamma - 1}) + \gamma (\rho \alpha_p r^{\gamma - 1} - \alpha_p r^{\gamma - 1}) - (\alpha_p \rho^{\gamma} - 1)]$$

$$= \frac{p_1 r}{(\gamma - 1)(r-1)} [r^{\gamma - 1} \{(\alpha_p - 1) + \gamma \alpha_p (\rho - 1)\} (\alpha_p \rho^{\gamma} - 1)] \quad ...(2.15)$$

2.10 BRAYTON CYCLE

The air standard Brayton or Joule cycle is a constant pressure cycle usd in gas turbine power plants. The *p-v* and *T-s* diagrams are shown in Fig. 2.12. It consists of the following processes :

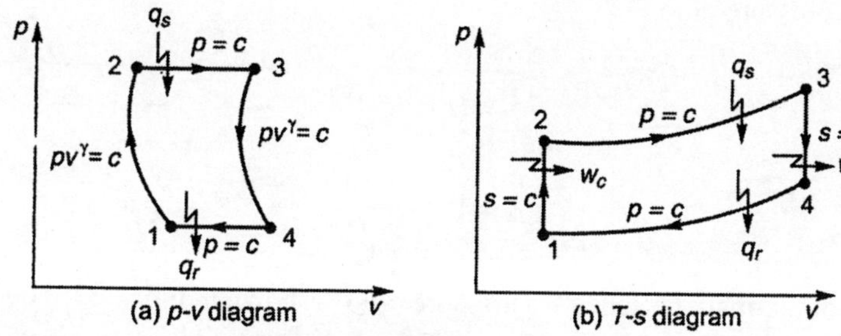

Fig. 2.12 Brayton (or Joule) cycle.

1-2 : Isentropic compression in the compressor

2-3 : Constant pressure heat addition

3-4 : Isentropic expansion of air in the turbine

4-4 : Constant pressure heat rejection.

Consider 1 kg of working fluid.

From first law of thermodynamics for steady flow, rejecting ΔKE and ΔPE, we have

$$\delta_q - \delta w = dh$$

Heat added, $$q_s = h_3 - h_2 = c_p (T_3 - T_2)$$

Hat rejected $$q_r = h_4 - h_1 = c_p (T_4 - T_1)$$

Net work done by turbine, $w_{net} = q_s - q_r = c_p[(T_3 - T_2) - (T_4 - T_1)]$

Also work done by turbine, $w_t = h_3 - h_4 = c_p (T_3 - T_4)$

Work consumed by compressor, $w_c = h_2 - h_1 = c_p (T_2 - T_1)$

$$w_{net} = c_p[(T_3 - T_4) - (T_2 - T_1)] = c_p[(T_3 - T_2) - (T_4 - T_1)]$$

Thermal efficiency, $$\eta_{th} = \frac{\text{Net work}}{\text{Heat added}} = \frac{w_{net}}{q_s}$$

$$= \frac{c_p[(T_3 - T_2) - (T_4 - T_1)]}{c_p[T_3 - T_2]} = 1 - \frac{T_4 - T_1}{T_3 - T_2}$$

From isentropic compression process 1 – 2, we have

$$\frac{T_2}{T_1} = \left(\frac{p_2}{P_1} \right)^{\frac{\gamma - 1}{\gamma}}$$

and from isentropic expansion process 3 – 4,

$$\frac{T_3}{T_4} = \left(\frac{p_3}{p_4}\right)^{\frac{\gamma-1}{\gamma}}$$

But $$p_2 = p_3 \text{ and } p_1 = p_4$$

$$\frac{T_2}{T_1} = \frac{T_3}{T_4} = \left(\frac{p_3}{p_4}\right)^{\frac{\gamma-1}{\gamma}} \text{ or } \frac{T_1}{T_2} = \frac{T_4}{T_3} = \frac{T_4 - T_1}{T_3 - T_2}$$

$\therefore$ $$\eta_{th} = 1 - \frac{T_2}{T_1} = 1 - \frac{T_4}{T_3}$$

Let $r_p = \frac{p_2}{p_1}$ = pressure ratio

$$\frac{T_2}{T_1} = (r_p)^{(\gamma-1)/\gamma}$$

$\therefore$ $$\eta_{th} = 1 - \frac{1}{(r_p)^{(\gamma-1)/\gamma}} \quad ...(2.16)$$

The variation of thermal efficiency v's pressure ratio is shown in Fig. 2.13. The thermal efficiency increases with increasing values of pressure ratio. This cycle is used in gas turbines.

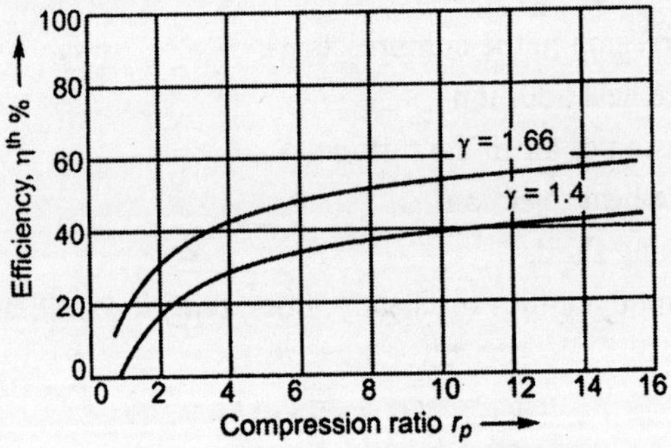

Fig. 2.13 Efficiency v's pressure ratio in simple Brayton cycle.

$$\text{Work ratio} = \frac{w_t - w_c}{w_t}$$

w_t = work done by turbine

w = work supplied to compressor

$$= \frac{(T_3 - T_4) - (T_2 - T_1)}{T_3 - T_4} = 1 - \frac{T_2 - T_1}{T_3 - T_4}$$

$$= 1 - \frac{T_1}{T_3}\left[\frac{r_p^{(\gamma-1)/\gamma} - 1}{1 - \frac{1}{r_p^{(\gamma-1)/\gamma}}}\right] = 1 - \frac{T_1}{T_3}(r_p)^{(\gamma-1)/\gamma} \quad ...(2.13)$$

Pressure ratio for maximum work

$$W = mc_p[(T_3 - T_2) - (T_4 - T_1)]$$
$$= mc_p[(T_3 - T_4) - (T_2 - T_1)]$$
$$\frac{T_3}{T_4} = r_p^{\frac{\gamma-1}{\gamma}} = \frac{T_2}{T_1}$$

Let $$z = \frac{\gamma - 1}{\gamma},\ mc_p = y, \text{ then}$$

$$W = y\left[T_3\left(1 - \frac{1}{r_p^z}\right) - T_1(r_p^z - 1)\right]$$

For W to me maximum, $dW/dr_p = 0$

$$\frac{dW}{dr_p} = [T_3 z r_p^{-(z+1)} - T_1 z r_p^{z-1}] = 0$$

$$r_p = \left(\frac{T_3}{T_1}\right)^{1/(2z)} = \left(\frac{T_3}{T_1}\right)^{\frac{\gamma}{2(\gamma-1)}} \quad \text{...(2.17)}$$

2.11 COMPARISON BETWEEN OTTO, DIESEL, AND DUAL CYCLES

The comparison parameters selected are :

1. Equal compression ratio and heat input
2. Constant maximum pressure and heat input
3. Constant maximum pressure and output
4. Constant maximum pressure and temperature
5. Equal compression ratio and heat rejection.

1. Equal compression ratio and heat input : The *p-v* and *T-s* diagrams for the three cycles are shown in Fig. 2.14. The cycles have been represented as follows :

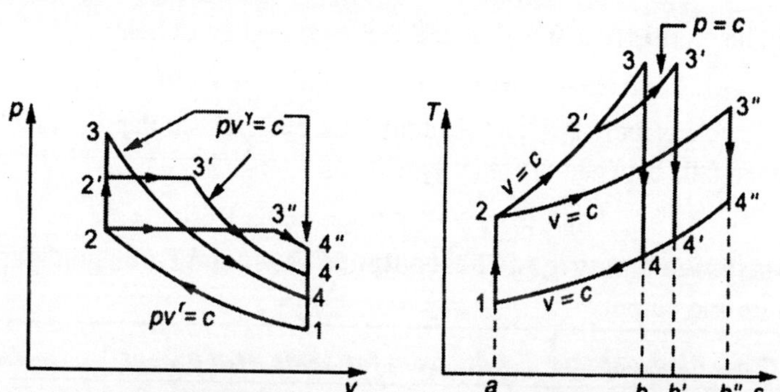

(a) *p-v* diagrams (b) *T-s* diagrams

Fig. 2.14 Comparison for equal compression ratio and heat input.

Cycle	*p-v diagram*	*T-s diagram for same heat input*	*Heat rejected*
Otto	$1-2-3-4$	$a-2-3-b$	$a-1-4-b$
Diesel	$1-2-3''-4''$	$a-2-3''-b''$	$a-1-4''-b''$
Dual	$1-2-2'-3'-4'$	$a-2-2'-3'-b'$	$a-1-4'-b'$

That cycle will be more efficient which rejects the least amount of heat after expansion, because

$$\eta_{th} = 1 - \frac{q_r}{q_s} = 1 - \frac{q_r}{\text{constant}}$$

Thus the order of efficiencies is :

$$\therefore \qquad (\eta_{th})_{Otto} > (\eta_{th})_{Dual} > (\eta_{th})_{Diesel} \qquad \text{...(2.18)}$$

2. Constant maximum pressure and heat input : The *p-v* and *T-s* diagrams for Otto and Diesel cycles are shown in Fig. 2.15 for constant maximum pressure and heat input.

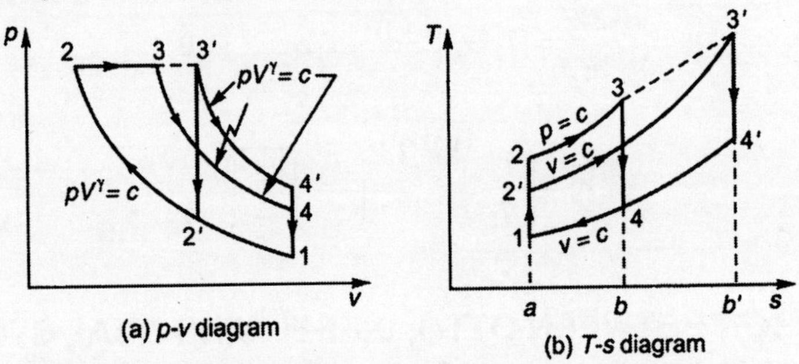

Fig. 2.15 Comparison for equal maximum pressure and heat input.

Cycle	*p-v diagram*	*T-s diagram for same heat input*	*Heat rejected*
Diesel	$1-2-3-4$	$a-2-3-b$	$a-1-4-b$
Otto	$1-2'-3'-4'$	$a-2'-3'-b'$	$a-1-4'-b'$

Area $a-2-3-b$ = Area $a-2'-3'-b'$ for same heat input

$$\therefore \qquad \eta_{th} = 1 - \frac{q_r}{q_s} = 1 - \frac{q_r}{\text{Constant}}$$

Thus cycle will be more efficient that rejects the least amount of heat.

$$(\text{Area } a-1-4-b) < (\text{Area } a-1-4'-b')$$

Thus the Diesel cycle is more efficient than the Otto cycle. A similar argument will show that the dual cycle performance falls between the other two. Thus

$$(\eta_{th})_{Diesel} > (\eta_{th})_{Dual} > (\eta_{th})_{Otto} \qquad \text{...(2.19)}$$

3. Constant maximum pressure and heat output : Th *p-v* and *T-s* diagrams ar shown in Fig. 2.16 for maximum pressure and output.

Cycle	*p-v diagram*	*T-s diagram for same heat output*	*Heat rejected*
Diesel	$1-2-3-4$	$1-2-3-4$	$a-1-4-b$
Otto	$1-2'-3'-4'$	$1-2'-3'-4'$	$a-1-4'-b'$

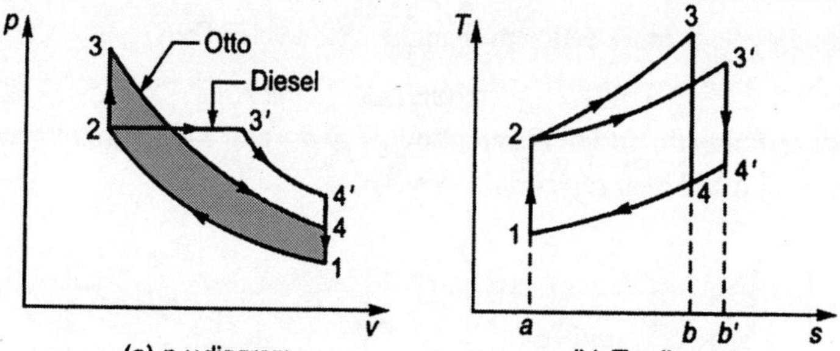

(a) *p-v* diagram (b) *T-s* diagram

Fig. 2.16 Comparison for maximum pressure and heat output.

Area 1 – 2 – 3 – 4 = Area 1 – 2′ – 3′ – 4′ for same heat output

$$\eta_{th} = \frac{w}{q_s} = \frac{w}{w + q_r} = \frac{\text{Constant}}{\text{Constant} + q_r}$$

Thus cycle will be more efficient which rejects least amount of heat.

$$(\text{Area } a - 1 - 4 - b) < (\text{Area } a - 1' - 4' - b')$$

Hence, the Diesel cycle is more efficient than the Otto cycle. Thus

$$\therefore \quad (\eta_{th})_{\text{Diesel}} > (\eta_{th})_{\text{Dual}} > (\eta_{th})_{\text{Otto}} \qquad \ldots(2.20)$$

4. Constant maximum pressure and temperature : The *p-v* and *T-s* diagrams are shown in Fig. 2.17 for same maximum pressure and temperature.

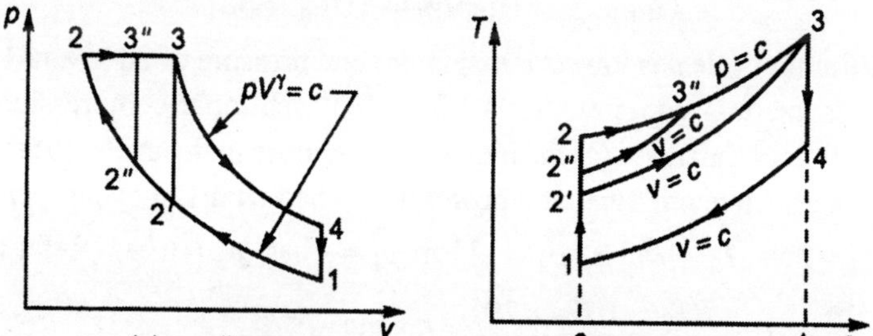

(a) *p-v* diagram

(b) *T-s* diagram

Fig. 2.17 Comparison for equal maximum pressure and temperature.

Cycle	*p-v diagram*	*T-s diagram for same maximum temperature*	*Heat rejected*
Diesel	1 – 2 – 3 – 4	1 – 2 – 3 – 4	$a - 1 - 4 - b$
Otto	1 – 2′ – 3 – 4	1 – 2′ – 3 – 4	$a - 1 - 4 - b$
Dual	1 – 2″ – 3″ – 3 – 4	1 – 2″ – 3″ – 3 – 4	$a - 1 - 4 - b$

Now
$$\eta_{th} = 1 - \frac{q_r}{q_s} = \frac{\text{constant}}{q_s}$$

All the cycles reject equal amount of heat.

Thus cycle will be more efficient which has higher heat addition.

Hence, the Diesel cycle is more efficient than the Otto cycle. Thus

$$\therefore \qquad (\eta_{th})_{Diesel} > (\eta_{th})_{Dual} > (\eta_{th})_{Otto} \qquad ...(2.21)$$

5. Equal compression ratio and heat rejection : The *p-v* and *T-s* diagrams are shown in Fig. 2.18, for equal compression ratio and heat rejection.

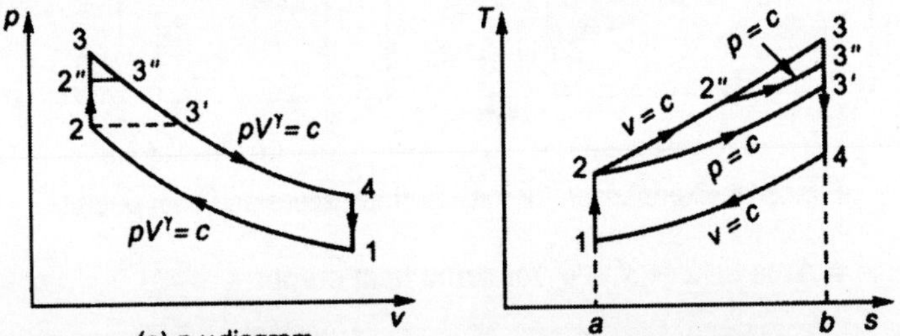

(a) *p-v* diagram

(b) *T-s* diagram

Fig. 2.18 Comparison for equal compression ratio and heat rejection.

Cycle	*p-v diagram*	*T-s diagram*	*Heat input*
Otto	1 – 2 – 3 – 4 – 1	1 – 2 – 3 – 4 – 1	*a* – 2 – 3 – *b*
Diesel	1 – 2 – 3′ – 4 – 1	1 – 2 – 3′ – 4 – 1	*a* – 2 – 3′ – *b*
Dual	1 – 2 – 2″ – 3″ – 4 – 1	1 – 2 – 2″ – 3″ – 4′ – 1	*a* – 2 – 2″ – 3″ – *b*

For same q_r, if q_s is more, η is more

$$\therefore \qquad (\eta_{th})_{Otto} > (\eta_{th})_{Dual} > (\eta_{th})_{Diesel} \qquad ...(2.22)$$

■ **Example 2.1** A Carnot cycle has lowest temperature and pressure of 20°C and 1 bar. The pressures are : 4 bar after isothermal compression; 12 bar after isentropic compression, and 6 bar after isothermal heat addition. Calculate (a) the highest temperature in the cycle, (b) thermal efficiency of the cycle, (c) mean effective pressure, and power developed with 150 cycles per minute.

Solution. Given : $p_4 = 1$ bar, $T_2 = 293$ K, $p_2 = 12$ bar, $p_1 = 4$ bar, $p_3 = 6$ bar (Refer to Fig. 2.2)

$$v_4 = \frac{RT_2}{p_4} = \frac{0.287 \times 293}{1 \times 10^2} = 0.8409 \text{ m}^3/\text{kg}$$

$$v_1 = \frac{p_4 v_4}{p_1} = \frac{1 \times 10^2 \times 0.8409}{4 \times 10^4} = 0.2102 \text{ m}^3/\text{kg}$$

$$v_2 = v_1 \left(\frac{p_1}{p_2}\right)^{1/\gamma} = 0.2102 \left(\frac{4}{12}\right)^{1/1.4} = 0.0958 \text{ m}^3/\text{kg}$$

$$v_3 = \frac{v_2 p_2}{p_3} = \frac{0.0958 \times 12}{6} = 0.1916 \text{ m}^3/\text{kg}$$

Stroke volume $v_s = v_4 - v_2 = 0.8409 - 0.0958 = 0.7451 \text{ m}^3/\text{kg}$

(a) Highest temperature in the cycle,

$$T_1 = T_2 \left(\frac{v_1}{v_2}\right)^{\gamma - 1} = 293 \left(\frac{0.2102}{0.0985}\right)^{0.4} = 401.2 \text{ K}$$

(b) Heat supplied, $q_s = p_2 v_2 \ln\left(\frac{v_3}{v_2}\right) = 12 \times 10^2 \times 0.0958 \ln\left(\frac{0.1916}{0.0958}\right)$

$$= 79.67 \text{ kJ/kg}$$

Heat rejected, $q_r = p_4 v_4 \ln\left(\frac{v_1}{v_4}\right) = 1 \times 10^2 \times 0.8409 \ln\left(\frac{0.2102}{0.8409}\right)$

$$= -58.07 \text{ kJ/kg}$$

Net work done, $w_{net} = q_s - q_r = 79.67 - 58.07 = 21.60$ kJ/kg

Thermal efficiency, $\eta_{th} = \frac{w_{net}}{q_s} = \frac{21.60}{79.67} = 0.2695$ or 26.95%

(c) Mean effective pressure $= \frac{w_{net}}{\text{Stroke volume}}$

$$= \frac{21.60}{0.7451} = 28.98 \text{ kg/m}^2 \text{ or } 0.2898 \text{ bar}$$

(d) Power developed $= w_{net} \times \text{rpm}/60 = 21.6 \times 150/60 = 54$ kW

■ **Example 2.2** In a Carnot cycle engine, the ratio of greatest to least volume is fixed as x. The lower temperature of the cycle is fixed but the isentropic compression ratio r is variable. Prove that for maximum work done, the following equation holds good :

$$(\gamma - 1) \ln\left(\frac{x}{r}\right) + \frac{1}{r^{\gamma - 1}} - 1 = 0$$

Solution. Refer to Fig. 2.19 and assume 1 kg of working fluid.

Work done, $w = q_s - q_r = p_2 v_2 \ln \frac{v_3}{v_2} - p_4 v_4 \ln \frac{v_4}{v_1}$

$$= p_2 v_2 \ln \frac{v_3}{v_2} + p_4 v_4 \ln \frac{v_1}{v_4}$$

$$= RT_1 \ln \frac{v_3}{v_2} + RT_2 \ln \frac{v_1}{v_4}$$

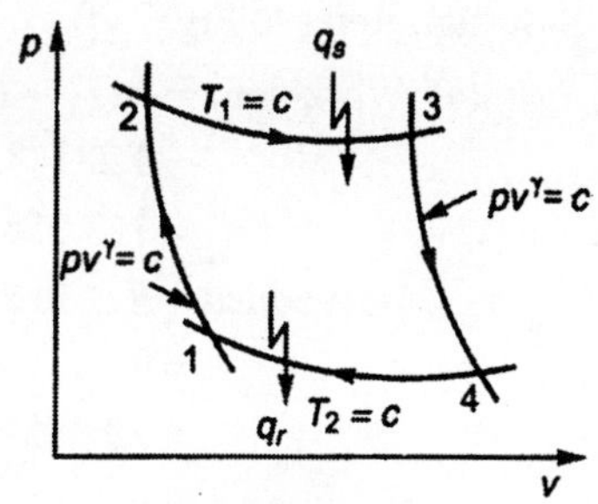

Fig. 2.19

Now, $T_1 = T_2 \left(\frac{v_1}{v_2}\right)^{\gamma - 1} = T_2 r^{\gamma - 1}; r = \frac{v_1}{v_2}$

$$T_1 = T_2 \left(\frac{v_4}{v_3}\right)^{\gamma - 1}$$

which give

$$\frac{v_1}{v_2} = \frac{v_4}{v_3} \text{ or } \frac{v_3}{v_2} = \frac{v_4}{v_1}$$

Now $x = \frac{v_4}{v_2}$ and $r = \frac{v_1}{v_2} = \frac{v_4}{v_3}$

$$\therefore \qquad \frac{v_3}{v_2} = \frac{v_3}{v_4} \times \frac{v_4}{v_2} = \frac{x}{r} = \frac{v_4}{v_1}$$

$$\therefore \qquad w = RT_1 \ln \frac{x}{r} + RT_2 \ln \frac{r}{x} = RT_2 r^{\gamma - 1} \ln \frac{x}{r} + RT_2 \ln \frac{r}{x}$$

$$= RT_2 (r^{\gamma - 1} - 1) \ln \frac{x}{r}$$

For maximum work, $\frac{dw}{dr} = 0$

$$RT_2 \left[\left(\ln \frac{x}{r} \right) \times (\gamma - 1) r^{\gamma - 2} + (r^{\gamma - 1} - 1) \times \frac{r}{x} \left(-\frac{x}{r^2} \right) \right] = 0$$

$$\left(\ln \frac{x}{r} \right) \times (\gamma - 1) r^{\gamma - 2} - r^{\gamma - 2} + \frac{1}{r} = 0$$

$$\left(\ln \frac{x}{r} \right) \times (\gamma - 1) r^{\gamma - 1} - r^{\gamma - 1} + 1 = 0$$

$$\ln \left(\frac{x}{r} \right) (\gamma - 1) + \frac{1}{r^{\gamma - 1}} - 1 = 0$$

$$(\gamma - 1) \ln \left(\frac{x}{r} \right) + \frac{1}{r^{\gamma - 1}} - 1 = 0 \qquad \text{Hence Proved.}$$

■ **Example 2.3** The following particulars refer to an engine working on Stirling cycle : Compression ratio = 3, lower temperature = 30°, speed = 50 rpm, higher temperature = 550°C, regenerative efficiency = 90%, initial pressure = 1 bar, heat added = 2300 kJ/min.

Calculate (a) work done per kg, (b) thermal efficiency, (c) mean effective pressure and (d) indicated power.

Solution. Refer to Fig. 2.20

(a) Work done $= R(T_1 - T_2) \ln r$

$= 0.287\ (823 - 300) \ln 3$

$= 163.956$ kJ/kg

Heat added $= RT_1 \ln r + (1 - \eta_{reg})\, c_v (T_1 - T_2)$

$= 0.287 \times 823 \times \ln 3 + (1 - 0.9) \times 0.718 \times (823 - 303)$

$= 296.829$ kJ/kg

(b) Thermal efficiency $= \frac{\text{Work done}}{\text{Heat added}}$

$= \frac{163.956}{286.829} = 0.5523$ or 55.23%

(c) Air flow rate $= \frac{2300}{296.829} = 7.7485$ kg/min

Working cycles/min $= \frac{50}{2} = 25$

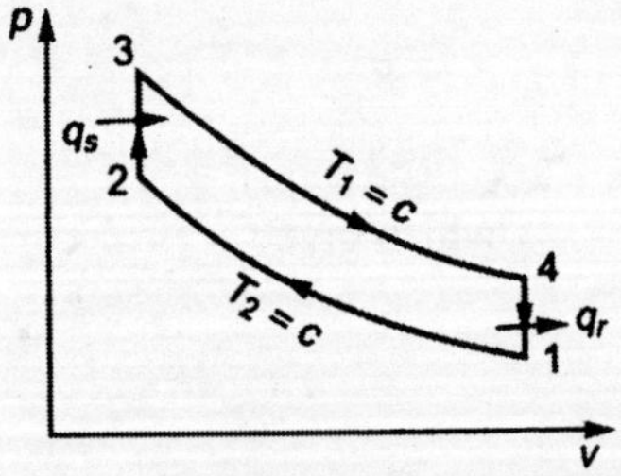

Fig. 2.20

Mass of air par working cycle, $m = \frac{7.7485}{25} = 0.31\,\text{kg}$

Volume of air at state 1, $V_1 = \frac{mRT_2}{p_1} = \frac{0.31 \times 0.287 \times 303}{1 \times 10^2} = 0.2694\,\text{m}^3$

$$V_2 = \frac{V_1}{r} = \frac{0.2694}{3} = 0.0898\,\text{m}^3$$

Stroke volume, $V_s = V_1 - V_2 = 0.2694 - 0.0898 = 0.1796\,\text{m}^3$

Mean effective pressure, $p_m = \frac{\text{Work done}}{\text{Stroke volume}} = \frac{163.956 \times 0.31}{0.1796 \times 10^2} = 2.978\text{ bar}$

(d) Indicated power $= \frac{p_m LAN}{60} = \frac{2.978 \times 0.1796 \times 50 \times 10^2}{60} = 42.337\,\text{kW}$

■ **Example 2.4** An engine working on Atkinson cycle takes in air at 1 bar and 25°C. Th air is compressed isentropically by a compression ratio of 6 : 1. The heat is added at constant volume increasing the final pressure to 20 bar. Now, the air is expanded isentropically to 1 bar. The heat is rejected at constant pressure. Calculate (a) pressure and temperature at various points, (b) work done per kg of air, and (c) cycle efficiency.

Solution. Refer to Fig. 2.21

(a) $p_1 = 1\text{ bar}$, $T_1 = 25 + 273 = 298\text{ K}$

$$T_2 = T_1 r^{\gamma - 1} = 298 \times 6^{0.4} = 610.2\text{ K}$$

$$p_2 = p_1 r^{\gamma} = 1 \times 6^{1.4} = 12.286\text{ bar}$$

$$p_3 = 20\text{ bar}$$

$$T_3 = T_2 \frac{p_3}{p_2} = 610.2 \times \frac{20}{12.286} = 993.3\text{ K}$$

$$p_4 = p_1 = 1\text{ bar}$$

$$T_4 = \frac{T_3}{\left(\frac{p_3}{p_4}\right)^{\frac{\gamma-1}{\gamma}}} = \frac{993.3}{20^{0.286}} = 422.7\text{ K}$$

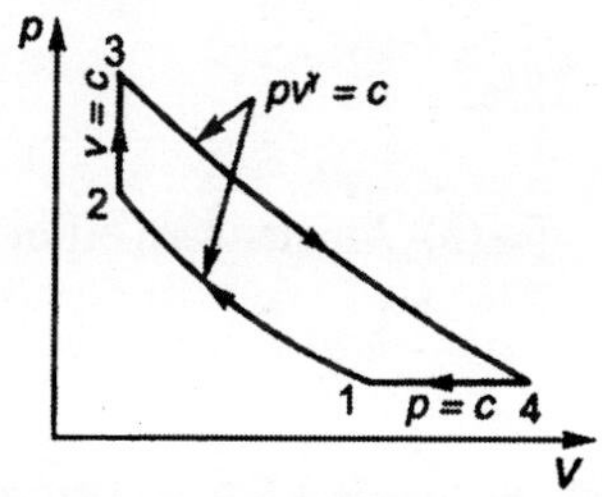

Fig. 2.21

Heat added, $q_s = c_v(T_3 - T_2) = 0.718\,(993.3 - 610.2) = 279.4\text{ kJ/kg}$

Heat rejected $q_r = c_p(T_4 - T_1) = 1.005(422.7 - 298) = 124.7\text{ kJ/kg}$

(b) Wore done, $w = q_s - q_r = 279.4 - 124.7 = 154.7\text{ kJ/kg}$

(c) Thermal efficiency $= \frac{w}{q_s} = \frac{154.7}{279.4} = 0.5536$ or 55.36%

■ **Example 2.5** An engine working on Otto cycle has the following conditions :

Pressure at the beginning of compression = 1 bar

Pressure at the end of compression = 10 bar

Calculate (a) the compression ratio, and (b) air standard efficiency of the engine.

Assume $\gamma = 1.4$.

Solution. Refer to Fig. 2.6.

(a) Compression ratio,

$$r = \frac{v_1}{v_2} = \left(\frac{T_2}{T_1}\right)^{\frac{1}{\gamma-1}} = \left(\frac{p_2}{p_1}\right)^{\frac{1}{\gamma}}$$

$$= 10^{1/1.4} = 10^{0.7143} = 5.18$$

(b) Air standard efficiency,

$$\eta_a = 1 - \frac{1}{r^{\gamma-1}} = 1 - \frac{1}{5.18^{0.4}} = 0.482 \text{ or } 48.2\%$$

■ **Example 2.6** In an engine working on ideal Otto cycle, the temperature at the beginning and end of compression are 45°C and 370°C. Find the compression ratio and air standard efficiency of the engine. Assume $\gamma = 1.4$.

Solution. Refer to Fig. 2.6

(a) Compression ratio

$$r = \frac{v_1}{v_2} = \left(\frac{T_2}{T_1}\right)^{\frac{1}{\gamma-1}}$$

$$T_1 = 45 + 273 = 318 \text{ K}$$

$$T_2 = 370 + 273 = 643 \text{ K}$$

$$r = \left(\frac{643}{310}\right)^{\frac{1}{0.4}} = 5.814$$

(b) Air standard efficiency,

$$\eta_a = 1 - \frac{1}{r^{\gamma-1}} = 1 - \frac{1}{5.814^{0.4}} = 0.505 = 50.5\%$$

■ **Example 2.7** In an Otto-cycle, air at 15°C and 1 bar is compressed adiabatically until the pressure is 15 bar. Heat is added at constant volume until the pressure rises to 40 bar. Calculate (a) the air standard efficiency, (b) the compression ratio, (c) the mean effective pressure for the cycle. Assume $c_v = 0.718$ kJ/kg.K, $\gamma = 1.4$ and $R = 8.314$ kJ/mol.K.

Solution. The Otto cycle is shown in Fig. 2.22

For isentropic process 1 – 2,

$$p_1 v_1^{\gamma} = p_2 v_2^{\gamma}$$

Compression ratio,

$$r = \frac{v_1}{v_2} = \left(\frac{p_2}{p_1}\right)^{\frac{1}{\gamma}} = 15^{\frac{1}{1.4}} = 6.92$$

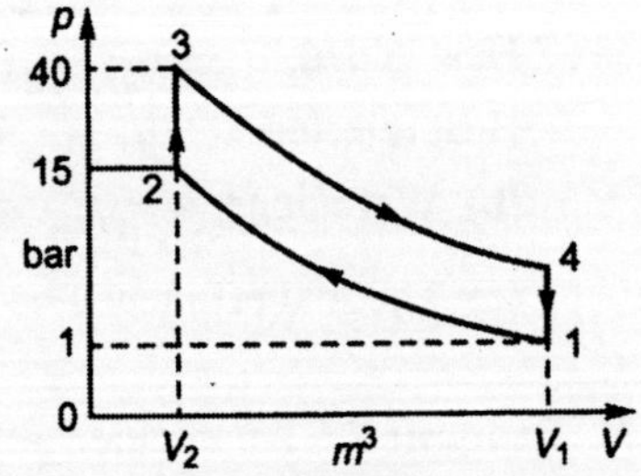

Fig. 2.22

Air standard efficiency,

$$\eta_a = 1 - \frac{1}{r^{\gamma-1}}$$

$$= 1 - \frac{1}{6.92^{0.4}} = 0.539 \text{ or } 59.9\%$$

$$T_1 = 15 + 273 = 288 \text{ K}$$

$$T_2 \left(\frac{p_2 v_2}{p_1 v_1}\right) T_1 = \frac{15 \times 288}{6.92} = 629.3 \text{ K}$$

For constant volume process 2 – 3 :

$$T_3 = \frac{p_3}{p_2} T_2 = \frac{40}{6.92} \times 629.3 = 1669.7 \text{ K}$$

Heat supplied. $q_s = c_v(T_3 - T_2) = 0.718(1669.7 - 629.3) = 747.03$ kJ/kg

Work done, $w = \eta_a q_s = 0.539 \times 747.03 = 402.65$ kJ/kg

Mean effective pressure,

$$p_m = \frac{\text{Work done}}{\text{Swept volume}} = \frac{w}{v_1 - v_2}$$

$$v_1 = \frac{V_1}{m} = \frac{mRT_1}{mp_1} = \frac{RT_1}{p_1} = \frac{8314 \times 288}{29 \times 1 \times 10^5} = 0.8256 \text{ m}^3/\text{kg}$$

$$v_2 = \frac{v_1}{r} = \frac{0.8257}{6.92} = 0.1193 \text{ m}^3/\text{kg}$$

Specific volume, $v_1 - v_2 = 0.7064 \text{ m}^3/\text{kg}$

$$p_m = \frac{402.65}{0.7064} \times 10^3 = 9.7 \times 10^5 \text{ N/m}^2 \text{ or } 9.7 \text{ bar}$$

■ **Example 2.8** A gas engine working on Otto cycle has a cylinder of diameter 220 mm and stroke 300 mm. The clearance volume is 1600 cc. Find the air standard efficiency. Assume $c_p = 1.004$ kJ/kg.K and $c_v = 0.718$ kJ/kg.K for air

Solution. Swept volume,

$$V_s = \left(\frac{\pi}{4}\right) d^2 L = \left(\frac{\pi}{4}\right) \times 22^2 \times 30 = 11404 \text{ cc}$$

Compression ratio, $\quad r = 1 + \frac{V_s}{V_c} = 1 + \frac{11404}{1600} = 8.13$

Adiabatic index, $\quad \gamma = \frac{c_p}{c_v} = \frac{1.004}{0.718} = 1.4$

Air standard efficiency,

$$\eta_a = 1 - \frac{1}{r^{\gamma - 1}} = \frac{1 - 1}{(8.13)^{0.4}} = 0.5675 \text{ or } 56.75\%$$

■ **Example 2.9** A gas engine working on ideal Otto cycle has a compression ratio of 6 : 1. The pressure and temperature at the commencement of compression are 1 bar and 27°C. The heat supplied during the constant volume combustion process is 1200 kJ/kg. Determine th peak pressure and temperature, work output per kg of air and air standard efficiency. Assume $c_v = 0.718$ kJ/kg and $\gamma = 1.4$ for air.

Solution. Isentropic compression process 1–2 (see Fig. 2.6)

$$\frac{p_1}{p_2} = \left(\frac{v_1}{v_2}\right)^{\gamma} = r^{\gamma} = 6^{1.4} = 12.28$$

$$p_2 = 1 \times 12.28 = 12.28 \text{ bar} = 12.28 \times 10^5 \text{ Pa}$$

$$\frac{T_2}{T_1} = \left(\frac{v_1}{v_2}\right)^{\gamma - 1} = r^{\gamma - 1} = 6^{0.4} = 2.048$$

$$T_2 = 2.048 \times 300 = 619.3 \text{ K} = 341.3°\text{C}$$

Constant volume heat addition process 2 – 3:

$$q_s = q_{2-3} = c_v(T_3 - T_2)$$

$$1200 = 0.718\,(T_3 - 619.3)$$

$$T_3 = 2289.6 \text{ K}$$

$$\frac{p_3}{p_2} = \frac{T_3}{T_2} = \frac{2285.6}{614.3} = 9.72$$

$$p_3 = 12.28 \times 10^5 \times 9.72 = 49.67 \times 10^5 \text{ Pa} = 49.67 \text{ bar}$$

Work output, w = area 1–2–3–4

$$= \frac{1}{\gamma - 1}[(p_3 v_3 - p_4 v_4) - (p_2 v_2 - p_1 v_2)]$$

$$= \frac{R}{\gamma - 1}[(T_3 - T_4) - (T_2 - T_1)]$$

$$\frac{T_3}{T_4} = \left(\frac{v_3}{v_4}\right)^{\gamma - 1} = r^{\gamma - 1} = 6^{0.4} = 2.048$$

$$T_4 = \frac{2285.6}{2.048} = 1116.0 \text{ K}$$

$$R = c_p - c_v = 1.004 - 0.718 = 0.286 \text{ kJ/kg-K}$$

$$w = \frac{0.286}{0.4}[(2289.6 - 1116.0) - (619.3 - 300)]$$

$$w = 611.54 \text{ kJ/kg}$$

Air standard efficiency,

$$\eta_a = 1 - \frac{1}{r^{\gamma - 1}} = 1 - \frac{1}{6^{0.4}} = 0.5116 \text{ or } 51.16\%$$

■ **Example 2.10** The pressure limits in an Otto cycle are 1 bar and 20 bar. The compression ratio is 5. Calculate (a) thermal efficiency and (b) mean effective pressure. Assume $\gamma = 1.4$ for air.

Solution. (a) $\eta_{th} = 1 - \dfrac{1}{r^{\gamma - 1}} = 1 - \dfrac{1}{5^{0.4}} = 0.4747 \text{ or } 47.47\%$

(b) From Fig. 2.6, we have

$$\frac{p_2}{p_1} = \left(\frac{v_1}{v_2}\right)^{\gamma} = r^{\gamma} = 5^{1.4} = 9.5182$$

$$p_2 = 1 \times 9.5182 = 9.5182 \text{ bar}$$

$$p_3 = 20 \text{ bar}$$

Pressure Ratio, $$\alpha_p = \frac{p_3}{p_2} = \frac{20}{9.5182} = 2.1$$

$$p_m = p_1 r \left[\frac{(\alpha_p - 1)(r^{\gamma-1} - 1)}{(r-1)(\gamma-1)}\right] = 1 \times 5 \left[\frac{(2.1-1)(5^{0.4}-1)}{(5-1)(1.4-1)}\right] = 9.1 \text{ bar}$$

■ **Example 2.11** An air standard Otto cycle has a compression ratio of 6. The temperature at the start of compression process is 25°C and the pressure is 1 bar. If the maximum temperature of the cycle is 1150°C, calculate (a) the heat supplied per kg of air, (b) net work done per kg of air, and (c) thermal efficiency of cycle. Assume $\gamma = 1.4$, $c_v = 0.718$ kJ/kg.K for air.

Solution. Process 1–2 (see Fig. 2.6)

$$\frac{T_2}{T_1} = \left(\frac{v_1}{v_2}\right)^{\gamma-1} = 6^{0.4} = 2.0476$$

$$T_1 = 25 + 273 = 298 \text{ K}$$

$$T_2 = 298 \times 2.0476 = 610.2 \text{ K}$$

Process 3–4 : $$\frac{T_4}{T_3} = \left(\frac{v_3}{v_4}\right)^{\gamma-1} = \left(\frac{1}{6}\right)^{0.4} = 0.4884$$

$$T_3 = 1150 + 273 = 1423 \text{ K}$$

$$T_4 = 1423 \times 0.4884 = 699.9 \text{ K}$$

(a) Hat supplied, $q_s = c_v(T_3 - T_2) = 0.718\,(1423 - 610.2) = 589.59$ kJ/kg

(b) Heat rejected, $q_r = c_v(T_4 - T_1) = 0.718\,(699.9 - 298) = 289.97$ kJ/kg

Work done per cycle, $w = q_s - q_r = 589.59 - 289.97 = 298.62$ kJ/kg

$$\eta_{th} = \frac{w}{q_s} = \frac{298.62}{584.59} = 0.5117 \text{ or } 51.17\%$$

■ **Example 2.12** A petrol engine with compression ratio of 5 develops 20 kW indicated power and consumes 8 litres of fuel per hour. The specific gravity of fuel is 0.78 and its calorific value is 44 MJ/kg. Calculate the indicated thermal efficiency and relative efficiency. Take $\gamma = 1.4$.

Solution. Air standard efficiency,

$$\eta_a = 1 - \frac{1}{r^{\gamma-1}} = 1 - \frac{1}{5^{0.4}} = 0.4747 = 47.47\%$$

Fuel consumption $= 8 \times 0.78 \times 1 = 6.24$ kg/h

$$= \frac{6.24}{3600} = 1.733 \times 10^{-3} \text{ kg/s}$$

Indicated thermal efficiency,

$$(\eta_{th})_i = \frac{\text{Indicated work}}{\text{Heat supplied}} = \frac{20}{44 \times 10^3 \times 1.733 \times 10^{-3}}$$

$$= 0.2622 = 26.22\%$$

Relative efficiency $= \dfrac{(\eta_{th})_i}{\eta_a} = \dfrac{0.2622}{0.4747} = 0.5524 \text{ or } 59.24\%$

■ **Example 2.13** A diesel engine has a compression ratio of 20 and cut-off takes place at 5% of the stroke. Find the air-standard efficiency. Assume $\gamma = 1.4$

Solution. Refer to Fig. 2.8.

$$\frac{v_1}{v_2} = r = 20$$

$$v_1 = 20\, v_2$$

$$v_s = v_1 - v_2 = 20v_2 - v_2 = 19\, v_2$$

$$v_3 = 0.05\, v_s + v_2 = 0.05 \times 19v_2 + v_2 = 1.95v_2$$

Cut-off ratio, $\rho = \dfrac{v_3}{v_2} = 1.95$

$$\eta_a = 1 - \frac{1}{r^{\gamma-1}}\left[\frac{\rho^{\gamma} - 1}{\gamma(\rho - 1)}\right]$$

$$\eta_a = 1 - \frac{1}{20^{0.4}}\left[\frac{1.95^{1.4} - 1}{1.4\,(1.95 - 1)}\right] = 0.649 \text{ or } 69.9\%$$

■ **Example 2.14** In an engine working on diesel cycle, inlet pressure and temperature are 1 bar and 20°C. Pressure at the end of adiabatic compression is 40 bar. The ratio of expansion after constant pressure heat addition is 9. Calculate the heat supplied, heat rejected and the efficiency of the cycle. Assume $c_p = 1.004$ kJ/kg-K and $c_v = 0.717$ kJ/kg-K.

Solution. Refer to Fig. 2.8

Adiabatic index, $\gamma = \dfrac{c_p}{c_v} = \dfrac{1.004}{0.717} = 1.4$

Process 1–2 : $r = \dfrac{v_1}{v_2} = \left(\dfrac{p_2}{p_1}\right)^{\frac{1}{\gamma}} = (40)^{1/1.4} = 19.942$

Cut-off ratio : $\rho = \dfrac{v_3}{v_2} = \dfrac{v_3}{v_1} \times \dfrac{v_1}{v_2} = \dfrac{14.942}{5} = 2.788$

$$\frac{T_2}{T_1} = \left(\frac{p_2}{p_1}\right)^{\frac{\gamma-1}{\gamma}} = (40)^{\frac{0.4}{1.4}} = 40^{0.286}$$

$$T_2 = 2.872 \times 293 = 841.5 \text{ K}$$

Process 2–3 : $T_3 = T_2\dfrac{v_3}{v_2} = 841.5 \times 2.788 = 2346.1 \text{ K}$

Process 3–4 : $T_4 = T_3\left(\dfrac{v_3}{v_2}\right)^{\gamma-1} = T_3\left(\dfrac{v_3}{v_1}\right)^{\gamma-1}$

$$= 2346.1\left(\frac{1}{5}\right)^{0.4} = 1232.4 \text{ K}$$

Heat supplied, $q_s = c_p(T_3 - T_2) = 1.004\,(2346.1 - 841.5)$

$$= 1510.62 \text{ kJ/kg}$$

Heat rejected, $q_r = c_v(T_4 - T_1) = 0.717(1232.4 - 293)$

$$= 673.55 \text{ kJ/kg}$$

$$\eta_a = \frac{q_s - q_r}{q_s} = 1 - \frac{673.55}{1510.62} = 0.5541 = 55.41\%$$

■ **Example 2.15** A diesel cycle operates at a pressure of 1 bar at the beginning of compression and the volume is compressed to $\frac{1}{15}$ th of the initial volume. Heat is supplied until the volume is twice that of the clearance volume. Calculate the mean effective pressure of the cycle. Assume $\gamma = 1.4$.

Solution. With reference to Fig. 2.8,

$$r = \frac{v_1}{v_2} = 15$$

$$v_1 = 15 v_2$$

$$v_3 = 2\,v_2$$

Swept volume, $v_s = v_1 - v_2 = (r - 1)v_2$

$$= (15 - 1)v_2 = 14 v_2$$

or

$$v_2 = \frac{v_s}{14}$$

Process 1–2 :

$$p_3 = p_1\left(\frac{v_1}{v_2}\right)^{\gamma} = 1 \times (15)^{1.4} = 49.31 \text{ bar}$$

$$p_2 = p_3 = 49.31 \text{ bar}$$

$$p_4 = p_3\left(\frac{v_3}{v_4}\right)^{\gamma} = p_3\left(\frac{v_3}{v_1}\right)^{\gamma} = p_3\left(\frac{v_3}{v_2} \times \frac{v_2}{v_1}\right)^{\gamma}$$

$$= 49.31\left(\frac{2}{15}\right)^{1.4} = 2.64 \text{ bar}$$

$$p_m = \frac{1}{v_s}\left[p_2(v_3 - v_2) + \frac{p_3 v_3 - p_4 v_4}{\gamma - 1} - \frac{p_2 v_2 - p_1 v_1}{\gamma - 1}\right]$$

$$= \frac{v_2}{v_s}\left[p_2\left(\frac{v_3}{v_2} - 1\right) + \frac{1}{\gamma - 1}\left(p_3 \frac{v_3}{2} - p_4 \frac{v_4}{v_2}\right) - \frac{1}{\gamma - 1}\left(p_2 - p_1 \frac{v_1}{v_2}\right)\right]$$

$$= \frac{1}{14}\left[49.31(2 - 1) + \frac{1}{0.4}(49.31 \times 2 - 2.64 \times 15) - \frac{1}{0.4}(49.31 - 1 \times 15)\right]$$

$$= 0.071\,[49.31 + 122.55 - 79.27] = 6.685 \text{ bar}$$

■ **Example 2.16** The mean effective pressure of an ideal diesel cycle is 10 bar. If the initial pressure is 1 bar and the compression ratio is 14, determine the cut-off ratio and the air-standard efficiency. Assume $\gamma = 1.4$.

Solution. See Fig. 2.8,

$$\text{Work output} = p_m v_s = p_2(v_3 - v_2) + \frac{p_3v_3 - p_4v_4}{\gamma - 1} - \frac{p_2v_2 - p_1v_1}{\gamma - 1}$$

$$r = \frac{v_1}{v_2} = 1 + \frac{v_s}{v_c} = 14$$

$$v_s = 13v_c$$

Now,

$$v_2 = v_c$$

$\therefore$

$$v_s = 13v_2$$

and

$$v_4 = v_1 = 14v_2 = 14v_c$$

$$v_3 = \rho v_2 = \rho v_c$$

$$\frac{p_2}{p_1} = r^\gamma = 14^{1.4} = 40.23$$

$$p_2 = 1 \times 40.23 = 40.23 \text{ bar}$$

$$\frac{p_3}{p_4} = \left(\frac{r}{\rho}\right)^\gamma = \frac{14^{1.4}}{\rho^{1.4}} = \frac{40.23}{\rho^{1.4}}$$

$$p_4 = \frac{40.23 \times \rho^{1.4}}{40.23} = \rho^{1.4}$$

Work done per cycle

$$= p_2(v_3 - v_2) + \frac{p_3v_3 - p_4v_4}{\gamma - 1} - \frac{p_2v_2 - p_1v_1}{\gamma - 1}$$

$$= \frac{1}{0.4}(40.23 \times \rho v_c - \rho^{1.4} \times 1.4v_c) - \frac{1}{0.4}(40.23\, v_c - 1 \times 14\, v_c)$$

$$p_m v_s = v_c\left[40.24\,(p-1) + \frac{1}{0.4}(40.23\,p - 14\,p^{1.4}) - \frac{1}{0.4}(40.23 - 14)\right]$$

$$10 \times 13v_c = v_c[40.23\,\rho - 40.23 + 100.6\rho - 35\rho^{1.4} - 100.6 + 35]$$

$$130 = 140.83\rho - 35\rho^{1.4} - 109.83$$

or

$$f(\rho) = \rho^{1.4} - 9.024\rho + 6.738 = 0$$

ρ	2	2.5	2.6	2.64
$f(\rho)$	1.329	0.285	0.086	0.007

We take

$$\rho = 2.64$$

$$\eta = 1 - \frac{1}{r^{\gamma-1}}\left[\frac{\rho^\gamma - 1}{\gamma(\rho - 1)}\right]$$

$$1 - \frac{1}{14^{0.4}}\left[\frac{2.64^{1.4} - 1}{1.4(2.64 - 1)}\right] = 0.5616 = 56.16\%$$

■ **Example 2.17** In an engine working on the diesel cycle, the air-fuel ratio by weight is 50 : 1. The temperature of air at the beginning of combustion is 40°C and the compression ratio is 19. What is the ideal efficiency of the engine? Calorific value of fuel is 42 MJ/kg. Assume $c_v = 0.717$ kJ/kg K, $c_p = 1.004$ kJ/kg K.

Solution. Refer to Fig. 2.8

$$\gamma = \frac{c_p}{c_v} = \frac{1.004}{0.717} = 1.4$$

$$\rho = \frac{v_3}{v_2} = \frac{T_3}{T_2}$$

Process 1–2 :

$$\frac{T_2}{T_1} = \left(\frac{v_1}{v_2}\right)^{\gamma-1} = r^{\gamma-1} = 15^{0.4} = 2.954$$

$$T_2 = 929.66 \text{ K}$$

Process 2–3 :

$$q_s = c_p(T_3 - T_2) = \frac{F}{A} \times CV$$

$$T_3 - T_2 = \frac{42000}{1.004 \times 50} = 836.66 \text{ K}$$

$$T_3 = 1761.32 \text{ K}$$

$$\rho = \frac{1761.32}{924.66} = 1.905$$

$$\eta = 1 - \frac{1}{r^{\gamma-1}}\left[\frac{\rho^\gamma - 1}{\gamma(\rho - 1)}\right]$$

$$= 1 - \frac{1}{15^{0.4}}\left[\frac{1.905^{1.4} - 1}{1.4\,(1.905 - 1)}\right]$$

$$= 0.6085 = 60.85\%$$

■ **Example 2.18** An air standard diesel cycle has a compression ratio of 16. The pressure at the beginning of compression stroke is 1 bar and the temperature is 20°C. The maximum temperature is 1430°C. Determine the thermal efficiency and the mean effective pressure for this cycle. Take $\gamma = 1.4$.

Solution. Refer to Fig. 2.8.

$$r = \frac{v_1}{v_2} = 16,\ p_1 = 1 \text{ bar},\ T_1 = 20 + 273 = 293 \text{ K}$$

$$T_3 = 1430 + 273 = 1703 \text{ K}$$

$$\frac{T_2}{T_1} = \left(\frac{v_1}{v_2}\right)^{\gamma-1} = r^{\gamma-1} = 16^{0.4} = 9.03$$

$$T_2 = 293 \times 9.03 = 887.79 \text{ K}$$

Process 2-3 : $\rho = \frac{v_3}{v_1} = \frac{T_3}{T_2} = \frac{1703}{887.79} = 1.918$

$$\eta_{th} = 1 - \frac{1}{\gamma r^{\gamma-1}}\left[\frac{\rho^\gamma - 1}{\rho - 1}\right] = 1 - \frac{1}{1.4 \times 16^{0.4}}\left[\frac{1.918^{1.4} - 1}{1.918 - 1}\right]$$

$$= 0.618 = 61.8\%$$

$$p_m = p_1 r\left[\frac{\gamma r^{\gamma-1}(\rho - 1) - (\rho^\gamma - 1)}{(r-1)(\gamma-1)}\right]$$

$$= 1 \times 16\left[\frac{1.4 \times 16^{0.4}(1.918 - 1) - (1.918^{1.4} - 1)}{(16-1)(1.4-1)}\right] = 6.414 \text{ bar}$$

■ **Example 2.19** A diesel engine has a compression ratio of 16. The fuel is cut-off at 0.08 stroke. The relative efficiency is 52%. Find the mass of the fuel of calorific value 41800 kJ/kg which would be required per kW per hour.

Solution. From Fig. 2.8, we have

$$r = \frac{v_1}{v_2} = 16, \; v_s = v_1 - v_2$$

$$v_3 - v_2 = 0.08\, v_s = 0.08\,(v_1 - v_2)$$

$$\frac{v_3}{v_2} - 1 = 0.08\left(\frac{v_1}{v_2} - 1\right) = 0.08\,(16 - 1) = 1.2$$

$$\rho = \frac{v_3}{v_2} = 1 + 1.2 = 2.2$$

$$\eta_a = 1 - \frac{1}{\gamma r^{\gamma-1}}\left[\frac{\rho^\gamma - 1}{\rho - 1}\right]$$

$$= 1 - \frac{1}{1.4 \times 16^{0.4}}\left[\frac{2.2^{1.4} - 1}{2.2 - 1}\right] = 0.5682 = 56.82\%$$

$$\text{Actual thermal efficiency} = \frac{\text{Actual work done}}{\text{Heat added}}$$

$$= \eta_r \times \eta_a = 0.52 \times 0.5682 = 0.2955$$

$$= \frac{\text{Work done}}{\text{C.V.} \times m_f} = \frac{\text{Work done}}{41800 \times m_f}$$

$$m_f = \frac{3600}{0.2955 \times 41800} = 0.2915 \text{ kg/kW}$$

■ **Example 2.20** An engine works on dual combustion cycle, the compression ratio being 11. The pressure at the commencement of combustion is 1 bar and the temperature is 90°C. The maximum pressure in the cycle is 50 bar and the constant pressure heat addition continues for $\frac{1}{20}$ of the stroke. Calculate the work done per kg. of air and the ideal thermal efficiency. Assume $c_v = 0.718$ kJ/kg.K, $c_p = 1.005$ kJ/kg.K.

Solution. $\gamma = \dfrac{c_p}{c_v} = \dfrac{1.005}{0.718} = 1.4$

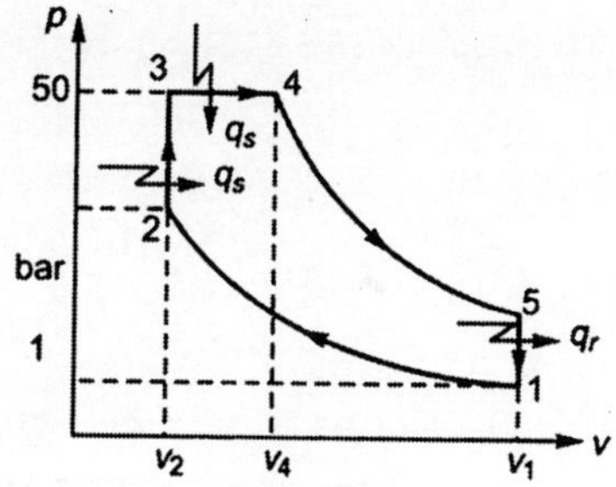

Fig. 2.23

$$T_1 = 90 + 273 = 363 \text{ K}$$

Refer to Fig. 2.23

$$r = \frac{v_1}{v_2} = 11$$

$$v_s = v_1 - v_2 = 10v_2$$

$$v_4 - v_3 = \frac{1}{20}(v_1 - v_2)$$

$$= \frac{v_2}{20}\left(\frac{v_1}{v_2} - 1\right)$$

$$= \frac{v_2}{20}(11-1)$$

$$= \frac{v_2}{2} = \frac{v_3}{2}$$

$$\frac{v_4}{v_3} - 1 = \frac{1}{2} \quad \text{or} \quad \frac{v_4}{v_3} = \frac{3}{2} = 1.5$$

Process 1-2 : $\quad T_2 = T_1 r^{0.4} = 363 \times 11^{0.4} = 947.24 \text{ K}$

$$p_2 = p_1 r^{\gamma} = 1 \times 11 = 28.704 \text{ bar}$$

Process 2-3 : $\quad T_3 = T_2 \dfrac{p_3}{p_2} = 947.24 \times \dfrac{50}{28.704} = 1650 \text{ K}$

Process 3-4 : $\quad T_4 = T_3\left(\dfrac{v_4}{v_2}\right) = 1650 \times 1.5 = 2475 \text{ K}$

Process 4-5 : $\quad T_5 = T_4\left(\dfrac{v_4}{v_5}\right)^{\gamma - 1} = 1119.46 \text{ K}$

$$w = q_s - q_r = c_v(T_3 - T_2) + c_p(T_4 - T_3) - c_v(T_5 - T_1)$$

$$= [0.718(1650 - 947.24) + 1.005(2475 - 1650)] - 0.718(1119.46 - 363)$$

$$= 1339.70 - 360.26 = 979.4 \text{ kJ / kg}$$

$$\eta_{th} = w/q_s = \frac{979.4}{1339.70} = 0.7298 = 72.98\%$$

■ **Example 2.21** A gas turbine operates on Brayton cycle with intake air at 1 bar and 15°C. The air is compressed to 15 bar and heated in a combustion chamber to 800°C. The hot air expands in the turbine to 1 bar. Find (a) the power developed, (b) thermal efficiency of the cycle. Take $\gamma = 1.4$ and $c_p = 1.005$ kJ/kg.K.

Solution. Refer to Fig. 2.12.

$$T_1 = 15 + 273 = 288 \text{ K}$$

$$T_2 = T_1 (r_p)^{\gamma - 1/\gamma} = 288\left(\frac{5}{1}\right)^{0.286} = 456.14 \text{ K}$$

Compressor work, $w_c = c_p(T_2 - T_1) = h_2 - h_1 = 1.005\,(456.14 - 288)$
$= 168.98$ kJ/kg

$$T_4 = \frac{T_3}{(r_p)^{\frac{\gamma-1}{\gamma}}} = \frac{273 + 800}{(5)^{0.286}} = 677.47 \text{ K}$$

Turbine work, $w_t = h_3 - h_4 = c_p\,(T_3 - T_4)$
$= 1.005(1073 - 677.47) = 397.5$ kJ/kg

Heat added, $q_s = q_{2-3} = h_3 - h_2 = c_p(T_3 - T_2)$
$= 1.005(1073 - 456.14) = 619.94$ kJ/kg

Net work, $w = w_t - w_c = 397.5 - 168.98 = 228.52$ kJ/kg

$$\eta_{th} = \frac{w}{q_s} = \frac{228.52}{619.94} = 0.3686 = 36.86\%$$

Power output = 228.52 kJ/kg

■ **Example 2.22** An air-standard dual cycle has a compression ratio of 10. The pressure and the temperature at the beginning of compression are 1 bar and 27°C. The maximum pressure reached is 40 bar and the maximum temperature is 1400°C. Determine (a) the temperature at the end of constant volume heat addition, (b) cut-off ratio, (c) work done per kg of air, and (d) the cycle efficiency. Assume $c_p = 1.004$ kJ/kg.K and $c_v = 0.717$ kJ/kg.K for air.

Solution. Refer to Fig. 2.12

$$\frac{v_s}{v_c} = \frac{v_1 + v_2}{v_c} = r - 1 = 10 - 1 = 9 \text{ or } v_s = 9v_c$$

$$\gamma = \frac{c_p}{c_v} = \frac{1.004}{0.717} = 1.4$$

Process 1–2 :

$$\frac{T_2}{T_1} = r^{\gamma-1} = 10^{0.4} = 2.512$$

$$T_1 = 27 + 273 = 300 \text{ K}$$

$$T_2 = 2.512 \times 300 = 759.6 \text{ K}$$

$$\frac{p_2}{p_1} = r^{\gamma} = 10^{1.4} = 29.12$$

$$p_2 = 1 \times 29.12 = 29.12 \text{ bar}$$

(a) Process 2–3 :

$$\frac{T_3}{T_2} = \frac{p_3}{p_2} = \frac{40}{25} = 1.592$$

$$T_3 = 1.592 \times 759.6 = 1200 \text{ K}$$

$$t_3 = 1200 - 273 = 927^\circ\text{C}$$

(b) Cut-off ratio, $\rho = \dfrac{T_4}{T_3} = \dfrac{1673}{1200} = 1.394$

(c) Work done per kg of air, $w = q_s - q_r$

$$q_s = c_v(T_3 - T_2) + c_p(T_4 - T_3)$$
$$= 0.717(1200 - 759.6) + 1.004\,(1673 - 1200) = 799.96 \text{ kJ/kg}$$

Process 4–5 :
$$\frac{T_4}{T_5} = \left(\frac{v_5}{v_4}\right)^{\gamma-1} = \left(\frac{r}{\rho}\right)^{\gamma-1} = \left(\frac{19}{1.394}\right)^{0.4} = 2.199$$

$$T_5 = \frac{1673}{2199} = 706.68 \text{ K}$$

$$q_r = c_v(T_s - T_1) = 0.717(760.68 - 300 = 330.3 \text{ kJ/kg}$$
$$w = 799.96 - 330.3 = 469.66 \text{ kJ/kg}$$

(d) $\eta_{\text{dual}} = \dfrac{w}{q_s} = \dfrac{464.66}{794.96} = 0.5845$ or 58.45%

■ **Example 2.23** Prove that the optimum pressure ratio r_p for maximum net work done between the temperature T_1 and T_3 for a Brayton cycle is given by :

$$r_p = \left(\frac{T_3}{T_1}\right)^{\frac{\gamma}{2(\gamma-1)}}$$

Solution. From Fig. 2.12, we have

$$\frac{T_3}{T_4} = \frac{T_2}{T_1} = \left(\frac{p_2}{p_1}\right)^{\frac{\gamma-1}{\gamma}} = (r_p)^{\frac{\gamma-1}{\gamma}} = x \text{ (say)}$$

Let
$$\frac{T_3}{T_1} = y$$

Net-work done/kg of air,

$$w = (h_3 - h_2) - (h_4 - h_1) = c_p(T_3 - T_2) - c_p(T_4 - T_1)$$
$$= c_p(T_3 - T_4) - c_p(T_2 - T_1) = c_pT_3(1 - 1/x) - c_pT_1(x - 1)$$

or
$$\frac{w}{c_pT_1} = \frac{T_3}{T_1}\left(1 - \frac{1}{x}\right) - (x-1) = y\left(1 - \frac{1}{x}\right) - (x-1)$$
$$= y - y/x - x - 1$$

$$\frac{d}{dx}\left(\frac{w}{c_pT_1}\right) = 0 \text{ for maximum work done.}$$

Here y is constant and x is variable

$$0 + \frac{y}{x^2} - 1 = 0 \quad \text{or} \quad x = \sqrt{y}$$

$$(r_p)^{\frac{\gamma-1}{\gamma}} = \sqrt{\frac{T_3}{T_1}}$$

$$r_p = \left(\frac{T_3}{T_1}\right)^{\frac{\gamma}{2(\gamma-1)}}$$

Hence Proved.

■ **Example 2.24** An air standard Brayton cycle has pressure ratio across the compression equal to 6. Air enters the compressor at 1 bar and 27°C. The maximum temperature of the cycle is 850°C. Calculate the specific output of the cycle. What will be the power developed by the unit for a mass flow rate of 10 kg/s? For air, $\gamma = 1.4$ and $c_p = 1.005$ kJ/kg.K.

Solution. From Fig. 2.12, we have

$$r_p = \frac{p_2}{p_1} = 6, T_1 = 27 + 273 = 300 \text{ K}$$

$$p_1 = 1 \text{ bar}, T_3 = 850 + 273 = 1123 \text{ K}$$

$$T_2 = T_1 r_p^{(\gamma-1)/\gamma} = 300 \times 6^{1/9.5} = 500.8 \text{ K}$$

$$\frac{T_3}{T_4} = \left(\frac{p_3}{p_4}\right)^{\frac{\gamma-1}{\gamma}} = \left(\frac{p_2}{p_1}\right)^{\frac{\gamma-1}{\gamma}} = 6^{1/3.5} = 1.669$$

$$T_4 = 1123/1.699 = 672.86 \text{ K}$$

Specific output,

$$w = c_p(T_3 - T_2) - c_p(T_4 - T_1)$$
$$= 1.005[(1123 - 500.8) - 1.005(672.86 - 300)]$$
$$= 250.58 \text{ kJ/kg}$$

Power developed

$$P = mw = 10 \times 250.58 = 2505.8 \text{ kW}$$

■ **Example 2.25** In an air standard Brayton cycle, the air enters the compressor at 300 K and 100 kPa. The pressure ratio is 8 and the maximum allowable temperature is 1300 K. Determine the temperature and pressure at each state in the cycle, the compressor work, the turbine work per kg of air and the cycle efficiency.

Solution. From Fig. 2.12, we have

$$p_1 = 100 \text{ kPa}, T_1 = 300 \text{ K}, \frac{p_2}{p_1} = r_p = 8$$

$$p_2 = 100 \times 8 = 800 \text{ kPa}$$

$$T_2 = T_1 r_p^{(\gamma-1)/\gamma} = 300 \times 8^{0.286} = 543.4 \text{ K}$$

$$p_3 = p_2 = 800 \text{ kPa}$$

$$T_3 = T_{max} = 1300 \text{ K}$$

$$T_4 = T_3 \left(\frac{p_4}{p_3}\right)^{\frac{\gamma-1}{\gamma}} = 1300 \times \left(\frac{1}{8}\right)^{0.286} = 716.7 \text{ K}$$

$$p_4 = p_1 = 100 \text{ kPa}$$

$$w_c = (h_2 - h_1) = c_p(T_2 - T_1) = 1.005(543.4 - 300)$$
$$= 244.6 \text{ kJ/kg}$$

$$w_t = h_3 - h_4 = c_p(T_3 - T_4) = 1.005(1300 - 716.7)$$
$$= 586.2 \text{ kJ/kg}$$

$$w = w_t - w_c = 586.2 - 244.6 = 341.6 \text{ kJ/kg}$$

$$q_s = h_3 - h_2 = c_p(T_3 - T_2) = 1.005(1300 - 543.3)$$
$$= 760.4 \text{ kJ/kg}$$

$$\eta = \frac{w}{q_s} = \frac{341.6}{760.4} = 0.4492 \text{ or } 49.92\%$$

■ **Example 2.26** In an air standard Otto cycle, the conditions at the beginning of the compression stroke are 100 kPa and 37°C. The maximum pressure and temperature in the cycle are 7 MPa and 2700°C respectively. Determine the compression ratio, thermal efficiency, and net work/kg of air.

Solution. Refer to Fig. 2.6.

$$r = \frac{p_3 T_1}{T_3 p_1} = \frac{7 \times 310}{2973 \times 0.1} = 7.3$$

$$\eta = 1 - \frac{1}{r^{\gamma - 1}} = 1 - \frac{1}{7.3^{0.4}} = 0.5485 \text{ or } 59.85\%$$

$$T_2 = T_1\left(\frac{v_1}{v_2}\right)^{\gamma - 1} = 310(7.3)^{0.4} = 686.6 \text{ K}$$

$$q_s = c_v(T_3 - T_2) = 0.717(2973 - 686.6) = 1639.3 \text{ kJ/kg}$$
$$w = \eta q_s = 0.5485 \times 1639.3 = 899.2 \text{ kJ/kg}$$

■ **Example 2.27** An engine operates on the air standard diesel cycle. The conditions at the start of the compression stroke are 323 K and 100 kPa while at the end of the stroke the pressure is 4 MPa. The energy absorbed is 600 kJ/kg of air. Calculate (a) the compression ratio, (b) the cut-off ratio, (c) the work done per kg air, and (d) the thermal efficiency.

Solution. From Fig. 2.8 we have

Isentropic compression process 1–2 :

$$r = \frac{v_1}{v_2} = \left(\frac{p_2}{p_1}\right)^{1/\gamma} = \left(\frac{4}{0.1}\right)^{1/1.4} = 1395$$

$$T_2 = \frac{p_2 v_2 T_1}{p_1 v_1} = \frac{4 \times 323}{0.1 \times 13942} = 926.7 \text{ K}$$

$$q_s = c_p(T_3 - T_2)$$
$$600 = 1.005(T_3 - 926.7)$$
$$T_3 = 1523.7 \text{ K}$$

Constant pressure process 2–3 :

$$p_2 = p_3 = 4 \text{ MPa}$$

$$\rho = \frac{v_3}{v_2} = \frac{1523.7}{926.7} = 1.6442$$

$$\eta = 1 - \frac{1}{\gamma r^{\gamma - 1}}\left[\frac{\rho^{\gamma} - 1}{\gamma - 1}\right]$$

$$= 1 - \frac{1}{1.4 \times (13.942)^{0.4}} \left[\frac{(1.6442)^{1.4} - 1}{1.6642 - 1} \right]$$

$$= 0.6112 \text{ or } 61.12\%$$

Work done, $w = \eta q_s = 0.6112 \times 600 = 366.72$ kJ/kg

■ **Example 2.28** An air standard Otto cycle has a compression ratio of 8. At the beginning of the compression stroke, the temperature is 300 K and pressure is 100 kPa. Thermal energy is supplied at 1840 kJ/kg. Determine (a) the pressure and temperature at the end of various processes, (b) the thermal efficiency, (c) the work done per kg air, and (d) the mean effective pressure. Take $\gamma = 1.4$, $c_v = 0.718$ kJ/kg.K.

Solution. From Fig. 2.6, we have

$$T_1 = 300 \text{ K}, \; p_1 = 100 \text{ kPa}, \; r = \frac{v_1}{v_2} = 8$$

(a) Process 1–2: $pv^{\gamma} = \text{Constant}$

$$T_2 = T_1 \left(\frac{v_1}{v_2} \right)^{\gamma - 1} = 300(8)^{0.4} = 689.2 \text{ K}$$

$$p_2 = p_1 \left(\frac{v_1}{v_2} \right)^{\gamma} = 100(8)^{1.4} = 1837.9 \text{ kPa}$$

Process 2–3:

$$q_{2-3} = c_v(T_3 - T_2)$$

$$1840 = 0.717(T_3 - 689.2)$$

$$T_3 = 3258.4 \text{ K}$$

$$p_3 = \frac{p_2 T_3}{T_2} = \frac{1837.9 \times 3259.4}{689.2} = 8681.3 \text{ kPa}$$

Process 3–4:

$$T_4 = T_3 \left(\frac{v_3}{v_4} \right)^{\gamma - 1} = 3259.4 \times \left(\frac{1}{8} \right)^{0.4} = 1.417 \text{ K}$$

$$p_4 = p_3 \left(\frac{v_3}{v_4} \right)^{\gamma} = 8681.3 \left(\frac{1}{8} \right)^{1.4} = 472.3 \text{ kPa}$$

(b)

$$\eta_{th} = 1 - \frac{1}{r^{\gamma - 1}} = 1 - \frac{1}{8^{0.4}} = 0.5647$$

(c)

$$w = \eta q_{2-3} = 0.5647 \times 1840 = 1039 \text{ kJ/kg}$$

(d)

$$v_1 = \frac{mRT_1}{p_1} = \frac{8.314 \times 300}{28.97 \times 10^5} = 0.861 \text{ m}^3/\text{kg}$$

$$v_2 = \frac{v_1}{r} = \frac{0.861}{8} = 0.1076 \text{ m}^3/\text{kg}$$

$$p_m = \frac{w}{v_1 - v_2} = \frac{1039}{0.861 - 0.1076} = 1379.1 \text{ kPa}$$

■ **Example 2.29** 0.5 kg of air (ideal gas) executes a Carnot power cycle having a thermal efficiency of 50 percent. The heat transfer to the air during the isothermal expansion is 40 kJ. At the beginning of the isothermal expansion the pressure is 7 bar and the volume is 0.12 m^3. Determine the maximum and minimum temperatures for the cycle, in K, the volume at the end of isothermal expansion, in m^3, and the work and heat transfer for each of the four processes,k in kJ. For air, $c_v = 0.721$ kJ/kg.K and $c_p = 1.008$ kJ/kg.K.

Solution. The Carnot cycle is shown in Fig. 2.24.

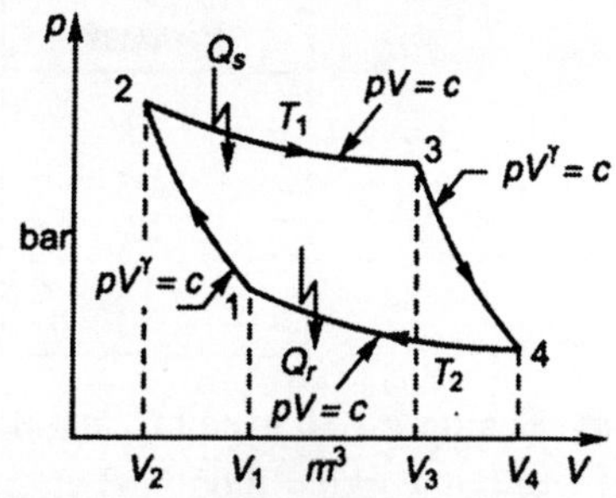

Fig. 2.24

$$m = 0.5 \text{ kg}, \eta_{th} = 0.5, Q_{2-3} = 40 \text{ kJ}$$

$$p_2 = 7 \text{ bar}, V_2 = 0.12 \text{ m}^3$$

$$\eta_{th} = 1 - \frac{T_2}{T_1} = 0.5$$

$$T_1 = 2T_2$$

Now

$$R = c_p - c_v = 1.008 - 0.721$$
$$= 0.287 \text{ kJ/kg-K}$$

$$p_2V_2 = mRT_1$$

$$T_1 = \frac{7 \times 10^5 \times 0.12}{0.5 \times 0.287 \times 10^3} = 585.4 \text{ K or } 312.4°\text{C}$$

$$T_2 = 292.7 \text{ K } 19.7°\text{C}$$

$$Q_{2-3} = mRT_1 \ln\left(\frac{V_3}{V_2}\right)$$

$$40 = 0.5 \times 0.287 \times 585.4 \ln\left(\frac{V_3}{0.12}\right)$$

$$V_3 = 0.1932 \text{ m}^3$$

Process 1–2 : Adiabatic compression

$$\Delta Q = 0$$

$$\Delta U = mc_v(T_1 - T_2) = 0.5 \times 0.721(585.4 - 292.7)$$
$$= 105.52 \text{ kJ}$$

$$\Delta W = -\Delta U = -105.52 \text{ kJ}$$

Process 2–3 : Isothermal expansion

$$\Delta Q = \Delta W + \Delta U$$

Since

$$\Delta U = 0$$

∴

$$\Delta W = \Delta Q = 40 \text{ kJ}$$

Process 3–4 : Adiabatic expansion

$$\Delta U = -mc_v(T_1 - T_2) = -0.5 \times 0.721\,(585.4 - 292.7) = -105.5 \text{ kJ}$$

$$\Delta W = -\Delta U = 105.52 \text{ kJ}$$

Process 4–1 : Isothermal compression

$$\eta_{th} = \frac{Q_s - Q_r}{Q_s}$$

$$0.5 = \frac{40 - Q_r}{40}$$

$$Q_r = 20 \text{ kJ or } \Delta Q = -20 \text{ kJ}$$

$$\Delta U = 0$$

$$\Delta W = \Delta Q = -20 \text{ kJ}$$

Process	*ΔW, kJ*	*ΔU, kJ*	*ΔQ, kJ*
1 – 2	– 105.52	105.52	0
2 – 3	40	0	40
3 – 4	105.52	–105.52	0
4 – 1	– 20	0	–20

■ **Example 2.30** In an I.C. engine operating on the dual cycle (limited pressure cycle), the temperature of the working fluid (air) at the beginning of compression is 27°C. The ratio of the maximum and minimum pressures of the cycle is 70 and the compression ratio is 15. The amounts of heat added at constant volume and at constant pressure are equal. Compute the air standard thermal efficiency of the cycle. State three main reasons why the actual thermal efficiency is different from the theoretical value.

Solution. The *p-v* diagram for the dual cycle is shown in Fig. 2.25.

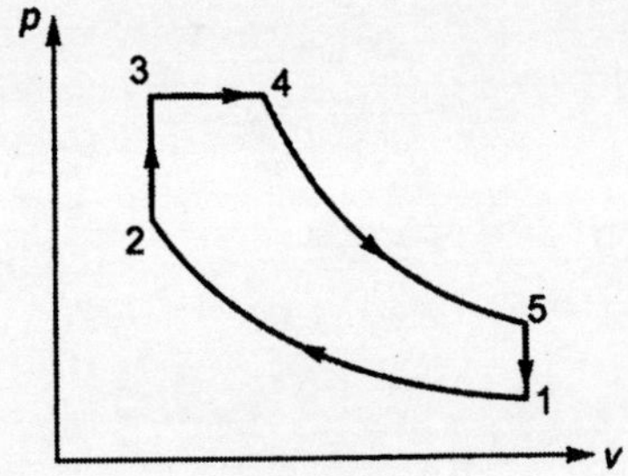

Fig. 2.25

$$T_1 = 273 + 27 = 300 \text{ K}$$

$$\frac{p_3}{p_1} = \frac{p_4}{p_1} = 70, \frac{v_1}{v_2} = \frac{v_1}{v_3} = 15$$

$$q_{2-3} = q_{3-4}$$

$$\frac{T_2}{T_1} = \left(\frac{v_1}{v_2}\right)^{\gamma - 1} = (15)^{0.4} = 2.954$$

$$T_2 = 300 \times 2.954 = 886.25 \text{ K}$$

$$\frac{p_2}{p_1} = \left(\frac{v_1}{v_2}\right)^{\gamma} = 15^{1.4} = 44.31$$

$$\frac{p_3}{p_2} = \frac{p_3}{p_1} \times \frac{p_1}{p_2} = \frac{70}{44.31} = 1.58$$

Process 2-3 : $$T_3 = T_2 \times \frac{p_3}{p_2} = 886.25 \times 1.58 = 1400 \text{ K}$$

Now, $$c_v(T_3 - T_2) = c_p(T_4 - T_3)$$

∴ $$T_4 = T_3 + \frac{T_3 - T_2}{\gamma} = 1400 + \frac{1400 - 886.25}{1.4} = 1767 \text{ K}$$

Process 3-4 : $$\frac{v_4}{v_3} = \frac{T_4}{T_3}, v_4 = \frac{v_3 \times 1767}{1400} = \frac{v_1}{15} \times \frac{1767}{1400} = 0.0841 v_1$$

Also $$v_5 = v_1$$

Process 4-5 : $$\frac{T_4}{T_5} = \left(\frac{v_5}{v_4}\right)^{\gamma-1} = \left(\frac{v_1}{0.0841 v_1}\right)^{0.4} = 2.6915$$

$$T_5 = \frac{1767}{2.6915} = 656.5 \text{ K}$$

Thermal efficiency, $$\eta_{th} = 1 - \frac{q_r}{q_s} = 1 - \frac{c_v(T_5 - T_1)}{c_v(T_3 - T_2) + c_p(T_4 - T_3)}$$

$$= 1 - \frac{T_5 - T_1}{(T_3 - T_2) + \gamma(T_4 - T_3)}$$

$$= 1 - \frac{656.5 - 300}{(1400 - 886.25) + 1.4(1767 - 1400)} = 0.653 \text{ or } 65.3\%$$

■ **Example 2.31** In an air-standard Brayton cycle the minimum and maximum temperature are 300 K and 1200 K, respectively. The pressure ratio is that which maximizes the net work developed by the cycle per unit mass of air flow. Calculate the compressor and turbine work, each in kJ/kg air, and the thermal efficiency of the cycle.

Solution. With reference to Fig. 2.26, $T_{max} = 1200$ K and $T_{min} = 300$ K

For maximum work, optimum a pressure ratio,

$$(r_p)_{opt} = \left[\frac{T_{max}}{T_{min}}\right]^{\frac{\gamma}{2(\gamma-1)}} = \left[\frac{1200}{300}\right]^{\frac{1.4}{0.8}} = 11.314$$

$$T_2 = T_1 (r_p)^{\frac{\gamma-1}{\gamma}} = 300\,(11.314)^{0.4/1.4} = 600 \text{ K}$$

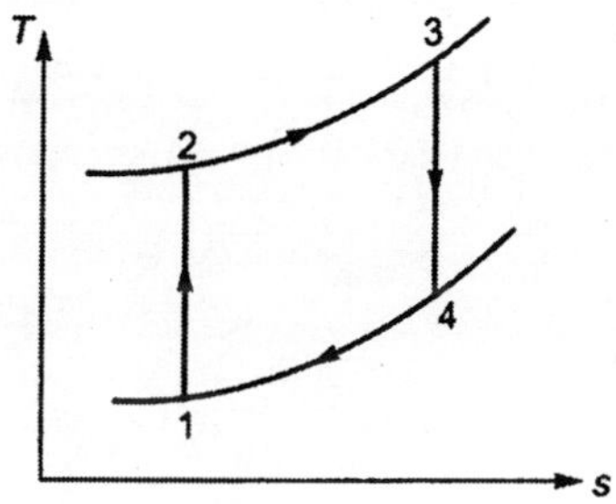

Fig. 2.26

Compressor work,

$$w_c = c_p(T_2 - T_1)$$
$$= 1.005(600 - 300) = 301.5 \text{ kJ/kg}$$

$$T_4 = \frac{T_3}{(r_p)^{\frac{\gamma-1}{\gamma}}} = \frac{1200}{(11.314)^{0.4/1.4}} = 600 \text{ K}$$

Turbine work, $$w_t = c_p(T_3 - T_4) = 1.005(1200 - 600)$$
$$= 603 \text{ kJ/kg}$$

Heat supplied, $$q_s = c_p(T_3 - T_2) = 1.005(1200 - 600)$$
$$= 603 \text{ kJ/kg}$$

Thermal efficiency, $$\eta_{th} = \frac{w_t - w_c}{q_s} = \frac{603 - 301.5}{603}$$
$$= 0.5 \text{ or } 50\%$$

■ **Example 2.32** Prove that for the same compression ratio the Otto cycle is more efficiency than the Diesel cycle. [IES, 1996]

Solution. The p-v diagrams are shown in Fig. 2.27 for same compression ratio.

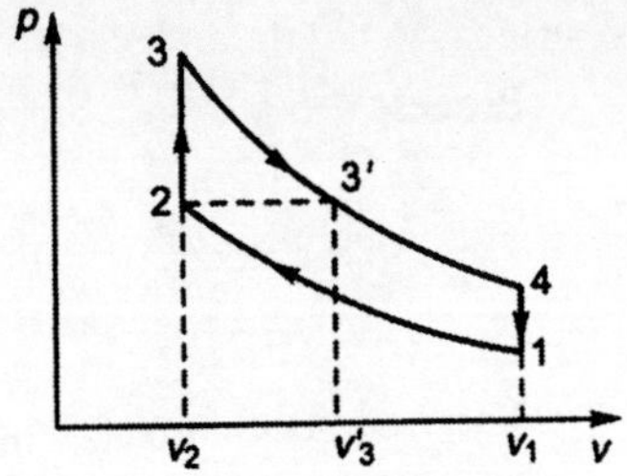

Fig. 2.27

1 – 2 – 3 – 4 – 1 : Otto cycle
1 – 2 – 3′ – 4 – 1 : Diesel cycle

Compression ratio,
$$r = \frac{v_1}{v_2}$$

$$e = \frac{v_1}{v_3'}, \rho = \frac{v_3'}{v_2} = \frac{v_3'}{v_1} \times \frac{v_1}{v_2} = \frac{r}{e}$$

where $e = r - \Delta r$, and Δr is a small quantity.

Air standard efficiency of Otto cycle,

$$\eta_0 = 1 - \frac{1}{r^{\gamma-1}}$$

Air standard efficiency of Diesel cycle,

$$\eta_d = 1 - \frac{1}{\gamma r^{\gamma-1}}\left[\frac{\rho^\gamma - 1}{\rho - 1}\right] = 1 - \frac{1}{\gamma r^{\gamma-1}}\left[\frac{\left(\frac{r}{e}\right)^\gamma - 1}{\left(\frac{r}{e}\right) - 1}\right]$$

Now,
$$\frac{r}{e} = \frac{r}{r - \Delta r} = \frac{1}{1 - \frac{\Delta r}{r}} = \left[1 - \frac{\Delta r}{r}\right]^{-1}$$

$$= 1 + \frac{\Delta r}{r} + \frac{\Delta r}{r^2} + \frac{\Delta r}{r^3} + \ldots$$

and
$$\left(\frac{r}{e}\right)^\gamma = \left(\frac{r}{r - \Delta r}\right)^\gamma = \left(1 - \frac{\Delta r}{r}\right)^\gamma$$

$$= 1 + \gamma . \frac{\Delta r}{r} + \frac{\gamma(\gamma + 1)}{2!} \cdot \left(\frac{\Delta r}{r}\right)^2 + \ldots$$

∴
$$\eta_d = 1 - \frac{1}{r^{\gamma-1}}\left[\frac{\frac{\Delta r}{r} + \frac{(\gamma+1)}{2!} \cdot \left(\frac{\Delta r}{r}\right)^2 + \ldots}{\frac{\Delta r}{r} + \frac{\Delta r}{r^2} + \ldots}\right]$$

The coefficients of $\frac{\Delta r}{r}, \left(\frac{\Delta r}{r}\right)^2$, … in the numerator are more than 1. Thus $\eta_0 > \eta_d$.

■ **Example 2.33** The volume ratios of compression and expansion for a Diesel engine as measured from an indicator diagram are 15.3 and 7.5 respectively. The pressure and temperature at the beginning of compression are 1 bar and 27°C. Assuming an ideal engine, determine the mean effective pressure, the ratio of maximum pressure to mean effective pressure and the cycle efficiency.

Also find the fuel consumption per kWh if the indicated thermal efficiency is 0.5 of ideal efficiency, mechanical efficiency 0.8 and the calorific value of oil 42000 kJ/kg. [IES, 1996]

Solution. Refer to Fig. 2.28

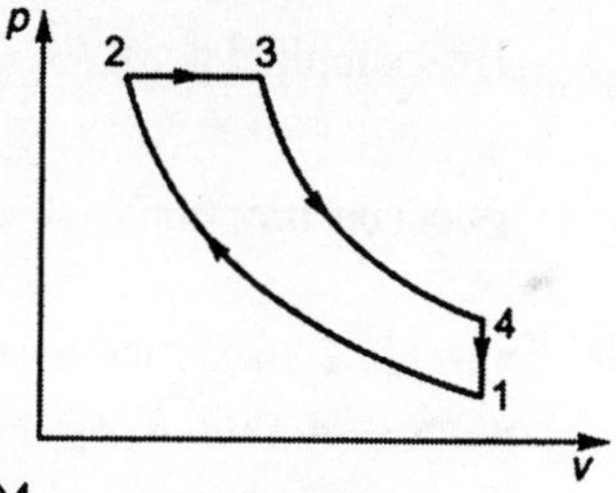

Fig. 2.28

Compression ratio, $r = \frac{v_1}{v_2} = 15.3$

Expansion ratio, $e = \frac{v_4}{v_3} = 7.5$

Cut-off ratio, $\rho = \frac{v_3}{v_2} = \frac{v_3}{v_1} \times \frac{v_1}{v_2} = \frac{v_3}{v_4} \times \frac{v_1}{v_2} = \frac{r}{e} = \frac{15.3}{7.5} = 2.04$

$p_1 = 1$ bar, $T_1 = 273 + 27 = 300$ K

$$\frac{(\eta_{th})_i}{\eta_{ideal}} = 0.5, \ \eta_{mech} = 0.8, \text{ C.V.} = 42000 \text{ kJ/kg}$$

Process 1–2 : $T_2 = T_1\left(\frac{v_1}{v_2}\right)^{\gamma-1} = T_1 r^{\gamma-1} = 300 \times 15.3^{0.4} = 893.3$ K

$$p_2 = p_1\left(\frac{v_1}{v_2}\right)^{\gamma} = p_1 r^{\gamma} = 1 \times 15.3^{1.4} = 45.56 \text{ bar}$$

Process 2–3 : $T_3 = T_2 \frac{v_3}{v_2} = T_2\rho = 893.3 \times 2.04 = 1822.33$ K

Process 3–4 : $T_4 = T_3\left(\frac{v_3}{v_4}\right)^{\gamma-1} = \frac{T_3}{e^{\gamma-1}} = \frac{1822.33}{7.5^{0.4}} = 813.95$ K

Let us consider 1 kg of working fluid.

Work done, $W = Q_s - Q_r = mc_p(T_3 - T_2) - mc_v(T_4 - T_1)$

$$= 1 \times 1.005(1822.33 - 893.3) - 1 \times 0.718(813.95 - 300)$$

$$= 564.66 \text{ kJ}$$

Swept volume, $V_s = V_1 - V_2 = R\left(\frac{T_1}{p_1} - \frac{T_2}{p_2}\right) = 0.287\left(\frac{300}{100} - \frac{893.3}{4556}\right)$

$$= 0.8047 \text{ m}^3$$

Mean effective pressure,

$$p_m = \frac{W}{V_s} = \frac{564.66 \times 10^3}{0.8047} = 701.7 \times 10^3 \text{ Pa or } 7.017 \text{ bar}$$

$$\frac{p_{max}}{p_m} = \frac{p_2}{p_m} = \frac{45.56}{7.017} = 6.493$$

Cycle efficiency $= \frac{W}{Q_s} = \frac{564.66}{1 \times 1.0056(1822.33 - 893.3)}$

$$= 0.6048 \text{ or } 60.48\%$$

Indicated thermal efficiency $= 0.5 \times 60.48 = 30.24\%$

Brake thermal efficiency $= 30.24 \times 0.8 = 24.19\%$

$$\text{Heat supplied per kWh} = \frac{3600}{0.2419} = 14882.2 \text{ kJ/kWh}$$

$$\text{Fuel consumption} = \frac{14882.2}{42000} = 0.3543 \text{ kg/kWh}$$

■ **Example 2.34** Derive an expression for air standard efficiency of the following cycle in terms of compression ratio R, c_v and γ.

(i) an isothermal compression, compression ratio, r.

(ii) an increase of pressure at constant volume.

(iii) an adiabatic expansion. [IES, 1997]

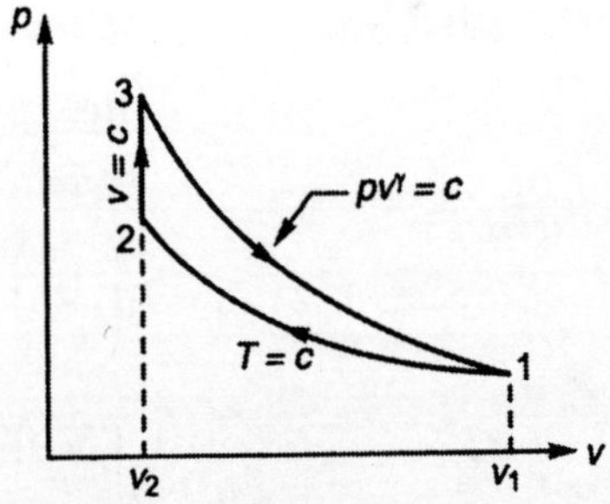

Fig. 2.29

Solution. The p-v diagram is shown in Fig. 2.29

Process 1–2 : Isothermal compression

$$W_{1-2} = Q_{1-2} = mRT_1 \ln \frac{V_1}{V_2} = mRT_1 \ln r$$

Process 2–3 : Constant volume process

$$W_{2-3} = \int_2^3 p \cdot dV = 0$$

$$Q_{2-3} = U_3 - U_2 = mc_v(T_3 - T_2) = mc_v(T_3 - T_1)$$

Process 3–1 : Adiabatic expansion

$$W_{3-1} = mc_v(T_3 - T_1)$$

$$Q_{3-1} = 0$$

Air standard efficiency,

$$\eta = 1 - \frac{Q_{2-3} + Q_{3-1}}{Q_{1-2}} = 1 - \frac{mc_v(T_3 - T_1) + 0}{mRT_1 \ln r}$$

$$= 1 - \frac{c_v(T_3 - T_1)}{RT_1 \ln r}$$

Now $$T_3 = T_1\left(\frac{v_1}{v_3}\right)^{\gamma-1} = T_1\left(\frac{v_1}{v_2}\right)^{\gamma-1} = T_1 r^{\gamma-1}$$

$\therefore$ $$\eta = 1 - \frac{c_v(r^{\gamma-1} - 1)}{R \ln r} = 1 - \frac{r^{\gamma-1} - 1}{(\gamma - 1) \ln r}$$

■ **Example 2.35** An air standard Otto cycle has a volumetric compression ratio of 6, the lowest cycle pressure of 0.1 MPa and operates between temperature limits of 27°C and 1569°C.

(a) Calculate the temperature and pressure after the isentropic expansion (Ratio of specific heats = 1.4).

(b) Since it is observed that values in (a) are well above the lowest cycle operating conditions, the expansion process was allowed to continue down to a pressure of 0.1 MPa. Which process is required to complete the cycle? Name the cycle so obtained.

(c) Determine by what percentage the cycle efficiency has been improved.

Solution. The p-v diagram is shown in Fig. 2.30

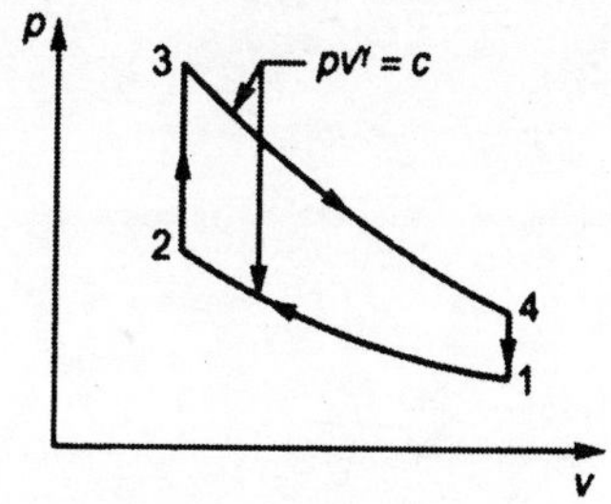

Fig. 2.30 Otto cycle.

$$\frac{v_1}{v_2} = 6$$

$$p_1 = 0.1 \text{ MPa}$$

$$T_1 = 27 + 273 = 300 \text{ K}$$

$$T_3 = 1569 + 273 = 1842 \text{ K}$$

(a) $p_2 = p_1\left(\frac{v_1}{v_2}\right)^{\gamma} = 0.1 \times 6^{1.4} = 1.2286 \text{ MPa}$

$$T_2 = T_1\left(\frac{v_1}{v_2}\right)^{\gamma - 1} = 300 \times 6^{0.4} = 614.3 \text{ K}$$

$$p_3 = \frac{p_2 T_3}{T_2} = \frac{1.2286 \times 1842}{614.3} = 3.684 \text{ MPa}$$

$$p_4 = p_3\left(\frac{v_3}{v_4}\right)^{\gamma} = p_3\left(\frac{v_2}{v_1}\right)^{\gamma} = \frac{3.684}{6^{1.4}} = 0.3 \text{ MPa}$$

$$T_4 = T_1\left(\frac{p_4}{p_1}\right) = 300 \times \frac{0.3}{0.1} = 900 \text{ K}$$

(b) Constant pressure scavenging is required to complete the cycle with expansion process to continue down to 0.1 MPa as shown in Fig. 2.31. The cycle is called Atkinson cycle.

(c) $\eta_{Otto} = 1 - \frac{1}{r^{\gamma - 1}} = 1 - \frac{1}{6^{0.4}} = 0.5116 \text{ or } 51.16\%$

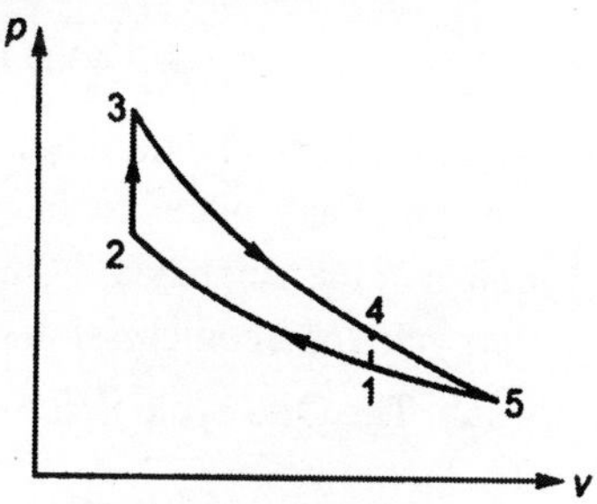

Fig. 2.31 Atkinson cycle.

$$\eta_{\text{Atkinson}} = 1 - \frac{\gamma\,(T_5 - T_1)}{T_3 - T_2}$$

$$T_5 = T_3\left(\frac{p_5}{p_3}\right)^{\frac{\gamma - 1}{\gamma}} = 1842\left(\frac{0.1}{3.684}\right)^{\frac{0.4}{1.4}} = 657.3 \text{ K}$$

$$\eta_{\text{Atkinson}} = 1 - \frac{1.4(657.3 - 300)}{(1842 - 614.3)} = 0.5925 \text{ or } 59.25\%$$

Improvement in cycle efficiency = 59.25 − 51.16 = 8.09%

■ **Example 2.36** Air enters the compressor of a gas turbine plant operating on Brayton cycle at 1 bar, 27°C. The pressure ratio in the cycle is 6. If $W_t = 2.5\ W_c$, where W_t and W_c are the turbine and compressor work respectively, calculate the maximum temperature and the cycle efficiency.

Solution. The Brayton cycle is shown in Fig. 2.32

$$p_1 = 1 \text{ bar} = 1 \times 10^5 \text{ N/m}^2$$

$$T_1 = 27 + 273 = 300 \text{ K}$$

$$\frac{p_2}{p_1}=6,\ p_2=6\times10^5\ \text{N/m}^2$$

$$\frac{T_2}{T_1}=\left(\frac{p_2}{p_1}\right)^{\frac{\gamma-1}{\gamma}}$$

$$T_2=300\times6^{0.4/1.4}=500.55\ \text{K}$$

$$\frac{T_3}{T_4}=\left(\frac{p_2}{p_1}\right)^{\frac{\gamma-1}{\gamma}}=6^{0.4/1.4}=1.6685$$

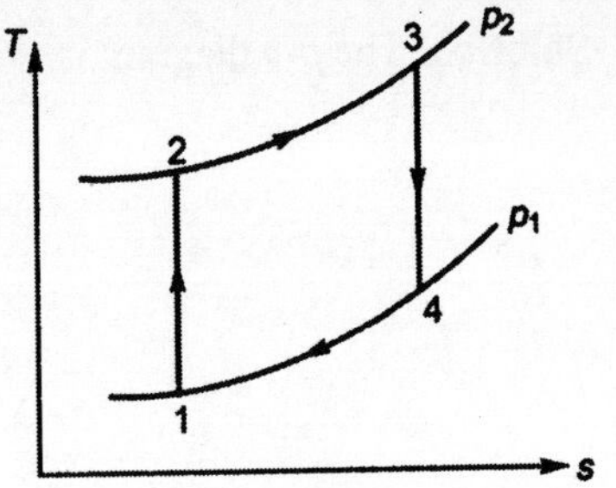

Fig. 2.32

$$W_c=mc_p\,(T_2-T_1)$$
$$W_t=mc_p(T_3-T_4)$$
$$mc_p(T_3-T_4)=2.5\,mc_p(T_2-T_1)$$
$$T_3-T_4=2.5(T_2-T_1)$$
$$T_3\left(1-\frac{1}{1.6685}\right)=2.5(500.55-300)$$
$$T_3=1251.4\ \text{K},\ T_4=750\ \text{K}$$

$$\text{Cycle efficiency}=\frac{W_t-W_c}{mc_p(T_3-T_2)}=\frac{(T_3-T_4)-(T_2-T_1)}{T_3-T_2}$$

$$=\frac{(1251.4-750)-(500.55-300)}{1251.4-500.55}=\frac{501.4-200.55}{750.85}$$

$$=0.4007\text{ or }40.07\%$$

■ **Example 2.37** The minimum pressure and temperature in an Otto cycle are 100 kPa and 27°C. The amount of heat added to the air-cycle is 1500 kJ/kg. Determine the pressures and temperatures at all points of the air standard Otto cycle. Also calculate the specific work and the thermal efficiency of the cycle for a compression ratio of 8 : 1. (Take c_v (air) = 0.72 kJ/kg.K and $(c_p/c_v)_{\text{air}}=1.4$).

Solution. The Otto cycle is shown in Fig. 2.33

$$p_1=100\ \text{kPa},\ T_1=27+273=300\ \text{K}$$
$$q_{2-3}=1500\ \text{kJ/kg},\ r=8$$
$$\frac{T_2}{T_1}=r^{\gamma-1},\ T_2=300\times8^{0.4}=689.22\ \text{K}$$
$$p_2=p_1r^{\gamma}=100\times8^{1.4}=1837.917\ \text{kPa}$$

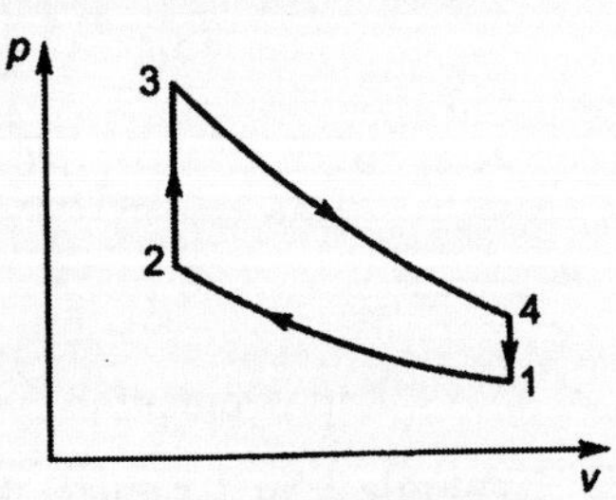

Fig. 2.33

$$q_{2-3}=c_v(T_3-T_2)$$
$$1500=0.72(T_3-689.22)$$
$$T_3=277255\ \text{K}$$
$$p_3=p_2\frac{T_3}{T_2}=1837.917\times\frac{2772.55}{689.22}$$
$$=7393.46\ \text{kPa}$$

$$T_4 = \frac{T_3}{r^{\gamma-1}} \frac{2772.55}{8^{0.4}} = 1206.82 \text{ K}$$

$$p_4 = \frac{p_3}{r^{\gamma}} = \frac{7393.46}{8^{1.4}} = 402.274 \text{ kPa}$$

$$\eta_{th} = 1 - \frac{1}{r^{\gamma-1}} = 1 - \frac{1}{8^{0.4}} = 0.5647 \text{ or } 56.47\%$$

$$q_{4-1} = c_v(T_4 - T_1) = 0.72\,(1206.82 - 300) = 652.91 \text{ kJ/kg}$$

Specific work, $w_{net} = q_{2-3} - q_{4-1} = 1500 - 652.91 = 847.09$ kJ/kg

■ **Example 2.38** A Brayton cycle (air standard) has a pressure ratio of 4 and inlet conditions of one standard atmospheric pressure and 27°C. Find the air flow rate required for 100 kW power output if the maximum temperature in the cycle is 1000°C. Assume $\gamma = 1.4$ and $c_p = 1.0$ kJ/kg.K.

Solution. Refer to Fig. 2.34

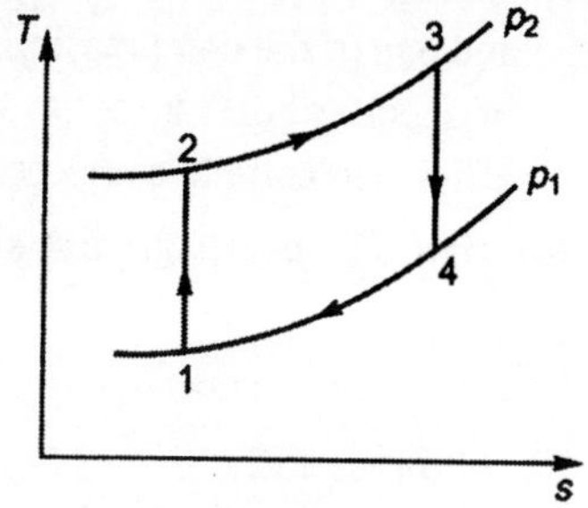

Fig. 2.34

$$\frac{p_2}{p_1} = 4, \; p_1 = 1 \text{ atm}$$

$$T_1 = 27 + 273 = 300 \text{ K}$$

$$T_3 = 1000 + 273 = 1273 \text{ K}$$

$$T_2 = T_1\left(\frac{p_2}{p_1}\right)^{\frac{\gamma-1}{\gamma}} = 300 \times 4^{0.4/1.4} = 445.8 \text{ K}$$

$$\frac{T_3}{T_4} = r^{\frac{\gamma-1}{\gamma}}, \; T_4 = \frac{1273}{4^{0.4/1.4}} = 856.66 \text{ K}$$

$$W_{net} = W_t - W_c$$

$$= mc_p[(T_3 - T_4) - (T_2 - T_1)]$$

$$= m \times 1[(1273 - 856.66) - (445.8 - 300)]$$

$$= 270.5345 \times m \text{ kW} = 100$$

$$m = 0.3698 \text{ kg/s}$$

■ **Example 2.39** Calculate the percentage loss in the ideal efficiency of a diesel engine with compression ratio 14 if the fuel cut-off is delayed from 5% to 8%.

Solution. The diesel cycle is shown in Fig. 2.35

Compression ratio, $r = 14$, Cut-off ratio, $\rho = 0.05$

Efficiency,

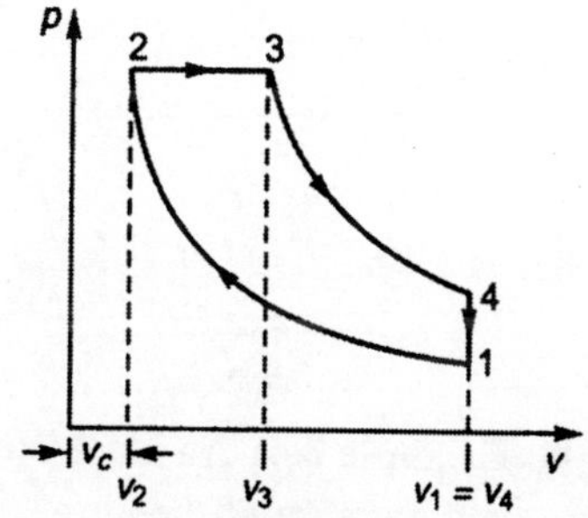

Fig. 2.35

$$\eta_1 = 1 - \frac{1}{r^{\gamma-1} \cdot \gamma}\left[\frac{\rho^{\gamma} - 1}{\rho - 1}\right]$$

$$= 1 - \frac{1}{14^{0.4} \times 1.4}\left[\frac{0.05^{1.4} - 1}{0.05 - 1}\right]$$

$$= 0.7423$$

For $\rho = 0.08$

$$\eta_2 = 1 - \frac{1}{r^{\gamma-1} \cdot \gamma}\left[\frac{\rho^{\gamma} - 1}{\rho - 1}\right]$$

$$= 1 - \frac{1}{14^{0.4} \times 1.4}\left[\frac{0.08^{1.4} - 1}{0.08 - 1}\right]$$

$$= 0.7377$$

$$\text{Change in efficiency} = \frac{\eta_1 - \eta_a}{\eta_1}$$

$$= \left[\frac{0.7423 - 0.7377}{0.7423}\right] = 0.006197 \text{ or } 0.6197\%$$

■ **Example 2.40** A reciprocating engine works on Otto cycle where the temperature at the beginning and end of the compression stroke are found to be 40°C and 360°C respectively. Assuming the ratio of specific heats at constant pressure and constant volume to be 1.4, determine the air standard efficiency and the compression ration.

Solution. The *p-v* diagram is shown in Fig. 2.36

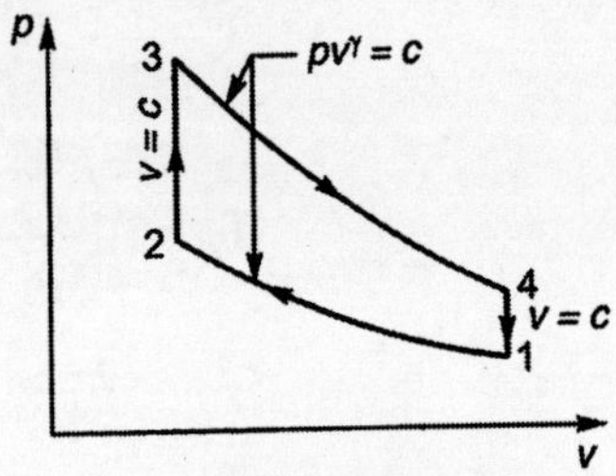

Fig. 2.36 *p-v* diagram for Otto cycle.

$$T_1 = 273 + 40 = 313 \text{ K}$$

$$T_2 = 273 + 360 = 633 \text{ K}$$

Process 1–2

Compression ratio,

$$r = \frac{v_1}{v_2} = \left(\frac{T_2}{T_1}\right)^{\frac{1}{\gamma-1}}$$

$$= \left(\frac{633}{313}\right)^{\frac{1}{0.4}} = 5.816$$

Also

$$v_3 = v_2 \text{ and } v_4 = v_1$$

$$\frac{v_4}{v_3} = \frac{v_1}{v_2} = r$$

$$\frac{T_4}{T_3} = \left(\frac{1}{r}\right)^{\gamma-1} = \frac{T_1}{T_2} \quad \text{or} \quad \frac{T_4}{T_1} = \frac{T_3}{T_2}$$

$$\text{Air standard efficiency} = \frac{\text{Heat added} - \text{Heat rejected}}{\text{Heat added}}$$

$$= \frac{c_v(T_3 - T_2) - c_v(T_4 - T_1)}{c_v(T_3 - T_2)} = 1 - \frac{T_4 - T_1}{T_3 - T_2} = 1 - \frac{T_1}{T_2}$$

$$= 1 - \frac{1}{r^{\gamma-1}} = 1 - \frac{1}{(5.816)^{0.4}} = 0.5055 \text{ or } 50.55\%$$

■ **Example 2.41** An ideal Diesel engine operates with air as the working substance. The temperature and pressure of the air at the beginning of the compression process are 25°C and 1.005 bar. The compression ratio is 18 and cut-off occurs at 6.5% of the expansion stroke.

(i) Draw the *p-v* and *T-s* diagrams for the cycle indicating clearly each of the process;

(ii) Determine the pressure and temperature at the end of each process;

(iii) Determine the air-standard efficiency of the engine assuming $\gamma = 1.4$ for air,

(iv) The work done per cycle (assume $c_p = 1.0$ kJ/kg.K); and

(v) The mean effective pressure.

Solution. (i) The *p-v* and *T-s* diagrams are shown in Fig. 2.37.

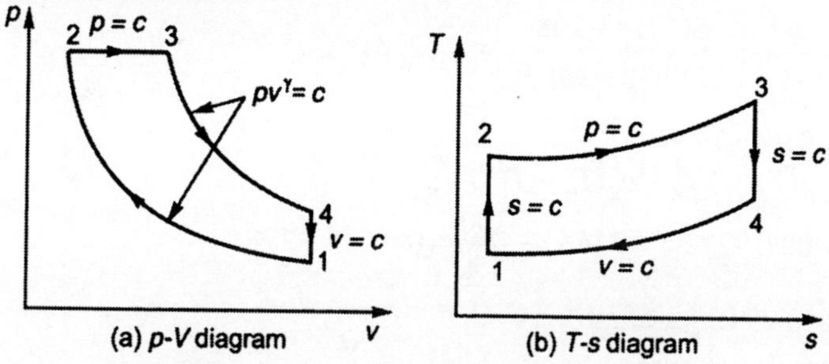

Fig. 2.37 *p-v* and *T-s* diagrams for a Diesel cycle.

(ii)

$$T_1 = 273 + 25 = 298 \text{ K}$$

$$p_1 = 1.005 \text{ bar}$$

$$r = \frac{v_1}{v_2} = 18$$

$$\rho = \frac{v_3}{v_2}$$

$$T_2 = T_1 r^{\gamma - 1} = 298 \times 18^{0.4} = 946.9 \text{ K}$$

$$p_2 = p_1 r^{\gamma} = 1.005 \times 18^{1.4} = 57.48 \text{ bar}$$

Now

$$v_3 - v_2 = 0.065(v_4 - v_3) = 0.065(v_1 - v_3)$$

$$\frac{v_3}{v_2} - 1 = 0.065\left(\frac{v_1}{v_2} - \frac{v_3}{v_2}\right)$$

$$1.065\rho = 0.065r + 1$$

$$\rho = \frac{0.065 \times 18 + 1}{1.065} = 2.0376$$

$$T_3 = T_2 \frac{v_3}{v_2} = T_2 \rho = 946.9 \times 2.0376 = 1929.36 \text{ K}$$

$$p_3 = p_2$$

$$T_4 = T_3\left(\frac{v_3}{v_4}\right)^{\gamma - 1} = T_3\left(\frac{v_3}{v_1}\right)^{\gamma - 1} = T_3\left(\frac{v_3}{v_2} \times \frac{v_2}{v_1}\right)^{\gamma - 1} = T_3\left(\frac{\rho}{r}\right)^{\gamma - 1}$$

$$= 1929.36\left(\frac{2.0376}{18}\right)^{0.4} = 807.14 \text{ K}$$

$$p_4 = p_3\left(\frac{T_4}{T_3}\right)^{\frac{\gamma}{\gamma-1}} = 57.48\left(\frac{807.14}{1929.36}\right)^{\frac{1.4}{0.4}} = 2.722 \text{ bar}$$

(iii) Air-standard efficiency

$$= \frac{\text{Work done}}{\text{Heat supplied}} = \frac{c_p(T_3 - T_2) - c_v(T_4 - T_1)}{c_p(T_3 - T_2)} = 1 - \frac{1}{\gamma}\left(\frac{T_4 - T_1}{T_3 - T_2}\right)$$

$$= 1 - \frac{1}{1.4}\left(\frac{807.14 - 298}{1929.36 - 946.9}\right) = 0.6298 \text{ or } 62.98\%$$

(iv) Work done per cycle,

$$w = c_p(T_3 - T_2) - c_v(T_4 - T_1)$$

$$= 1.00\,(1929.36 - 946.9) - \frac{100}{1.4} = (807.14 - 298)$$

$$= 618.79 \text{ kJ/kg of air}$$

(v) Mean effective pressure

$$p_m = \frac{\text{Work done}}{\text{Swept volume}} = \frac{w}{v_s}$$

$$R = c_p - c_v = 1.00 - \frac{1}{1.4} = 0.286 \text{ kJ/kg.K}$$

$$v_1 = \frac{RT_1}{p_1} = \frac{0.286 \times 298}{1.005 \times 10^2} = 0.848 \text{ m}^3/\text{kg}$$

$$v_s = v_1 - v_2 = v_1 - \frac{v_1}{18} = 0.848 - \frac{0.848}{18} = 0.801 \text{ m}^3/\text{kg}$$

$$p_m = \frac{618.79}{0.801} = 772.52 \text{ kPa}$$

■ **Example 2.42** An engine working on the ideal Otto cycle takes in air at 1 bar and 30°C which is compressed to 15 bar at the end of the compression stroke. The temperature attained at the end of constant volume heat addition is 900°C. Assuming adiabatic index to be 1.4, determine (i) the compression ratio, (ii) the thermal efficiency, (iii) the temperature at the end of compression, (iv) the pressure at the end of constant volume heating, and (v) the mean effective pressure.

Solution. The *p-v* diagram is shown in Fig. 2.38

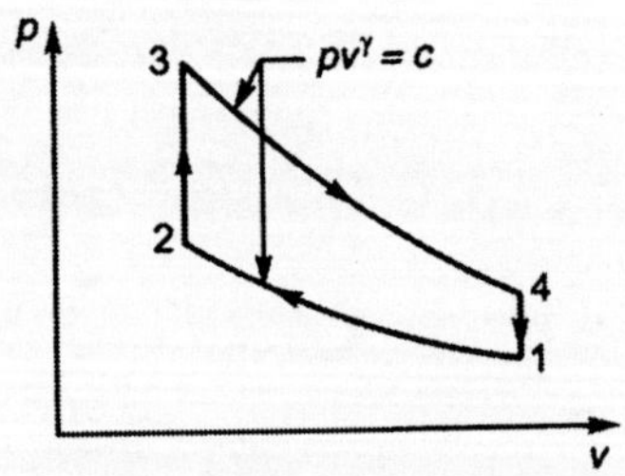

Fig. 2.38 *p-v* diagram for Otto cycle.

$$p_1 = 1 \text{ bar}, T_1 = 273 + 30 = 303 \text{ K}$$

$$p_2 = 15 \text{ bar}, T_3 = 273 + 900 = 1173 \text{ K}, \gamma = 1.4$$

(i) Compression ratio,

$$r = \frac{v_1}{v_2} = \left(\frac{p_2}{p_1}\right)^{\frac{1}{\gamma}} = \left(\frac{15}{1}\right)^{\frac{1}{1.4}} = 6.92$$

(ii) Thermal efficiency

$$\eta_{th} = 1 - \frac{1}{r^{\gamma-1}} = 1 - \frac{1}{(6.92)^{0.4}} = 0.5387 \text{ or } 53.87\%$$

(iii) $T_2 = T_1 r^{\gamma-1} = 303 \times (6.92)^{0.4} = 656.88$ K

(iv) $p_3 = p_2 = \frac{T_3}{T_2} = 15 \times \frac{1173}{656.88} = 26.8$ bar

(v) Mean effective pressure,

$$p_m = \frac{\text{Work done}}{\text{Swept volume}} = \frac{w}{v_s}$$

$$w = c_v[(T_3 - T_2) - (T_4 - T_1)]$$

$$T_4 = T_3\left(\frac{v_3}{v_4}\right)^{\gamma-1} = T_3\left(\frac{v_2}{v_1}\right)^{\gamma-1} = \frac{T_3}{r^{\gamma-1}} = \frac{1173}{(6.92)^{0.4}} = 541.1\text{K}$$

$$v_s = v_1 - v_2 = R\left(\frac{T_1}{p_1} - \frac{T_2}{p_2}\right) = 0.287\left(\frac{303}{1} - \frac{656.88}{15}\right) \times 10^{-2}$$

$$= 0.74392 \text{ m}^3/\text{kg}$$

$$w = 0.718[(1173 - 656.88) - (541.1 - 303)] = 199.618 \text{ kN.m/kg}$$

$$p_m = \frac{199.618}{0.74392} = 268.33 \text{ kPa}$$

■ **Example 2.43** The diameter and stroke of a gas engine cylinder are 18 cm and 30 cm respectively. The ratio of expansion is 5. The pressure and temperature of the mixture at the beginning of compression are 1 bar and 100°C respectively. Find the index of the compression process and the weight of the mixture in the cylinder, if the pressure at the end of compression is 7 bar. Also calculate the work done and heat transferred during the process, indicating the direction of flow. Assume $R = 0.287$ kJ/kg.K and ratio of specific heats equal to 1.4 for the mixture. Take 1 bar $= 100$ kPa.

Solution. The p-V diagram is shown in Fig. 2.39

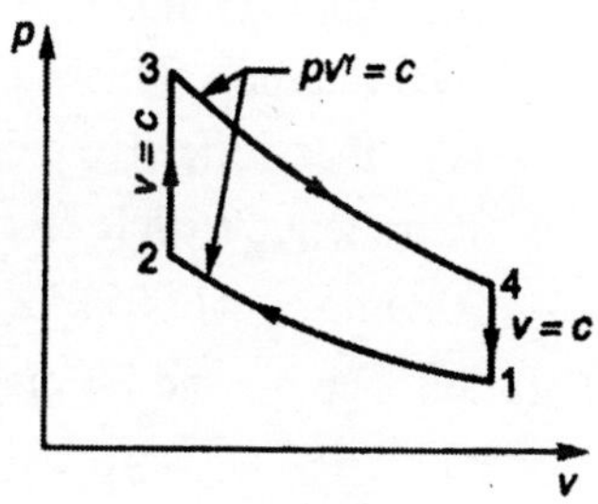

Fig. 2.39

$$d = 18 \text{ cm}, L = 30 \text{ cm}, r_e = 5 = \frac{V_4}{V_3} = \frac{V_1}{V_2}$$

$$p_1 = 1 \text{ bar}, T_1 = 100 + 273 = 373 \text{ K}$$

$$p_2 = 7 \text{ bar}, \gamma = 1.4, R = 0.287 \text{ kJ/kg.K}$$

$$V_1 = \frac{\pi}{4} d^2 L = \frac{\pi}{4} \times 18^2 \times 30 = 7634 \text{cm}^3$$

$$m = \frac{p_1 V_1}{R T_1} = \frac{1 \times 10^2 \times 7634 \times 10^{-6}}{0.287 \times 373}$$

$$= 71.312 \times 10^{-4} \text{ kg}$$

$$\frac{p_2}{p_1} = \left(\frac{V_1}{V_2}\right)^n$$

$$\ln \frac{p_2}{p_1} = n \ln \frac{V_1}{V_2}, \quad n = \frac{\ln\left(\frac{7}{1}\right)}{\ln 5} = 1.209$$

Work done $= \frac{p_1V_1 - p_2V_2}{n-1} = \frac{\left(1\times 1734 - 7\times\frac{7634}{5}\right)\times 10^{-6}\times 10^2}{1209-1}$

$= -1.461\,\text{kJ}$

$$T_2 = \frac{T_1}{\left(\frac{p_1}{p_2}\right)^{\frac{n-1}{n}}} = \frac{373}{\left(\frac{1}{7}\right)^{\frac{0.209}{1.209}}} = 522.16\,\text{K}$$

Heat added, $Q = mc_v\left(\frac{\gamma - n}{1-n}\right)(T_2 - T_1)$

Now $c_v = \frac{R}{\gamma - 1} = \frac{0.287}{0.4} = 0.718\,\text{kJ/kg.K}$

$$Q = 71.312\times 10^{-2}\times 0.718\times\left(\frac{14-1.209}{1-1.209}\right)(522.16-373)$$

$= -0.6979\,\text{kJ/cycle}$

–ve sign indicates that heat is rejected by the system.

■ **Example 2.44** A four-stroke limited pressure cycle (diesel) engine draws 1.2 kg/s of air at 1 atm and 27°C. Compression ratio of the cycle is 16. Pressure ratio during constant volume heat addition is 2.0. Total heat added is equal to 2310 kJ/kg of air in the cylinder. Determine

(i) pressure, volume and temperature at all salient points,

(ii) % of heat added during constant pressure process,

(iii) cut-off ratio,

(iv) thermal efficiency, and

(v) mean effective pressure.

Represent the cycle on *p-v* and *T-s* planes.

Assume $c_p = 1.005$ kJ/kg.K and $c_v = 0.718 =$ kJ/kg.K.

Solution. The *p-v* and *T-s* diagrams are shown in Fig. 2.40

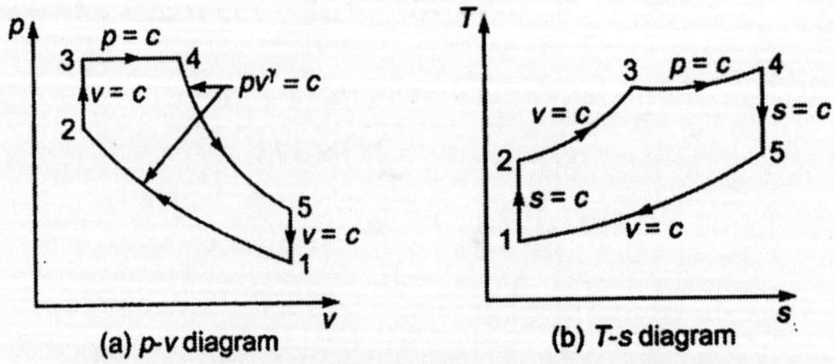

(a) *p-v* diagram (b) *T-s* diagram

Fig. 2.40 Limited pressure cycle (diesel) 4-stroke engine.

$$p_1 = 1\,\text{atm} = 1.03\times 10^5\,\text{Pa}$$

$$T_1 = 27 + 273 = 300\,\text{K}$$

$$\frac{p_3}{p_2} = 2, r = \frac{v_1}{v_2} = 16, m_a = 1.2 \text{ kg/s}$$

$$q = 2310 \text{ kJ/kg}$$

$$R = c_p - c_v = 1.005 - 0.718 = 0.287 \text{ kJ/kg.K}$$

$$\gamma = \frac{c_p}{c_v} = \frac{1.005}{0.718} = 1.4$$

(i)
$$v_1 = \frac{RT_1}{p_1} = \frac{0.287 \times 300}{1.03 \times 10^2} = 0.8359 \text{ m}^3\text{/kg}$$

$$p_2 = p_1 \left(\frac{v_1}{v_2}\right)^{\gamma} = p_1 r^{\gamma} = 1.03 \times 10^2 \times 16^{1.4}$$

$$= 49.958 \times 10^2 \text{ kPa}$$

$$T_2 = T_1 \left(\frac{p_2}{p_1}\right)^{\frac{\gamma - 1}{\gamma}} = 300 \left(\frac{49.958}{1.03}\right)^{\frac{0.4}{1.4}} = 909.43 \text{ K}$$

$$v_2 = \frac{v_1}{16} = \frac{0.8359}{16} = 0.0522 \text{ m}^3\text{/kg}$$

$$p_3 = 2p_2 = 2 \times 49.958 \times 10^2 = 99.916 \times 10^2 \text{ kPa}$$

$$v_3 = v_2 = 0.522 \text{ m}^3\text{/kg}$$

$$T_3 = T_2 \frac{p_3}{p_2} = 909.43 \times 2 = 1818.86 \text{ K}$$

Heat added,
$$q_s = c_v (T_3 - T_2) + c_p (T_4 - T_3)$$

$$2310 = 0.718\,(1818.86 - 909.43) + 1.005(T_4 - 1818.86)$$

$$T_4 = 3467.64 \text{ K}$$

$$v_4 = v_3 \frac{T_4}{T_3} = 0.0522 \times \frac{3467.64}{1818.86} = 0.09952 \text{ m}^3\text{/kg}$$

$$T_5 = T_4 \left(\frac{v_4}{v_5}\right)^{\gamma - 1} = 3467.64 \left(\frac{0.09952}{0.8359}\right)^{0.4} = 1480.25 \text{ K}$$

$$p_5 = p_4 \left(\frac{v_4}{v_5}\right)^{\frac{1}{\gamma}} = p_3 \left(\frac{v_4}{v_1}\right)^{\frac{1}{\gamma}} = 99.916 \times 10^2 \times \left(\frac{0.09952}{0.8359}\right)^{\frac{1}{1.4}}$$

(ii) Heat added during constant pressure process,

$$q_p = c_p (T_4 - T_3) = 1.005\,(3467.64 - 1818.86) = 1657.024 \text{ kJ/kg}$$

Percentage of heat added $= \dfrac{q_p}{q_s} = \dfrac{1657.024}{2310} = 0.7173$ or 71.73%

(iii) Cut-off ratio, $\rho = \dfrac{v_4}{v_3} = \dfrac{0.09952}{0.0522} = 1.906$

(iv) Thermal efficiency, $\eta_{th} = \dfrac{\text{Heat added} - \text{Heat rejected}}{\text{Heat added}}$

Heat rejected during process 5 – 1

$$= c_v(T_5 - T_1) = 0.718\,(1480.25 - 300) = 847.42 \text{ kJ/kg}$$

$$\eta_{th} = \frac{2310 - 847.42}{2310} = 0.6331 \text{ or } 63.31\%$$

(v) Mean effective pressure,

$$p_m = \frac{\text{Work done}}{\text{Swept volume}} = \frac{2310 - 847.42}{0.8359 - 0.0522}$$

$$= 1866.25 \text{ kPa or } 18.119 \text{ bar}$$

■ **Example 2.45** A reversible cycle using an ideal gas as the working substance consists of an isentropic compression from an initial temperature to 555 K, a constant volume process from 555 K to 835 K, a reversible adiabatic expansion to 555 K, a constant pressure expansion from 555 K to 835 K, followed by constant volume process to the initial temperature. Draw the cycle on *p-v* and *T-s* diagrams and determine the initial temperature. Take γ = 1.40. Also compute the work done.

Solution. The *p-v* and *T-s* diagrams are shown in Fig. 2.41 (a) and (b) respectively.

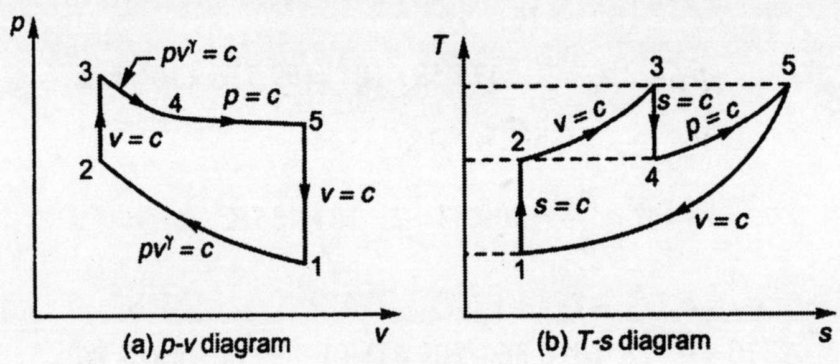

Fig. 2.41

Process 1–2 : $\dfrac{v_2}{v_1} = \left(\dfrac{T_1}{T_2}\right)^{\frac{1}{\gamma-1}} = \left(\dfrac{T_1}{55}\right)^{2.5} = \dfrac{v_3}{v_5}$

Process 3–4 : $\dfrac{v_3}{v_4} = \left(\dfrac{T_3}{T_4}\right)^{\frac{1}{\gamma-1}} = \left(\dfrac{835}{555}\right)^{2.5} = 2.7764$

Process 4–5 : $\dfrac{v_4}{v_5} = \dfrac{T_4}{T_5} = \dfrac{555}{835} = 0.66467$

Now $\dfrac{v_3}{v_5} = \dfrac{v_3}{v_4} \times \dfrac{v_4}{v_5} = 2.7764 \times 0.66467 = 1.8454$

$\therefore$ $$\left(\frac{T_1}{555}\right)^{2.5} = 1.8454$$

$$T_1 = 709.13 \text{ K}$$

Work done $= \frac{R}{(\gamma-1)}(T_1 - T_2) + 0 + \frac{R}{(\gamma-1)}(T_3 - T_4) + R(T_5 - T_4) + 0$

$$= \frac{0.287}{0.4}(709.13 - 555) + \frac{0.287}{0.4}(835 - 555) + 0.287(835 - 555)$$

$$= 110.5883 + 200.9 + 80.36 = 391.848 \text{ kJ/kg}$$

■ **Example 2.46** In an ideal Otto cycle if T_3 and T_1 represent the maximum and minimum temperatures respectively, show that for maximum work done in the cycle

$$T_2 = \sqrt{T_1 T_3} = T_4$$

where T_2 and T_4 are respectively the temperature after compression and expansion.

Solution. The *p-v* diagram for Otto cycle is shown in Fig. 2.42

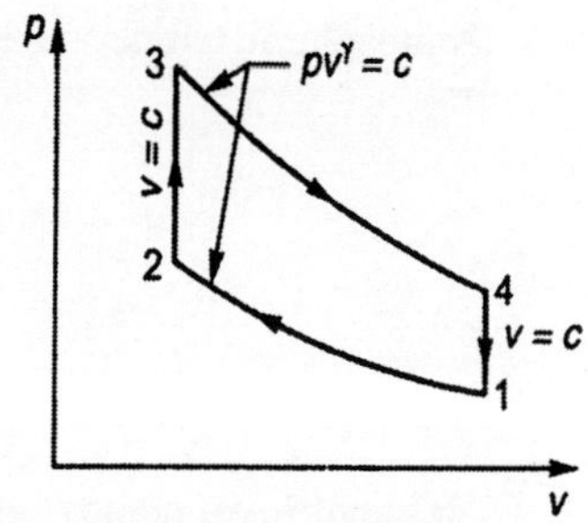

Fig. 2.42 *p-v* diagram for Otto cycle.

Work done, $w = c_v(T_3 - T_2) - c_v(T_4 - T_3)$

Now $T_2 = T_1 r^{\gamma - 1}$ and $T_3 = T_4 r^{\gamma - 1}$

$$w = c_v\left[T_3 - T_1 r^{\gamma-1} - \frac{T_3}{r^{\gamma-1}} + T_1\right]$$

For w to be maximum, $\frac{dw}{dr} = 0$

$$-(\gamma - 1) T_1 r^{\gamma-2} + (\gamma - 1)\frac{T_3}{r^{\gamma}} = 0$$

or $T_1 r^{\gamma-2} = \frac{T_3}{r^{\gamma}}$

$$r^{2(\gamma-1)} = \frac{T_3}{T_1}$$

$$r = \left(\frac{T_3}{T_1}\right)^{\frac{1}{2(\gamma-1)}}$$

$$T_2 = T_1\left(\frac{T_3}{T_1}\right)^{\frac{1}{2}} = \sqrt{T_1 T_3}$$

$$T_4 = \frac{T_3}{\left(\frac{T_3}{T_1}\right)^{\frac{1}{2}}} = \sqrt{T_1 T_3}$$

∴ $T_2 = T_4 = \sqrt{T_1 T_3}$

SOLVED QUESTIONS FROM UNIVERSITY EXAMINATIONS

■ **Example 2.47** A petrol engine has a cylinder diameter of 60 mm and stroke 100 mm. If the mass of the charge admitted per cycle is 0.0002 kg, find the volumetric efficiency of the engine. Assume $R = 287$ J/kg.K. [MDU, Dec., 2007]

Solution. Given : $d = 60$ mm, $L = 100$ mm, $m = 0.0002$ kg/cycle, $R = 287$ J/kg.K.

Swept volume per stroke, $V_s = \frac{\pi}{4} d^2 L = \frac{\pi}{4} (60)^2 \times 100 \times 10^{-9}$

$$= 0.28274 \times 10^{-3} \text{ m}^3$$

Let atmospheric pressure,

$$p = 1 \text{ bar or } 1 \times 10^5 \text{ N/m}^2$$

Atmospheric temperature, $T = 20°C$ or 293 K

$$V_a = \text{theoretical swept volume, m}^3$$

$$pV_a = \text{mRT}$$

$$1 \times 10^5 \times v_a = 0.0002 \times 287 \times 293$$

$$V_a = 0.1682 \times 10^{-3} \text{ m}^3$$

Volumetric efficiency, $\eta_{vol} = \frac{V_a}{V_s} = \frac{0.1682 \times 10^{-3}}{0.28274 \times 10^{-3}}$

$$= 0.5949 \quad \text{or} \quad 59.49\%$$

■ **Example 2.48** Derive an expression for the efficiency of a dual cycle. [MDU, Dec., 2007]

Solution. Refer to art 2.9.

■ **Example 2.49** An oil engine working on a dual compaction cycle has a compression ratio 10 and cut-off takes place 1/10 of stroke. If the pressure at the beginning of compression is 1 bar and maximum pressure is 40 bar, determine the air standard efficiency of the cycle. Take $\gamma = 1.4$. [MDU, Dec. 2007]

Solution. Given : $r = 10$, $v_4 - v_2 = \frac{1}{10} v_s$, $p_1 = 1$ bar, $p_3 = 40$ bar, $\gamma = 1.4$

Refer to Fig. 2.10.

$$r = \frac{v_1}{v_2} = \frac{v_1}{v_3} = 10$$

$$v_4 = v_2 + \frac{1}{10} v_s = v_2 + \frac{1}{10}(v_1 - v_2)$$

$$= v_2 + \frac{1}{10}(10v_2 - v_2) = 1.9v_2$$

Cut-off ratio, $\rho = \frac{v_4}{v_2} = 1.9$

Pressure ratio, $\alpha_p = \frac{p_3}{p_2} = \frac{T_3}{T_2}$

Now, $$p_1 v_1^\gamma = p_2 v_2^\gamma$$

$$\left(\frac{v_1}{v_2}\right)^\gamma = \frac{p_2}{p_1}$$

or $$p_2 = p_1 r^\gamma = 1 \times 10^{1.4} = 25.119 \text{ bar}$$

$$\alpha_p = \frac{p_3}{p_2} = \frac{40}{25.119} = 1.592$$

Air standard efficiency, $$\eta_a = 1 - \frac{1}{r^{\gamma-1}}\left[\frac{(\alpha_p \rho^\gamma - 1)}{(\alpha_p - 1) + \gamma\alpha_p(\rho - 1)}\right]$$

$$= 1 - \frac{1}{10^{0.4}}\left[\frac{1.592 \times (1.9)^{1.4} - 1}{(1.592 - 1) + 1.4 \times 1.592\,(1.9 - 1)}\right]$$

$$= 1 - \frac{1}{2.51188}\left[\frac{2.9102}{2.59792}\right]$$

$$= 0.5540 \quad \text{or} \quad 55.40\%$$

■ **Example 2.50** Derive an expression for efficiency and MEP for the Otto cycle. [MDU, May, 2008]

Solution. Refer to art. 2.7

■ **Example 2.51** Two I.C. engines are to operate on Diesel and Otto cycles with the following data :

Maximum temperature = 1500 K

Exhaust temperature = 700 K

Ambient conditions = 1 bar and 300 K

Compute the compression ratio, maximum pressure and efficiency of the two engines. [MDU, May, 2008]

Solution. Given : $p_1 = 1$ bar, $T_1 = 300$ K, $T_4 = 700$ K, $T_3 = 1500$ K

Otto cycle, Refer to Fig. 2.6

Adiabatic process 3–4 :

$$\frac{T_3}{T_4} = \left(\frac{v_4}{v_3}\right)^{\gamma-1} = \left(\frac{v_1}{v_2}\right)^{\gamma-1} = r^{\gamma-1}$$

Compression ratio, $$r = \left(\frac{T_3}{T_4}\right)^{\frac{1}{\gamma-1}} = \left(\frac{1500}{700}\right)^{\frac{1}{0.4}} = 6.722$$

Adiabatic process 1–2 :

$$p_1 v_1^\gamma = p_2 v_2^\gamma$$

$$p_2 = p_1\left(\frac{v_1}{v_2}\right)^\gamma = p_1 r^\gamma = 1 \times (6.722)^{1.4} = 14.4 \text{ bar}$$

$$\frac{T_2}{T_1} = \left(\frac{p_2}{p_1}\right)^{(\gamma-1)/\gamma} = (14.4/1)^{\frac{0.4}{1.4}} = 2.143$$

Constant volume process 2–3 :

$$T_2 = 300 \times 2.143 = 642.8 \text{ K}$$

$$\frac{p_2}{T_2} = \frac{p_3}{T_3}$$

Maximum pressure, $p_3 = \left(\frac{T_3}{T_2}\right) p_2 = \left(\frac{1500}{642.8}\right) \times 14.4 = 33.603$ bar

Thermal efficiency, $\eta_{th} = 1 - \frac{1}{r^{\gamma-1}} = 1 - \frac{1}{(6.722)^{0.4}}$

$$= 0.5333 \text{ or } 53.33\%$$

Diesel cycle : Refer to Fig. 2.8

Adiabatic process 3–4 :

$$\frac{T_3}{T_4} = \left(\frac{v_4}{v_3}\right)^{\gamma-1} = \left(\frac{v_1}{v_3}\right)^{\gamma-1}$$

$$\frac{v_1}{v_3} = \left(\frac{1500}{700}\right)^{\frac{1}{0.4}} = 6.722$$

Constant volume process 4–1 :

$$\frac{p_4}{T_4} = \frac{p_1}{T_1}$$

$$p_4 = 1 \times \frac{700}{300} = \frac{7}{3} \text{ bar}$$

Also from process 3–4,

$$\frac{T_3}{T_4} = \left(\frac{p_3}{p_4}\right)^{\frac{\gamma-1}{\gamma}}$$

or

$$p_3 = p_4 \left(\frac{T_3}{T_4}\right)^{\frac{\gamma}{\gamma-1}} = \frac{7}{3}\left(\frac{1500}{700}\right)^{\frac{1.4}{0.4}} = 33.6 \text{ bar}$$

Now $\quad p_2 = p_3 = 33.6$ bar

Adiabatic process 1–2 :

$$p_1 v_1^\gamma = p_2 v_2^\gamma$$

$$r = \frac{v_1}{v_2} = \left(\frac{p_2}{p_1}\right)^{\frac{1}{\gamma}} = \left(\frac{33.6}{1}\right)^{\frac{1}{1.4}} = 12.31$$

∴ Compression ratio, $r = 12.31$

Maximum pressure, $p_2 = p_3 = 33.6$ bar

Cut-off ratio, $\rho = \frac{v_3}{v_2} = \frac{v_3}{v_1} \times \frac{v_1}{v_2} = \frac{1}{6.722} \times 12.31 = 1.83$

Thermal efficiency, $\eta_{th} = 1 = \frac{1}{r^{\gamma-1}} = \left[\frac{\rho^{\gamma}-1}{\gamma(\rho-1)}\right]$

$$= 1 - \frac{1}{(12.31)^{0.4}}\left[\frac{(1.83)^{1.4}-1}{1.4\,(1.83-1)}\right]$$

$$= 0.5806 \text{ or } 58.06\%$$

■ **Example 2.52** An engine uses air as the working substance. At the beginning of compression the pressure is 90 kN/m^2 and the temperature is 40°C. During the adiabatic compression, the volume is reduced to one-sixth of its value at the beginning of the compression stroke. Heat is then added at constant pressure until the temperature is 1400°C. The stroke is completed by adiabatic expansion until the initial volume is reached, when the cycle is closed by a constant volume process.

(a) Sketch the *p-V* diagram of the engine.

(b) Calculate the temperature and pressure at all the corners of the diagram.

(c) Determine the thermal efficiency of the cycle. [MDU, Dec. 2008]

Solution. Given : $p_1 = 90 \text{ kN/m}^2, T_1 = 273 + 40 = 313 \text{ K}, \frac{v_1}{v_2} = 6,$

$$T_3 = 273 + 1400 = 1673 \text{ K}$$

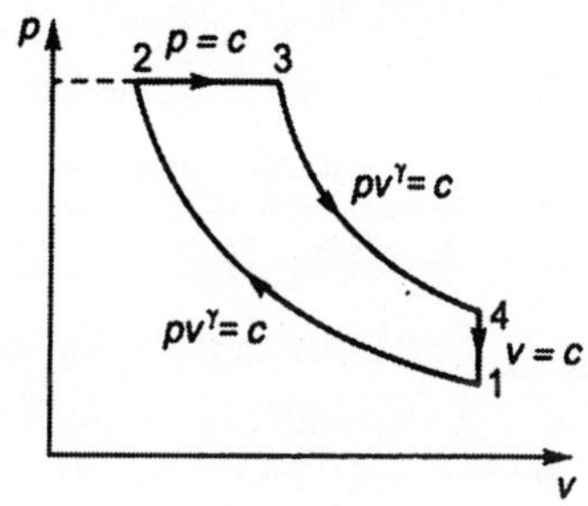

Fig. 2.43

(a) The *p-v* diagram is shown in Fig. 2.43.

(b) Adiabatic compression process 1–2 :

$$p_1 v_1^{\gamma} = p_2 v^{\gamma}$$

$$p_2 = p_1\left(\frac{v_1}{v_2}\right)^{\gamma} = p_1 r^{\gamma} = 90(6)^{1.4} = 1105.75 \text{ kN/m}^2$$

$$\frac{T_2}{T_1} = \left(\frac{p_2}{p_1}\right)^{\frac{\gamma-1}{\gamma}} = \left(\frac{1105.74}{90}\right)^{\frac{0.4}{1.4}} = 2.0476$$

$$T_2 = 313 \times 2.0476 = 640.9 \text{ K}$$

Isobaric process 2–3 :

$$p_3 = p_2 = 1105.74 \text{ kN/m}^2$$

$$T_3 = 1673 \text{ K}$$

$$\frac{v_2}{T_2} = \frac{v_3}{T_3}$$

$$\frac{v_3}{v_2} = \frac{T_3}{T_2} = \frac{1673}{640.9} = 2.61$$

$$\rho = \frac{v_3}{v_2} = 2.61$$

Adiabatic expansion process 3–4 :

Now $v_4 = v_1$ and $\frac{v_4}{v_3} = \frac{v_1}{v_3} = \frac{v_1}{v_2} \times \frac{v_2}{v_3}$

$$= 6 \times \frac{1}{2.61} = 2.3$$

$$p_3 v_3^\gamma = p_4 v_4^\gamma$$

$$p_4 = p_3 \left(\frac{v_3}{v_4} \right)^\gamma = 1105.74 \left(\frac{1}{2.3} \right)^{1.4} = 344.53 \text{ kN/m}^2$$

$$\frac{T_3}{T_4} = \left(\frac{p_3}{p_4} \right)^{\frac{\gamma - 1}{\gamma}} = \left(\frac{1105.74}{344.53} \right)^{\frac{0.4}{1.4}} = 1.395$$

$$T_4 = \frac{1673}{1.395} = 1199 \text{ K}$$

(c) Thermal efficiency of cycle, η_{th}

$$= 1 - \frac{1}{r^{\gamma - 1}} \left[\frac{\rho^\gamma - 1}{\gamma(\rho - 1)} \right]$$

$$= 1 - \frac{1}{6^{0.4}} \left[\frac{(2.61)^{1.4} - 1}{1.4\,(2.61 - 1)} \right]$$

$$= 0.3867 \text{ or } 38.67\%$$

■ **Example 2.53** In an ideal constant volume cycle, the pressure and temperature at the beginning of compression are 100 kN/m^2 and 50°C. The volume ratio of compression is 5 : 1. The heat supplied during the cycle is 950 kJ/kg of working fluid. Determine :

(a) The maximum temperature attained in the cycle.
(b) The thermal efficiency of the cycle.
(c) The work done during the cycle / kg of working fluid.

Assume $\gamma = 1.4$ and $c_v = 0.717$ kJ/kg.K. [MDU, May, 2009]

Solution. Given : $p_1 = 100 \text{ kN/m}^2, T_1 = 273 + 50 = 323 \text{ K}, r = 5, q_s = 950 \text{ kJ/kg}, \gamma = 1.4,$
$c_v = 0.717$ kJ/kg.K.

Refer to Fig. 2.6.

(a) Adiabatic compression process 1–2 :

$$p_1 v_1^\gamma = p_2 v_2^\gamma$$

$$p_2 = p_1 \left(\frac{v_1}{v_2} \right)^\gamma = p_1 r^\gamma = 100 \times 5^{1.4} = 951.8 \text{ kN/m}^2$$

$$\frac{T_2}{T_1} = \left(\frac{v_1}{v_2} \right)^{\gamma - 1} = r^{\gamma - 1} = 5^{0.4} = 1.9036$$

$$T_2 = 323 \times 1.9036 = 614.9 \text{ K}$$

Constant volume process 2–3 :

$$q_s = c_v(T_3 - T_2)$$
$$950 = 0.717\,(T_3 - 614.9)$$
$$T_3 = 1939.86\text{ K}$$

Adiabatic expansion process 3–4

$$\frac{T_3}{T_4} = \left(\frac{v_4}{v_3}\right)^{\gamma-1} = \left(\frac{v_1}{v_2}\right)^{\gamma-1} = r^{\gamma-1} = 5^{0.4} = 1.9036$$

$$T_4 = \frac{1939.86}{1.9036} = 1019.05\text{ K}$$

Heat rejected, $q_r = c_v(T_4 - T_1) = 0.717\,(1019.05 - 323) = 499$ kJ/kg

(c) Work done, $w_{net} = q_s - q_r = 950 - 499 = 451$ kJ/kg

(b) Thermal efficiency of cycle, $\eta_{th} = \dfrac{w_{net}}{q_s}$

$$= \frac{451}{950} \times 100 = 47.47\%$$

■ **Example 2.54** For an engine working on the ideal dual cycle, the compression ratio is 10 and the maximum pressure is limited to 70 bar. If the heat supplied is 1680 kJ/kg, find the pressure and temperature at the various salient points of the cycle and the cycle efficiency.The pressure and temperature of air at the commencement of the compression are 1 bar and 100°C respectively. Assume, $c_p = 1.004$ kJ/kg and $c_v = 0.717$ kJ/kg/.K for air. [MDU, Dec. 2009]

Solution. Given $r = 10$, $p_3 = p_4 = 70$ bar, $q_s = 1680$ kJ/kg, $p_1 = 1$ bar, $T_1 = 273 + 100 = 373$ K, $c_p = 1.004$ kJ/kg.K, $c_v = 0.717$ kJ/kg.K.

Refer to Fig. 2.10.

Adiabatic compression process 1–2 :

$$p_1 v_1^{\gamma} = p_2 v_2^{\gamma}$$

$$p_2 = p_1\left(\frac{v_1}{v_2}\right)^{\gamma} = p_1 r^{\gamma} = 1 \times 10^{1.4} = 25.119\text{ bar}$$

$$\frac{T_2}{T_1} = \left(\frac{v_1}{v_2}\right)^{\gamma-1} = r^{\gamma-1} = 10^{0.4} = 2.5119$$

$$T_2 = 373 \times 2.5119 = 936.9\text{ K}$$

Constant volume process 2–3 :

$$\frac{p_2}{T_2} = \frac{p_3}{T_3}$$

$$T_3 = T_2 \times \frac{p_3}{p_2} = 936.9 \times \frac{70}{25.119} = 2611\text{ N}$$

Now
$$q_s = c_v(T_3 - T_2) + c_p(T_4 - T_3)$$
$$1680 = 0.717\,(2611 - 936.9) + 1.004(T_4 - 2611)$$
$$T_4 = 3088.76\text{ K}$$

Isobaric process 3–4 :

$$p_4 = p_3 = 70\text{ bar}$$
$$T_4 = 3088.76\text{ K}$$
$$\frac{v_3}{T_3} = \frac{v_4}{T_4}$$

Cut-off ratio,
$$\rho = \frac{v_4}{v_3} = \frac{T_4}{T_3} = \frac{3088.76}{2611} = 1.183$$

Adiabatic expansion process 4–5 :

$$p_4 v_4^{\gamma} = p_5 v_5^{\gamma}$$
$$\frac{p_5}{p_4} = \left(\frac{v_4}{v_5}\right)^{\gamma} = \left(\frac{v_4}{v_1}\right)^{T} = \left(\frac{1.183 v_3}{v_1}\right)^{\gamma} = \left(\frac{1.183\, v_2}{v_1}\right)^{\gamma}$$
$$= \left(\frac{1.183}{10}\right)^{1.4} = 0.0504$$
$$p_5 = 70 \times 0.0504 = 3.526\text{ bar}$$
$$\frac{T_5}{T_4} = \left(\frac{v_4}{v_5}\right)^{\gamma - 1} = \left(\frac{1.183}{10}\right)^{0.4} = 0.4258$$
$$T_5 = 3088.76 \times 0.4258 = 1315.1\text{ K}$$

Heat rejected,
$$q_r = c_v(T_5 - T_1) = 0.717(1315.1 - 373)$$
$$= 675.5\text{ kJ/kg}$$

Net work done,
$$w_{\text{net}} = q_s - q_r = 1680 - 675.5$$
$$= 1004.5\text{ kJ/kg}$$

Cycle efficiency,
$$\eta_{\text{th}} = \frac{w_{\text{net}}}{q_s} = \frac{1004.5}{1680} = 0.5979 \text{ or } 59.79\%$$

■ **Example 2.55** In an engine working on Diesel cycle inlet pressure and temperature are 1 bar and 17°C respectively. Pressure at the end of adiabatic compression is 35 bar. The ratio of expansion, *i.e.*, after constant pressure heat-addition is 5. Calculate the heat-addition, heat-rejection and efficiency of the cycle.

Assume, $\gamma = 1.4$, $c_p = 1.004$ kJ/kg.K, $c_v = 0.717$ kJ/kg.K [MDU, Dec. 2010]

Solution. Given : $p_1 = 1$ bar, $T_1 = 273 + 17 = 290$ K, $p_2 = 35$ bar, $\frac{v_4}{v_3} = 5$, $\gamma = 1.4$, $c_p = 1.004$ kJ/kg.K, $c_v = 0.717$ kJ/kg.K.

Refer to Fig. 2.8.

Adiabatic compression process 1–2 :

$$\frac{T_2}{T_1} = \left(\frac{p_2}{p_1}\right)^{\frac{\gamma-1}{\gamma}} = \left(\frac{35}{1}\right)^{\frac{0.4}{1.4}} = 2.7616$$

$$T_2 = 290 \times 2.7616 = 800.86\text{ K}$$

$$p_1 v_1^\gamma = p_2 v_2^\gamma$$

$$r = \frac{v_1}{v_2} = \left(\frac{p_2}{p_1}\right)^{\frac{1}{\gamma}} = (35)^{\frac{1}{1.4}} = 12.67$$

Isobaric process 2–3 :

$$p_2 = p_3 = 35\text{ bar}$$

Adiabatic expansion process 3–4 :

$$p_3 v_3^\gamma = p_4 v_4^\gamma$$

$$p_4 = p_3\left(\frac{v_3}{v_r}\right)^\gamma = 35\left(\frac{1}{5}\right)^{1.4} = 3.677\text{ bar}$$

$$\frac{T_3}{T_4} = \left(\frac{v_4}{v_3}\right)^{\gamma-1} = 5^{0.4} = 1.9$$

Constant volume process 4–1 :

$$\frac{p_4}{T_4} = \frac{p_1}{T_1}$$

$$T_4 = T_1\left(\frac{p_4}{p_1}\right) = 290\left(\frac{3.677}{1}\right) = 1066.3\text{ K}$$

Then $T_3 = 1.9\,T_4 = 1.9 \times 1066.3 = 2026\text{ K}$

Heat added, $q_s = c_p(T_3 - T_2) = 1.004(2026 - 800.86) = 1230\text{ kJ/kg}$

Heat rejected, $q_r = c_v(T_4 - T_1) = 0.717\,(1066.3 - 290) = 556.6\text{ kJ/kg}$

Net work done, $w_{net} = q_s - q_r = 1230 - 556.6 = 673.4\text{ kJ/kg}$

Thermal efficiency, $\eta_{th} = \frac{w_{net}}{q_s} = \frac{673.4}{1230} = 0.5475$ or 54.75%

■ **Example 2.56** (a) Derive an expression for the efficiency and mean effective pressure of a diesel cycle.

(b) The compression ratio of an ideal air standard cycle is 15. The heat transfer is 1465 kJ/kg of air. Find the pressure and temperature at the end of each process and determine the cycle efficiency.

What is the mean effective pressure of the cycle, if the inlet conditions are 1 bar and 300 K ?

[KU, Dec., 2007]

Solution. (a) Refer to art. 2.8.

(b) Given : $r = 15$, $q_s = 1465\text{ kJ/kg}$, $p_1 = 1\text{ bar}$, $T_1 = 300\text{ K}$

Refer to Fig. 2.6.

Adiabatic compression process 1–2 :

$$p_1 v_1^\gamma = p_2 v_2^\gamma$$

$$p_2 = p_1\left(\frac{v_1}{v_2}\right)^\gamma = p_1 r^\gamma = 1\times 15^{1.4} = 44.31\text{ bar}$$

$$\frac{T_2}{T_1} = \left(\frac{v_1}{v_2}\right)^{\gamma-1} = r^{\gamma-1}$$

$$T_2 = 300\times 15^{0.4} = 886.25\text{ K}$$

Constant volume heat addition process 2–3 :

$$q_s = c_v(T_3 - T_2)$$

$$1465 = 0.717(T_3 = 886.25)$$

$$T_3 = 2929.48\text{ K}$$

$$\frac{p_2}{T_2} = \frac{p_3}{T_3}$$

$$p_3 = p_2\left(\frac{T_3}{T_2}\right) = 44.31\times\frac{2929.48}{886.25} = 146.47\text{ bar}$$

Adiabatic expansion process 3–4 :

$$p_3 v_3^\gamma = p_4 v_4^\gamma$$

or

$$p_4 = p_3\left(\frac{1}{r}\right)^\gamma = 146.47\left(\frac{1}{15}\right)^{1.4} \qquad [\because v_3 = v_2 \text{ and } v_4 = v_1]$$

$$= 3.3\text{ bar}$$

$$\frac{T_4}{T_3} = \left(\frac{v_3}{v_4}\right)^{\gamma-1} = \left(\frac{1}{r}\right)^{\gamma-1} = \left(\frac{1}{15}\right)^{0.4} = 0.3385$$

$$T_4 = 991.64\text{ K}$$

Cycle efficiency, $\eta_a = 1 - \dfrac{1}{r^{\gamma-1}} = 1 - \dfrac{1}{(15)^{0.4}} = 0.6615$ or 66.15%

Heat rejected, $q_r = c_v(T_4 - T_1) = 0.717\,(991.64 - 300) = 495.9$ kJ/kg

Net work, $w_{net} = q_s - q_r = 1465 - 495.9 = 969.1$ kJ/kg

Swept volume, $v_s = v_1 - v_2$

$$p_1 v_1 = RT_1$$

$$v_1 = \frac{287\times 300}{1\times 10^5} = 0.861\text{ m}^3/\text{kg}$$

$$v_2 = \frac{v_1}{15} = \frac{0.861}{15} = 0.0574\text{ m}^3/\text{kg}$$

Mean effective pressure, $p_m = \frac{w_{net}}{v_s} = \frac{969.1}{(0.861 - 0.0574)}$

$$= 1205.95 \text{ kN/m}^2$$

■ **Example 2.57** Derive an expression for the mean effective pressure of the Otto cycle.

[KU, Dec., 2008]

Solution. Refer to art. 2.7.

■ **Example 2.58** A single cylinder four stroke diesel engine working on dual combustion cycle has a compression ratio of 15 : 1. The engine draws in air at 1 bar and 27°C and maximum pressure in the cylinder is limited to 55 bar. If heat transfer at constant volume is twice that at constant pressure, determine (a) the pressure ratio of constant volume process. (b) the cut-off ratio and (c) η_{th} of the cycle. Take $c_p = 1.005$ kJ/kg.K and $c_v = 0.718$ kJ/kg.K, $\gamma = 1.4$. [KU. Dec., 2008]

Solution. Given : $r = 15$, $p_1 = 1$ bar, $T_1 = 273 + 27 = 300$ K, $p_3 = p_4 = 55$ bar, $(q_s)_v = 2\,(q_s)_p$, $c_p = 1.005$ kJ/kg.K, $c_v = 0.718$ kJ/kg.K, $\gamma = 1.4$

Adiabatic process 1–2 :

$$p_1 v_1^{\gamma} = p_2 v_2^{\gamma}$$

$$p_2 = p_1 \left(\frac{v_1}{v_2}\right)^{\gamma} = p_1 r^{\gamma} = 1 \times 15^{1.4} = 44.31 \text{ bar}$$

$$\frac{T_2}{T_1} = \left(\frac{v_1}{v_2}\right)^{\gamma - 1} = r^{\gamma - 1}$$

$$T_2 = 300 \times 15^{0.4} = 886.25 \text{ K}$$

$$(q_s)_v = c_v(T_3 - T_2) = 0.718(T_3 - 886.25)$$

$$(q_s)_p = c_p(T_4 - T_3) = 1.005(T_4 - T_3)$$

$$\frac{T_3}{T_2} = \left(\frac{p_3}{p_2}\right)^{\frac{\gamma - 1}{\gamma}}$$

$$T_3 = 886.25 \left(\frac{55}{44.31}\right)^{\frac{0.4}{1.4}} = 942.7 \text{ K}$$

∴ $$0.718(942.7 - 886.25) = 2 \times 1.005(T_4 - 942.7)$$

$$T_4 = 962.86 \text{ K}$$

Constant pressure process 3–4 :

$$\frac{v_3}{T_3} = \frac{v_4}{T_4}$$

∴ $$\rho = \frac{v_4}{v_3} = \frac{T_4}{T_3} = \frac{962.86}{942.7} = 1.02$$

(a) Pressure ratio of constant volume process,

$$\alpha_p = \frac{p_3}{p_2} = \frac{55}{44.31} = 1.24$$

(b) Cut-off ratio, $\rho = \dfrac{v_4}{v_3} = 1.02$

(c) Thermal efficiency of the cycle,

$$\eta_{th} = 1 - \frac{1}{r^{\gamma-1}}\left[\frac{\alpha_p\rho - 1}{(\alpha_p - 1) + \gamma\alpha_p(\rho - 1)}\right]$$

$$= 1 - \frac{1}{(15)^{0.4}}\left[\frac{1.24 \times 1.02 - 1}{(1.24 - 1) + 1.4 \times 1.24\,(1.02 - 1)}\right]$$

$$= 1 - \frac{1}{2.954}\left[\frac{0.2648}{0.2747}\right]$$

$$= 0.6737 \text{ or } 67.37\%$$

■ **Example 2.59** (a) Explain assumptions made in air standard cycle.

(b) Derive expressions for thermal efficiency, work output and mean effective pressure for an air standard cycle. [KU, Sept. 2009]

Solution. (a) Refer to art. 2.1.

(b) Refer to art. 2.7.

SHORT ANSWER TYPE QUESTIONS

1. Define an air-0standard cycle.
2. Write four important assumptions for air-standard cycle.
3. Define air-standard efficiency and relative efficiency.
4. Differentiate between cylinder volume and swept volume.
5. What is compression ratio? Give its range for Otto cycle and Diesel cycle.
6. Define clearance ratio and cut-off ratio.
7. Define mean effective pressure.
8. Write the processes involved in the Carnot cycle.
9. Draw the Carnot cycle on *p-V* and *T-s* diagrams.
10. What are the processes involved in the Stirling cycle?
11. Draw the *p-V* and *T-s* diagrams for the Stirling cycle.
12. What is the principle of working of Stirling cycle?
13. What are the processes involved in the Ericsson cycle?
14. Draw the Ericsson cycle on *p-V* and *T-s diagrams.*
15. What are the processes involved in the Atkinson cycle?
16. Draw the *p-V* and *T-s* diagrams for the Atkinson cycle.
17. What are the processes involved in the Otto cycle?

18. Draw the *p-V* and *T-s* diagrams for the Otto cycle.
19. Explain the effect of compression ratio and index of compression on thermal efficiency of Otto cycle.
20. Write the processes involved in the Diesel cycle.
21. Draw the *p-V* and *T-s* diagrams for the Diesel cycle.
22. What is the effect of cut-off ratio on the thermal efficiency of diesel cycle?
23. What are the processes involved in the dual cycle?
24. Draw the *p-V* and *T-s* diagrams for the dual cycle.
25. What is the application of Brayton cycle?
26. Draw the *p-V* and *T-s* diagrams for the Brayton cycle.
27. Write the processes involved in the Brayton cycle.
28. How does the thermal efficiency of Otto, Diesel and Dual cycles compare for constant maximum pressure and heat input?

EXERCISES

1. A diesel cycle operates at pressure of 1 bar at the beginning of compression and the volume is compressed to1/15 of the initial volume. Heat is then supplied until the volume is twice that of the clearance volume. Determine the mean effective pressure. Take $\gamma = 1.4$. **[Ans.** 6.69 bar]
2. A semi diesel engine works on dual combustion cycle. The pressure and temperature at the beginning of the compression is 1 bar and 27°C respectively and the compression ratio being 12. If the maximum pressure is 50 bar and heat received at constant pressure is for 1/30 th of the stroke, find the work done per kg of air and the thermal efficiency. Take $c_p = 1.004$ kJ/kg.K, $c_v = 0.717$ kJ/kg.K. **[Ans.** 537 kJ/kg, 69.27%]
3. A compression ignition engine has compression ratio of 10 and 2/3 of heat of combustion is liberated at a constant volume and the remainder at constant pressure. The pressure and temperature at the beginning are 1 bar and 27°C and the maximum pressure is 40 bar. Find the temperature at the end of compression and expansion, if it follows the law $pV^{1.35}$ = constant, and $\gamma = 1.4$. **[Ans.** (i) 398.6°C, (ii) 347.3°C]
4. An Otto cycle engine having a clearance volume of 250 cc has a compression ratio of 8. The ratio of pressure at constant volume is 9. If the initial pressure is 1 bar, find the work done per cycle and the theoretical mean effective pressure. Take $\gamma = 1.4$. **[Ans.** (i) 1946.1 J/cycle, (ii) 11.12 bar]
5. Find the m.e.p. for the ideal air standard Otto cycle having a maximum pressure of 40 bar and minimum pressure of 1 bar. The compression ratio is 5 : 1. Take $\gamma = 1.4$. **[Ans.** 9.04 bar]
6. An engine working on Otto cycle, the ratio of temperature at the beginning of compression is 300 K. If the ideal air standard efficiency = 0.5, calculate the compression ratio of the engine. If the peak temperature of the cycle is 1150 K, calculate the temperature when the piston is at BDC during expansion stroke. **[Ans.** (i) 9.66, (ii) 575 K]
7. A petrol engine working on Otto cycle has a maximum pressure of 50 bar. If the pressure during combustion is 12.286 bar, find the compression ratio and also the ratio of peak temperature to inlet temperature. Take $p_1 = 1$ bar and $t_1 = 27°$ C. **[Ans.** (i) 6, (ii) 8.34]
8. An Otto cycle takes in air at 300 K. The ratio of maximum to minimum temperature is 6. Find out the optimum compression ratio for the maximum work output of the cycle. **[Ans.** 9.39]

9. The pressure and temperature of a diesel cycle at the start are 1 bar and 20°C respectively and the compression ratio is 19. The pressure at the end of expansion is 2.5 bar. Find the percentage of working stroke at which heat is supplied and heat supplied per kg of air. Assume $\gamma = 1.4$ and $c_p = 1.004$ kJ/kg.K.
[Ans. (i) 7.11%, (ii) 781.46 kJ/kg]

10. An oil engine works on diesel cycle, the compression ratio being 19. the temperature at the start of compression is 17°C and 700 kJ of heat is supplied at constant pressure per kg of air and attains a temperature of 417°C at the end of adiabatic expansion. Find the air-standard efficiency of the cycle. What would be the theoretical work done per kg of air. Take $c_v = 0.717$ kJ/kg.K and $\gamma = 1.4$. **[Ans.** (i) 61.32% (ii) 419.20 kJ]

11. An internal combustion engine works on Diesel cycle with a compression ratio of 8 and expansion ratio of 9. Calculate the air-standard efficiency. Assume $\gamma = 1.4$. **[Ans.** 51.76%]

12. A diesel engine works on Diesel cycle with a compression ratio of 16 and cut-off ratio of 1.8. Calculate the thermal efficiency assuming $\gamma = 1.4$. **[Ans.** 62.38%]

13. A compression-ignition engine works on dual combustion cycle. The pressure and temperature at the beginning of compression are 1 bar and 27° respectively and the pressure at the end of compression is 25 bar. If 420 kJ of heat is supplied per kg of air during constant volume heating and the pressure at the end of adiabatic expansion is found to be 3 bar, find the ideal thermal efficiency. Assume $c_p = 1.004$ kJ/kg.K and $c_v = 0.717$ kJ/kg.K. **[Ans.** 58.27%]

14. The cycle of an internal combustion engine with isochoric heat supply is performed with the compression ratio equal to 8. Find heat supplied to the cycle and the useful work, if the removed heat is 500 kJ/kg and the working fluid is air. **[Ans.** (i) 1148.63 kJ/kg (ii) 648.63 kJ/kg]

□□□

Chapter 3

Carburetion, Fuel Injection and Ignition Systems

Inside This Chapter

PART - I : CARBURETION

3.1 CARBURETION

In S.I. engines, the air and fuel are mixed outside the engine cylinder and partly evaporated mixture is supplied to the engine. The process of preparing this mixture is called **carburetion.** The device used for this purpose is known as **carburettor**. The carburettor atomises the fuel and mixes it with air. This complicated process is achieved in the induction system, which is shown in Fig. 3.1. The pipe that carries the prepared mixture to the engine cylinder is called the intake manifold.

During the suction stroke vacuum is created in the cylinder which causes the air to flow thorough the carburettor and the fuel to be sprayed from the fuel jets.. Because of the high volatility of the S.I. engine fuels, most of the fuel vaporises and forms a combustible fuel-air mixture. However, some of the larger droplets may reach the cylinder in the liquid form and must be vaporised and mixed with air during the compression stroke before ignition takes place by the electric spark.

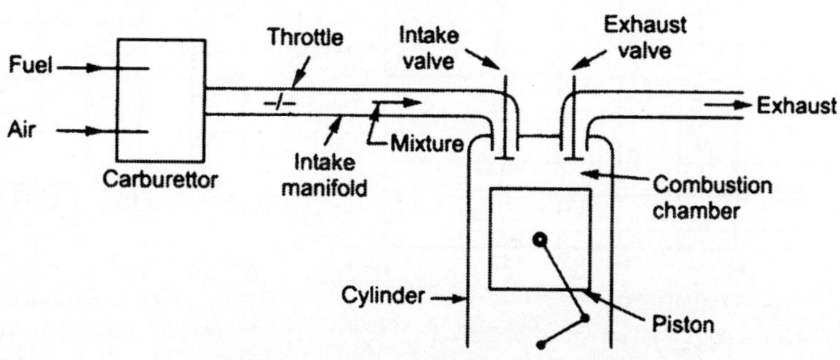

Fig. 3.1 Induction system for S.I. engine.

3.1.1 Factors Affecting the Process of Carburetion

The important factors which significantly affect the process of carburetion are :

1. The time available for the preparation of the mixture.
2. The temperature of incoming air in the intake manifold.
3. The quality of the fuel supplied.
4. The design of the induction system and combustion chamber.

Atomisation, mixing and vaporisation, are the processes which require a finite time to occur. In high speed engines the time available for mixture formation is very small (less than 0.02s). Because of the short time available complete and efficient mixing, vaporisation, and distribution is difficult to achieve.

Temperature is a factor which controls the vaporization process of the fuel. A high temperature results in high rate of vaporization. The temperature of the mixture can be increased by heating the induction manifold but this would reduce the power output due to reduction in mass flow.

The volatility of the fuel affects the vaporization and distribution of fuel. The design of the induction manifold, etc., affect the uniform distribution of the mixture among the cylinders.

3.1.2 Mixture Requirements in S.I. Engines

There is a limited range of air-fuel ratios, in a homogeneous mixture, which can be ignited in a SI engine. These limits are about 7 : 1 A/F by mass (0.14 : 1 F/A) on the rich side, and about 20 : 1 A/F (0.05 : 1 F/A) on the lean side in single cylinder engines (See Fig 3.2). The properties of different air-petrol mixtures in

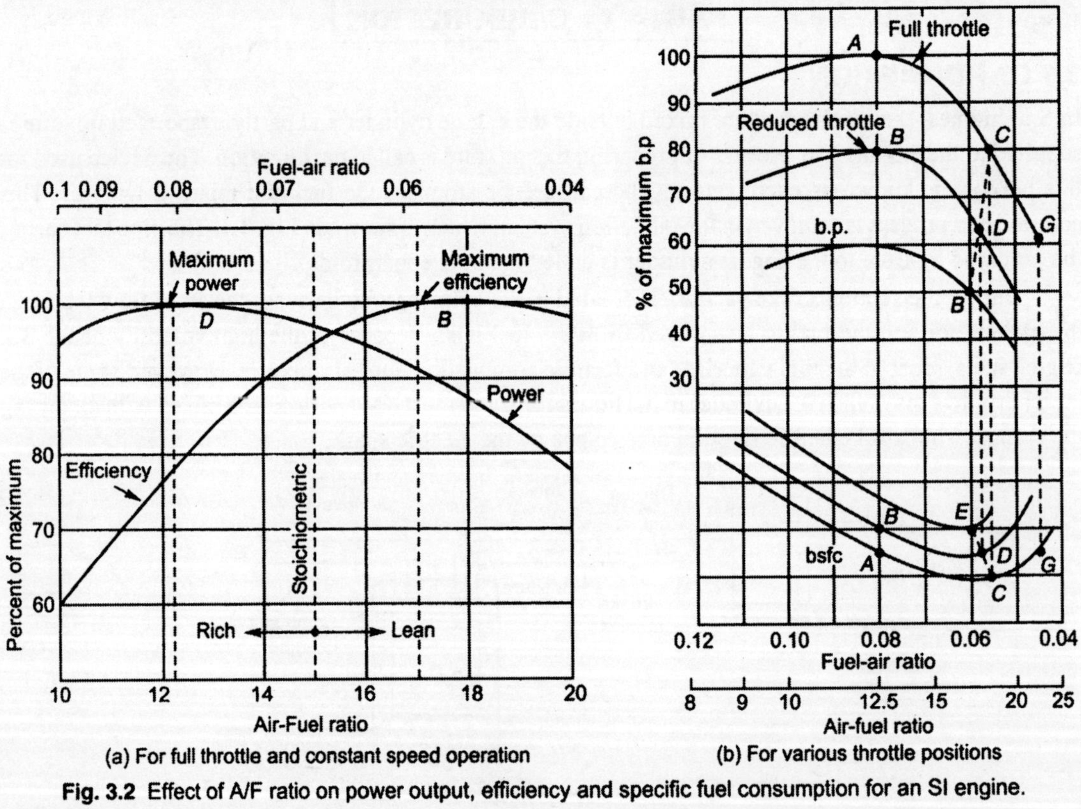

(a) For full throttle and constant speed operation

(b) For various throttle positions

Fig. 3.2 Effect of A/F ratio on power output, efficiency and specific fuel consumption for an SI engine.

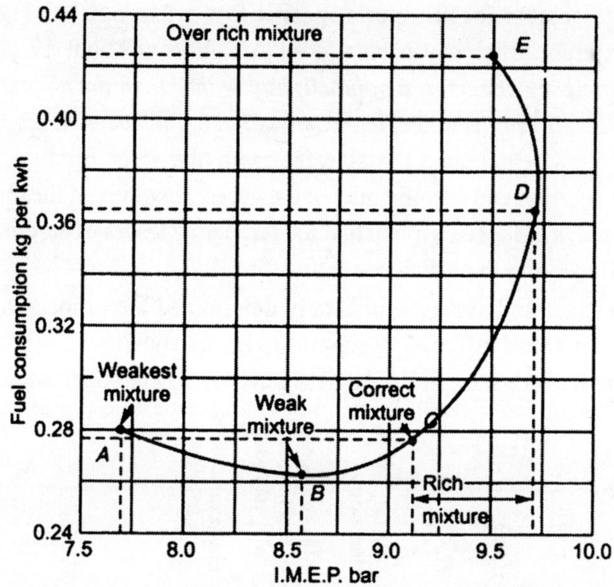

Fig. 3.3 Fuel consumption and imep curve for various A/F ratios.

this range are shown in Figs. 3.2 to 3.4. Fig 3.2 shows the effect of A/F ratio on power output and specific fuel consumption for a SI engine for full throttle and constant speed condition. Fig. 3.3 shows a typical fuel consumption loop for various A/F ratios (see Table 3.1 also). The most important hint to note from these figures is that the A/F ratio for maximum power is not the same as the A/F ratio for maximum economy. Further, the A/F ratio for maximum power and maximum economy varies with load as shown in Fig. 3.4.

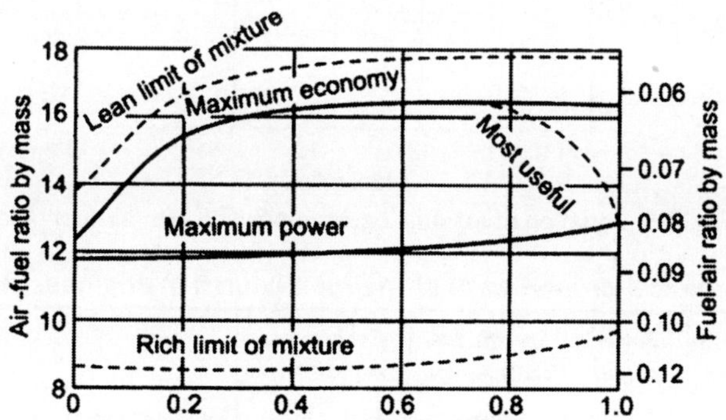

Fig. 3.4 Ranges of air-fuel ratio at varying load.

(a) Mixture requirements for maximum power : Fig. 3.2 shows that the maximum power is obtained at about 12.5 : 1 A/F (0.08 : 1 F/A). *Maximum energy is released when slightly excess fuel is introduced so that all the oxygen present in the cylinder is utilized.* More fuel than this does no help. In fact it is disadvantageous because the combustion of a large excess of fuel with the same amount of oxygen results in smaller energy release due to partial combustion, and more carbon monoxide is formed.

Mechanical efficiency is maximum at maximum power position.

(b) Mixture requirement for minimum specific fuel consumption : Fig. 3.2 shows that at full throttle, maximum efficiency occurs at an A/F ratio of about 17 : 1 (0.06 : 1F/A ratio). *Maximum efficiency occurs at a point slightly leaner than the chemically correct A/F ratio because excess air requires complete combustion of fuel when mixing is not perfect;* and the power maximum temperatures assoicated with the inlet mixture favourably affect the chemical equilibrium and specific inlet of the gases. However, if the mixture is made too lean, the flame speed is reduced so much that the large time losses overcome the above-mentioned beneficial effects, and the efficiency falls off.

Various investigators have experimentally determined the properties of air-petrol mixtures which are summarized in Table 3.1 (also see Fig. 3.3 and Fig. 3.5). For practical purposes the portion of the fuel consumption loop between *B* and *D* is important.

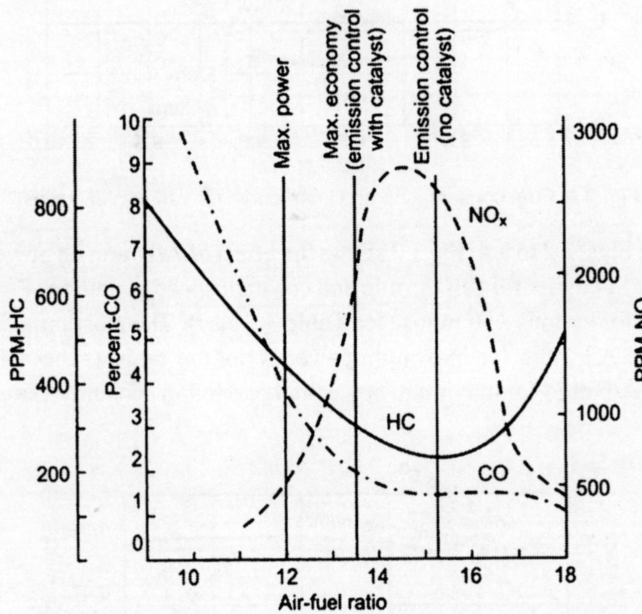

Fig. 3.5 Variation of exhaust gas composition with air-fuel ratio.

Table 3.1 : Principal properties of air-petrol mixtures in engines. (Refer Fig. 3.3)

S. No.	*Air/petrol (by mass)*	*Designation of mixture*	*Power output*	*Specific fuel consumption*	*Exhaust gas composition, %*	*Remarks*
1.	18-22 (Point A)	Very weak	Minimum about 40% less than maximum	Low	84.0 N_2 8.0 CO_2 8.0 O_2	Minimum power. Occasional backfire and popping in carburettor. Slow burning. Erratic running with overheating.
2.	16-18 (Point B)	Weak	About 10% less than maximum	Minimum value for all mixtures	84.5 N_2 12.0 CO_2 3.5 O_2	Most economical mixture. Will take greatest spark advance. Less power than that for chemically correct mixture.
3.	15 (Approx.) (Point C)	Chemically correct	About 4% less than maximum	About 4% more than minimum	86.8 N_2 13.2 CO_2 Practic ally no O_2	Best compromise between maximum fuel economy and maximum power.

4.	12-14 (Point D)	Rich	Gives maximum power	About 25% to 30% more than minimum	84.5 N_2 10.5 CO_2 5.0 CO_2 No O_2	Best for full power running. No oxygen in exhaust. Maximum temperature and flame propagation rate.
5.	9-11 (Point E)	Very rich	10% to 20% less than maximum	From 35% to 50% more than minimum	82.0 N_2 6.0 CO_2 13.0 CO with carbon for richest mixtures	Poor result. Low power. High fuel consumjption. Low flame propagation rate. Much carbon deposited in exhaust.

3.1.3 Mixture Requirements for Stedy State Operation

A carburettor is a mechanical device designed to fulfil the following conditions :

(1) Meter the liquid fuel in such quantities as to produce the air-fuel ratio required by the engine at all speeds and loads (steady as well as transient conditions).

(2) Atomise the fuel and mix it homogeneously with the air.

In stationary engines the desired air-fuel ratio is that which gives the maximum economy. Automotive engines, operating under variable speed and load conditions, present the most difficult requirements to carburettors. The air-fuel ratios must change depending on whether maximum economy or maximum power is required. Also, required air-fuel ratio must be provided for transient conditions like starting, warm-up, and acceleration. Further, in all conditions the consideration of minimum exhaust emission should be kept in mind.

For automotive engines there are three main areas of steady state operation requiring different air-fuel ratios. (Steady state operation means continuous operation at a given speed and power output with normal engine temperature). These are given in Table 3.2 and shown in Fig. 3.6, and explained in the following paragraphs.

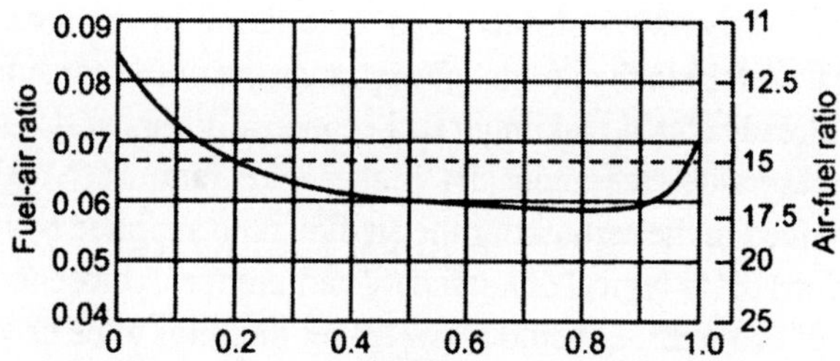

Fig. 3.6 Main areas of automotive engine operation.

Table 3.2 : Major areas of operation of automotive engines

Rangs of Operation	*% of rated power*	*Governing factor*	*A/F**	*F/A**	*Remarks*
1. Idling	0–20	Dilution of mixture by products of combustion	12.5	0.08	Rich
2. Normal power	20–75	Economy	16.5	0.06	Slightly lean
3. Maximum power	75–100	Full utilization of air	13	0.076	Rich

* Approximate

1. **Idling and low load :** (From no load to about 20% of rated power). The no load running mode of the engine is called *idling*. During idling, the air supply is restricted by the nearly closed throttle and the suction pressure is very low. This condition of low pressure gives rise to backflow of exhaust gases and air leakage from the various parts of the engine intake system.

   At idling and during part load operation backflow during the valve overlap period occurs since the exhaust pressure is higher than the intake pressure. This increases the amount of residual gases. During the suction process the residual gases expand, thereby, reducing the fresh mixture inhaled. Increased dilution causes the combustion to be erratic or even impossible. Irregular and slow combustion, so obtained, results in poor thermal efficiency and higher exhaust emissions.

   The problem of dilution by residual gases becomes more pronounced at low loads and idling because the exhaust temperature reduces with decreasing load, i.e., the density and hence mass of the residual gases increase.

   Further dilution of the charge occurs due to air leakage past valves, etc., at low inlet manifold pressures obtained at idling and low loads.

   The above two phenomena require that the air-fuel ratios used for idling and low loads, say up to about 20 per cent of full load, should be rich for smooth engine operation (*F/A* ratio 0.08 or *A/F* ratio $\approx$ 12.5 : 1). Up to this point the amount of fuel burned is quite small and, hence, fuel economy is not important.

   How rich mixture imporves combustion needs some explanation. The amount of fresh charge broght in during idling is much less than that during full throttle operation. The presence of exhaust gas tends to obstruct the contact of fuel and air particles – a requirement necessary for satisfactory combustion. *The richening of mixture increases the probability of contact between fuel and air particles and thus improves combustion.*

2. **Normal power range or cruising range :** (From abou 20% to 75% of rated power). As the load is increased above 20% of rated load the dilution by residual gases as well as leakage decreases and therefore in the normal power range the prime consideration is usually the fuel economy. As already stated maximum fuel economy occurs at *A/F* ratio of about 17 : 1 (*F/A* $\approx$ 0.06). Practical values are about 5% higher (*A/F* ratio 16.7 : 1). Due to manufacturing tolerances provided in the carburettor the air-fuel ratio supplied by it varies during operation. A value of abour ±6% is typical of a standard carburettor. A closer control of these tolerances, say to a value of abour ±3%, would allow leaner mixtures to be used and thereby, improving both the fuel economy as well as the exhaust levels. (The mixture ratios for best economy are very near to the mixture ratios for minimum emissions).

3. **Maximum power range :** (From abour 75% to 100% rated power). As already stated the mixture requirement for maximum power is a rich mixture, of *A/F* about 14 : 1 or (*F/A* $\approx$ 0.07). Besides providing maximum power, a rich mixture also prevents overheating of exhaust valve at high load and inhibits detonation. At high load there is greater heat transfer to engine parts. Enriching the mixture reduces the flame temperature and the cylinder temperature, thereby reducing the cooling problem and lessening the chances of damaging the exhust valves. Also, reduced temperature tends to reduce detonation. Aircraft engines have elaborate arrangement for enrichment of mixture, as detonation can wreck the engine in a matter of seconds. (In automobile engines elaborate arrangement for enrichment is not

required as these engines generlly operate well below full power. Further in these eingines detonation is audible as 'knock' and the driver can take corrective measure by decreasing the load or increasing the engine speed by shifting to a lower gear ratio).

3.1.4 Distribution

In multi-cylinder engines the carburettor supplies the air-fuel maxture to each of several cylinders. Idealy each cylinder should receive mixture of the same air-fuel ratio. However, in practice it is very difficult to achieve this ideal condition.

In the carburettor, complete atomization and vaporization of the fuel is not achieved. Therefore, the mixture passing through the intake manifold generally contains a certain amount of petrol in the droplet form. These droplets have greater inertia than the gaseous mixture and, hence, whenever the direction of flow is changed abruptly, the droplets tend to continue in their original direction of movement, as shown in Fig. 3.7. As a result, there is a variation in the *A/F* ratio between cylinders, the outer cylinders getting richer mixture than the inner cylinders. There is another factor which contributes to this uneven distribution—the existence of thin film of liquid fuel adhering to the inner walls of the intake manifold.

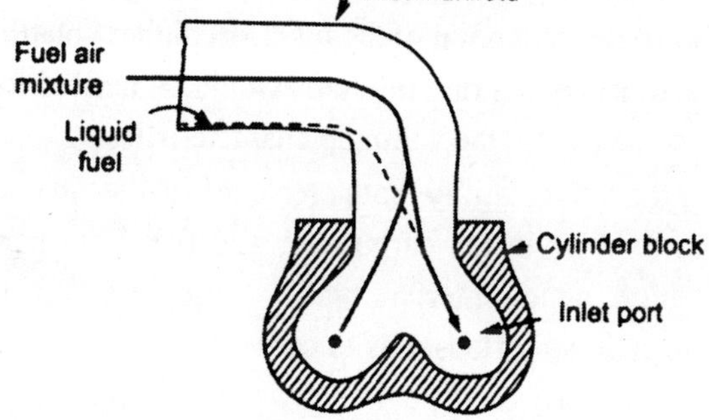

Fig. 3.7 Tendency of liquid fuel droplet in a multi-cylinder engine.

The remedies adopted to partially correct this imbalance in *A/F* ratio in different cylinders are the following :

(i) The mixture in the intake manifold is heated to vaporise the liquid droplets. Howerver, this reduces the mass of the charge with consequent reduction in power output.

(ii) A rich overall air-fuel mixture is supplied so that the leanest cylinder receives the required *A/F* ratio. Of course, this would result in other cylinders getting a richer mixture than necessary.

3.1.5 Transient Mixture Requirements

Besides providing a suitable mixture for steady-running a carburettor has to provide mixture for transient conditions under which speed, load temperature, or pressure change rapidly. The principle transient conditions of operation are starting, warming up, acceleration, and deceleration.

The transient mixture requirements are different from steady running mixture requirements because in the transient case the evaporation of the fuel may be incomplete, the quantity of liquid fuel in the inlet manifold may be increasing or decreasing, and the distribution of fuel to various cylinders may be different.

To understand why evaporation may be incomplete it is important to remember that the petrol is a mixture of hydrocarbons having different boiling temperatures. Those which have high vapour pressures and low boiling points are called 'light ends' and those which are less volatile are called 'heavy ends'. The temperature of the vapour when 10 per cent of fuel is vaporised is called the '10 per cent point', and the temperature of the vapour when complete fuel is vaporised is called the 'end point'.

1. **Starting and warm up requirements :** While starting from cold the speeds as well as engine temperatures are low, hence much of the 'heavy ends' supplied by the carburettor do not vaporise and remain in liquid form. Also, the vaporised fuel may recondense on coming in contact with cold cylinder walls and piston head. Therefore even when the fuel-air ratio at the carburettor is well within the normal combustion limits or petrol-air mixtures, the ratio of *evaporated fuel* to air in the cylinder may be too lean to ignite. Therefore during starting a very rich mixture must be supplied, as much as 5 to 10 times the normal amount of petrol (*A/F* ratio 3 : 1 to 1.5 : 1 or *F/A* ratio 0.3 to 0.7), so that enough 'light ends' are available for proper ignition. As the engine warms up the amount of evaporated fuel increases and hence the mixture ratio should be progressively made leaner to avoid too rich *evaporated* fuel-air ratio.

   The volatility of the fuel significantly affects the starting and warm up characteristics of the engine. Lower the 10 per cent point of the fuel (*i. e.*, higher volatility), more is the quantity of evaporated fuel and hence less rich mixtures would be needed for starting. The 50 per cent point dictates, to some extent, the warm up characteristics.

   Note. Too high or too low volatility, both create difficulties in operation. Too high volatility may form vapour bubbles in the carburettor and fuel lines particularly when the engine temperatures are high, which interfere with the supply and metering of the fuel and may distrub the fuel-air ratio so seriously as to cause the engine to stop. Too low volatility may cause the petrol to condense on the cylinder walls, diluting and removing the lubricating oil film. Eventually the petrol may reach the crankcase past the piston rings and dilute the engine oil. Condensation of petrol on cylinder walls also causes carbon deposits. Thus petrol should be carefully made to suit the type of the engine and the climate of the place.

2. **Acceleration requirements :** The term 'acceleration' with regard to engines, is generally used to refer to an increase in engine speed resulting from opening the throttle. However, the main purpose of opening the throttle is to provide an increase in torque and whether or not an increase in speed follows depends on the nature of the load.

   The fuel evaporated in the intake manifold moves much faster than the liquid film formed on the induction system walls. Under steady running conditions this does not cause any problem but when the throttle is *suddenly* opened, *e.g.*, during acceleration, the liquid fuel lags behind and, temporarily, the cylinder receives a lean mixture whilst actually a rich mixture is needed to produce more instantaneous power for acceleration. Therefore, to compensate for the temporary leaning of the mixture and to provide rich mixture needed during the accelerating period, additional fuel must be supplied by a suitable mechanism.

   A summary of the F : A ratio requirements for various running conditions is given in Table 3.3.

Table 3.3 F : A ratios for various running conditions of S.I. engines.

	F : A ratio
1. Starting	0.2 : 1
2. Warming	0.15 : 1
3. Idling	0.085 : 1
4. Running with maximum thermal efficiency	(0.06 – 0.07) : 1
5. Running with maximum power (80% throttle)	(0.0775 – 0.08) : !
6. Full throttle	0.085 : 1
7. Acceleration	0.1 : 1

3.1.6 The Simple Carburettor

The details of a simple carburettor are shown in Fig. 3.8. It mainly consists of a float chamber, fuel discharge nozzle and a metering orifice, a venturi, a throttle valve and a choke. A float and a needle valve system maintains a constant level of gasoline in the float chamber. If the amount of fuel in the float chamber falls below the designed level, the float goes down, thereby opening the fuel supply valve and admitting fuel. When the designed level has been reached, the float closes the fuel supply valve thus stopping additional fuel flow from the supply valve thus stopping additional fuel flow from the supply system. The float chamber is vented either to the atmosphere or to the upstream side of the venturi.

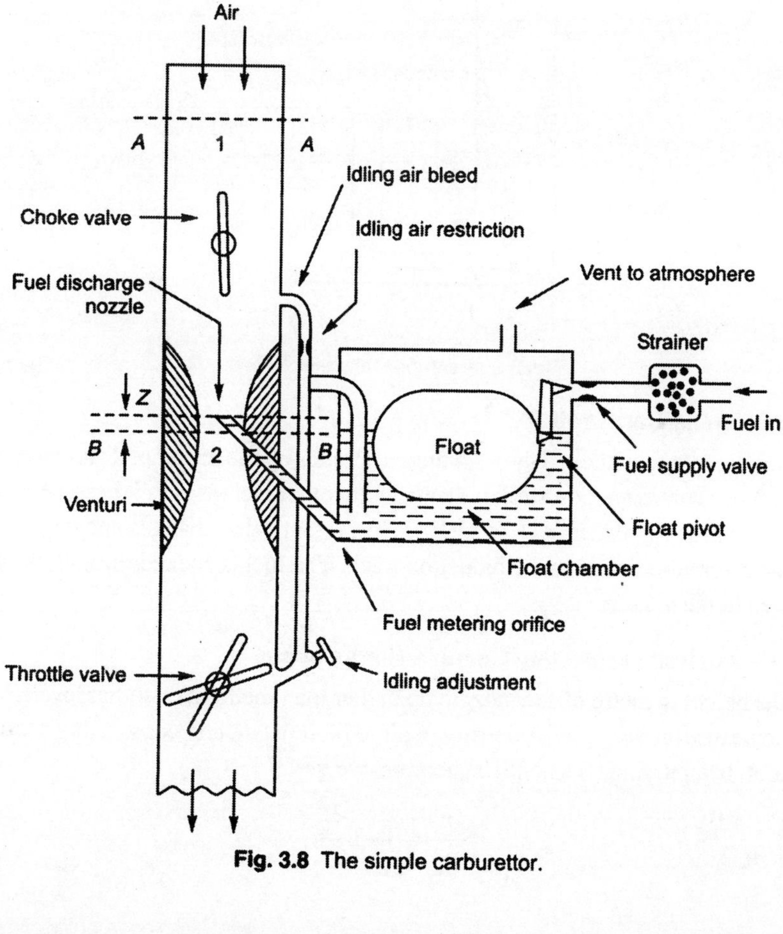

Fig. 3.8 The simple carburettor.

During suction stroke air is drawn through the venturi (or choke tube). As the air passes through the venturi the velocity increases reaching a maximum at the venturi throat and pressure decreases to minimum. From the float chamber, the fuel is fed to a discharge jet, the tip of which is located in th throat of the venturi, fuel is discharged into the air stream. To avoid overflow of fuel through the jet, the level of the liquid in the float chamber is maintained at a level slightly below the tip of the discharge jet.

The throttle valve controls the amount of charge delivered to the cylinder, which is situated after the venturi tube. As the throttle is closed less air flows through the venturi tube and less is the quantity of air-fuel mixture delivered to the cylinder and hence power output is reduced. The reverse takes place when the throttle is opened.

3.1.7 Compensating Jet

The function of the compensating jet device is to make the mixture leaner as the throttle opens progressively. As shown in Fig. 3.9, the compensating jet is connected to the compensating well which is vented to the atmosphere. The compensating well is supplied with fuel from the main float chamber through a restricting orifice. With the increase in air flow rate, there is decrease of fuel level in the compensating well, with the result that fuel supply through the compensating jet decreases. The compensating jet thus progressively makes the mixture leaner as the main jet progressively makes the mixture richer.

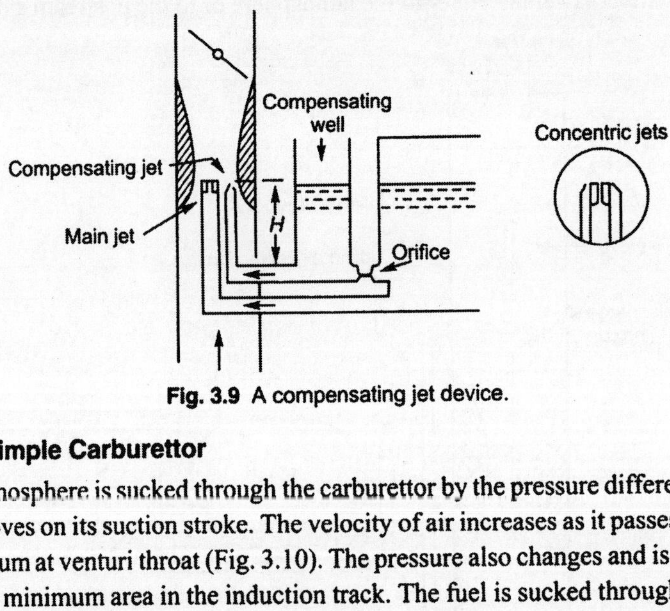

Fig. 3.9 A compensating jet device.

3.1.8 Theroy of Simple Carburettor

The air from the atmosphere is sucked through the carburettor by the pressure difference across it created when the piston moves on its suction stroke. The velocity of air increases as it passes through the venturi and reaches maximum at venturi throat (Fig. 3.10). The pressure also changes and is maximum at section 2-2, because this is minimum area in the induction track. The fuel is sucked through the nozzle because of suction created in the venturi.

(*a*) Approximate Analysis (neglecting Compressibility of air)

Let z be the height in metre of fuel nozzle tip higher than the float chamber level. Assuming initial velocity of air to be negligible ($c_1 = 0$), density of air to be constant and considering sections at entrance and venturi throat, by applying, Bernoulli's theorem, we get

$$\frac{p_1}{\rho_a} = \frac{p_2}{\rho_a} + \frac{c_2^2}{2}$$

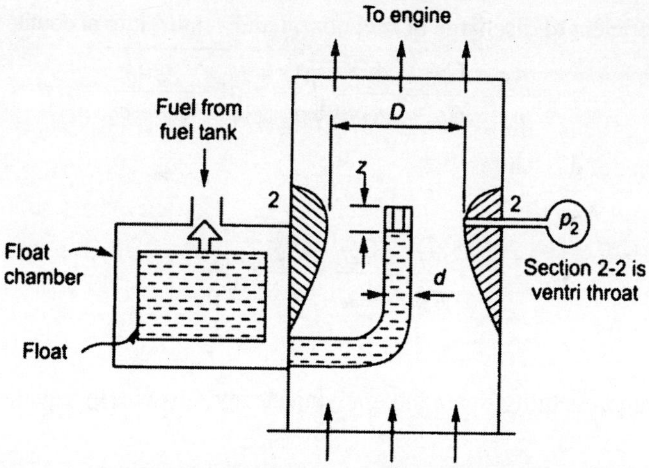

Fig. 3.10 Principles of a simple carburettor.

where p_1, p_2 = pressure at sections 1 and 2 respectively, N/m^2

c_2 = velocity of air at section 2, m/s

ρ_a = density of air, kg/m^3

$$\therefore \qquad c_2 = \sqrt{\frac{2(p_1 - p_2)}{\rho_a}}$$

Mass of air per second,

$$m_a = A_2 c_2 \rho_a = A_2 \sqrt{2\rho_a (p_1 - p_2)} \qquad \ldots(3.1)$$

where A_2 = area of venturi throat, m^2

Similarly, for the flow of fuel, we have

$$\frac{p_1}{\rho_f} = \frac{p_2}{\rho_f} + \frac{c_f^2}{2} + zg$$

where ρ_f = constant density of fuel, kg/m^3

c_f = velocity of flow of fuel, m/s

$$\therefore \qquad c_f = \sqrt{\frac{2(p_1 - p_2 - zg\rho_f)}{\rho_f}}$$

Mass of fuel per second,

$$m_f = A_f c_f \rho_f = A_f \sqrt{2\rho_f (p_1 - p_2 - zg\rho_f)} \qquad \ldots(3.2)$$

where A_f = cross-sectional area of fuel nozzle, m^2

Fuel-air ratio,

$$\frac{m_f}{m_a} = \frac{A_f}{A_2} \cdot \sqrt{\frac{\rho_f}{\rho_a}} \times \sqrt{1 - \frac{zg\rho_f}{p_1 - p_2}} \qquad \ldots(3.3)$$

Taking the coefficient of discharge of fuel nozzle and venturi into account, we have

$$\frac{m_f}{m_a} = \frac{C_{df}}{C_{da}} \times \frac{A_f}{A_2} \times \sqrt{\frac{\rho_f}{\rho_a}} \times \sqrt{1 - \frac{zg\rho_f}{p_1 - p_2}} \quad \text{...(3.3a)}$$

where C_d = coefficient of discharge.

If $z = 0$.

$$\frac{m_f}{m_a} = \frac{C_{df}}{C_{da}} \times \frac{A_f}{A_2} \times \sqrt{\frac{\rho_f}{\rho_a}} \quad \text{...(3.3b)}$$

(b) Exact Analysis

Considering compressibility of air and applying steady flow energy equation to sections 1 and 2, we get

$$Q - W = h_2 - h_1 + \frac{c_2^2 - c_1^2}{2}$$

where h_1, h_2 = enthalpies at sections 1 and 2 respectively

Since $Q = 0$, $W = 0$ and $c_1 = 0$

$$\therefore \quad c_2 = \sqrt{2(h_1 - h_2)}$$

$$= \sqrt{2c_p(T_1 - T_2)} = \sqrt{2c_pT_1\left(1 - \frac{T_2}{T_1}\right)} \quad \text{...(3.4)}$$

For isentropic flow between the atmosphere and venturi throat, we have

$$\frac{T_2}{T_1} = \left(\frac{p_2}{p_1}\right)^{\frac{\lambda - 1}{\lambda}} \quad \text{...(3.5)}$$

$$\therefore \quad c_2 = \sqrt{2c_pT_1\left[1 - \left(\frac{p_2}{p_1}\right)^{\frac{\gamma-1}{\gamma}}\right]} \quad \text{...(3.6)}$$

Now

$$\dot{m}_a = \frac{A_1c_1}{v_2} = \frac{A_2c_2}{v_2} \quad \text{...(3.7)}$$

where v = specific volume

Since

$$p_1v_1^\gamma = p_2v_2^\gamma$$

$$v_2 = v_1\left(\frac{p_1}{p_2}\right)^{\frac{1}{\gamma}} = \frac{RT_1}{p_1}\left(\frac{p_1}{p_2}\right)^{\frac{1}{\gamma}} \quad \text{...(3.8)}$$

Eq. (3.7) becomes

$$\dot{m}_a = \frac{A_2p_1}{RT_1}\left(\frac{p_2}{p_1}\right)^{\frac{1}{\gamma}} \sqrt{2c_pT_1\left[1 - \left(\frac{p_2}{p_1}\right)^{\frac{\gamma-1}{\gamma}}\right]}$$

$$= \frac{A_2 p_1}{R\sqrt{T_1}} \sqrt{2c_p \left[\left(\frac{p_2}{p_1} \right)^{\frac{2}{\gamma}} - \left(\frac{p_2}{p_1} \right)^{\frac{\gamma+1}{\gamma}} \right]} \qquad \text{...(3.9)}$$

$$\frac{m_a}{m_f} = \frac{C_{da}}{C_{df}} \times \frac{A_2}{A_f} \times \frac{p_1}{R\sqrt{T_1}} \times \frac{\sqrt{2c_p \left[\left(\frac{p_2}{p_1} \right)^{\frac{2}{\gamma}} - \left(\frac{p_2}{p_1} \right)^{\frac{\gamma+1}{\gamma}} \right]}}{\sqrt{2\rho_f (p_1 - p_2 - zg\rho_f)}} \qquad \text{...(3.10)}$$

3.1.9 Limitations of Single Jet Carburettor

The drawbacks of single jet carburettor are listed below :

1. A single jet carburettor cannot provide a very rich mixture as required at the time of starting the engine. This is because, at low speed (starting or idling) the pressure difference causing the fuel flow is very small, (as the throttle is nearly closed). It is not possible to discharge fuel to make the mixture considerably rich.
2. Also, it cannot provide very rich mixture required for sudden acceleration of the engine.
3. For gradually increasing pressure difference over the jet (at higher speed of the engine), the weight of the petrol discharged from a single jet increases at a greater rate than does the air supply. Hence, a single jet carburettor gives a progressively richer mixture as the air speed increases when set to given a correct mixture at low air speeds.
4. It cannot reduce the quantity of air flow during Starting as required in cold weather conditions.
5. The automatic control of air and fuel according to the required conditions is not possible.

The carburettor used with a variable speed engine must fulfil all the following requirements :

1. It must automize the fuel and mix it homogeneously with the air.
2. It must be able to run the engine smoothly without hunting or fuel wastage.
3. It must provide rich mixture during starting and idling.
4. It must provide a constant air-fuel ratio during normal running of the engine which is the maximum period.
5. It must provide rich mixture required for quick acceleration of the engine.
6. It must be able to start the engine even in very cold weather conditions (during snowfall).
7. All operations should be automatic.

To fulfil the above requirements, the following devices are introduced.

1. A starting or a pilot jet (to start engine).
2. Compensating devices (to provide constant A : F ratio during normal operation conditions).
3. An automatic control of choke valve (to start the engine in cold weather).

3.1.10 Different Devices Used to Meet the Requirements of an Ideal Carburettor

The main function of the carburettor is to vaporise the petrol in the current of air created by means of engine suction and supply the required quantity of air and petrol mixture in proper proportion in accordance with the load on the engine and its speed.

In the design of a carburettor, the following points should be considered :

1. It is seen that for a gradually increasing pressure difference over the fuel jet, the mass of petrol discharged from a single jet increases at a greater rate than does the air supply. Hence a carburettor of this type will give a progressively rich mixture as air speed is increased. There must be some device to maintain A : F ratio constant over a wide range of engine speed.
2. To ensure rapid and complete combustion, it is necessary that the fuel should be finely divided and mixed intimately with air supply. This can be done by proper design of venturi and inlet manifold.
3. For starting and accelerating, a rich mixture must be supplied momentarily but the supply, should come to correct, mixture strength automatically when engine attains the desired speed.
4. There must be provision to supply extra rich mixture to start the engine in a very cold weather (where temperature falls to zero or sub-zero temperature).
5. The float in the float chamber of a carburettor must maintain the fuel level constant irrespective of load or speed of the engine.

3.1.11 Complete Carburettor

In order to satisfy the demands of an engine under all conditions of operation the following additional systems are added to the simple carburettor :

1. Main metering system
2. Idling system
3. Power enrichment by economizer system
4. Acceleration pump system
5. Choke.

1. Main Metering System

The main metering system of a carburettor is designed to supply a nearly constant basis fuel-air ratio over a wide range of speeds and loads. This mixture corresponds approximately to best economy at full throttle (A/F ratio $\approx$ 15.6 or F/A ratio 0.064). Since a simple or elementary carburettor tends to enrich the mixture at higher speeds automatic compensating devices are incorporated in the main metering system to correct this tendency. These devices are :

(i) Use of a *compensating jet* that allows an increasing flow of air through a fuel passage as the mixture flow increases.

(ii) Use of *emulsion tube* for air bleeding. In this device the emphasis is on air bleeding alone.

(iii) Use of a tapered metering pin that is arranged to be moved in and out of the main or auxiliary fuel orifice either manually or by means of some automatic mechanism changing the quantity of fuel drawn into the air charge.

(iv) Back-suction control or pressure reduction in the float chamber.

(v) changing the position or jet in the venturi. The suction action is highest at the venturi throat, therefore by raising the venturi the nozzle relatively moves to points with smaller suction and the flow of fuel is decreased.

(vi) Use of an auxiliary air valve or port that automatically admits additional air as mixture flow increases.

The main devices are explained below in detail.

(a) *Compensating jet device.* This device is shown in Fig. 3.11. In this device, in addition to the main jet, a compensating jet is provided which is in communication with a compensating

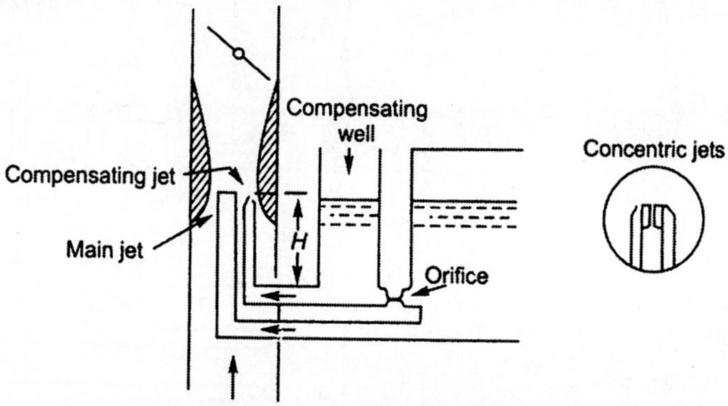

Fig. 3.11 A compensating jet device.

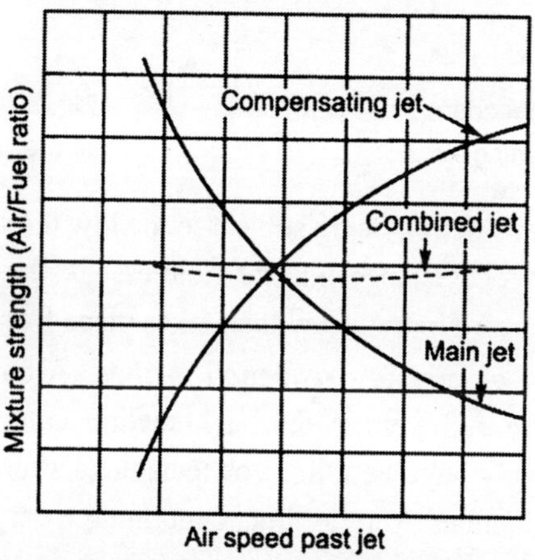

Fig. 3.12 Variation of air-fuel ratio vs. air flow with main and compensating jet.

well. The compensating well is open to atmosphere and gets its fuel supply from the float chamber through a restricting orifice. As the air flow increases, the level of fuel in the well decreases, thus reducing the fuel supply through the compensating jet. The compensating jet thus tends towards leanness as the main jet tends towards richness, the sum of the two remaining constant as shown in Fig. 3.12. At even higher rates of air flow, when the compensating jet has been emptied, air is bled through the compensating jet to continue the leanness effect, and incidentally, to assist in fuel atomization.

(b) *Emulsion tube or air bleeding device.* In the modern carburettors the mixture correction is done by air bleeding alone. In this arrangement the main metering jet is fitted about 25 mm below the petrol level and it is called a submerged jet (see Fig. 3.13). The jet is situated at the bottom of a well, the sides of which have holes which are in communication with the atmosphere. Air is drawn through the holes in the well, the petrol is emulsified, and the pressure difference across the petrol column is not as great as that in the simple or elementary carburettor. Initially, the petrol in the well is at a level equal to that in the float chamber. On

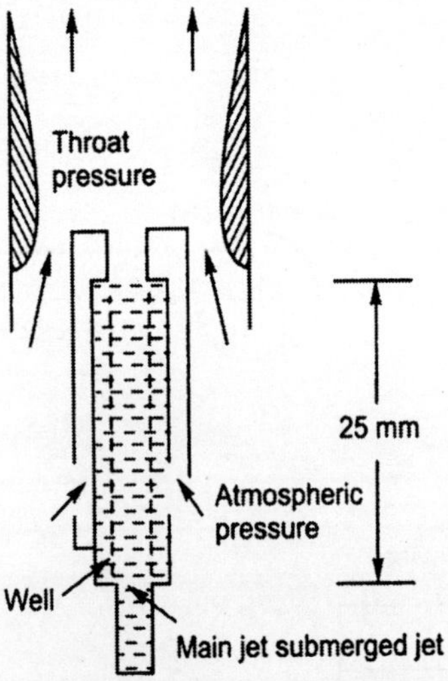

Fig. 3.13 Correction in modern carburettors by air bleeding.

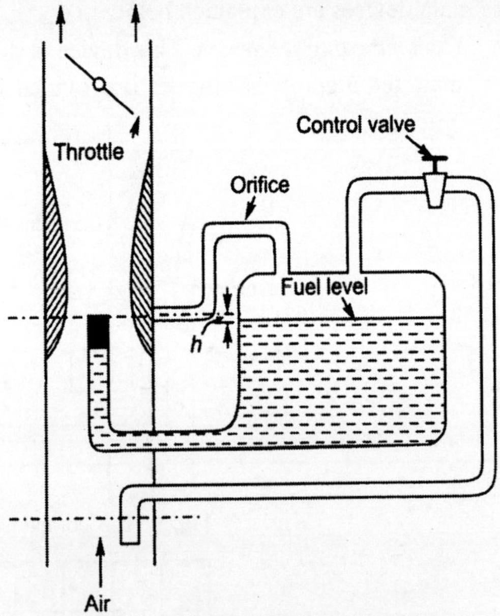

Fig. 3.14 Back-suction control or pressure reduction method.

opening the throttle this petrol, being subject to the low throat pressure, is drawn into the air. This continues with decreasing mixture richness as the boles in the central tube are progressively uncovered. Normal flow then takes place from the main jet.

(c) *Back-suction control or pressure reduction method.* A common method of changing the air-fuel ratio in large carburettors the back-suction control shown in Fig. 3.14. In this arrangement a relatively large vent line connects the carburettor entrance (say point 1) with the top of the float chamber. Another line, containing a very small orifice line, connects the top of the float chamber with the venturi throat (say point 2). A control valve is placed in the large vent line. When the valve is wide open, the vent line is unrestricted the pressure in the float chamber is equal to p_1, and the pressure difference acting on the fuel orifice is $(p_1 - p_2)$. If the valve is closed, the float chamber communicates only with the venturi throat and the pressure on the fuel surface will be p_2. Then Δp_f will be zero, and no fuel will flow. By adjusting the control valve, any pressure between p_1 and p_2 may be obtained in the float chamber, thus changing the quantity of fuel discharged by the nozzle.

(d) *Auxiliary valve carburettor.* An auxiliary valve carburettor is illustrated in Fig. 3.15. With an increase of engine load, the vacuum at the venturi throat also increases. This causes the valve spring to lift the valve admitting additional air and the mixture is prevented from becoming over-rich.

(e) *Auxiliary part carburettor.* An auxiliary port carburettor is illustrated in Fig. 3.16. By opening the butterfly valve, additional air is admitted and at the same time the depression at the venturi throat is reduced, decreasing the quantity of fuel drawn in. This method is used in aircraft carburettors for altitude compensation.

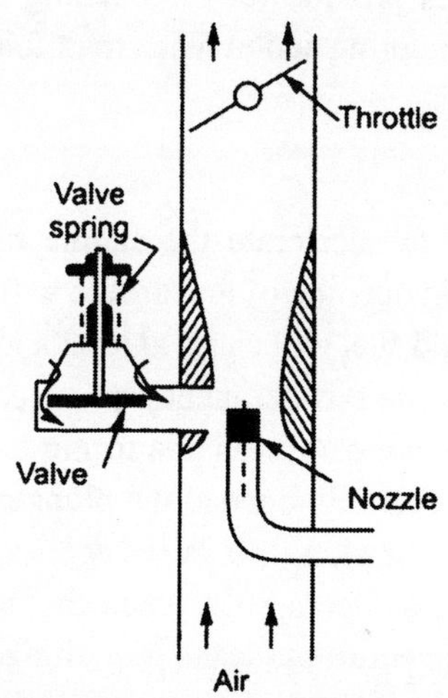

Fig. 3.15 An auxiliary valve carburettor.

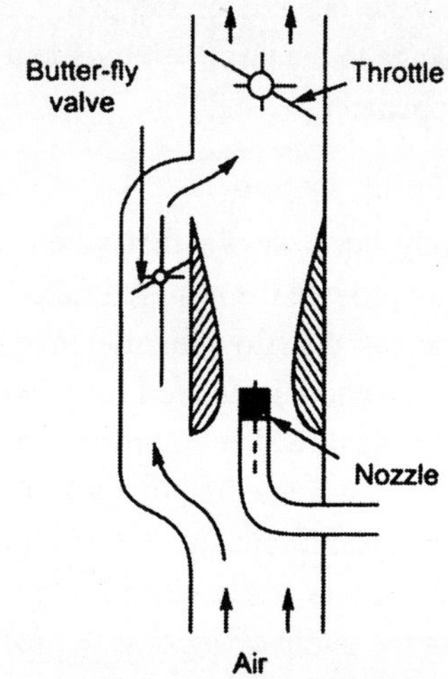

Fig. 3.16 An auxiliary port carburettor.

2. Idling System

It has already been shown that at idling and low lead the engine requires a rich mixture (about *A/F* 12 : 1). However, the main metering system not only fails to enrich the mixture at low air flows but also supplies no fuel at all at idling. For this reason, a separate idling jet must be added to the basic carburettor. An example of idling jet is shown in Fig. 3.17. It consists of a small fuel line from the float chamber to a point a little on the engine side of the throttle. This line contains a fixed fuel orifice. When the throttle is practically closed, the full manifold suction operates on the outlet to this jet. In addition, the very high velocity past be throttle plate increases the suction locally. Fuel can therefore be lifted by the additional height up to the discharge point, but this occurs only at very low rates of air flow. As the throttle is opened, the main jet gradually takes over while the idle jet becomes ineffective.

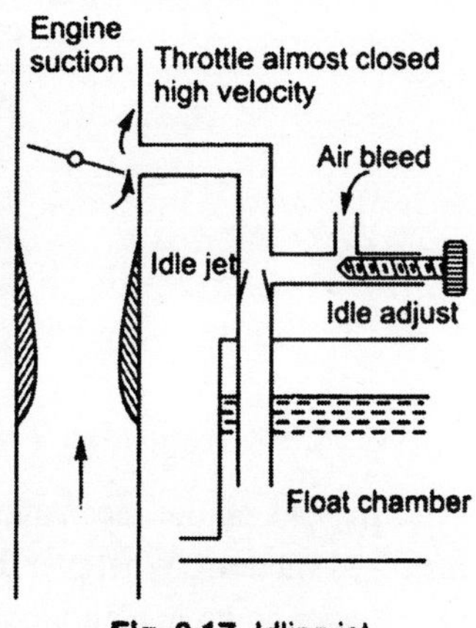

Fig. 3.17 Idling jet.

The desired air-fuel ratio for the idling jet is regulated manually to *idle adjust,* which is a needle valve controlling the air bleed.

3. Power Enrichment of Economizer System

As the maximum power range of operation (75% to 100% load) is approached, some device must allow richer mixture (*A/F* about 13 : 1, *F/A* 0.08) to be supplied despite the compensating leanness. Such a device is the *meter rod economiser* shown in Fig. 3.18. The name economiser is rather misleading. It stems from the fact that such a device provides a rich uneconomical mixture at high load demand without interferring with economical operation in the normal power range. The meter rod economizer shown in Fig. 3.18, simply provides a large orifice opening to the main jet as the throttle is opened beyond a certain point.

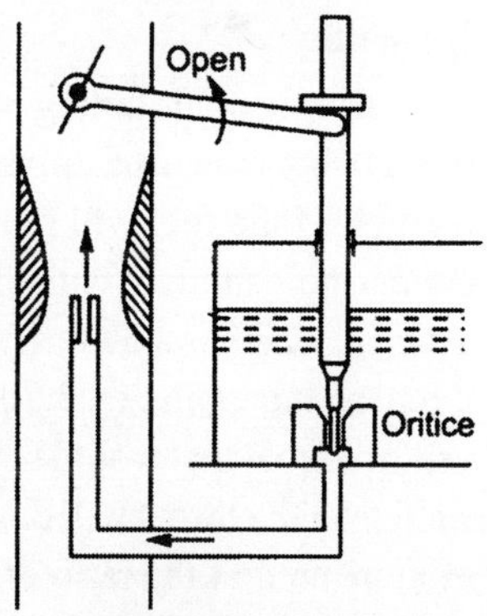

Fig. 3.18 Meter rod economiser.

The rod may be tapered or stepped. Other examples provide for the opening of auxiliary jets through some linkage to the throttle movement or through a spring action when manifold vacuum is lost as the throttle is opened.

4. Acceleration Pump System

It has already been shown that when it is desired to accelerate the engine rapidly, a simple carburettor will not provide the required rich mixture. Rapid opening of the throttle will be immediately followed by an increased airflow, but the inertia of the liquid fuel will cause at least a momentarily lean mixture just when richness is desired for power. To overcome this deficiency an *acceleration pump* is provided, one example of which is shown in Fig. 3.19. The pump consists of a spring-loaded plunger. A linkage mechanism is provided so that when the throttle is rapidly opened the plunger moves into the cylinder and forces an additional jet of fuel into the venturi. The plunger is raised again against the spring force when the throttle is partly closed. Arrangement is provided so that when the throttle is opened slowly, the fuel in the pump cylinder is not forced into the venturi but leaks past plunger or some holes into the float chamber.

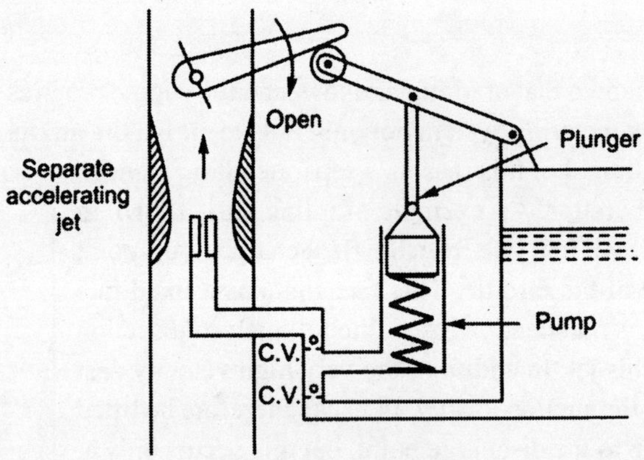

Fig. 3.19 Acceleration pump.

Instead of the mechanical linkage shown some carburettors have a pump plunger held up by manifold vacuum. Wherever that vacuum is reduced by the rapid opening of the throttle, a spring forces the plunger down in a pumping action identical to that of the pump illustrated.

5. Choke

During cold starting period, at low cranking speed and before the engine has warmed up, a mixture mach richer than usual mixtures (almost 5 to 10 times more fuel) must be supplied simply because a large fraction of the fuel will remain liquid even in the cylinder, and only the vapour fraction can provide a combustible mixture with the air. The most common means of obtaining this rich mixture is by the use of a *choke*, which is a butterfly type of valve placed between the entrance to the carburettor and the venturi throat as shown in Fig. 3.20. By partially closing the choke, a large pressure drop can be produced at the venturi throat that would normally result from the amount of air flowing through the venturi. This strong suction at the throat will draw large quantities of fuel from the main nozzle and supply a sufficiently rich mixture so that the ratio of evaporated fuel to air in the cylinders is within combustible limits. Choke valves are sometimes made with a spring loaded by pass so that high pressure drops and excessive choking will not result after the engine has started and has attained a higher speed. some manufacturers

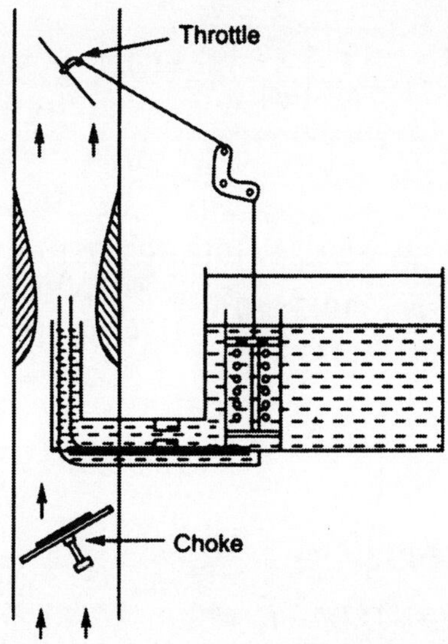

Fig. 3.20 Choke valve with spring-loaded bypass.

make the choke operate automatically by means of a thermostat such that when the engine is cold the choke is closed by a bimetallic element. After starting and as the engine warms up the bimetallic element gradually opens the choke to its fully open position.

An alternative to the choke is the provision of auxiliary fuel jets that are opened manually or automatically only as required.

■ **Example 3.1** A single jet carburettor has to supply 5 kg of air per minute. The air is at a pressure of 1.013 bar and at a temperature of 27°C. Calculate the throat diameter of the choke for air flow velocity of 90 m/s. Take velocity coefficient to be 0.8. Assume isentropic flow. Assume the flow to be incompressible. [IES, 1992]

Solution : $m_a = 5\text{ kg/min},\ p_1 = 1.013\text{ bar},\ T_1 = 273 + 27 = 300\text{ K},\ c_2 = 90\text{m/s},\ C_{da} = 0.8$

Applying Bernoulli's theorem between inlet and throat, we have

$$h_1 + \frac{c_1^2}{2} = h_2 + \frac{c_2^2}{2}$$

Taking $c_1 = 0.$

$$c_2 = \sqrt{2(h_1 - h_2)} = \sqrt{2c_p(T_1 - T_2)}$$

Actual velocity,

$$c_2 = C_{da}\sqrt{2c_pT_1\left(1 - \frac{T_2}{T_1}\right)}$$

$$= C_{da}\left[2c_pT_1\left\{1 - \left(\frac{p_2}{p_1}\right)^{\frac{\lambda-1}{\lambda}}\right\}\right]^{\frac{1}{2}}$$

$$90 = 0.8\sqrt{2 \times 1.005 \times 10^3 \times 300\left[1 - \left(\frac{p_2}{p_1}\right)^{\frac{0.4}{1.4}}\right]}$$

$$1-\left(\frac{p_2}{p_1}\right)^{\frac{1}{3.5}} = 0.021$$

$$\frac{p_2}{p_1} = 0.9284$$

$$p_2 = 0.9284 \times 1.013 = 0.9405 \text{ bar}$$

Density of air at inlet, $\rho_{a1} = \frac{p_1}{RT_1} = \frac{1.013 \times 10^5}{287 \times 300} = 1.1765 \text{ kg/m}^3$

Density of air at throat, $\rho_{a2} = \rho_1 \left(\frac{p_2}{p_1}\right)^{\frac{1}{\gamma}} = 1.1765\,(0.9284)^{\frac{1}{14}} = 1.1157 \text{ kg/m}^2$

Mass flow rate at throat, $\dot{m}_a = \rho_{a2} A_2 c_2$

$$\frac{5}{60} = 1.1157 \times \frac{\pi}{4} d_2^2 \times 90$$

throat dia, $d_2 = 0.0325$ m or 32.5 mm

■ **Example 3.2** A simple jet carburettor is required to supply 6 kg of air per minute and 0.45 kg of fuel of density 740 kg/m^3. The air is initially at 1.013 bar and 27°C. Calculate the throat diameter of the choke for a flow velocity of 92 m/s. Take velocity coefficient equal to 0.8.

If the pressure drop across the fuel metering orifice is 0.75 of that at the choke, calculate orifice diameter assuming $C_{da} = 0.60$.

Solution :

(*a*) Velocity of air at venturi throat,

$$c_2 = C_{da}\sqrt{2c_pT_1\left[1-\left(\frac{p_2}{p_1}\right)^{\frac{\gamma-1}{\gamma}}\right]}$$

$$92 = 0.8\sqrt{2 \times 1.005 \times 10^3 \times 300\left[1-\left(\frac{p_2}{p_1}\right)^{\frac{1}{3.5}}\right]}$$

$$\frac{p_2}{p_1} = 0.92539$$

$$p_2 = 0.92530 \times 10.13 = 0.937 \text{ bar}$$

Now $p_1 v_1^\gamma = p_2 v_2^\gamma$

$$v_2 = v_1\left(\frac{p_1}{p_2}\right)^{\frac{1}{\gamma}} = \frac{RT_1}{p_1}\left(\frac{p_1}{p_2}\right)^{\frac{1}{\gamma}} = \frac{287 \times 300}{1.013 \times 10^5}\left(\frac{1}{0.92539}\right)^{\frac{1}{1.4}}$$

$$= 0.898 \text{m}^3/\text{kg}$$

Throat area, $$A_2 = \frac{\dot{m}_a v_2}{c_2} = \frac{6 \times 0.898}{60 \times 92} = 9.76 \times 10^{-4}\,\text{m}^2$$

$$d_2 = 35.25\text{ mm}$$

(*b*) Pressure drop at venturi $= p_1 - p_2 = 1.013 - 0.937 = 0.076$ bar

Pressure drop in jet $= 0.75 \times 0.076 = 0.057$ bar

Now $$\dot{m}_f = A_f\, C_{df}\, \sqrt{2\rho_f\,(p_1 - p_2)}$$

$$= A_f\, C_{df}\, \sqrt{2 \times 740 \times 0.057 \times 10^5} = A_f C_{df} \times 2904$$

$$A_f = \frac{m_f}{C_{df} \times 2904} = \frac{0.45 \times 10^4}{60 \times 0.6 \times 2904} = 0.043\text{ cm}^2$$

$$d_f = 23.4\text{ mm}$$

■ **Example 3.3** An automobile carburettor having its float chamber vented to the atmosphere is tested at sea level conditions in the factory without an air cleaner. The main metering system of this carburettor is found to yield a fuel-air ratio of 0.065. The venturif/throat pressure is 0.84 bar. This carburettor is now installed in an automobile and an air cleaner is placed on the inlet to the carburettor. The air flow with and without the air filter is 230 kg/hr. The pressure drop through the filter is found to be 0.035 bar at sea level conditions. Assuming $z = 0$ and orifice coefficient to be constant, calculate

(i) The venturi/throat pressure with the air cleaner,

(ii) Fuel-air ratio with the air cleaner.

Assume incompressible flow.

Solution :

(i) Mass of air flow rate,

$$\dot{m}_a = C_{da}\, A\,[2\rho_a \,.\, \Delta p_a]^{0.5}$$

Without air cleaner, $\Delta p_a = 1.013 - 0.840 = 0.173$ bar

With air cleaner, let p_1 = throat pressure

$$\Delta p'_a = (1.013 - 0.035 - p_1) = (0.978 - p_1)\text{ bar}$$

For the same air flow and constant co-efficients,

$$\Delta p_a = \Delta p'_a$$

$$0.173 = 0.978 - p_1$$

$$p_1 = 0.805\text{ bar}$$

(ii) Without air cleaner,

$$\Delta p_f = \Delta p_a = p_1 = 0.805\text{ bar}$$

With air cleaner, $\Delta p_f = 1.013 - 0.805 = 0.208$ bar

As Δp_f has increased, more fuel will flow making the mixture richer.

Now $$A/F\text{ ratio} = \left(\frac{1}{0.065}\right)\left[\frac{0.173}{0.208}\right]^{0.5} = 14.03$$

or $$F/A\text{ ratio} = \frac{1}{14.03} = 0.07127$$

■ **Example 3.4** A six-cylinder, 4-stroke engine, 80 mm in diameter and 120 mm stroke length runs at 3600 rpm. The volumetric efficiency of the engine is 0.8. If the maximum head causing the flow is limited to 11.765 cm of mercury, find the throat diameter of the venturi required. Also find the diameter of the nozzle orifice if the desired A : F ratio is 15 : 1. Take $C_{da} = 0.9$, $C_{df} = 0.7$, $\rho_a = 1.3\ \text{kg/m}^3$ and $\rho_f = 720\ \text{kg/m}^3$.

Solution : Given $i = 6$, $d = 0.08$ m, $L = 0.120$ m, $N = 3600$ rpm, $\eta_{vol.} = 0.8$, $h_{Hg} = 11.765$cm

$$\text{Volume flow rate of air, } \dot{V}_a = \frac{\pi}{4} d^2 L\, \eta_{vol} \times i \frac{N}{2 \times 60}$$

$$= \frac{\pi}{4}(0.08)^2 \times 0.120 \times 0.8 \times 6 \times \frac{3600}{2 \times 60} = 0.0868\ \text{m}^3/\text{s}$$

Mass flow rate of air, $\dot{m}_a = \rho_a \dot{V}_a = 1.3 \times 0.0868 = 0.113$ kg/s

$$h_w = h_{Hg} \times \gamma_{Hg} = 11.765 \times 13.6 = 160 \text{ cm of water}$$

$$m_a = 14 C_{da} A_a \sqrt{\rho_a h_w}$$

$$0.113 = 14 \times 0.9 \times \frac{\pi}{4}(0.01\, d_a)^2 \sqrt{13 \times 160}, \quad d_a \text{ in cm}$$

$$d_a = 2.8 \text{ cm}$$

$$\frac{m_a}{m_f} = \frac{C_{da}}{C_{df}} \cdot \frac{A_a}{A_f} \sqrt{\frac{\rho_a}{\rho_f}} = \frac{C_{da}}{C_{df}} \left(\frac{d_a}{d_f}\right)^2 \sqrt{\frac{\rho_a}{\rho_f}}$$

$$\frac{15}{1} = \frac{0.9}{0.7}\left(\frac{2.8}{d_f}\right)^2 \sqrt{\frac{13}{720}}$$

$$d_f = 0.169 \text{ cm or } 1.69 \text{ mm}$$

■ **Example 3.5** The throat diameter of a carburettor is 80 mm and nozzle diameter is 5.5 mm. The $C_{da} = 0.85$ and $C_{df} = 0.7$. The nozzle lip is 6 mm. The pressure differences causing the flow is 0.1 bar. Find (*a*) *A* : *F* ratio supplied by the carburettor neglecting nozzle lip. (*b*) *A* : *F* ratio comsidering nozzle lip and (*c*) the minimum velocity of air required to start the fluid flow. Neglect air compressibility. Take $\rho_a = 1.2\ \text{kg/m}^3$ and $\rho_f = 750\ \text{kg/m}^3$.

Solution : Given $d_a = 80$ mm, $d_f = 5.5$ mm, $z = 6$ mm, $\Delta p = 0.1$ bar,

$$C_{da} = 0.85,\ C_{df} = 0.7,\ \rho_a = 1.2\ \text{kg/m}^3,\ \rho_f = 750\ \text{kg/m}^3$$

$$\text{(a)}\ \frac{m_a}{m_f} = \frac{C_{da}}{C_{df}} \cdot \frac{A_a}{A_f} = \sqrt{\frac{\rho_a}{\rho_f}} = \frac{0.85}{0.7}\left(\frac{80}{5.5}\right)^2 \sqrt{\frac{1.2}{750}} = 10.2:1$$

$$\text{(d)}\ \frac{m_a}{m_f} = \frac{C_{da}}{C_{df}} \cdot \frac{A_a}{A_f} = \sqrt{\frac{\rho_a}{\rho_f}\left(\frac{h_w}{h_w - h_{wl}}\right)}$$

$$h_w = \Delta p = 0.1 \times 10^5\ \text{N/m}^2 = \frac{0.1 \times 10^5}{98} = 102 \text{ cm of } H_2O$$

$$h_{wl} = z\gamma_f = 6\times10^{-1}\times0.75 = 0.45 \text{ cm}$$

$$\frac{m_a}{m_f} = \frac{0.85}{0.7}\times\left(\frac{80}{5.5}\right)^2\sqrt{\frac{1.2}{750}\left(\frac{102}{102-0.45}\right)} = 10.28:1$$

(c) $c_a = 14\sqrt{\frac{h_{wl}}{\rho_a}} = 14\sqrt{\frac{0.45}{1.2}} = 8.57 \text{ m/s}$

■ **Example 3.6** $A:F$ ratio of a mixture supplied to an engine by a carburettor is 15. The fuel consumption of the engine is 7.5 kg/h. The diameter of the venturi is 2.2 cm. Find the diameter of fuel nozzle if the lip of the nozzle is 4 mm.

Take $\rho_f = 750 \text{ kg/ m}^3$, $C_{da} = 0.82$, $C_{df} = 0.7$,

atmospheric pressure = 1.013 bar and temperature = 25° C.
Neglect compressibility effect of air.

Solution : Given $A:F = 15$, $m_f = 7.5$ kg/h, $d_a = 2.2$ cm,

$$z = 4 \text{ mm}, \rho_f = 750 \text{ kg/ m}^3, C_{da} = 0.82, C_{df} = 0.7,$$

$$p = 1.013 \text{ bar}, T = 273 + 25 = 298 \text{ K}$$

Density of air, $$\rho_a = \frac{p}{RT} = \frac{1.013\times10^5}{287\times298} = 1.185 \text{ kg/m}^3$$

Mass flow rate of air, $$\dot{m}_a = \frac{\dot{m}_f \times A:F \text{ ratio}}{3600}$$

$$= \frac{7.5\times15}{3600} = 0.03125 \text{ kg/s}$$

Now $$\dot{m}_a = A_a\, C_{da}\sqrt{2\rho_a\cdot\Delta p}$$

$$0.03125 = \frac{\pi}{4}(2.2\times10^{-2})^2\times0.82\sqrt{2\times1.185\times\Delta p}$$

$$\Delta p = 4260 \text{ N/m}^2$$

$$m_f = A_f\, C_{df}\sqrt{2\rho_f(\Delta p - zg\cdot\rho_f)}$$

$$\frac{7.5}{3600} = \frac{\pi}{4}(d_f\times10^{-3})^2\times0.7\sqrt{2\times750\,(4260-4\times10^{-3}\times9.81\times750}$$

$$d_f = 1.245\text{mm}$$

■ **Example 3.7** A 4-Stroke petrol engine of 1710 cm^3 capacity is to be designed to develop maximum power at 5400 rpm, when the mixture $A:F$ ratio supplied to the engine is 13:1. Two carburettors are to be fixed. The maximum velocity of air is limited to 107 m/s. Find the diameters of venturi and fuel nozzle jet, if η_{vol} (engine) = 0.7, $C_{da} = 0.85$, $C_{df} = 0.66$, nozzle lip = 6 mm and $\rho_f = 750 \text{ kg/m}^3$.

Atmospheric pressure and temperature are 1.013 bar and 300 K. Take $c_p = 1$ kJ /kg. K.

Solution : Given $V_e = 1710 \text{ cm}^3$, N = 5400 rpm, $A:F = 13$, $C_a = 107$ m/s, $d_a = ?$, $d_f = ?$, $\eta_{vol} = 0.7$, $C_{da} = 0.85$, $C_{df} = 0.66$, $z = 6$ mm, $\rho_f = 750 \text{ kg/m}^3$ $p = 1.013$ bar, $T = 300$K, $c_p = 1$ kJ/kg. K

Volume of air supplied to engine at STP,

$$\dot{V} = V_e \times \frac{N}{2 \times 60} \times \eta_{vol} = 1710 \times 10^{-6} \times \frac{5400}{2 \times 60} \times 0.7 = 0.054\ \text{m}^3/\text{s}$$

Air flow through each carburettor,

$$\dot{V}_a = \frac{\dot{V}}{2} = \frac{0.054}{2} = 0.027 \text{m}^3/\text{s}$$

Mass flow rate of air,

$$\dot{m}_a = \frac{p\dot{V}_a}{RT} = \frac{1.013 \times 10^5 \times 0.027}{287 \times 300} = 0.032\ \text{kg/s}$$

Air velocity through ventrue,

$$c_a = \left[2c_p\, T_1 \left\{ 1 - \left(\frac{p_2}{p_1} \right)^{\frac{\gamma - 1}{\gamma}} \right\} \right]^{1/2}$$

$$107 = \left[2 \times 1 \times 10^3 \times 300 \left\{ 1 - \left(\frac{p_2}{p_1} \right)^{\frac{1}{3.5}} \right\} \right]^{1/2}$$

$$\frac{p_2}{p_1} = 0.935$$

$$p_2 = 0.935 \times 1.013 = 0.947\ \text{bar}$$

Assuming the flow to be isentropic,

$$p_1 \dot{V}_1^{\gamma} = p_2 \dot{V}_2^{\gamma}$$

$$\dot{V}_2 = \dot{V}_1 \left(\frac{p_1}{p_2} \right)^{\frac{1}{\gamma}} = 0.027 \left(\frac{1}{0.935} \right)^{\frac{1}{1.4}} = 0.0287\ \text{m}^3/\text{s}$$

Volume of air passing through each carburettor,

$$\dot{V}_2 = A_a c_a C_{da}$$

$$0.0287 = \frac{\pi}{4} (d_a \times 10^{-2})^2 \times 107 \times 0.85,\ d_a \text{ in cm}$$

$$d_a = 2 cm$$

Mass flow rate of fuel per carburettor,

$$\dot{m}_f = \frac{\dot{m}_a}{A:F} = \frac{0.032}{13} = 0.00246\ \text{kg/s}$$

Now

$$\dot{m}_f = A_f \cdot C_{df} \sqrt{2\rho_f\, (\Delta p - zg\rho_f)}$$

$$\Delta p = p_1 - p_2 = 1.013 - 0.947 = 0.066\ \text{bar}$$

$$0.00246 = \frac{\pi}{4} (d_f \times 10^{-3})^2 \sqrt{2 \times 750\, (0.066 - 6 \times 10^{-3} \times 9.81 \times 750)}$$

where d_f is in mm.

$$d_f = 1.23\ \text{mm}$$

PART-II : FUEL INJECTION

3.2 PETROL INJECTION SYSTEMS

Although, the modern carburettor is cheap and reliable, it has a number of inherent disadvantages that make the supply of correct $A : F$ mixture at all times difficult. The problem is further accentuated when a single carburettor has to supply the mixture to multicylinder engine. In addtion to this, the throat restricts the flow of air to the engine and the maximum power. The maximum power can be increased by using a large throat as this affects the economy at low speeds because of low air speed and worst control on the petrol spary and automization. Multiple carburettors assist the distribution problem but it also increases fuel consumption due to low air velocity at low engine speed and capital cost.

For these reasons, many attempts have been made to design a satisfactory petrol injection system, in which, each cylinder is supplied with its correct quantity of petrol for each working cycle under all operating conditions. It is also claimed that the system devised to do this, allow the engine to produce more power and gives less vaporization troubles.

The adavntages opf fuel injection system over conventional carburettor have been appreciated for many years. The quest for improved engine performance linked with fuel economy together with legislation regarding the control of exhaust emission has enormously increased the use of fuel injection system. The incorporation of electronic control systems has also considerably helped in the development of very efficient and commerically viable systems.

There are two petrol injection systems which are mechanically operated.

1. Combustion Chamber Injection.
2. Continuous Port Injection.

The combustion chamber injection system is just similar to CI-engines. But this system is not adopted as todays emission and fuel economy requirements make it impractical.

1. Continuous Port Injection System (Lucas Mechanical Petrol Injection System)

A simplified line diagram of the above mentioned system is shown in Fig. 3.21. The petrol is sucked from the tank by a pump and pressurised petrol at 3 bar is supplied through a distributor to the fuel injector to a particular cylinder. The relief valve shown in figure maintains the pressure and allows excess petrol to return to the tank. The pump may be driven by engine or by a separate electric motor. The

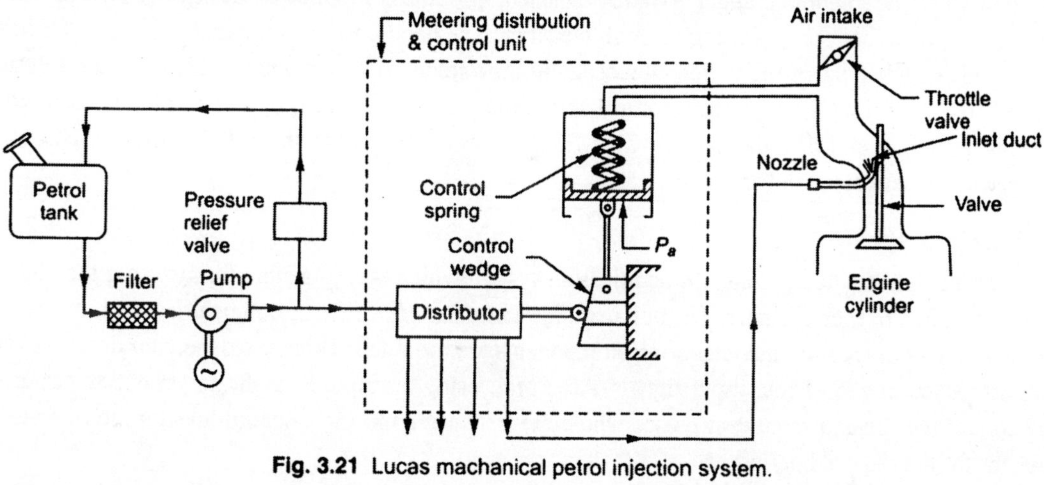

Fig. 3.21 Lucas machanical petrol injection system.

latter is preferred because it starts pumping at its normal running speed and pressurise as soon as ignition is switched on.

In this system, the petrol is injected continuously into the inner port as shown in figure at a varied rate. The distribution may be made by having a separate metering pump for each cylinder, all are timed by the arrangement of a series of cams on one cam shaft or by having one pump operated by a single cam with a lob for each cylinder (similar to ignition system) that feeds to distribution unit which passes the petrol to each inlet port in turn. The unit may be driven by shaft, chain or V-belt.

Now-a-days electronically controlled fuel injection system are commonly used as it functions rapidly and responds automatically to the change in manifold air pressure, engine speed, crankshaft angle and many other secondary factors. The electronic control unit assesses data (manifold pressure, engine speed, crank angle) received from a various sensing devices and then adjust the A : F supply for the best performance of the engine.

This system has to contain a means of supplying extra fuel for cold starting, during warming and enriching the mixture during acceleration.

Ice formation is virtually impossible with this system and the danger of vaporization is minimised because the petrol is under pressure right upto the injection point.

2. Electronic Fuel Injection System

The amount of fuel supplied to the engine is controlled by three factors, injector cross sectional area, fuel pressure and duration of injection.

Injector orifice size is fixed for a particular design. Fuel pressure supplied is controlled by electric pump and regulator. Therefore, the only variable factor in electronic fuel injection is to control the period of injection. This is done by translating the data supplied to the electronic control unit from the various sensors into electric pulses which are in turn relayed to the solenoids which operate the injectors thus determining the moment and period of injection.

The main parts of the system is injector and electronic control unit.

Injectors

The solenoid operated fuel injector is shown in Fig. 3.22 (a). It consists of a valve body and needle valve to which the solenoid plunger is rigidly attached. The fule is supplied to the injector under pressure from the electric fuel pump passing through the filter. The needle valve is pressed against a seat in the valve body by a helical spring to keep it closed until the solenoid winding is energised. When a current pulse is received from the electronic control unit, a magnetic field builds up in the solenoid coil which attracts a plunger and lifts the needle valve from its seat. This opens the path to pressurised fuel to emerge as a finely atomised spray.

Electronic Control Unit

This unit is the heart of electronic injection system which is responsible for metering the quantity of fuel supplied to each cylinder. The unit contains a number of printed circuit boards on which, a series of transistors, diodes and other electronic components are mounted. This makes the vital data analysing circuits responding to various input signals. After processing the input data, the power output circuits in the control unit generates current pulses which are transmitted to the solenoid injectors to operate the injector for the required period.

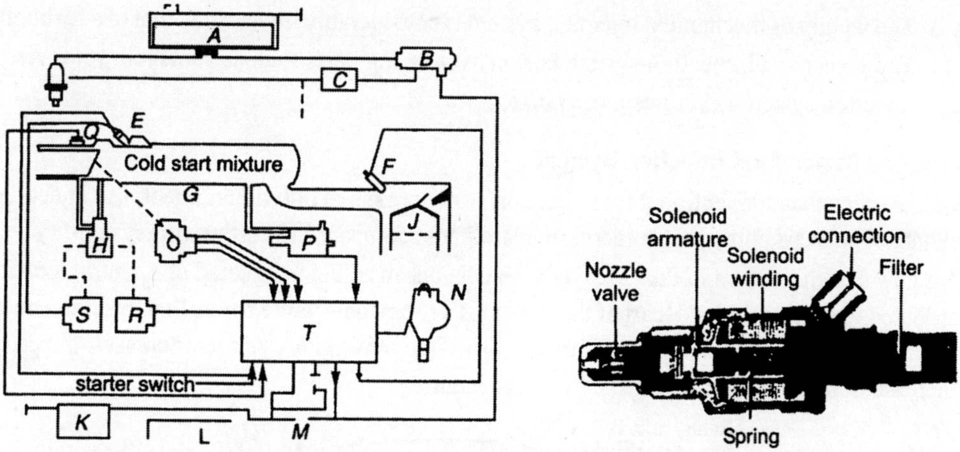

(a) (b)

Fig. 3.22 (a) Solenoid operated injection valve (b) Electronically controlled fuel injection system.

All the electrical units of this system are connected to a cable terminating in a 25 pole plug to the mating socket. Hence, as soon as the ignition switch is actuated, the fuel pump is switched on and the fuel injection system becomes operational.

The arrangement of this system is shown in Fig. 3.22 (b).

Advantages and Disadvantages of Petrol Injection System

Advantages :

1. Fuel injection system can precisely match fuel delivery to engine requirements under all load and speed conditions. This reduces fuel consumption with no loss of engine performance.
2. The manifold in an injection system carries only air, so there is no problem of the air and fuel separation and design of manifold becomes simple.
3. Due to absence of venturi which abstracts the air passage, the petrol injection results in better volumetric efficiency and increased power.
4. Fuel injection system is relatively free from icing and surge when tilted, cornering and braking.
5. Starting and acceleration is much pattern than carburetor system.
6. It provides identical A : F ratio mixture to all the cylinders and maintain better balancing. In addition to this, the engine can work more economically closer to its lean limit whereas in case of carburettor engine, one or two ratios richer than best economy mixture is to be used to compensate for driveability. This gives lower specific fuel consumption.
7. By maintaining a precise A : F ratio according to engine requirements, exhaust emissions are lowered. The improved air fuel flow in injection system also helps to reduce emission levels.

Disadvantages :

1. The major disadvantage is its high capital cost due to precise and complicated components of electronic circuit.
2. The maintenance of this system is difficult and costly as this system contains 4 times the components of mechanical system. A Junker 12 cylinder engine has 1576 parts against 433 parts used with 12 cylinder Mercedes Carburettor system.

3. The weight of mechanical injection system is considerably higher than that of a carburettor.
4. The mechanical injection system has many wearing parts such as camshaft, rotor etc.
5. Injection system generates more noise.

3. Rotary Gate Meter Fuel Injection System

We have seen that conventional fuel injection system requires individual injector to inject fuel into each cylinder. The extra components increase overall cost compared to carburettor system.

This new system, shown in Fig. 3.23 reduces cost because fuel is injected at a central location and then distributed to each cylinder. Heart of the system is a rotary gate valve controlled by engine intake air to regulate fuel injection. As the volume of the air taken into the engine cylinder increases, the gate valve tips to increase the amount of fuel injected into the engine.

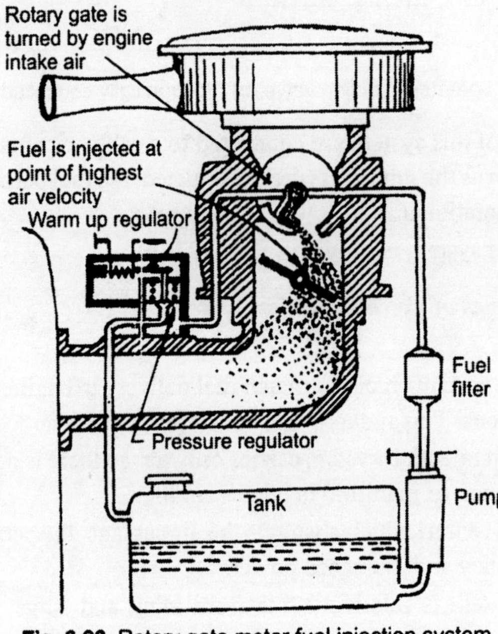

Fig. 3.23 Rotary gate meter fuel injection system.

Robert Bosch Corporation, Germany, reported that because of this system, fuel atomization and distribution are optimized as the fuel is injected into the air stream at the point of highest velocity.

3.3 FUEL INJECTION IN C.I. ENGINES

Unlike petrol engines, the fuel in C.I. engines is supplied at a very high pressure partly during the compression stroke and partly during the power stroke. The air is taken in during the suction stroke and compressed to a high pressure (30 to 70 bar) and high temperature (500 to 700°C) according to the compression ratio used (12 to 20). The high temperature of air at the end of stroke is sufficient to ignite the fuel.

As the fuel is injected in a high pressure air, the pressure of fuel injected lies between 100 to 200 bar. During the process of injection, the fuel is broken into very fine droplets. The droplets vaporise taking the heat from the hot air and form a combustible mixture and starts burning. As the burning starts, the varporization of fuel is accelerated as more heat is available. As the combustion advances, the amount of oxygen available for burning reduces and therefore heat release rate is reduced.

The period between the injection and ignition of fuel is known as delay period (ignition delay). This lies between 0.001 to 0.002 sec according to the speed of the engine. This counts for nearly 25° crank rotation. The whole performance of the engine is totally dependent on the delay period. The delay period is less, better is the formance of the engine.

The main functions of the injection system are

1. To supply the correct quantity of fuel to be injected as per the load of the engine and increase in speed for automobile engines.
2. To supply the fuel within a precisely defined period of the cycle.
3. To control the rate of fuel injection, such that it should result in the desired heat release pattern.
4. To atomic the fuel into very fine particles.
5. To distribute the fuel uniformly in the combustion chamber of the engine and results in rapid mixing of fuel and air.
6. Injection starts and stops sharply. There should not be any after injection.

Figure 3.24 shows the pattern of variation of pressure, temperature and heat release rates with respect to crank angle (lift of fuel needle). It is obvious from the fuel pattern, the fuel injection starts 22° before TDC and ignition starts at 12.5° before TDC (delay period 22 – 12.5 = 9.5°C). The heat release rate at the beginning of combustion is very high as it takes 2.5° crank angle from minimum to maximum which occurs at 10.5° before TDC. The combustion rate and heat release rate after this decrease and continued till 5° after TDC and again starts increasing till 18° crank angle after TDC (during a period of 13° rotation of crank).

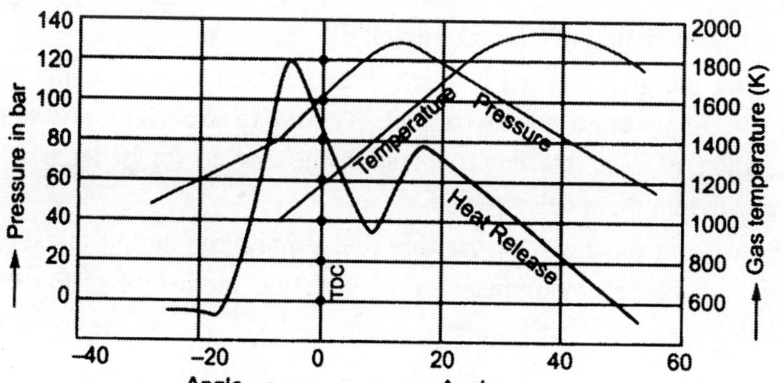

Fig. 3.24

The effect of this pattern of heat release on pressure and temperature are also represented. The temperature continuously rises. The pressure starts increasing and reaches maximum at 5° after TDC and then starts falling due to the movement of piston. But the temperature continues to increase even after this point as heat release is continued and peak of temperature occurs at 35° after TDC.

3.3.1 Requirements of an Ideal Injection System

The amount of oil injected per cycle depends upon the load of the engine. To obtain a complete combustion of fuel at full load, the supply must be about 50% greater than the minimum theoretical amount required. In engines, with high turbulence combustion chambers, the excess air may be reduced to about 30%.

If more fuel is injected then can be burned completely, the specific fuel combustion and cost of power goes up. If less fuel injected, the MEP decreases and small power will be developed per cm^3 of the displacement again increasing the cost of power. Therefore, exact amount of fuel giving the lowest cost of power for a given engine must be determined by testing and then only injection system must be designed.

For efficient injection system, the following factors are to be considered in more detail.

1. Automization

The automization is defined as formation of smallest fuel particles in the engine during fuel injection.

Better automization is essential for better mixing and complete combustion and ultimately better performance of the engine.

The automization of fuel in the engine depends upon rate of injection; injection pressure, air density, nozzle diameter and swirl provided to the air.

The factors which are responsible for automization as mentioned above are discussed in detail.

(a) Rate of Injection : The rate of injection is defined as (dm_f/dt) or $(dm_f/d\theta)$ during the rotation of crank is very important for automization. The total fuel supply during a cycle is given by

$$m_f = \int_{\theta_1}^{\theta_2} f(\theta) \cdot d\theta$$

where m_f is a function of θ (crank angle).

Considering the graphs *a* and *b* which show constant rate of supply and variable rate of supply and supply same quantity of fuel. We have to select a most desirable curve from the consideration of desirable environment in the engine for better evaporation and better burning with minimum delay period.

It is observed in practice that variable fuel supply rate during the supply period is most preferable as better environment exists with the motion of piston towards TDC. (See Fig. 3.25).

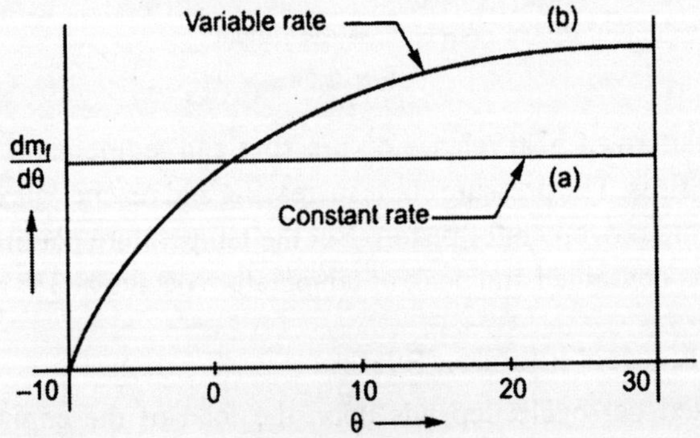

Fig. 3.25

The exact curve of $(dm_f/d\theta)$ depends upon, many factors as compression ratio, swirl produced and nozzle diameter.

Some information concerning the fuel injection rate is available as shown in Fig. 3.26. The injection period extends over 40° after TDC and combustion extends upto 78° after TDC. The horizontal distance between the two curves gives the time lag or delay period.

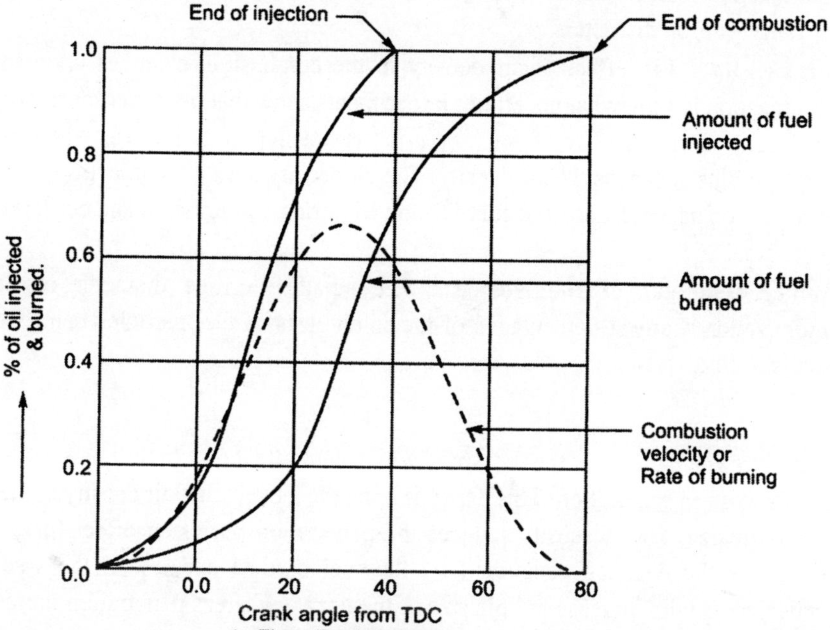

Fig. 3.26

The duration of combustion in an oil engine should not be more than 40° of crank travel and considering unavoidable after burning not over 80°. This means, the time available for burning the fuel is 0.067 sec at 200 R.P.M. and only 0.0067 sec at 2000 RPM. This exceedingly short time requires a very fine automization.

(b) Injection Pressure : The degree of automization is indicated by the smallness of the size of the particles in the spray. The fuel particle velocity is most important factor affecting the degree of automization. The automization is a function of $\sqrt{p_2 - p_1}$ where p_2 is the oil pressure and p_1 is the air pressure during the injection of fuel.

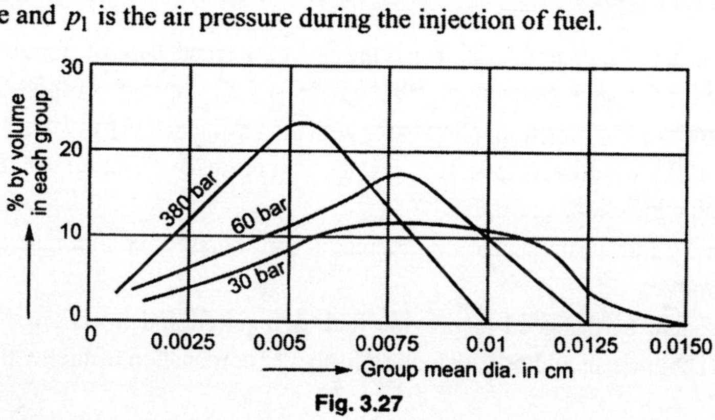

Fig. 3.27

The graph shown in Fig. 3.27 shows that increasing the injection pressure reduces the mean diameter of the particles. It is also noticeable from the figure that the percentage of particles with the large diameters decreases markedly with increasing injection pressure.

These automization curves show both the degree of fineness and the uniformity of the automization. Closer the curve to the vertical axis, the finer is the automization and smaller the range of drop diameters.

(c) **Air Density :** The effect of air density in the combustion chamber upon the degree of automization is somewhat uncertain. Experiments show that a smaller mean drop size as the air density increases. The effect $(dr_0/d\rho_a)$ is not definitely known, where r_0 is the fuel droplet radius and p_a is the air density. The air density during the injection and at the end of compression depends upon the curve followed during compression and compression ratio of the engine.

(d) **Orifice Diameter of the Nozzle :** The smallest orifice diameter results in better automization. Largest percentages of the small sizes of fuel particles being obtained with smalls orifice.

2. Penetration

The distance to which the tip of the fuel spray will peneratrate the air in the combustion chamber in the given time is known as penetration. This depends upon the jet velocity, air density in the combustion chamber and nozzle size. Fuel viscosity reduces the penetration but exact effect $(dL_p/d\mu_f)$ is not known, where L_p is the distance penetrated by the fuel particle and μ_f is the viscosity of oil. Automization decreases with increase in hole size of the nozzle whereas penetration increases with the increase in hole size. Therefore, the size of the hole must be determined as a compromise between proper automization and sufficient penetration.

The following factors effect the penetration :

(a) **Injection pressure :** The effect of injection pressure on penetration is listed below :

Injection pressure (bar)	135	270	540
Penetration (cm)	8.75	11.25	15

But these is limit for injection pressure, beyond which increase in injection pressure would not increase the penetration because the atomised drops are too light to penetrate the dense air.

(b) **Air Density :** Higher air density resists the penetration and therefore requires higher injection pressure for the same penetration. But high density air improves evaporation.

(c) **Fuel Density :** The penetration increases with increasing fuel density. Heavy oil with specific gravity of 0.9 would penetrate 10% greater than ordinary diesel oil of 0.85 specific gravity under same injection pressure.

(d) **Injection Value :** The penetration increases considerably with an increase in the angle of spiral groove.

The nozzle should be designed to provide fuel as required and injection should be such, the atomization and penetration should be satisfactory for better combustion of fuel without any danger of knocking.

3. Dispersion

Dispersion is defined as the distribution of fuel particles in the mass of air. The dispersion of the fuel spray depends mainly upon the nozzle design. The type of the nozzle that gives a large spray cone produces better dispersion of fuel than does a small spray cone. However, increase in dispersion cone decreases the penetration. Therefore, a compromise should be made between dispersion and penetration. The dispersion characteristics of the spray can be improved by providing turbulent flow.

4. Turbulence

The turbulence in the combustion chamber aids fuel dispersion and helps for better evaporation of fuel. This lowers fuel consumption and increases the engine power. The design of the intake valve, combustion space and piston top can be designed to create swirl and provides additional turbulence which is difficult to measure directly. But its influence can be measured by measuring the power increase and fuel consumption.

The additional turbulence can increase MEP upto 17% and lowers the fuel consumption by 15%.

5. Timing of Injection

In solid injection systems, there is always time lag between the injection and starting of combustion of injected fuel. This is an inherent property of solid injection regardless of the shape and construction of the spray nozzle. Solid injection generally gives a coarse automization than air injection and therefore requires an earlier injection timing from 10 to 30° before TDC. This is essential as there is time lag between the injection and combustion. The proper angle of advance for injection depends upon the following factors : (i) Fuel characteristics, (ii) Method of injection, (iii) Speed of engine, (iv) Combustion chamber design and (v) Load on the engine.

The angle of injection advance in open combustion system varies from 10 to 30° and in precombustion chamber, it varies from 30 to 45°.

In majority injection systems, the angle of advance remains constant. Some pumps are designed where the cycle of advance is gradually increased with increase in load. In variable speed engines, provision is also made for automatic increase of the angle of advance with increasing engine speed. These systems used for changing the angle of advance are not common and their details are out of the scope of this text.

3.3.2 Types of Injection System

Injection system may be divided into two general types, as follows :

1. Air injection system
2. Air less, or solid or mechanical injection.

1. Air Injection System

It was first developed by Rudlof Diesel. The arrangement of the system is shown in Fig. 3.28. In this system, air and fuel both are injected into the cylinder during the supply of fuel. The required pressure of the air for injecting the fuel is about 70 bar or higher.

A fuel pump is driven by the engine itself. (A cam shaft operates the fuel pump through a cam and the power required to rotate the cam shaft is taken from the main shaft of the engine with the help of gears and discharges a definite quantity of fuel into the injection valve as shown in Fig. 3.28. The injection valve is mechanically opened and high pressure air drives the fuel charge and some air into the

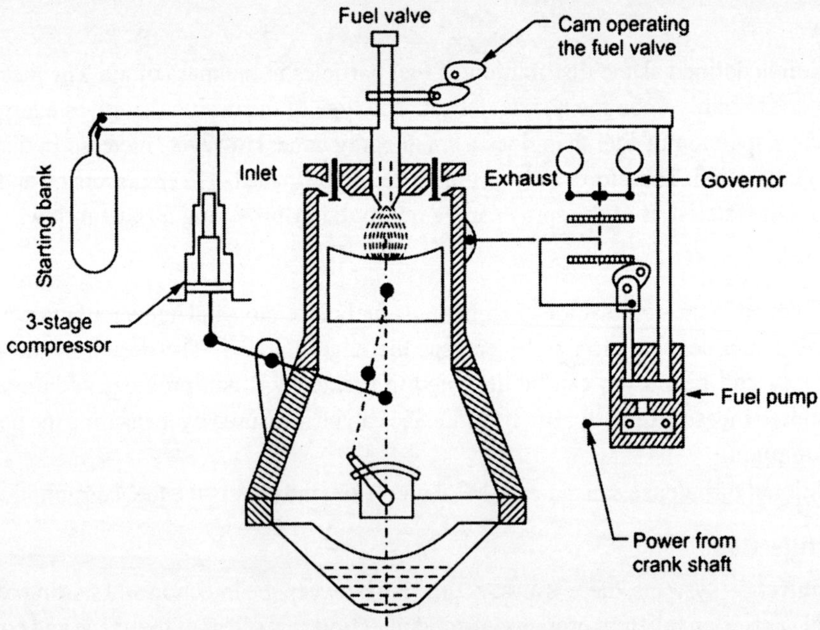

Fig. 3.28 Airless injection system.

combustion chamber. The amount of fuel delivered is under the control of oil pump suction valve, which is operated by a governor.

The air pressure is raised to about 70 bar by a three stage compressor (as shown in Fig. 3.28) providing intercooling. The compressor is also operated by the engine. The high pressure air projects the fuel into the combustion chamber and atomises it.

This type of system is rarely used now-a-days in diesel engines.

The advantages and disadvantages of this system are listed below :

Advantages :

1. It provides better automization and distribution of fuel.
2. As the combustion is more complete, the BMEP is higher than with other types of injection systems.
3. It allows to use the inferior fuels.

Disadvantages :

1. It requires complicated mechanism to run the compressor.
2. The weight of the engine increases.
3. Part of the power is used to drive the compressor and so the B.H.P. of the engine is reduced.

2. Air-less or Solid Injection

In this system, the fuel is supplied at a very high pressure (150 bar) from the fuel pump to the fuel injector from where it is injected to the combustion chamber with the help of an injector. The main parts of this system are fuel pump and fuel injector. The fuel pump is operated by a cam which is mounted on cam shaft. The power required to operate the cam is taken from the engine crank shaft. Depending upon the location of the fuel pumps and injectors, and upon the method used to meter the fuel, solid injection may further be classified as follows :

(a) Common Rail system

(b) Distributor system and Unit injector

(c) Individual pump and nozzle system.

The arrangement and working of each system are given below.

(a) Common Rail System

In this system, there is single high pressure pump, which supplies high pressure fuel to a common header (rail). The accumulator is connected to different cylinders by separate fuel line through the fuel nozzle as shown in Fig. 3.29. The pressure in the accumulator is maintained constant with the help of pressure relief valve. The fuel is supplied to each cylinder by operating the respective fuel valve with the help of cam mechanism driven by the engine crank shaft.

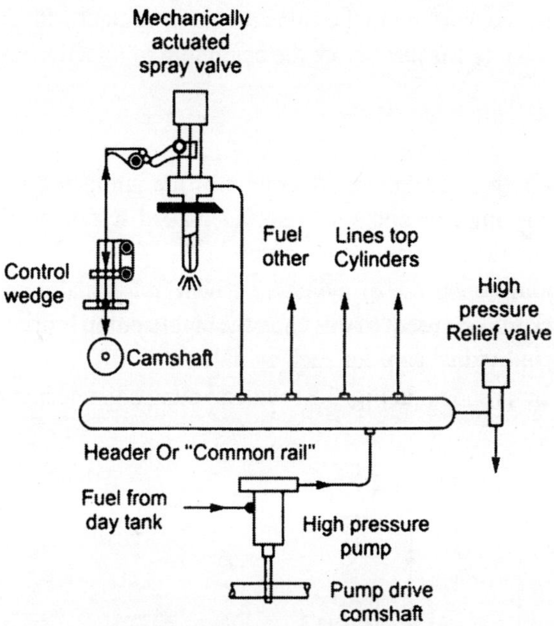

Fig. 3.29 Common rail system.

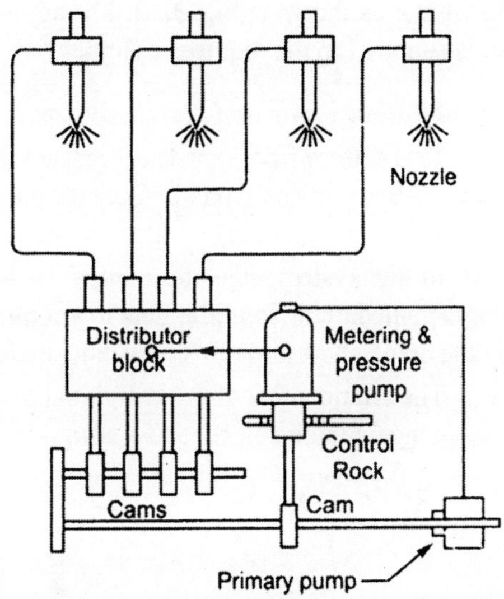

Fig. 3.30 Distributor system.

The quantity of fuel injected and the timing of injection are controlled by fuel valve and not by the injection pump. The arrangement of fuel valve (metering meter) and fuel nozzle is shown in figure. This system uses spring loaded injection valves which open and close by mechanical means.

The pressure used in this system ranges between 110 to 300 bar according to the compression ratio used for the engine design.

The advantages and disadvantages of the system are listed below.

Advantages :

1. It fulfills the requirements of either, (a) the constant load with variable speed or (b) constant speed with variable load.
2. Only one pump is sufficient for multi-cylinder engine.
3. Variation is pump supply pressure will effect all the cylinders uniformly.
4. The arrangement of the system is very simple and maintenance cost is less.

Disadvantages :

1. Very accurate design and workmanship are required.
2. There is tendency to develop leaks in the injection valve.

(b) Distributor System

This system, like common rail system, employs a single high pressure pump as shown in Fig. 3.30. The high pressure pump in this system is used for metering and compressing the fuel and then the fuel is delivered to the common rotating distributor. The fuel is supplied to each cylinder by the distributor. In every cycle, the injection strokes of the pump are equal to the number of cylinders. The quantity of fuel supplied and timing of fuel supply is done by single plunger (main pump) therefore equal amount of fuel is supplied to each cylinder and at the same point in the cycle. The function of the distributor is merely to select the cylinder to receive the fuel.

The distributor block selects a particular cylinder according to the cam coming in contact with the distributor as shown in Fig. 3.30. The appropriate valve opens just before the beginning of injection and oil is supplied to the required cylinder.

(c) Individual Pump and Nozzle System

This differs from constant pressure injection both in design and operation of the pump and fuel injector. Each injector has a separate pump and the injector contains a spring loaded hydraulically operated automatic plunger valve. No separate mechanism is required to operate it

In this system, separate pumps, each (depending upon no. of cylinder) driven individually or a single pump having four plungers in a common block may be used. In this case, the single pump is driven by the crank shaft through single cam shaft having individual cam for each cylinder.

The arrangement of the individual pump system is shown in Fig. 3.31 with all pumps in one block, four plungers in one barrel and a common cam shaft.

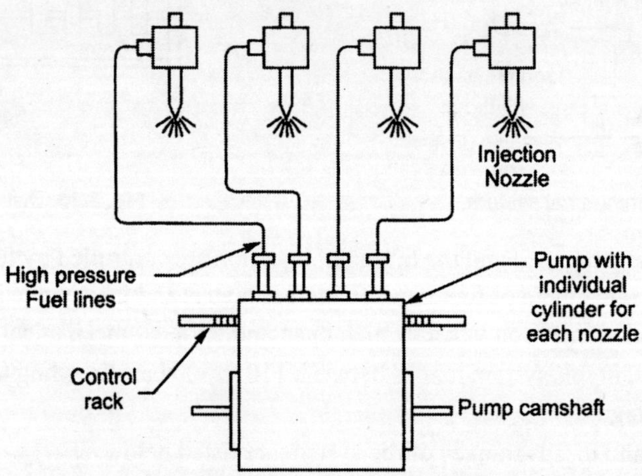

Fig. 3.31 Individual pump system.

The design of this type of pump must be very accurate and precise as the volume of fuel injected per cycle is 1/20,000 of the engine displacement at full load and 1/100,000 of the engine displacement during idling. The time allowed for injecting such a small quantity of fuel is very limited (about 1/450 second at 1500 r.p.m. of the engine providing injection through 20° angle). The pressure requirements vary from 100 to 300 bar.

The comparison of various fuel injection system is given in Table 3.3.

Table 3.3 : Comparison of fuel Injection Systems

S. No.	*Job*	*Air injection system*	*Solid injection system*		
			Individual pump	*Common rail*	*Distributor*
1.	Metering	Pump	Pump	Injection valve	Pump
2.	Timing	Fuel cam	Pump cam	Fuel cam	Fuel cam
3.	Injection rate	Spray valve	Pump cam	Spray valve	Fuel cam
4.	Atomisation	Spray valve	Spray tip	Spray tip	Spray tip
5.	Distribution	Spray valve	Spray tip	Spray tip	Spray tip

3.3.3 Design of Fuel Nozzle

Fuel consumed per cycle, $$m_f = \frac{\text{BSFC} \times \text{Power}}{\text{Cycles/h}} \quad \text{...(3.11)}$$

where cycles/h $= \dfrac{\text{RPM}}{2} \times 60$ for 4-stroke engine ...(3.12a)

$= \text{RPM} \times 60$ for 2-stroke engine ...(3.12b)

Duration of injection in seconds,

$$t = \frac{\theta \times 60}{\text{RPM} \times 360} \text{ s} \quad \text{...(3.13)}$$

Mass of fuel supplied per second,

$$m_f = \frac{m_f}{t} = C_{df}\, A_f \sqrt{2\rho_f \times \Delta p} \quad \text{...(3.14)}$$

Velocity of fuel passing through the orifice,

$$c_f = \sqrt{\frac{2(p_i - p_a}{\rho_f}} \quad \text{...(3.15)}$$

where p_i, p_a = injection pressure and air pressure in the cylinder respectively.

ρ_f = density of fuel injected.

Volume of fuel injected per second.

$$Q_f = \left[\left(\frac{\pi}{4} d^2 \times n\right) \times c_f \times \left(\frac{\theta}{360} \times \frac{1}{N_i}\right) \times \left(\frac{N}{60}\right)\right] C_{df} \quad \text{...(3.16)}$$

where d = orifice diameter

n = number of orifices

θ = crank angle during which the fuel is supplied

N_2 = number of injections per minute

$= \dfrac{N}{2}$ for a 4-stroke engine ...(3.17a)

$= N$ for a 2-stroke engine ...(3.17b)

$N =$ rpm

$C_{df} =$ coefficient of discharge of fuel nozzle.

■ **Example 3.8** A 6-cylinder, 4-stroke oil engine develops 200 kw at 1200 rpm and consumes 0.3 kg/kwh. Determine the diameter of a single orifice inject if the injection pressure is 200 bar and combustion chamber pressure is 40 bar. The injection is carried and for 30° rotation of a crank. Take $\rho_f = 900 \text{ kg/m}^3$, and $C_{df} = 0.7$. Each nozzle on a cylinder is provided with a single orifice.

Solution : Fuel consumed per cycle, $m_f = \dfrac{\text{BSFC} \times \text{Power}}{\text{Cycles/h}}$

$$= \frac{0.3 \times 200}{\frac{1200}{2} \times 60} = 1.67 \times 10^{-3} \text{ kg}$$

Duration of injection $t = \dfrac{30}{1} \times \dfrac{60}{1200 \times 360} = 4.17 \times 10^{-3}$ s

$$\dot{m}_f = \frac{m_f}{t} = \frac{1.67 \times 10^{-3}}{4.17 \times 10^{-3}} = 0.4 \text{ kg/s}$$

$$\dot{m}_f = C_{df} A_f \sqrt{\rho_f \times \Delta p}$$

$$0.4 = 0.7 \times A_f \sqrt{900 \times (200 - 40) \times 10^5}$$

$$A_f = 0.0476 \times 10^{-4} \text{ m}^2 \text{ or } 4.76 \text{ mm}^2$$

Area of each nozzle $= \dfrac{A_f}{6} = \dfrac{4.76}{6} = 0.793 \text{ mm}^2$

$$\frac{\pi}{4} d_f^2 = 0.793, \; d_f = 1.005 \text{ mm}$$

■ **Example 3.9** An engine having a simple single jet carburettor consumes 6.5 kg/h of fuel. The level of fuel in the float chamber is 3 mm below the top of jet when engine is not running. Ambient conditions are 76 cm of Hg paessure and 17°C temperature. The jet diameter is 1.25 mm and $C_{df} = 0.6$. The coefficient of discharge for air is 0.85 and $A : F$ ratio $= 15$. Determine the critical air velocity and the throat diameter. Express the pressure depression in cm of water. Take $\rho_f = 700 \text{ kg/m}^3$.

Solution : Given $\dot{m}_f = 6.5$ kg/h, $z = 3$ mm, $p_1 = 76$ cm of Hg,

$$T_1 = 273 + 17 = 290\text{K}, \; d_f = 1.25\text{mm}, \; C_{df} = 0.6,$$

$$C_{da} = 0.85, \; A : F = 15, \; \rho_f = 700 \text{ kg/m}^3$$

Air density, $\rho_a = \dfrac{p}{RT}$

$$p = 76 \text{ cm of Hg} = 1.013 \text{ bar}$$

$$\rho = \frac{1.013 \times 10^5}{287 \times 290} = 1.22 \text{ kg/m}^3$$

Critical air velocity, $(C_a)_{crit} = \sqrt{\dfrac{2zg\,\rho_f}{\rho_a}}$

$$= \sqrt{\frac{2 \times 2 \times 10^{-3}\ 9.81 \times 700}{1.22}}$$

$$= 5.8 \text{ m/s}$$

Mass flow rate of fuel,

$$\dot{m}_f = C_{df}\ A_f\ \sqrt{2\rho_f\ (\Delta p - zg\rho_f)}$$

$$\frac{6.5}{3600} = 0.6 \times \frac{\pi}{4} \times (1.25 \times 10^{-3})^2\ \sqrt{2 \times 700\ (\Delta P - 3 \times 10^{-3} \times 9.81 \times 700}$$

$$\Delta p = 4329 \text{ N/m}^2$$

Head causing flow in cm of water,

$$h_w = \frac{\Delta p}{98} = \frac{4329}{98} = 442 \text{ cm} \qquad [\because 1 \text{ cm of } H_2O = 98 \text{ N/m}^2]$$

Now

$$\dot{m}_a = \frac{\pi}{4}(d_a^2) \times c_a\ \rho_a C_{da} = \frac{\dot{m}_f \times 15}{3600}$$

where $c_a = \sqrt{2\rho_a \times \Delta p} = \sqrt{1 \times 1.22 \times 4329} = 102.8$ m/s

$$\frac{\pi}{4}(d_a \times 10^{-3})^2 \times 102.8 \times 1.22 \times 0.85 = \frac{6.5 \times 15}{3600}$$

$$d_a = 18 \text{ mm}$$

PART III : FUEL IGNITION

3.4 INTRODUCTION

The ignition of fuel is concerned only with starting the combustion and not with the behaviour of combustion flame. For starting the burning of fuel, it is necessary to raise the temperature of the air-fuel mixture to its ignition temperature. The energy required for this purpose is supplied through an electric spark. Within the range of mixtures normally used (12 : 1 to 15 : 1) in S.I. engines a spark energy of 10 kJ is sufficient to start the combustion process.

3.4.1 Requirement of Ignition System

The important requirements of the spark ignition system are listed below :

1. The voltage across the spark plug electrodes should be sufficiently large to produce an arc required to initiate the combustion. The voltage necessary to overcome the resistance of the spark gap and to release enough energy to initiate the self-propagating flame front in the combustible mixture is about 10,000 to 20,000 volts.
2. The intensity of spark should lie in a specified limit because too high intensity may burn the electrodes and too low intensity may not ignite the mixture properly.
3. The volume of the mixture (clearance volume) at the end of compression should not too large, otherwise the spark produced may not be sufficient to ignite the whole charge. There is definite relation between the size of the spark and clearance volume.

4. There should be no missing cycle due to failure of spark.
5. In a multi-cylinder engine, there must be arrangement (distributor) to carry this voltage to the right cylinder at the right time.

3.4.2 Ignition Systems

The basic ignition systems in use are :

1. Battery ignition system—conventional, transistor assisted.
2. Magneto ignition system—low temperature, high temperature.
3. Electronic ignition system.

Battery and magneto ignition systems differ only in the source of electrical energy. Battery ignition system uses a batter while magneto ignition system uses a magneto to supply low voltage, all other system components being similar.

3.4.3 Battery Ignition System

The function of battery ignition system is to produce high voltage spark and to deliver it to the spark plugs at regular intervals and at the correct time with respect to the crank position. The required components of the system are listed below :

1. A battery of 6 to 12 volts
2. Ignition coil
3. Contact breaker
4. Condenser
5. Distributor
6. Spark-plugs

The arrangement of all the components of battery ignition system for 4-cylinder engine is shown in Fig. 3.32.

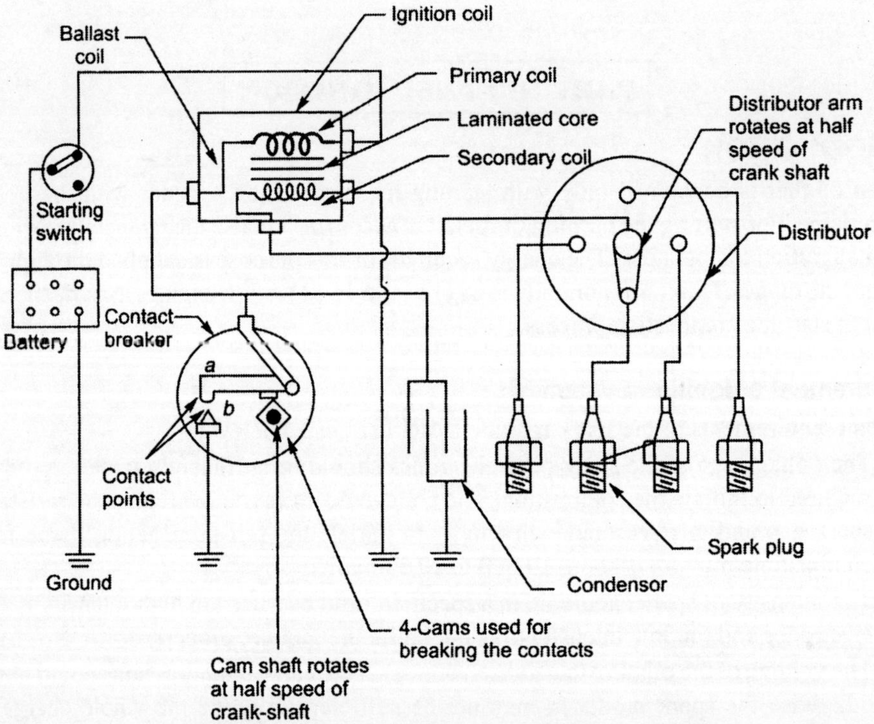

Fig. 3.32 Battery ignition system for multi-cylinder engines.

The source of current is the storage battery and it is connected to the primary of the induction coil through starting switch as shown in the figure. The other end of the primary coil is connected to the breaker and through it to the ground, when the breaker contact points are closed. (In the figure, the breaker contact points are shown in open position). As one terminal of the battery is grounded, the circuit is closed by passing the current from the battery through the starting switch, primary coil, contact breaker, ground and back to the battery when contact points are closed.

The induction coil consists of primary winding usually 100 to 200 turns and a secondary winding usually 10,000 turns. Both windings are mounted on soft iron core.

The contact breaker consists of contact points, camshaft on which a cam is mounted which is used to break and make the contacts between the contact points.

The distributor consists of distributor arm, as shown in figure. The arm is mounted on a cam shaft and is rotated at half the speed of crankshaft. The funetion of the arm is to make the contact with each spark plug as shown in figure.

The distributor unit generally includes the contact breaker also to make the unit more compact, as both are driven by the same cam-shaft.

A condenser is included in the circuit as shown in the figure.

(i) Principle of Induction

An E.M.F. is produced in the coil due to the relative movement of magnetic lines and coil because the magnetic field lines are cut by the coil. The E.M.F. produced depends on the relative movement between the two, higher the movement greater the E.M.F.

The principle of induction from one coil into another is shown in Fig. 3.33. When the current is allowed to flow through the primary coil, a magnetic field is set up and this field passes through the secondary coil and induces E.M.F. sending a current through a closed circuit. The current in the primary quickly attains a steady value and magnetic field is stabilized. The E.M.F. is not induced in the secondary as there is no relative movement between field (established by primary) and secondary.

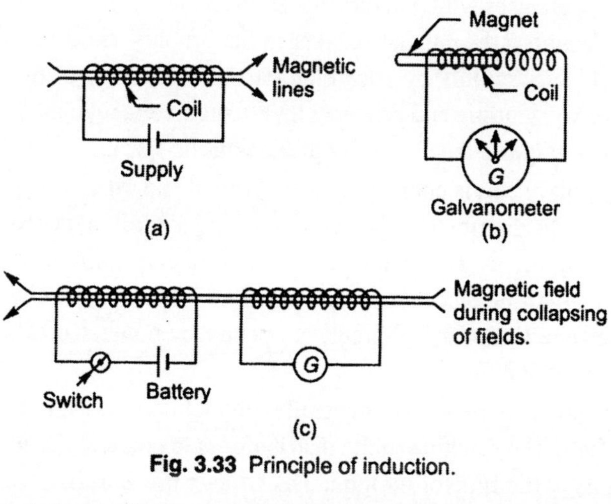

Fig. 3.33 Principle of induction.

If the primary is switched off, the establised magnetic field collapses and E.M.F. once again is induced in the secondary in the opposite direction. The greater E.M.F. is induced when the circuit is broken because the collapse is more rapid.

The E.M.F. induced in the secondary can be further increased by collapsing the established field more rapidly.

The magnetic effect in the secondary coil (E.M.F. produced) is intensified by winding the primary and secondary coils on a common soft iron bar or ring. The lines of magnetic field are concentrated around the bar or ring and magnetic effects are intensified.

A higher E.M.F. at the secondary terminal can be obtained by suitable proportioning of primary and secondary coil around a common iron coil (1 : 100).

(ii) Principle of Ignition

A high voltage can be introduced at the terminals of the secondary coil by collapsing the field established by primary and proper proportioning of the turns of primary and secondary as mentioned earlier. If this high voltage is connected to the two points providing an air gap between them as shown in Fig. 3.34 a spark is produced. Ignition can take place in the compressed charge of petrol engine if the spark is produced in the charge at the end of compression.

Fig. 3.34 Principle of ignition.

(iii) Working of Battery Ignition System

When the primary circuit is closed by the contact breaker (shown in open position in Fig. 3.1) a current begins to flow through the primary coil and magnetise core of the coil. The E.M.F. is induced in the secondary as the current in the primary increases. The E.M.F. induced in the secondary coil is proportional to the rate at which the magnetic flux increases. The E.M.F. produced in the secondary due to the growth of current in the primary is not sufficient to produce a spark at the spark plug because the primary circuit has to establish the magnetic flux.

When the primary circuit is opened by the contact breaker, the magnetic field collapses. Electromotive force is induced in the secondary which is directly proportional to the rate at which the magnetic field of the core collapses which in turn depends on the rate of decrease of the primary current. A condenser is connected across the contact breaker in the primary circuit as shown in Fig. 3.32. This helps to collapse the field very rapidly by absorbing part of the energy of the magnetic field which is thrown back into the primary winding and produces a very high voltage in the secondary. This E.M.F. in the secondary is sufficient to ignite the charge by producing the spark.

One end of the secondary coil is connected to the ground and other end is connected to the central terminal of the distributor. The distributor connects the secondary coil in turn to the different spark plugs of the engine in their firing order. The spark plug of a particular cylinder is placed in circuit of the secondary coil with the help of the distributor when the time comes for the charge in that cylinder to be ignited and at the same time the primary circuit is opened by contact breaker. A spark is produced between the points of the spark plug.

The distributor and contact breaker are generally mounted on the same cam-shaft which rotates at half speed of the crankshaft. The function of the distributor is to connect the secondary to each cylinder of a multi-cylinder engine at the time of ignition. The contact breaker also works simultaneously with distributor and its function is to disconnect the primary circuit exactly at the same time when the spark in the particular cylinder is required. The distributor arm connects 4 spark plugs in one rotation of the cam shaft and therefore four contact points are required in 4 cylinder engines. The contact breaker has to break the contacts 4-times in four cylinder engines so it requires 4 cams as shown in Fig. 3.32. If there are n cylinders, then the contact points and cams required are also n in number.

In a single cylinder engine, the distributor is not required as in scooter engine, and single cam is sufficient for giving the spark. A ignition system used in single cylinder petrol engine is shown in Fig. 3.35. Instead of battery, magneto is used in this system.

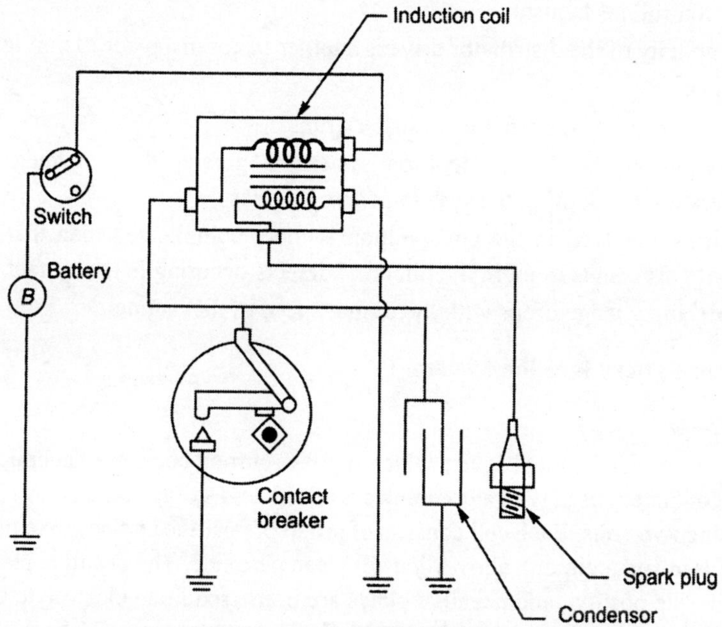

Fig. 3.35 Battery ignition system for single cylinder engine.

Cam is mounted on the crankshaft only as breaking of circuit during each rotation is required in two stroke engine and there is no necessity of cam-shaft.

(iv) Number of Sparks

Number of sparks produced must be equal to the number of working strokes in a single cylinder engine. If there are N_c, cylinders then the number of sparks produced for that engine are :

$$N_s, \text{(Number of spark)} = n \cdot N_c$$

where n = number of working strokes and N_c = number of cylinders

Further, $n = \frac{N}{2}$ for 4-stroke cycle engine

$= N$ for 2-stroke cycle engine where n is r.p.m. of the engine.

Thus, for 4-stroke engine, $N_s = \frac{NN_c}{2}$ and for 2-stroke engine, $N_s = NN_c$.

(v) Advantages and Disadvantages of Battery ignition System

Advantages :

1. Its initial cost is low compared with magneto. This is the main reason for the adoption of coil ignition on cars and commercial vehicles.

2. It provides better sparks at low speeds of the engine during starting and idling. This is because the maximum current is available throughout the engine speed range including starting.
3. The maintenance cost is negligible except for the battery.
4. The spark efficiency (intensity) remains unaffected by advance and retard positions of the timing control mechanism.
5. The simplicity of the distributor drive is another factor in favour of coil ignition.

Disadvantages :

1. The engine cannot be started if the battery runs down.
2. The weight of the battery ignition system is greater than magneto which is major consideration in adopting the system in aero-engines.
3. The wiring involved in the coil ignition is more complicated than that used in magneto ignition. This results in more likelihood of defects occuring in the system.
4. The sparking voltage drops with increasing speed of the engine.

(vi) Components of Battery Ignition System

1. Battery and cut-out

A battery is an electro-chemical device which supplies current because of chemical reaction inside. The common type of battery used with automobiles is a load acid storage battery. A 12 V battery has six cells each generating two volts. Each cell consists of group of positive and negative plates. The positive plate has a grid of lead and antimony alloy filled with lead peroxide. The negative plate has a grid filled with spoongy lead. The positive and negative plates are immersed in an electrolyte of dilute sulphuric acid. The plates are separated from each other by PVC or rubber separators. The positive and negative plate grids are connected to lead antimony strops. These strops connect separately the positive and negative sparks in series and then to the positive and negative terminals. The plates are placed in a battery container made of hard rubber which is acid proof. The plate tops are supported by insulating and acid resisting cell covers. Holes are provided in each cell to fill and unfill electrolyte or distilled water. The filter holes are covered with plugs containing small holds for allowing gases to escape.

The battery is charged continuously by a dynamo directly mounted on the crankshaft. To avoid over charging or discharging of the battery, an electric switch is introduced between the battery and dynamo, which is known as a cut-out.

2. Ignition coil

The purpose of the ignition coil is to step up 6 V or 12 V battery to 5000 V, which is sufficient to generate the spark. This coil consists of two insulated conducting coils having primary and secondary windings. The primary winding is connected to the battery through an ignition switch and contact breaker and secondary winding is connected to the spark plugs through the distributor.

3. Contact Breaker

As the number of sparks produced increases with an increase in engine speed, the time available for building up the magnetic field of the spark coil becomes shorter. As a result, the maximum value attained by the primary coil gets lesser and lesser with increasing speed and magnetic field produced becomes weaker and weaker, until finally not enough voltage is induced in the secondary winding to produce the spark. Contact breaker is a device to increase the time during which the primary circuit remains closed.

4. Condenser

For obtaining the highest voltage in the secondary circuit, a quick collapse of the magnetic field is essential. In addition to this, it is also necessary to prevent the arcing and consequently burning of contact points.

This is achieved by providing a condenser.

The condenser is a device which will absorb and held an electric charge when electromotive force is applied to its terminal. In its simplest form, it consists of two sheets of conducting material separated by a layer of insulating material.

The condenser capacity is given by, $C = A \cdot x \cdot p$ where A is the area of conducting material, x is the distance between the sheets of conducting material and P is the specific inductive capacity of the insulating material. The parts of the condenser are shown in Fig. 3.36.

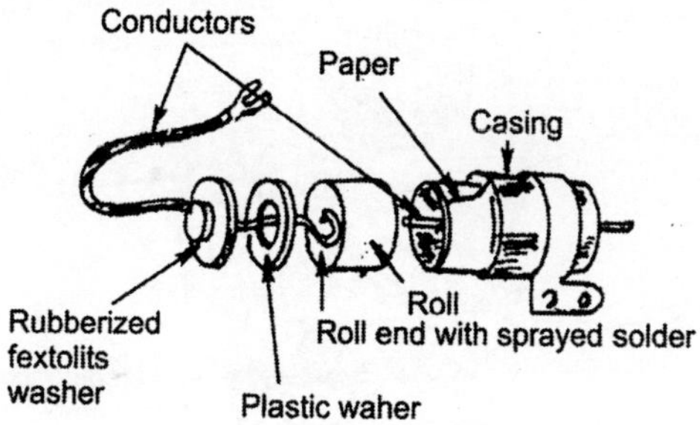

Fig. 3.36 Condenser.

In order to get large capacity with small space, instead of two sheets of conducting material, two sets of such sheets are used and they are separated by an insulating material. The conducting material used is "TINFOIL" while the insulating material is mica in condenser used with magneto and wax impregnated paper for battery system. All conducting sheets of a set are electrically connected. The arrangement for the condenser is shown in Fig. 3.37.

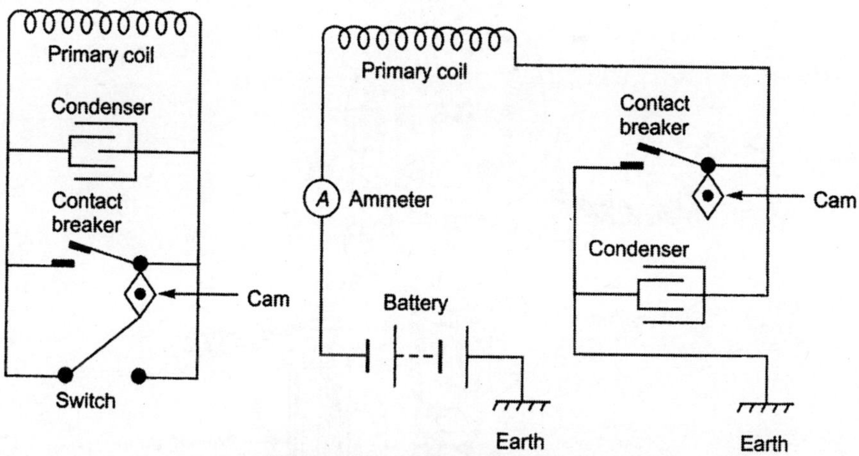

Fig. 3.37 Condenser arrangement.

The operation of the condenser is described as given below. The condenser is connected across the contact breaker. When the contacts points open, the current instead of passing across the points in the form of arc, flows into the condenser and stored by it and becomes charged. The charge in the condenser discharges back immediately in the primary circuit in the direction reverse to the flow of a battery current, thus assisting in a quicker collapse of magnetic field when the contact points open.

5. Distributor

The main function of the distributor is to distribute the high voltage surge to different plugs of multicylinder engine at the right time.

The high tension current first goes to the control electrode of the distributor and then to the rotor. As the rotor rotates, it passes this high tension current to metal electrodes which are connected to the spark plugs through high tension wiring according to the firing order of the engine.

These are two types of distributors as Brush type and Gap type. The blow up of distributor parts is shown in Fig. 3.38. In brush type, a carbon brush carried by a rotor arm slides over a metallic segments

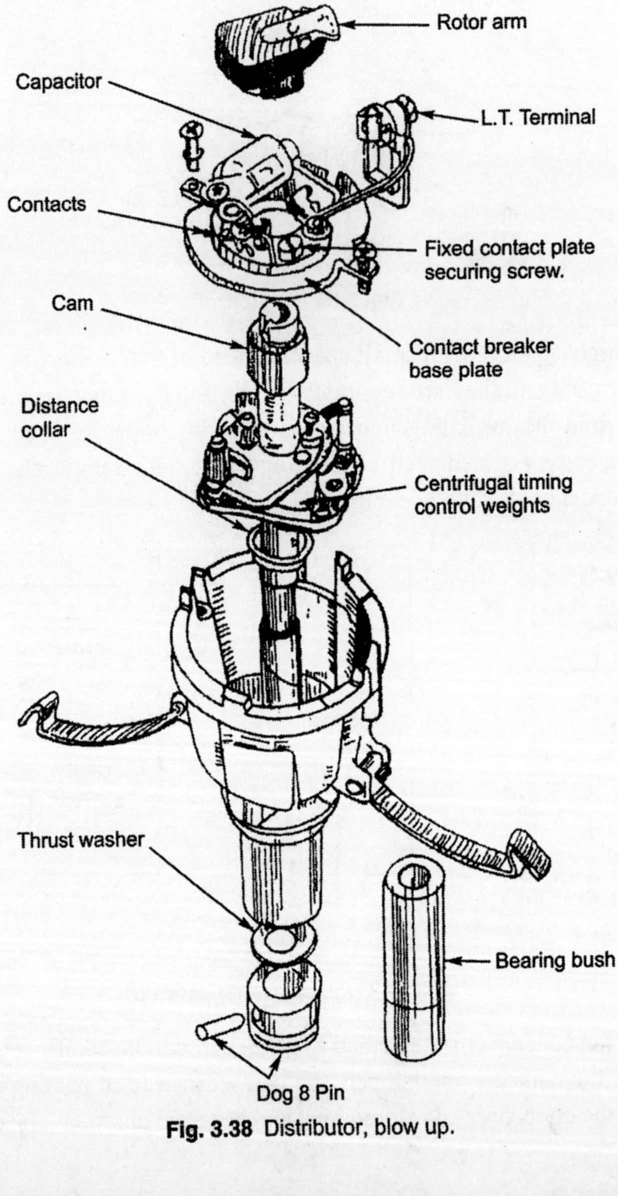

Fig. 3.38 Distributor, blow up.

embedded in the distributor cap of moulded insulating material, thus establishing electrical connection between the secondary winding of the coil and the spark plug.

In a gap type, the electrode of the rotor arm passes close to, but does not actually contact the segments in the distributor cap. The advantage of this is, there is no appreciable wear and misfiring due to fouled spark plug.

When the surface of the spark plug insulator inside combustion chamber becomes fouled with conducting carbon and oxides, there is considerable leakage of current through this conducting layer which pravents the voltage in the secondary from building up.

In actual practise, distributor, contact breaker, rotating shaft with breaker cam, condenser and ignition advance mechanism are housed together.

6. Spark Plug

The function of the spark plug is to generate the spark in the combustion using a high voltage communicated by the secondary. The spark plug provides two electrodes with a propery gap across which, high potential is discharged and spark is generated.

A sectional view of a conventional spark plug is shown in Fig. 3.39. It consists of a steel shell, an insulator, and two electrodes. The high voltage supply from secondary is given to the central electrode which is insulated with procelain. The other electrode is welded to the steel shell of the plug and thereby automatically grounded when the plug is fitted in the cylinder head of the engine. The electrodes are made of high nickel alloy to withstand severe corrosion and erosion to which they are subjected.

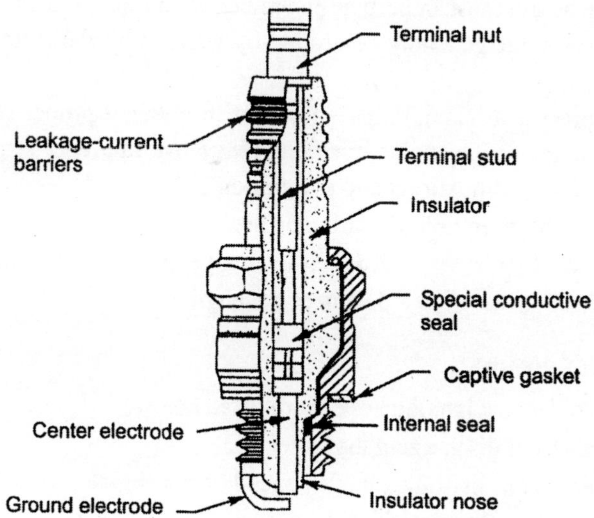

Fig. 3.39 Schematic of a typical spark plug.

The tips of central electrode and insulation are exposed to the burned gases. This results high thermal stresses and the insulator may crack. As the tips are subjected to high temperature (2000 – 25000°C), the heat must flow from the insulator and tip to the surrounding shell in order to cool the electrodes and prevent preignition.

The spark plugs are classified as hot plug and cold plug depending upon the temperature at the tip of the electrodes. The operating temperature of the tip depends upon the amount of heat transferred and it depends upon the path followed by the heat to flow. A cold plug has a short heat flow path where as hot plug follows a long flow path for the heat to flow as shown in Fig. 3.40.

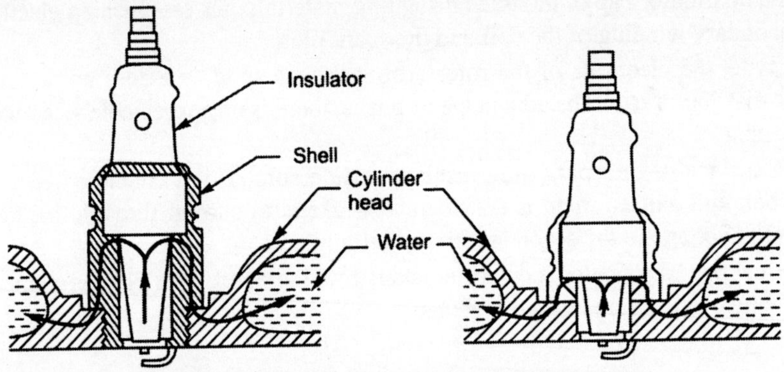

Fig. 3.40 Heat transfer path of hot and cold spark plug.

The hot plug is used to avoid cold fouling where combustion chamber temperatures are relatively low as during low power operation and continuous idling.

A spark plug which runs satisfactory, the temperature at cruising speed may run cool at idling speed and tips will be fouled by unburned carbon deposits or excess lubricating oil. The carbon deposits burns at 350°C where as lubricating deposits burn at 550°C. If the spark plug runs hot at idling speed to prevent carbon deposits, it may run too hot at high speed. This may cause undesirable preignition. If the plug runs above 800°C, then preignition generally occurs.

Insulator tip length is the most important parameter which controls the operating temperature. Therefore, the tip temperature is generally controlled by varying insulator tip position and electrode material.

It is necessary in practice to compromises in order to obtain a proper spark plug which would operate satisfactorily throughout the engine operating range. An improper spark plug has remained a major source of engine trouble as misfiring and preignition.

The factors affecting the operation of spark plug are :

1. Compression ratio
2. Electrode temperature
3. Speed of the engine
4. Mixture strength
5. Run of the engine

3.4.4 Magneto Ignition System

Some of the draw backs of battery ignition system are listed below:

1. There are chances of discharging the battery.
2. There are chances of misfiring at higher speed of the engine.
3. It requires complicated wiring.
4. There are chances of failure.
5. There are many mechanical complications in the operation of the system.

These difficulties experienced with battery ignition system can be avoided by using magneto ignition system.

Magneto Ignition System — Working

The arrangement of the Magneto Ignition system is shown in Fig. 3.41. The only difference between battery and Magneto system is that the battery is replaced by the rotating magnet. As the magneto rotates, the voltage and current is generated in the primary and circuit is completed passing the

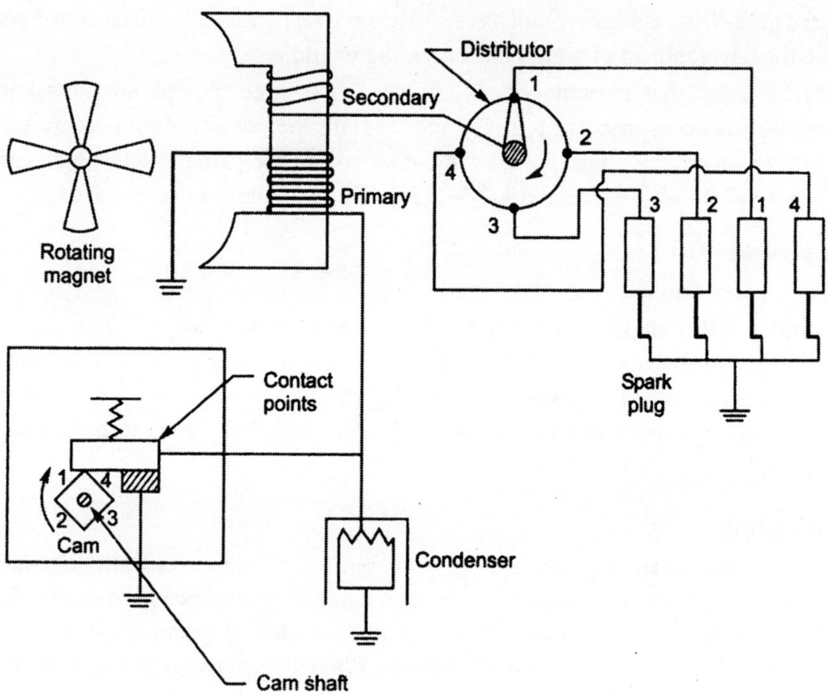

Fig. 3.41

current through the contact breaking point and through the ground. As the current passes through the primary coil through the contact breaker, the circuit is completed through ground. As the camshaft rotates, the cam 1 opens the contact breaker and interrupts the flow of current in the primary. This causes the decay in the magnetic field lines and cuts the lines of magnetic field in the secondary, and a high voltage is generated in the secondary. The process of generating the high voltage in the secondary is known as induction phenomenon. The voltage generation in the secondary depends upon the ratio of number of turns in the secondary and primary

$$e \propto \left(\frac{N_S}{N_P}\right)$$

Further consider the effect of firing sequence on engine cooling. When the first cylinder is fired its temperature increases. When the next cylinder fired is number 2, the portion of the engine between the cylinder number 1 and 2 gets overheated. If the third cylinder fired is cylinder number 3, this over-heating is shifted to the portion between the cylinders 2 and 3. Thus we that the task of the cooling system becomes very difficult because it is then, required to cool more at a place than at other places and this place too changing position with time. If we fire the third cylinder after the first the overheating problem can be mitigated.

Next, consider the flow of exhaust gases in the exhaust pipe. After firing the first cylinder, exhaust gases flow out to the exhaust pipe. If the next cylinder fired is the cylinder number 2, we find that before that gases exhausted from the first cylinder goes out to the exhaust pipe the gases exhausted from the second cylinder overtake them. This would require that the exhaust pipe be made bigger. Otherwise the back pressure in it would increase and the danger of back flow will arise. If instead of firing cylinder number 2, cylinder number 4 is fired then by the time the gases exhausted by cylinder 4 come into the exhaust pipe, the gases from cylinder 1 will have sufficient time to travel the distance between 1 cylinder and 4 and thus, the development of a high back pressure would be avoided.

It should be noted that to some extent all the above three requirements are conflicting and therefore, a trade-off is necessary. For four cylinder engines the possible firing orders are 1-2-4-3 and 1-3-4-2. The latter is in common use. For a six cylinder engine the firing orders which can be used are 1-5-3-6-2-4, 1-5-4-6-2-3, 1-2-4-6-5-3 and 1-2-3-6-5-4. The first one is in common use.

Other Firing Orders

For three cylinder engine 1-3-2

Eight cylinder in-line engine 1-6-2-5-8-3-7-4

Eight cylinder *V* shape engine 1-5-4-8-6-3-7-2, 1-8-4-3-6-5-7-2, 1-6-2-5-8-3-7-4, 1-8-7-3-6-5-4-2, 1-5-4-2-6-3-7-8. Cylinder No. 1 is taken from front of the in-line engines where as in *V* shape front cylinder on right side-bank is considered cylinder No. 1 for fixing H.T. leads according to engine firing order.

3.4.5 Ignition Timing

In order to obtain maximum power from an engine, the compressed mixture must deliver its maximum pressure at a time when the piston is about to commence its downward stroke. Since there is a time lag between occurrence of spark and the burning of the mixture the spark must take place before the piston reaches top dead centre on its compression stroke. Usually it should occur at about 15 degree *btdc*. If the spark occurs too soon combustion will take place before combustion stroke is completed and the pressure so developed will oppose the piston motion and thereby, reduce the engine power. If the spark occurs too late, the piston will have already completed a certain part of the expansion stroke before the pressure rise occurs and a corresponding amount of engine power will be lost.

The correct instant for the introduction of spark is mainly determined by the ignition lag. The ignition lag depends on many factors, such as compression ratio, mixture strength, throttle opening engine temperature, combustion chamber design and speed. Some of the important factors affecting the ignition timings are :

1. **Engine speed :** When an engine is rotating at 2000 rpm even one thousandth of a second represents 12° of crank rotation. So the spark must occur, say, 20° before the top dead centre for maximum power and economy. As the speed of the engine increases the combustion time

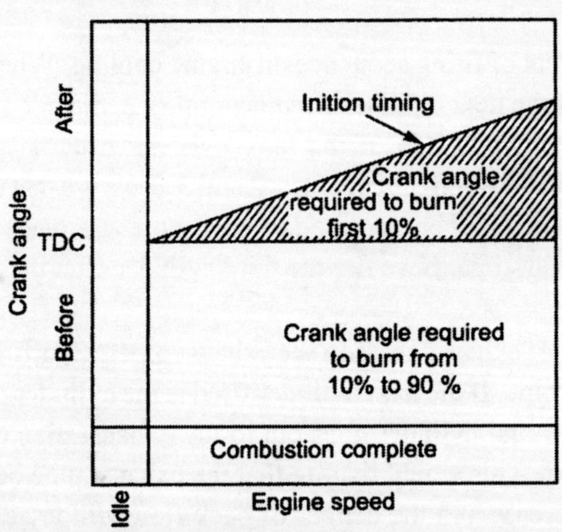

Fig. 3.42 Combustion time in terms of engine speed and degrees of crankshaft rotation.

increases in terms of crank angle degrees so the spark must advance accordingly, *i.e.,* the angle of advance must increase as the speed increases (see Fig. 3.42).

2. **Mixture strength :** In general, rich mixtures burn faster. So, as the mixture is made richer the optimum spark advance must be retarded, *i.e.,* the number of degress of crank angle before *tdc* is decreased and the spark occurs closer to *tdc.*
3. **Part load operation :** Part load operation of a spark ignition engine is affected by throttling. Due to throttling a smaller amount of charge enters the cylinder and the dilution due to residual gases is greater. In addition to this, higher air-fuel, ratio required for part load operation causes the combustion time to increase. So at part loads the spark advance must be increased.
4. **The type of fuel :** Ignition delay will depend on the type of the fuel used in engine. For maximum power and economy a slow burning fuel will need a higher spark advance than a fast burning fuel.

It is obvious from the above that the point in the cycle where the spark occurs must be regulated to ensure maximum power and economy at different speeds and loads and this must be automatic. Most of the engines are fitted with a mechanism which is integral with the distributor and automatically regulates the optimum spark advance to account for change of speed and load.

SOLVED QUESTIONS FROM UNIVERSITY EXAMINATIONS

■ **Example 3.10** Explain with neat sketch the theory of a simple Carburettor, and prove that

$$\frac{m_f}{m_a} = \frac{C_{df}}{C_{da}} \times \frac{A_f}{A_2}\sqrt{\frac{\rho_f}{\rho_a}}$$

[KU, Dec., 2006]

Solution : Refer to Art. 3.1.8.

■ **Exapmle 3.11** What is the function of a carburettor in a S.I. engine? Briefly explain with a neat sketch the operation of a simple float type carburettor? [Ku, Dec., 2006]

Solution : Refer to Art. 3.1.6.

■ **Example 3.12** (a) What is the function of a carburettor in S.I. engine? Briefly explain the working with neat sketch, the operation of a simple float type carburettor.

(b) A petrol engine consumes 7.5 kg of petrol per hr. The specific gravity of the fuel is 0.75. The air temperature is 25° C. The air fuel ratio is 15. The choke tube has a diameter of 22 mm. Calculate the diameter of the fuel jet of a simple carburettor. Top of the jet is 4 mm above the petrol level in the float chamber, take coeff. of discharge as 0.82 and 0.7 for air and fuel respectively. Atmospheric pressure $= 1.013$ bar. [Ku. Dec., 2005]

Solution : (a) Refer to Art. 3.1.6.

(b) Refer to Example 3.6.

■ **Example 3.13** With the help of neat sketch express the working of a simple carburettor. [MDU, May, 2008]

Solution : Refer to Art. 3.1.6.

■ **Example 3.14** Derive an expression for the calculation of air – fuel ratio in the carburetor. [MDU, Dec., 2009]

Solution : Refer to Art. 3.1.8.

■ **Example 3.15** Example construction details and working of elementary carburretor. [KU, Sept., 2009]

Solution : Refer to Art. 3.1.6

■ **Example 3.16.** What is Air-fuel mixture?

Explain-Rich Mixture

Stoichometric mixture

Lean Mixture [MDU, Dec., 2010]

Solution : Refer to Art. 3.1.2 (a) and Fig. 3.2. (a)

■ **Example 3.17** Explain constructional details and working of Carburettor used in S.I. engine. [KU, Dec, 2010]

Solution : Refer to Art. 3.1.6

■ **Example 3.18** What do you understand by air injection and solid injection system gnerally employed with the diesel engine? [MDU, May, 2008]

Solution : Refer to Art. 3.3.2.

■ **Example 3.19** What is S.I. injection? Name various solid injection systems. [MDU, Dec., 2008]

Solution : Refer to Art. 3.2 and Art. 3.3.2.

■ **Example 3.20** Explain the working of common rail system of injection. State its advantages and disadvantages. [MDU, Dec., 2008]

Solution : Refer to Art. 3.3.2.

■ **Example 3.21** State the essential requirements of a diesel injection system. [MDU, May, 2009]

Solution : Refer to Art. 3.3.1.

■ **Example 3.22** Describe the Battery ignition system with the help of neat sketch. What are the main disadvantages of a battery ignition system? How these can be overcome? [KU, Dec., 2007]

Solution : Refer to Art. 3.4.3.

■ **Example 3.23** (a) Explain ignition timing.

(b) Write a note on Spark plug. [KU, Dec., 2009]

Solution : (a) Refer to Art. 3.4.8.

(b) Refer to Art. 3.4.3(6).

■ **Example 3.24** What are the requirements of a good ignition system? [MDU, Dec., 2007]

Solution : Refer to Art. 3.4.1.

■ **Example 3.25** With the help of sketch, explain the electric circuit used for battery ignition in a four cylinder engine. Explain the function of each component. [MDU, Dec., 2007]

Solution : Refer to Art. 3.4.3.

■ **Example 3.26** Explain long and short plug through neat sketch and state where and when it is preferred. [MDU, May, 2009]

Solution : Refer to Art. 3.4.3 (6).

■ **Example 3.27** Sketch a schematic diagram of a spark plug. Explain its function and state causes of failure. [MDU, May, 2009]

Solution : Refer to Art. 3.4.3 (6).

■ **Example 3.28** State the essential requirements of an ignition system. [MDU, Dec., 2009]

Solution : Refer to Art. 3.4.1.

■ **Example 3.29** What is the function of a carburettor in an S.I. engine? Briefly explain with a neat sketch the operation of a simple flood type carburettor. [KU, Dec., 2007]

Solution : Refer to Art. 3.1 and Art. 3.1.6.

SHORT ANSWER TYPE QUESTIONS

1. What are the essential requirements for an ideal carburettor?
2. What do you understand by compensation? Explain clearly.
3. Explain with a neat diagram how the power and efficiency of a S.I. engine vary with A : F ratio.
4. Why rich mixtures are required for starting and during idling of an engine? Explain with the help of a neat sketch.
5. Draw a simple type of carburettor and explain its working.
6. What do you understand by stoichiometric A : F ratio?
7. What are the requirements of a good injection system?
8. What is solid injection? What are its advantages over air-injection system?
9. What are the basic requirements of an ideal ignition system?
10. Explain the working of a battery ignition system with the help of a neat sketch.
11. What are the main drawbacks of a battery ignition system? How these can be overcome?
12. Explain the working of a magneto ignition system with the help of a neat sketch.
13. Compare battery and magneto ignition systems.
14. Describe the working of an electronic ignition system.

EXERCISES

1. The venturi of a single jet carburettor has throat diameter of 75mm and fuel nozzle diameter of 5 mm. Find the $A : F$ ratio, of the mixture supplied to the engine for the following date :

$$C_{da} = 0.93, C_{df} = 0.68, \Delta p \text{ across venturi} = 0.15 \text{ bar},$$

$\rho_a = 1.29 \text{ kg/m}^3$, $\rho_f = 720 \text{ kg/m}^3$. Neglect nozzle lip and air compressibility effect. [**Ans.** 13.3 : 1]

2. A 4-cylinder, 4-stroke petrol engine of 82.5 mm diameter and 115 mm stroke has 80% volumetric efficiency, when running at 3000 rpm. The pressure head causing the flow is 11.03 cm of Hg. Determine :

(a) The size of the venturi, and

(b) The fuel nozzle diameter if the $A:F$ ratio of mixture supplied is 14 : 1.
Take $C_{da} = 0.84, C_{df} = 0.7, \rho_a = 1.29\,\text{kg/m}^3, \rho_f = 700\,\text{kg/m}^3$. **[Ans.** 22.3 mm, 4.15 mm]

3. The diameter of a simple venturi is 75 mm and fuel nozzle diameter is 5 mm. Nozzle lip = 5 mm, and $C_{da} = 0.85$,

$$C_{df} = 0.7, \rho_a = 1.29\,\text{kg/m}^3, \text{ and } \rho_f = 720\,\text{kg/ m}^3.$$

Determine :

(a) $A:F$ ratio supplied by the carburettor is the pressure drop causing air flow is 0.14 bar. Neglect nozzle lip.

(b) $A:F$ ratio when nozzle lip is considered.

4. A simple carburettor filled on a S.I. engine consumes 6.4 kg /h. The nozzle lip = 3 mm, $d_f = 1.27$ mm, $C_{da} = 0.8$ and $C_{df} = 0.6$. The atmospheric air pressure and temperature are 1.013 bar and 15.5°C and $\rho_f = 760$ kg/m^3. The $A:F$ ratio supplied by the carburettor = 15 : 1. Determine :

(a) The venturi diameter, and (b) Drop in venturi pressure.

[Ans. 21.5 mm, 37 cm of H_2O]

5. A single jet carburettor is to be designed to give $A:F$ ratio of 15 : 1 and fuel supply of 24.5 kg/h. The atmospheric air pressure and temperature are 1.013 bar and 288 K. Calculate :

(a) The venturi diamter if the air velocity is limited to 97.5 m/s, and

(b) The fuel nozzle diameter.

Take $C_{da} = 0.84$ and $C_{df} = 0.66$ **[Ans.** 33.8 mm, 2 mm]

6. A 4-stroke, 8-cylinder C.I. engine develops 500 Kw at 1200 spm and consumes 0.175 kg/hwh of fuel whose specific gravity is 0.89. Determine the diameter of a single orifice injector if the injection pressure is 160 bar and combustion chamber preassue is 40 bar. The period of injection is 30° crank rotation. Take $C_{df} = 0.9$.

7. A 4-stroke, 6-cylinder oil engine, 115 mm diameter and 140 mm stroke operates with $A:F = 16:1$. The suction condition in the engine are 1 bar and 300 K. The volumetric efficiency of the engine is 76%. Determine the maximum amount of fuel supplied per hour to the engine running at 1500 apm. The injection pressure is 125 bar and combustion chamber pressure is 40 bar. The fuel is supplied for 20° rotation of the crank. Also calculate the diemeter of the orifice assuming each nozzle has single orifice. Take

$\rho_f = 760\,\text{kg/m}^3$ and $C_{df} = 0.64$. **[Ans.** 21.7 kg/h, 1.76 mm]

❑❑❑

Chapter 4

Combustion in I.C. Engines

Inside This Chapter

4.1 INTRODUCTION

The combustion is defined as a rapid chemical reaction between the H_2 and C with an oxygen in the air and liberates energy in the form of heat.

It is absolutely essential to burn the fuel supplied completely for the economical working of the engine and also for the safety of the engine. Therefore, the mixture supplied to the engine should possess required A : F ratio otherwise, combustion cannot be initiated or if initiated it cannot be sustained. In addition to this, there must be some means to initiate the combustion and the generated flame should be able to propagate through the mixture and burn the mixture completely.

It is a known fact that the fuel with any A : F ratio cannot be burned. There is a igntion limit for any fuel to start the combustion and sustain it till the complete fuel burns by the flame generated with spark plug. In addition to this, the temperature of the mixture to initiate the ignition is equally important. It is also known that the flame will propagate if the temperature of the burnt gases exceeds 1500 K for S.I. engine (gasoline) fuels. The ignition limits of hydrocarbon fuel when temperature of mixture reaches to 1500 K are shown in Fig. 4.1.

The upper and lower limit of A : F ratio for ignition depends upon the temperature of a particular fuel. The limit becomes wider at higher temperatures because of higher reactin rate and higher thermal

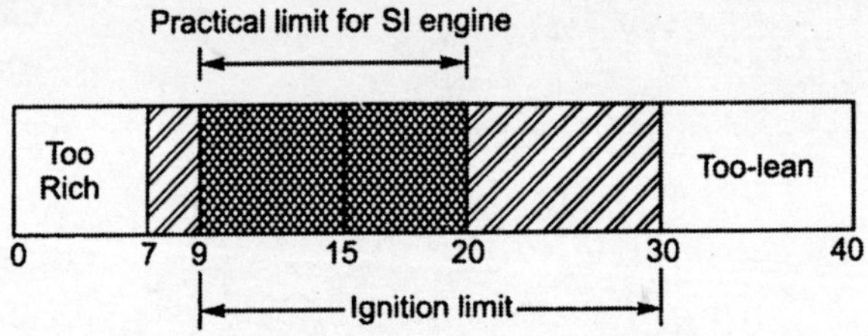

Fig. 4.1 Ignition limit of Hydro-carbon fuels at 1500 K.

diffusivity coefficients of the mixture. Therefore, it is very essential to see that the A : F ratio of the mixture supplied to the engine should lie in the practical limit as shown in Fig. 4.1.

4.2 STAGES OF COMBUSTION IN S.I. ENGINES

It is assumed that the heat is added instantaneously at constant volume in the ideal air-cycle of S.I. engines. To achieve this, the burning of the fuel in the S.I. engine should be instantaneous. In actual engines, combustion occurs over a finite period of time as the flame starting around the spark plug has to propagate through the entire mixture of air and fuel.

A high intensity spark is produced for few degrees of crank angle before TDC at the end of the compression stroke. The function of spark is to povide a source of heat to the combustible mixture. The temperature of the combustible mixture surrounding the spark is raised to a sufficient temperature level to start the combustion. The flame propagation is continued through the combustible mixture provided the release of heat from the initial sources is sufficient to heat the adjoining portion of the mixture to a temperature at which the heat from reacton is sufficient to overcome the heat losses. The heat liberated by burning portion 'flame front' prepares the adjacent portion of the unburned charge for the combustion reaction.

It has been observed experimentally that there is a certain time interval between the instant of spark and the instant when there is a noticeable rise in pressure due to combustion. This means that the combustion starts some time after the spark. This time lag is known as 'Ignition Lag'. Ignition lag is a time interval in the process of chemical reaction during which the molecules of the fuel get heated up to the self-ignition temperature, get ignited and produce a self-propagating nucleus of flame.

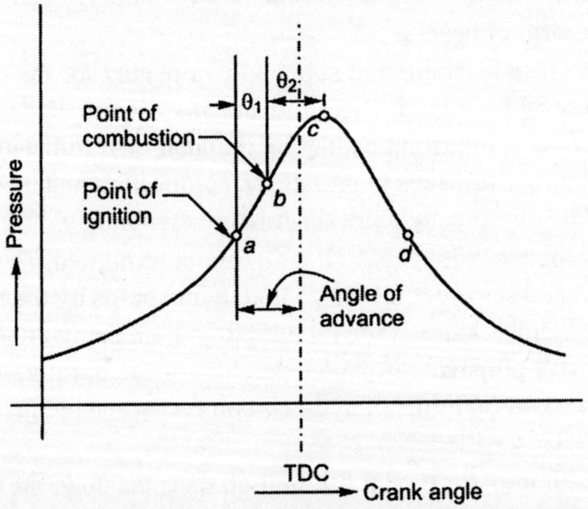

Fig. 4.2 Crank angle v's pressure variation in S.I. engine.

The pressure variation in the S.I. engine combustion chamber during the crank rotation is shown in Fig. 4.2. This is really an unfold *p-v* diagram.

The ignition is timed to take place at the point '*a*' but the burning commences only at the point '*b*'. The time interval between the points '*a*' and '*b*' is known as 'Ignition Lag'. This ignition lag is generally expressed in terms of crank angle and this is given by θ_1 as shown in Fig. 4.2. The time required for the crank to rotate through an angle θ_2 is known as combustion period during which the propagaton of flame takes place.

Many times, the combustion is not complete at the point '*c*' (when the mixture is too rich) and the fuel may burn after the point *c* during the expansion stroke. This burning of fuel after the point of maximum pressure is known as 'After Burning'. This is not desirable.

The major disadvantages of the ignition lag is to reduce the power developed. If the ignition lag is long, the peak pressure occurs during the expansion stroke, and therefore full advantage of expansion is not achieved. The time lag may vary from 10° to 50° according to the type of the fuel used and several other factors affecting the time lag. To maintain the 'time lag' constant at variable speed of the engine, it is necessary to change the angle of advance. The angle of advance must increase with increase in speed. This is commonly done in automobile engines by an automatic advancing mechanism.

All considerations should be taken into account in the design of the combustion chamber and selecting the fuel used to reduce the *Ignition Lag*.

The theoretical diagram of combustion is shown in Fig. 4.3 (a).

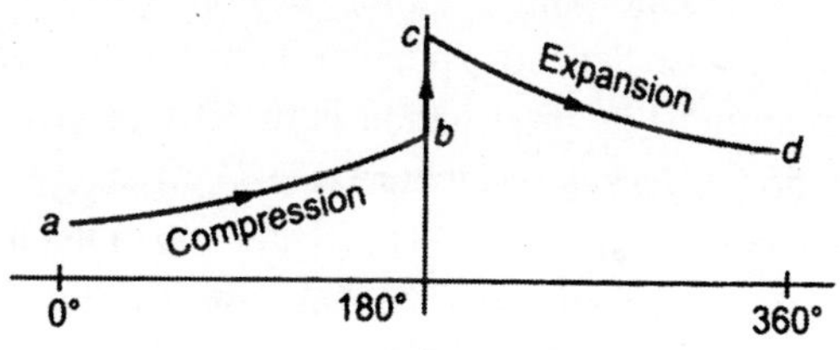

Fig. 4.3 (a) Ignition limit of hydro-carbhon fuels at 1500 K.

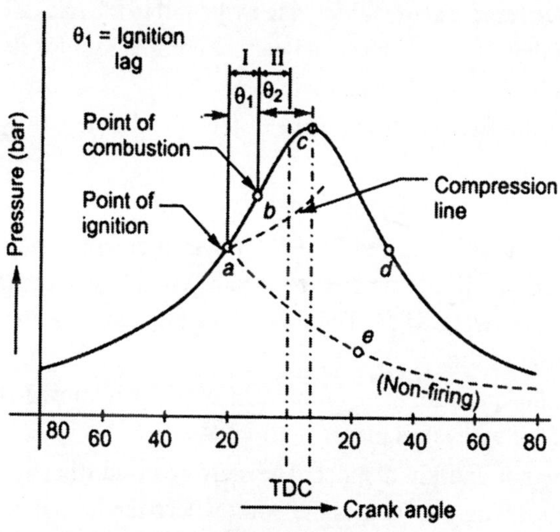

Fig. 4.3 (b)

But actual process differs from theoretical as instant combustion is not possible as shown by *bc*. The combustion to start takes small time after giving the spark as the surrounding mixture is to be heated upto ignition temperature and then formed nucleus of flame starts propagating through the surrounding mixture.

The pressure variation in the engine with respect to crank-angle is shown in Fig. 4.3 (b).

There are mainly three stages of combustion in S.I. engines as shown in Fig. 4.3 (b).

1. First Phase

This phase is considered between the point of ignition and point of combustion. As shown in figure, ignition is timed at point '*a*' and combustion starts at point '*b*'. The time interval between the points '*a*' and '*b*' is known as *Ignition lag*. The ignition lag is generally expressed in terms of crank angle (θ_1). This period of *Ignition lag* is very small and lies between 0.00015 to 0.002 seconds. An ignition lag of 0.002 seconds correspond to 35° crank rotation when the engine is running at 3000 RPM and crank angle required (which is known as angle of advance) increases with an increase in engine speed.

2. Second Phase

Once the flame formed at the point '*b*', it shoul be self-sustained and must be able to propagate through the mixture. This is possible when the rate of heat generaton (Q_g) by burning the surrounding mixture of the flame nucleus must be higher than the heat lost (Q_l) by the flame to the surrounding. As the difference between ($Q_g - Q_l$) is higher, the rate of flame propagation is higher and complete combustion will occur earliest possible which is most desirable requirement of combustion in S.I. engines. The propagation of flame also depends upon the flame temperature as well as temperature and density of the surrounding mixture as its propagation is directly proportional to these factors. Weak spark and low compression ratio (as density of mixture is less) gives low propagation of the flame.

After the point '*b*', the flame propagation is abnormally low at the beginning as heat loss is more than heat generated. Therefore the pressure rise is also slow as mass of mixture burned is small. Therefore, it is necessary to provide angle of advance 30 to 35° if the peak pressure is to be attained 5-10° after TDC.

After the point '*c*', the pressure starts falling due to the fall in the rate of heat release when the flame reaches the wall in the last part of combustion and cannot compensate for its fall due to gas expansion, and heat transfer to the walls.

The time required for the flame to travel 95% of the chamber length with respect to speed of the engine is shown in Fig. 4.4.

When $\theta_1 = 30°$ and $A:F = 13$

It is obvious, the crank angle required for 95% travel increases with increasing RPM (the time available is decreased), therefore, if the combustion is to be completed at point '*c*', the angle of advance (θ) must be increased with increasing RPM. The flame speed increases with increasing RPM because of increase in turbulence of the mixture.

The time required for the crank to rotate through an angle θ_2 is known as combustion period during which the propagation of the flame takes place.

The stage-I and II are not entirely distinct. The starting point of stage-II is measurable as rise in pressure can be seen on $p-\theta$ diagram. This is the point where the line of combustion departs from the line of compression.

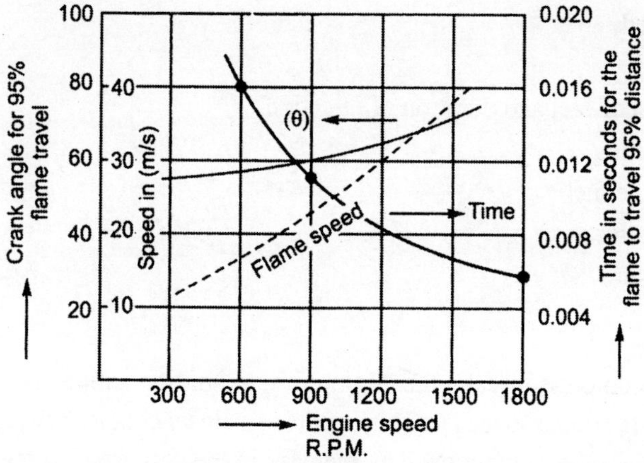

Fig. 4.4

3. Third Phase

Although the point '*c*' represents the end of flame travel, it does not assure the complete combustion of fuel. In this case, the comustion still continues after attaining the peak pressure also and this combustion is known as *After Burning.* This is continued throughout the expansion stroke. This generally happens when the rich mixture is supplied to the engine.

4.3 IGNITION LAG (OR DELAY) IN S.I. ENGINES

The ignition delay period is the phase during which some fuel has already been admitted but has not yet ignited. The fuel does not ignite immediately upon injection into the combustion chamber. There is a definite period of inactivity between the time when the first droplet of fuel hits the hot air in the combustion chamber and the time it starts through the actual burning phase. This period is known as the ignition delay period. In Fig. 4.5 the delay period is shown on pressure crank angle (or time) diagram between points '*a*' and '*b*'. Point '*a*' represents the time of injection and point '*b*' represents the time at which the pressure curve first separates from the motoring curve.

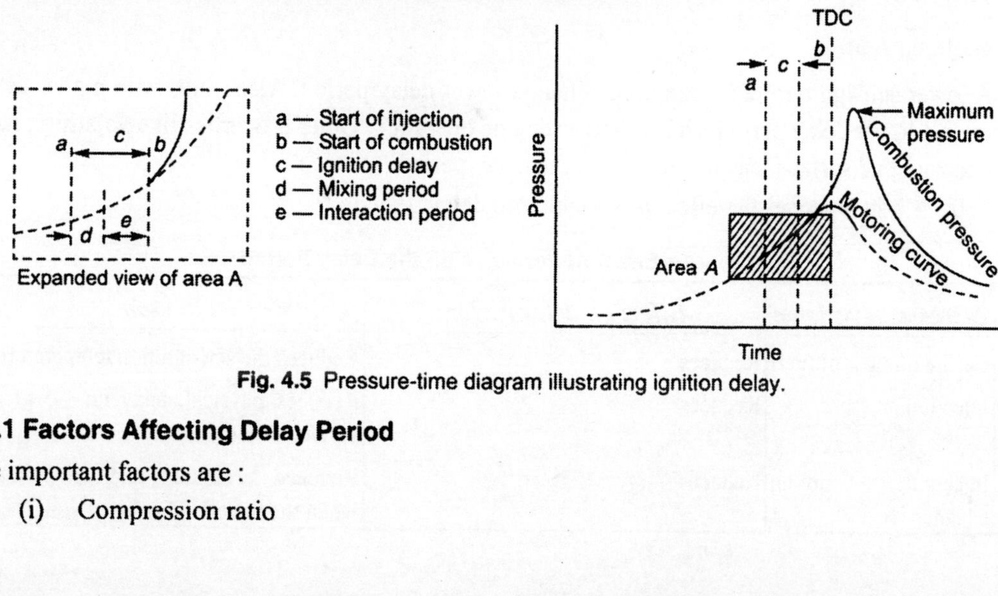

Fig. 4.5 Pressure-time diagram illustrating ignition delay.

4.3.1 Factors Affecting Delay Period

The important factors are :

(i) Compression ratio

(ii) Engine speed
(iii) Output
(iv) Atomization of fuel and duration of injection
(v) Injection timing
(vi) Quality of the fuel
(vii) Intake temperature
(viii) Intake pressure

(i) Compression ratio

With increase in compression ratio the minimum auto ignition temperature of a fuel decreases due to increased density of the compressed air. This results in a closer contact between the molecules of fuel and oxygen reducing the time of reaction. The increase in the compression temperature as well as the decrease in the minimum auto ignition temperature decreases the delay period. The peak pressure during the combustion proces is only marginally affected by the compression ratio (because delay period is shorter with higher compression ratio and hence the pressure rise is lower).

One of the practical disadvantages of using a very high compression ratio is that the mechanical efficiency tends to decrease due to increase in weight of the reciprocating parts. Therefore, in practice the engine designers always try to use a lower compression ratio which helps in easy cold starting and light load running at high speeds.

(ii) Engine speed

The delay period decreases with increase in engine speed in a variable speed operation with a given fuel. With increase in engine speed, the loss of heat during compression decreases, resulting in the rise of both the temperature and pressure of the compressed air thus reducing, the delay period.

(iii) Output

With an increase in engine output the air-fuel ratio decreases, operating temperatures increase and hence delay period decreases.

(iv) Quality of fuel

A lower self-ignition temperature results in a lower delay period. Also, fuels with higher cetane number give lower delay period, and smoother engine operation. Other properties like volatility, latent heat, viscosity and surface tension also affect the delay period.

Table 4.1 summarises the effect of variables on delay period

Table 4.1 : Effect of Variables on the Delay Period

	Increase in variable	*Effect on delay period*	*Reason*
1.	Cetane number of fuel	Reduces	Reduces the self-ignition temperature.
2.	Injection pressure	Reduces	Reduces physical delay due to greater surface-volume ratio.
3.	Injection timing advance	Reduces	Reduced pressures and temperatures when the injection begins.

4.	Compression ratio	Reduces	Increases air temperature and pressure and reduces autoignition temperature.
5.	Intake temperature	Reduces	Increases air temperature.
6.	Jacket water temperature	Reduces	Increases wall and hence air temperature.
7.	Fuel temperature	Reduces	Increases chemical reaction due to better vaporization.
8.	Intake pressure (supercharging)	Reduces	Increases density and also reduces autoignition temperature.
9.	Speed	Increase in terms of crank angle. Reduces in terms of milliseconds	Reduces loss of heat.
10.	Load (fuel-air ratio)	Decreases	Increases the operating temperature.
11.	Engine size	Decreases in terms of crank angle. Little effect in terms of milliseconds.	Larger engines operate normally at low speeds.
12.	Type of combustion chamber	Lower for engines with precombustion chamber.	Due to compactness of the chamber.

4.4 FACTORS AFFECTING THE FLAME PROPAGATION

After the ignition, the rate of flame propagation affects the combustion process in S.I. engines.

Better the propagaton, higher is the comustion efficiency and higher the economy. The propagation depends upon the flame velocity. Unfortunately, the flame velocities for most of the fuels range between 10 to 30 m/s. Therefore all steps should be taken in the design and operation of the engine that the flame velocity should be as maximum as possible.

The study of flame propagation rate and the factors affecting the same is important for two reasons.

1. Its effect on the rate of pressure rise in the cylinder.
2. Its effect in connection with abnormal combustion (knocking).

The factors which affect the flame propagation are discussed below :

1. **A : F ratio :** The mixture strength influences the rate of combustion and amount of heat generated. The maximum flame speed for all hydrocarbon fuels occurs at nearly 10% rich miture. Flame speed is reduced both for lean as well as for rich mixture.

   Lean mixture releases less heat resulting lower flame temperature and lower flame speed. Very rich mixture results incomplete combustion ($C \rightarrow CO$ instead of CO_2) and also results in production of less heat and flame speed remains low.

   The effects of A : F ratio on p-v diagram and p-θ diagram are shown in Fig. 4.6.

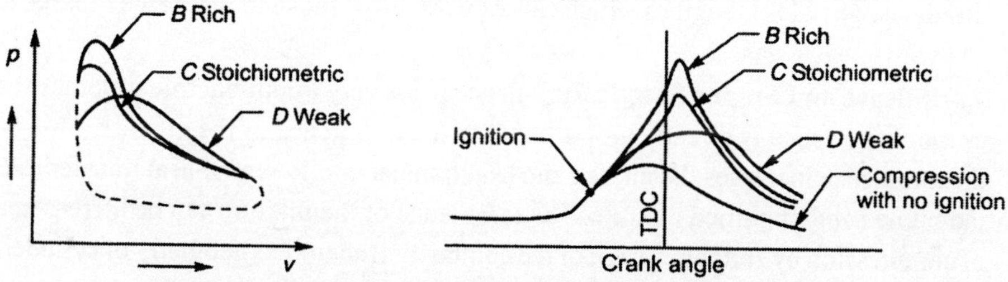

Fig. 4.6 Indicator diagrams for stoichiometric and weak mixtures.

2. **Compression Ratio (r_c) :** The higher compression ratio increases the pressure and temperature of the mixture and also decreases the concentration of residual gases. All these factors which are in favour reduces the ignition lag and helps to speed up the second phase of combustion. The maximum pressure of the cycle as well as mean effective pressure of the cycle will increase with increase in r_c. Fig. 4.7 shows the effect of compression ratio on pressure (indirectly on the speed of combustion) with respect to crank angle for same A : F ratio and same angle of advance.

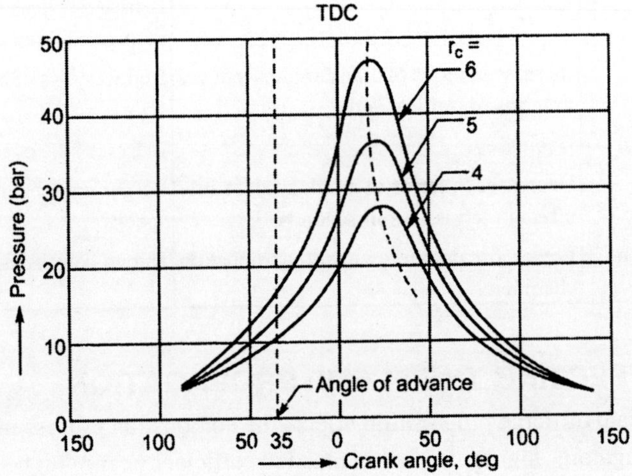

Fig. 4.7

Higher compression ratio increases the surface to volume ratio and thereby increases the part of the mixture which afterburns in the third phase.

3. **Load on Engine :** With increase in load, the cycle pressures increase and the flame speed also increases.

   In S.I engine, the power developed by an engine is controlled by throttling. At lower load and higher throttle, the initial and final pressure of the mixture after compression decrease and mixture is also diluted by the more residual gases. This reduces the flame propagation and prolongs the ignition lag. This is the reason, the advance mechanism is also provided with change in load on the engine. This difficulty can be partly overcome by providing rich mixture at part loads but this definitely increases the chances of after-burning. The after burning is prolonged with richer mixture.

   In fact, poor combustion at part loads and necessity of providing richer mixture are the main disadvantages of S.I. engines which causes wastage of fuel and discharge of large amount of CO with exhaust gases.

4. **Turbulence and Engine Speed :** Turbulence plays very important role in combustion of fuel as the flame speed is directly proportional to the turbulence of the mixture. This is because, the turbulence increases the mixing and heat transfer coefficient or heat transfer rate between the burned and unburned mixture. The turbulence of the mixture can be increased at the end of compression by suitable design of the combustion chamber (geometry of cylinder head and piston crown).

Insufficient turbulence provides low flame velocity and incomplete combustion and reduces the power output. But excessive turbulence is also not desirable as it increases the combustion rapidly and leads to detonation. Excessive turbulence causes to cool the flame generated and flame propagation is reduced.

Moderate turbulence is always desirable as it accelerates the chemical reaction, reduces ignition lag, increases flame propagation and even allows weak mixture to burn efficiently.

The turbulence of the mixture increases with an increase in engine speed. Therefore, the effects of increase in engine speed are similar to increase in turbulence.

5. **Other Factors :** Among the other factors, the factors which increase the flame speed are supercharging of the engine, spark timing and residual gases left in the engine at the end of exhaust stroke.

   The air humidity also affects the flame velocity but its exact effect is not known. Anyhow, its effect is not large compared with A : F ratio and turbulence.

4.5 PHENOMENA OF KNOCKING IN S.I. ENGINES

In a SI engine combustion is initiated between the spark plug electrodes and then spreads across the combustible mixture. A definite flame front which separates the fresh mixture from the products of combustion travels from the spark plug to the other end of the combustion chamber. Heat release due to combustion increases the temperature and pressure of the burned part of the mixture above those of the unburned mixture. In order to effect pressure equalization the burned part of the mixture will expand and compress the unburned part adiabatically thereby increasing the pressure and temperature of unburned part further. If the temperature of unburned mixture exceeds the self-ignition temperature of the fuel and remains at or above this temperature during the ignition lag, spontaneous ignition (or auto-ignition) occurs at various pin-point locations. This phenomenon is called knocking.

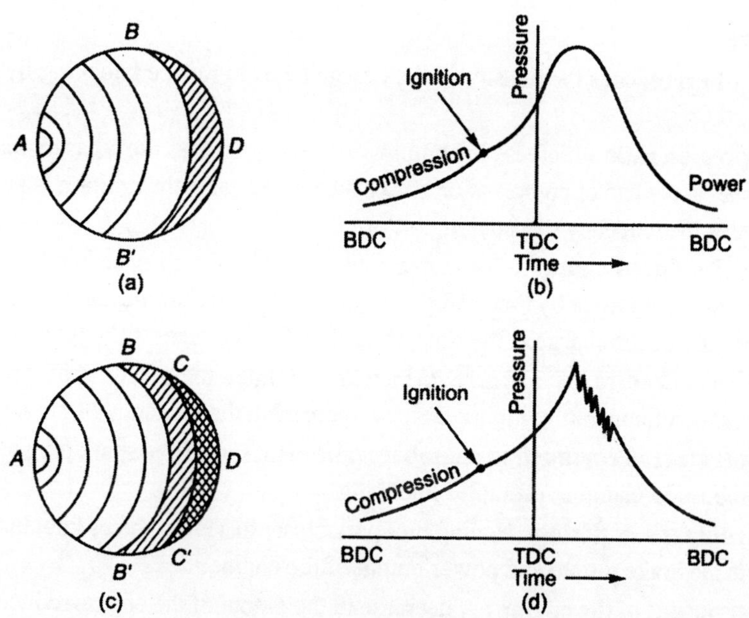

Fig. 4.8 Normal and abnormal combustion.

The phenomenon of knocking may be explained by referring to Fig. 4.8 (a) which shows the cross-section of the combustion chamber with flame advancing from the spark plug location A without knock whereas Fig. 4.8 (c) shows the combustion process with knock. In the normal combustion, the flame travels across the combustion chamber from A towards D. The advancing flame front compresses the end charge *BB′ D* farthest from the spark plug, thus raising its temperature. The temperature is also increased due to heat transfer from the hot advancing flame front and preflame oxidation in the end charge. If the temperature of the end charge does not reach its self-ignition temperature, the charge would not auto-ignite and flame will advance further and consume the charge *BB′ D*. This is the normal combustion process which is illustrated by means of the pressure-time diagram in Fig. 4.8 (b).

If the end charge *BB′ D* reaches its auto-ignition temperature and remains for some length of time equal to the time of preflame reactions the charge will auto-ignite, leading to knocking, combustion. In Fig. 4.8 (c), it is assumed that when the flame has reached the position *BB′*, the charge ahead of it has reached critical auto-ignition temperature. During the preflame reaction period if the flame front could move from *BB′* to only *CC′* then the charge ahead of *CC′* would auto-ignite. Because of the auto-ignition, another flame front starts travelling in the opposite direction to the main flame front. When the two flame fronts collide, a severe pressure pulse is generated. The gas in the chamber is subjected to compression and rarefaction along, the pressure pulse until pressure equilibrium is restored. This disturbance can force the walls of the combustion chamber to vibrate at the same frequency as the gas (≈ 5000 Hz). The pressure-time trace of such a situation is shown in Fig. 4.8 (d).

4.5.1 Factors Influencing Knocking

The factors influencing knocking are :

1. Density factors
2. Time factors
3. Composition factors
4. Temperature factors

1. Density factors

Any factor which reduces the density of the charge tends to reduce knocking by providing lower energy release.

(i) Compresson ratio : Increase in compression ratio increases the pressure and temperature of the gases at the end of compression stroke which decreases the ignition lag of the end gas and increases the tendency of knocking.

(ii) Mass of induced charge : An increase in mass of induced charge into the cylinder of an engine by throttling or by increasing the amount of supercharging increases both temperature and density of the charge at the time of ignition. This increases the tendency for knocking.

(iii) Inlet temperature of mixture : Increase in inlet temperature of mixture makes the compression temperature higher thereby, increasing the tendency of knocking.

(iv) Temperature of combustion chamber walls : Hot spots (like spark plug and exhaust valve) in combustion chamber promotes knocking.

(v) Retarding spark timing : Having the spark closer to TDC reduces knocking. However, this effects the brake torque and power output of the engine.

(vi) Power output of the engine : A decrease in the output of the engine reduces the tendency to knock.

2. Time Factors

The following factors reduce the possibility of knocking :

(i) **Turbulence :** Turbulence increase the engine speed and reduces the time available for the end charge to attain auto-ignition conditions thereby decreasing the tendency to kncok.

(ii) **Engine speed :** An increase in engine speed increases turbulence and decreases the tendency to knock.

(iii) **Flame travel distance :** Shortening the time required for the flame front to traverse the combustion chamber reduces the tendency for knocking. Flame travel distance is governed by combustion chamber size and spark plug position.

(iv) **Engine size :** A large engine (combustion chamber size) has a greater tendency for knocking as the flame requires a longer time to travel across the combustion chamber.

(v) **Combustion chamber shape :** The combustion chamber should be such which promotes turbulence to reduce knocking. Spherical shaped chambers minimizes knocking tendency.

(vi) **Location of spark plug :** A centrally located spark plug are multiple spark plugs minimizes flame travel time and reduces knocking.

3. Composition Factors

(i) **Fuel :** Air ratio-F/A ratio affects the reaction time or ignition delay. When the mixture is nearly 10% richer than stoichiometric (F/A ratio = 0.08) ignition lag of the end gas is minimum and velocity of flame propagation is maximum and knocking tendency is maximum. A rich mixture is effective in decreasing the knocking tendency due to longer delay period and lower temperature of compression.

(ii) **Octane rating of fuel :** Higher the octane number, lesser is the tendency for knocking. Paraffin series have the maximum and aromatic series the minimum tendency to knock. The knocking characteristics of a fuel can be decreased by adding small amounts of additives called dopes.

(iii) **Humidity of air :** Increasing humidity of atmosphere air decrease the tendency to knock.

4. Temperature Factors

Increasing the temperature of the unburned mixture by any factor in design or operation will increase the possibility of knocking in SI engine. The temperature of the unburned mixture is increased by the following factors :

(i) *Raising the compression ratio* : For a given engine setting and fuel there will be a critical compression ratio above which knocking occurs. This compression ratio is called the highest useful compression ratio (HUCR).

(ii) Supercharging.

(iii) Raising the inlet temperature.

(iv) Raising the coolant temperature.

(v) Increasing the load (opening the throttle).

(vi) Raising the temperature of the cylinder and combustion chamber walls.

(vii) Advancing the spark timing.

The sumary of variable affecting knocking in SI engines is given in Table 4.2

Table 4.2 : Summary of Variables Affecting Knock in an SI Engine

Increase in variable	*Major effect on unburned reduce charge*	*Action to be taken to knocking*	*Can operator usually control?*
1. Compression ratio	Increases temperature and pressure	Reduce	No
2. Mass of charge inducted	Increases pressure	Reduce	Yes
3. Inlet temperature	Increases temperature	Reduce	In some cases
4. Chamber temperature	Increases temperature	Reduce	Not ordinarily
5. Spark advance	Increases temperature and pressure	Retard	In some cases
6. A/F ratio	Increases temperature and pressure	Make very rich	In some cases
7. Turbulence	Decreases time factor	Increase	Somewhat (through engine speed)
8. Engine speed	Decreases time factor	Increase	Yes
9. Distance of flame travel	Increases time factor	Reduce	No

4.5.2 Methods for Suppressing Knocking

(i) **Proper location of spark Plug :** A compact combustion chamber with proper central location of the spark plug reduces the path of the flame front travel from the spark plug to the remotest part of the combustion chamber, and thereby, reduces the tendency of knocking. The same result could be obtained by multi-fuel injection system.

(ii) **Proper selection of material for piston and cylinder head :** The end charge cooling is much better when piston and cylinder are made of aluminium alloys as they have good thermal conductivity. A better cooling of end charge reduces the tendency for knocking.

(iii) **Injecting water into the intake manifold :** Injection of water in the cylinder reduces the temperature of the end gas and increases the delay, period, thereby decreasing the tendency to knock.

(iv) **Retarding the spark timings :** By retarding the spark timing, the peak pressures are reached only during the power stroke and are of lower magnitude. This reduces knock.

(v) **Extremely rich or lean mixture :** By using extremely rich or lean mixture the flame temperature can be kept low thus eliminating considerably the tendency of knock.

(vi) **Squish recesses in combustion chamber :** The provision of squish recess in combustion chamber cools the last position of the charge and reduces the tendency of knock.

The characteristics to reduce knocking are given in Table 4.3.

Table 4.3 : Characteristics Tending to Reduce Detonation or knock

Sl. No.	*Characteristics*	*SI Engines*	*CI Engines*
1.	Ignition temperature of fuel	High	Low
2.	Ignition delay	Long	Short

3.	Compression ratio	Low	High
4.	Inlet temperature	Low	High
5.	Inlet pressure	Low	High
6.	Combustion wall temperature	Low	High
7.	Speed rpm	High	Low
8.	Cylinder size	Small	Large

4.5.3 Effects of Knocking

1. **Mechanical damage :** Knocking creates very high pressure wave of large amplitude. This increases the rate of wear of almost all mechanical parts like piston, cylinder head and valves. The engine parts are also subjected to very high temperatures due to auto-ignition and the piston is damaged by overheating.
2. **Noise :** When the intensity of knock is high, a loud pulsating noise is created because of high intensity pressure wave which vibrates back and forth across the cylinder. The high vibrating motion of the gases causes crankshaft vibrations and engine runs rough.
3. **Heat transfer rate :** The heat transfer rate increases as more heat is lost to the coolant.
4. **Power output :** The power output slightly decreases due to rapid burning of last part of the charge.
5. **Pre-ignition :** Pre-ignition is the ignition of the charge in the absence of spark as it comes in contact with hot surface. This adds negative work by the piston during compression which increases high peak pressure and temperature in the cylinder.

4.6 COMBUSTION CHAMBERS FOR S.I. ENGINES

The design of combustion chamber involves the shape of the combustion chamber, location of the spark plug and dispotion of inlet and exhaust valves.

4.6.1 Basic Requirements of a Good Combustion chamber

The three main requirements of a SI engine combustion chamber are : high power output with minimum octane requirement, high thermal efficiency, and smooth engine operation.

(a) Higher power ouput requires the following :

1. High compression ratio.
2. Small or no excess air.
3. No dead pockets.
4. An optimum degree of turbulence.
5. High volumetric efficiency.

(b) High thermal efficiency require the following :

1. High compression ratio.
2. Small heat loss during combustion.
3. Good scavenging of exhaust gases.

(c) Smooth engine operation requires the following :

1. Moderate rate of pressure rise during combustion.
2. Absence of detonation : Compact combustion chamber, proper location of spark plug and exhaust valve, and satisfactory cooling of spark plug points.

4.6.2 Combustion Chamber Design Principles

1. To achieve high volumetric efficiency the largest possible inlet valve should be accommodated with ample clearance around the valve heads.
2. To prevent detonation the length of flame travel from the spark plug to the farthest point in the combustion space should be as short as possible.
3. To reduce the possibility of detonation the exhaust valve should be near the spark plug.
4. The exhaust valve should be kept small due to its hot surface, but to compensate for this, high lift should be employed.
5. Correct amount of turbulence should be created in the combustion chamber by properly positioning the inlet valve. Preferably turbulence should be created by squish.
6. The shape of combustion chamber should be such that the largest mass of the charge burns as soon as possible after ignition with progressive reduction in the mass of charge burnt towards the end of combustion.
7. To ensure high thermal efficiency and less air pollution the surface-volume ratio of chamber should be minimum in the beginning.
8. In the end gas region, surface-volume ratio should be large to achieved good cooling in the 'detonation zone'.
9. The spark plug should be so positioned so that it will be scoured of any residual exhaust products by the incoming charge.
10. The exhaust valve head should be well cooled as it is the hottest region of combustion chamber.
11. The sparking plug points should be sufficiently cooled to avoid pre-ignition effects.
12. There should be good scavenging of exhaust gases.
13. Thickness of wells must be uniform for uniform expansion.
14. It is desirable to employ a plain flat topped piston on manufacturing grounds.
15. To achieve maximum thermal efficiency the highest possible compression ratio should be used.

4.6.3 Combustion Chamber Designs

The various types of combustion chambers are shown in Fig. 4.9.

1. T-Head Combustion Chamber

In the earliest engine T-head design shown in Fig. 4.9 (a) was used. Its disadvantages are : (i) having two camshafts, and (ii) being very prone to detonation.

2. L-Head or Side Valve Combustion Chamber

This is shown in Fig. 4.9 (b). Its advantages are :

(i) A side valve engine is easy to manufacture.
(ii) It is easy to enclose and lubricate the valve mechanism.
(iii) In head can be removed without disturbing the valve-gear and pipe work for decarbonising.
(iv) It provides compact layout.
(v) Cooling of exhaust by inlet charge is more effective as both are situated side by side.

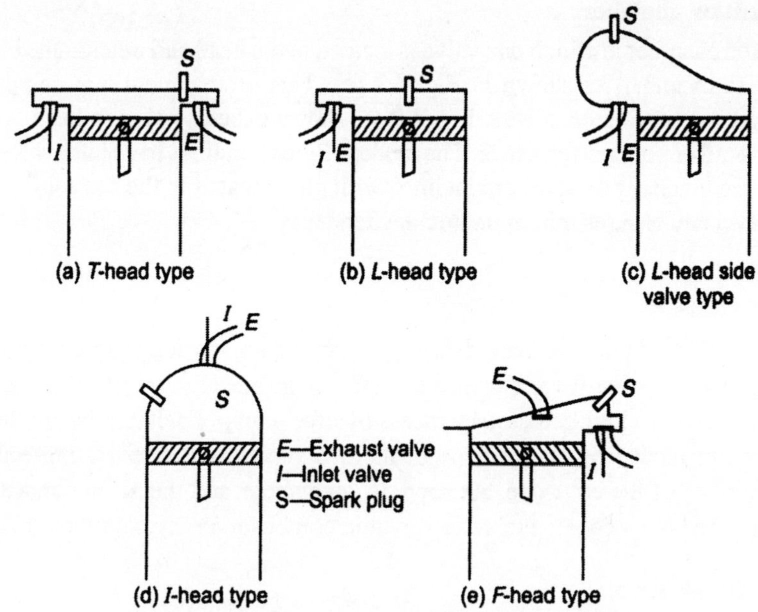

Fig. 4.9 Examples of typical combustion chambers.

Its shortcomings are :

(i) Lack of turbulence in the incoming charge.

(ii) Extremely prone to knocking as the flame has to travel long distance.

(iii) Extremely sensitive to ignition timing due to slow combustion.

3. Ricardo Turbulent L-Head side valve Design

This arrangement is shown in Fig. 4.9 (c). This gives higher flame speed and to reduce the knocking tendency. The entire combustion space is concentrated over the inlet and exhaust valves. This provides sufficient turbulence during the compression stroke and increased flame velocity. The ratio of area of volume of the space above the piston is large which provides sufficient quench area essential to avoid knocking.

4. Overhead valve or I-Head combustion chamber

In this arrangement, both the valves are located in the cylindrical head as shown in Fig. 4.9 (d). This arrangement is superior to side valve or L-head arrangement when high compression ratios are used for the following reasions :

(i) Lower pumping losses and higher volumetric efficiency from better breathing of the engine from larger valves or valve lifts and more direct passage ways.

(ii) Less distance for the flame to travel and therefore greater freedom from knock.

(iii) Less force on the head bolts and therefore less possibility of leakage.

(iv) Absence of exhaust valve from block results in more uniform cooling of cylinder and piston.

(v) Lower surface-volume ratio and, therefore, less heat loss and less air pollution.

(vi) Easier to cast and hence lower casting cost.

5. F-head combustion chambers

A combustion chamber in which one valve is located in the head and other in the block is known as F-head combustion chamber, as shown in Fig. 4.9 (e). This arrangement is a compromise between L-head and I-head combustion chambers, Fig. 4.9 (e) shows exhaust valve in head and inlet valve in block. This gave a rather poor performance. The modern F-head engines have inlet valve in the head and exhaust valve in the block. The valves are inclined with plug located in the flat roof. This provides the shortest flame travel and ensures minimum kocking tendency.

4.7 COMBUSTION IN C.I. ENGINES

In the C.I. engines, air alone is compressed through a large compression ratio during the compression stroke raising highly its temperature and pressure. At this stage one or more jets of full are injected in the liquid state, compressed to a high pressure by means of a fuel pump. Each minute droplet as it enters the hot air is quickly surrounded by an envelope of its own vapour and after an appreciable interval, is inflamed at the surface of the envelope. As soon as this vapour and the air in contact with it reach a certain temperature and the local air-fuel ratio is within combustion range, ignition takes place.

4.7.1 Stages of Combustion

The stages of combustion in the CI engine are shown in Fig. 4.10.

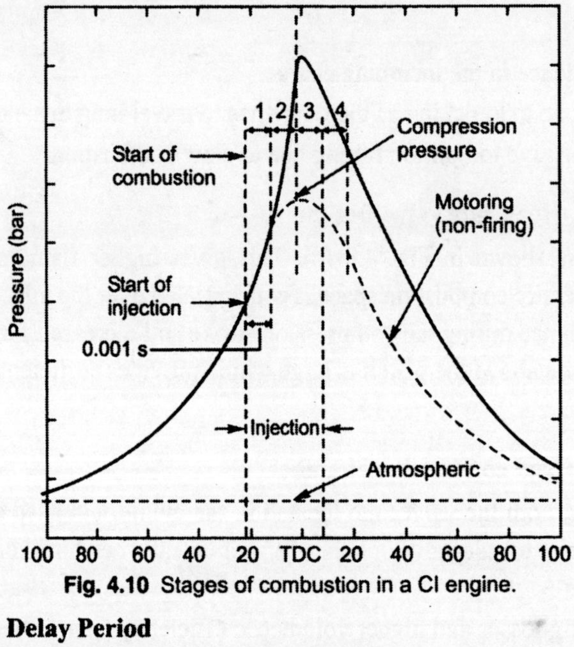

Fig. 4.10 Stages of combustion in a CI engine.

1. First stage : Ignition Delay Period

During the first stage some fuel has been admitted but has not yet been ignited. The ignition delay is counted from the start of injection to the point where the pressure-time (crankshaft rotation) curve separates from the pure air compression curve. The delay period is a sort of preparatory phase.

2. Second stage : Rapid or uncontrolled combustion (premixed flame)

In the second stage, the pressure rise is rapid because during the delay period the fuel droplets have had time to speed themselves over a wide area and they have fresh air all around them. This period is counted from the end of delay period to the point of maximum pressure on the indicator diagram. About one-third of the heat is evolved during this period.

3. Third stage : Controlled combustion (diffusion flame)

At the end of second stage the temperature and pressure are so high that the fuel droplets injected during the last stage burn almost as they enter and any further pressure rise can be controlled by the injection rate. The period of controlled combustion is assumed to end at maximum cycle temperature. About 70 to 80 percent of total heat of the fuel is supplied during this period.

4. Fourth stage : After burning

Because of poor distribution of the fuel particles, combustion continues during part of the remainder of the expansion stroke. This after-burning can be called the fourth stage of combustion. The after-burning period is about 70° to 80° of crank angle from TDC. The total heat evolved during the end of entire combustion process is 95 to 97% and 3 to 5% of heat goes to unburnt fuel.

4.7.2 Delay Period or Ignition Delay

In Fig. 4.11, the delay period is shown on pressure-crank angle (or time) diagram between points '*a*' and '*b*'. Point '*a*' represents the point of injection and point '*b*' represents the time at which the pressure curve (caused by combustion) first separates from the compression curve (non-firing or motoring curve). This ignition delay period can be roughly divided into two parts.

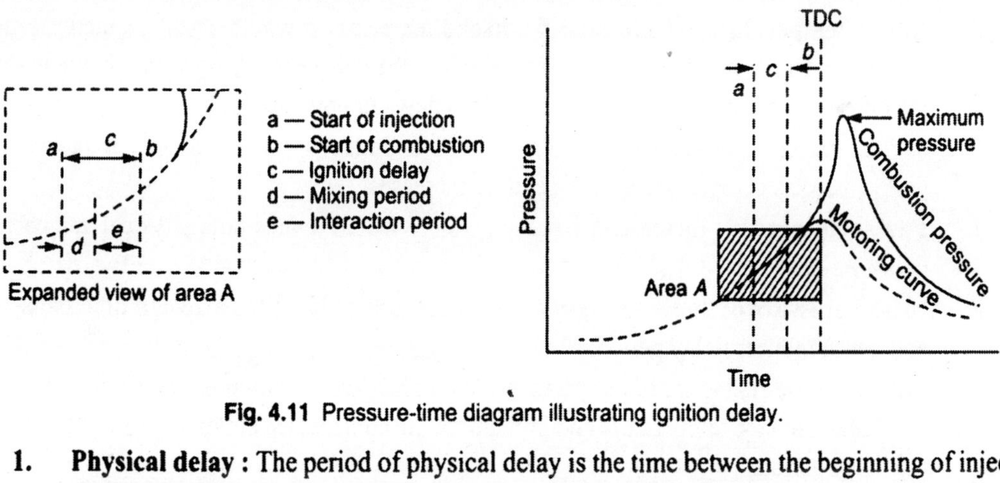

Fig. 4.11 Pressure-time diagram illustrating ignition delay.

1. **Physical delay :** The period of physical delay is the time between the beginning of injection and the attainment of chemical reaction conditions. In the physical delay period, the fuel is atomized, vaporized, mixed with air, and raised in temperature.
2. **Chemical delay :** In this period, the preflame reactions start slowly and then accelerate until local ignition takes place. Generally chemical delay is longer than the physical delay. The delay period refers to the sum of physical and chemical delay. The ignition lag in SI engine is basically equivalent to the chemical delay in CI engine. There is no component like physical delay in SI engine.

4.7.3 Variables Affecting Delay Period

1. **Fuel :** A lower pre-ignition temperature means a wider margin between it and the temperature of the compressed air and hence delay period. A higher cetane number means a lower delay period and smoother engine operation. Other properties of fuel which affects delay period are : volatility, latent heat, viscosity and surface tension.
2. **Injection pressure or size of droplet :** The smaller the size and greater the number of droplets the larger will be aggregate area of inflammation and therefore the greater the uncontrolled pressure rise. The size of the droplets depend on the injection pressure. Therefore, lower the injection pressure the lower the rate of pressure rise during the uncontrolled phase and smoother the running.
3. **Injection advance angle :** The delay period increases with increase in injection advance angle because the pressures and temperatures are lower when the injection begins. When the injection advance angles are small, the delay period reduces and operation of the engine is smoother but the power is reduced because larger amount of fuel burns during after-burning. The optimum angle of injection advance is 12° to 20° before TDC.
4. **Compressure ratio :** Increase in compression ratio reduces the delay period as it raises both temperature and density. However, higher compression ratio decreases the volumetric efficiency and power. It also decreases mechanical efficiency. In practice, the engine designer uses the lowest compression ratio which would satisfy the needs of easy cold starting and light load running at high speeds.
5. **Intake temperature :** Increasing the intake temperature would result in increase in the compressed air temperature, which would reduce the delay period. However, it would reduce the density of air and hence volumetric efficiency and power output.
6. **Jacket water temperature :** Delay period is reduced with increase in jacket water temperature as compressed air temperature increases.
7. **Fuel temperature :** Increase in fuel temperature would reduce both physical and chemical delay periods.
8. **Intake pressure or super charging :** Supercharging reduces the auto-ignition temperature and hence reduces delay period.
9. **Speed :** As the engine speed increases, the loss of heat during compression decreases with the result that both the temperature and pressure of the compressed air tend to rise, thus reducing the delay period.
10. **Air-fuel ratio (load) :** With increase in air-fuel ratio the combustion temperatures are lowered and hence the delay period increases. With increase in load, air-fuel, ratio decreases operating temperature increase and hence, delay period decreases.
11. **Engine size :** The engine size has little effect on the delay period in milliseconds but crank angle decreases.
12. **Type of combustion chamber :** A pre-combustion chamber gives shorter delay period as compared to an open type of combustion chamber.

The summary of the effect of delay period in C.I. engine is given in Table 4.4.

Table 4.4 : Effect of variables on delay period in C.I. engine

Sl. No.	*Increases in variable*	*Effect on delay period*	*Reason*
1.	Cetane number of fuel	Reduces	Reduces self-ignition temperature
2.	Injection pressure	Reduces	Greater surface volume ratio hence less physical delay.
3.	Injection advance angle	Increases	Pressures and temperatures lower when injection begins.
4.	Compression ratio	Reduces	Increases air temperature and pressure and reduces auto-ignition temperature.
5.	Intake tamperature	Reduces	Increases air temperatures.
6.	Jacket water temperature	Reduces	Increases wall and hence air temperature.
7.	Fuel temperature	Reduces	Better vaporisation and increases chemical reaction.
8.	Intake pressure (Supercharging)	Reduces	Increases in density reduces autoignition temperature.
9.	Speed	Reduces in milliseconds	Less loss of heat more crank angle in a given time.
10.	Load (fuel-air-ratio)	Increase in crank angle	
11.	Engine size	Little effect in	Operating temperature increases.
12.	Type of combustion chamber	Lower for precombustion chamber	Low rpm.

4.7.4 Diesel Knock

If the delay period is long a large amount of fuel will be injected and accumulated in the chamber. The auto-ignition if this large amount of fuel may cause high rate of pressure rise and high maximum pressure which may cause knocking in diesel engines, usually called diesel knock. The methods for controlling diesel knock are :

1. **Cetane Number of the Fuel Used :** Cetane number is a scale for comparing the ignition delay period of various diesel fuels. A higher Cetane number means a lower delay period, and smoother engine operation, because it reduces the self-ignition temperature. Some compounds called additives or dopes are used to improve combustion performance of fuels. Various additives used for disel fuels are : Isopropyl nitrate, Amyl nitrate, Heptyl peroxide and Butyl peroxide. These dopes delay the auto-ignition but they increase NO_x drastically.
2. **Compression Ratio :** Increase in compression ratio reduces the delay period as it raises both temperature and density. Any the same time the minimum auto-ignition temperature decreases which reduces the time of reaction when fuel is injected. Thus the delay period decreases. With increase of compression ratio the unused air would be much more decreasing

the volumetric efficiency and power. It further decreases the mechanical efficiency due to increase in weight of reciprocating parts.

3. **Air-Fuel Ratio :** With decrease in air-fuel ratio (richer mixture) the combustion temperatures are increased and cylinder wall temperatures are increased and hence the delay period decreases. With increases of load air-fuel ratio decreases, operating temperature increases and hence delay period decreases.
4. **Injection Pressure :** Lower the injection pressure, the lower the rate of pressure rise during the uncontrolled phase and smoother the running of engine. Lower injection pressure gives rise to fuel droplets of smaller size to obtain largest surface-volume ratio resulting in less physical delay. Large droplets of fuel results in too low subsequent burning.

However smaller droplets have lesser momentum resulting in partial suffocation by its own products of combustion. It also results in greater uncontrolled pressure rise.

4.8 DETONATION IN C.I. ENGINES

The fuel injection takes place over a definite period. Consequently, the first droplets injected are passing through the ignition delay, additional droplets are being injected into the chamber. If the ignition delay of the injected fuel is small the first droplets start burning in a relatively short time after injection and relatively small amount of fuel will be accumulated in the chamber. Therefore, the rate of pressure rise will be smooth as shown in Fig. 4.12 (a). But if the delay period is longer, the accumulation of fuel will be larger as the actual burning of the first droplets is delayed. When actual burning starts, the accumulated fuel and additional fuel injected starts burning and can cause too rapid rate of pressure rise. This rapid rate of pressure rise causes pulsating combustion and creates heavy noise. This pulsating variation of pressure due to sudden burning of accumulated fuel is shown in Fig. 4.12 (b). This type of abnormal combustion is known as Detonation. This detonation occurs during the initial period over the delay period when the burning of the first portion of the fuel is uncontrolled, thus giving rise to an excessive rate of pressure rise.

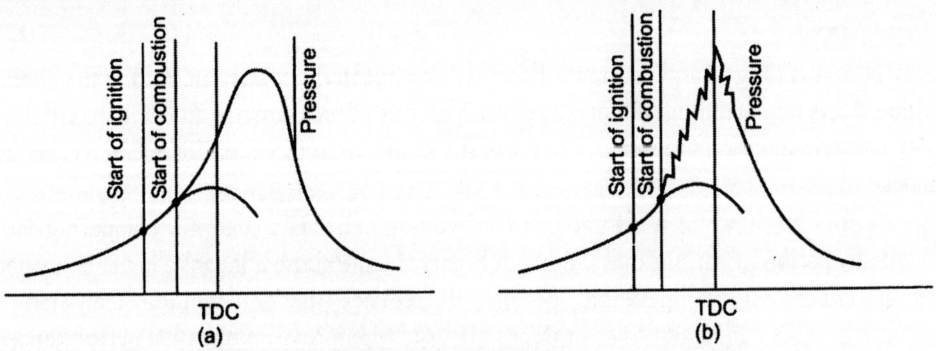

Fig. 4.12 Knocking combustion in C.I. and S.I. engines.

4.8.1 Factors Affecting the Detonation in C.I. Engines

The factors which influence the delay period also influence the detonation tendency of C.I. engines as listed below.

1. If the injection of fuel is too far advanced, the rate of ppressure rise during auto-ignition is very high and cause detonation.

2. Inferior fuels (having lower Cetane number) promote Diesel knock but this can be avoided by using better types of fuel (higher Cetane number).
3. Fuel injection parameters (better penetration and distribution) better combustion chamber design (split type) also influence detonation tendency of the engine.
4. Initial condition of the air (hgiher temperature or higher pressure in case of supercharged engine or both) influence detonation tendency. Higher pressure and temperature reduces detonation tendency.
5. The fuel having longer *delay period* and higher *self-ignition* temperature lead to detonation.

With proper design of combustion chamber and injecting the fuel uniformly over a hot surface of the combustion chamber, the detonation can be avoided.

4.8.2 Controlling the Detonation

We have discussed the factors which are responsible for the detonation in the previous article. If these factors are controlled, then the detonation can be avoided.

The detonation can be controlled by reducing the delay period which is fully responsible for detonation.

1. **Using a better fuel :** Higher CN fuel has lower delay period and reduces detonation tendency.
2. **Controlling the Rate of Fuel Supply :** It can be reduced by controlling the initial rate of fuel delivery without affecting efficient combustion condition.

   A small amount of fuel should be supplied till the combustion starts and then more amount of fuel should be supplied as inside combustion chamber condition is suitable (because of high temperature inside) to burn more fuel.

   In this case, cam shape is so designed which gives low initial injection followed by main injection at higher rate as shown in Fig. 4.13.

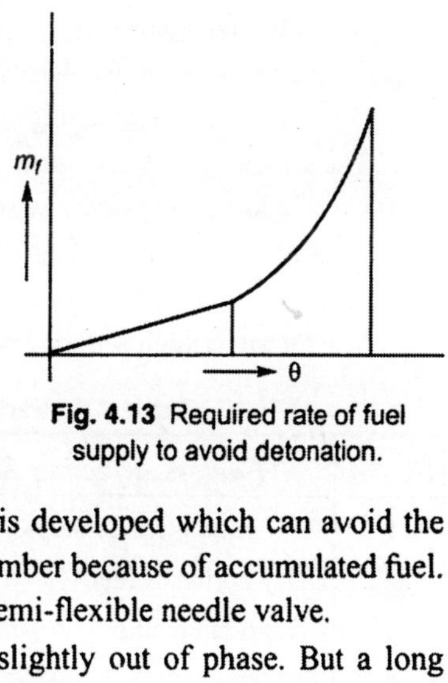

Fig. 4.13 Required rate of fuel supply to avoid detonation.

3. **Knock Reducing Fuel Injector :** This type of injector is developed which can avoid the sudden increase in the pressure inside the combustion chamber because of accumulated fuel. In this case, injection pressure of 100 bar is used with a semi-flexible needle valve.

   Doxford controlled detonation by using two injectors, slightly out of phase. But a long combustion period reduces the thermal efficiency of the engine.

4.8.3 Comparison of Knocking in S.I. and C.I. Engines

It is interesting to compare the knocking phenomenon in S.I. and C.I. engines. This phenomenon in both engine is fundamentally similar. Both are processes of *Auto-ignition* subject to the ignition-lag characteristics of A : F mixture supplied to the engine.

The differences in knocking phenomenon in both engines are listed below :

1. In S.I. engines, the knocking occurs near the end of combustion whereas in C.I. engine, this occurs at the beginning of combustion as shown in Fig. 4.14.

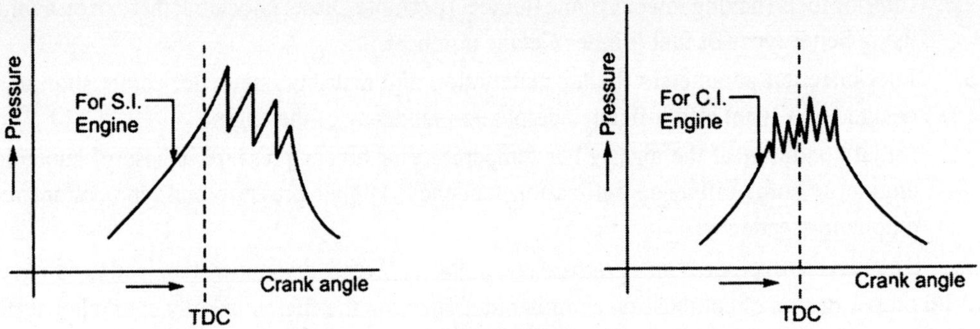

Fig. 4.14

2. The knocking in S.I. engine takes place in an homogeneous mixture, therefore, the rate of pressure rise and maximum pressure is considerably high. In case of C.I. engine, the mixture is not homogeneous and hence the rate of pressure is lower than in S.I. Engine.
3. The question of preignition does not arise in C.I. engines as the fuel is supplied only near the end of compression stroke.
4. It is more easy to distinguish between knocking and non-knocking operation in S.I. engine as the human ear easily finds the difference. However, in C.I. engine, the normal ignition is itself by auto-ignition and rate of pressure-rise under normal operation is considerably higher (10 bar against 2.5 bar for S.I. engines) and causes high noise. The noise level becomes excessive under detonation condition. Therefore, there is no definite distinction between *normal* and *knocking* combustion.

The knocking in S.I. engines is because of auto-ignition of the last part of the charge. To avoid this, the fuel must have *long-delay period* and high self-ignition temperature (SIT).

To avoid knock in C.I. engine, the *delay period* should be as small as possible and fuel self-ignition temperature should be as low as possible.

The factors which reduce knock in S.I. and C.I. engines are given in Table 4.5.

Table 4.5 : Factors which reduce knock in S.I. and C.I. Engines.

Sl. No.	*Factors Affecting Knock*	*S.I. Engines*	*C.I. Engines*
1.	Self ignition temperature	High	Low
2.	Delay period of fuel	Long	Short
3.	Compression ratio	Low	High
4.	Inlet temperature	Low	High
5.	Inlet pressure	Low	High
6.	Speed	High	Low
7.	Cylinder size	Small	Large
8.	Combustion chamber wall temperature	Low	High

Because of this dissimilarity in the time of starting of the knock in S.I. and C.I. engines, the conditions which reduce the knock tendency in S.I. engine increase the knock tendency in C.I. engines. The fuels which are better from the point of view of avoiding detonation in S.I. engine may promote detonation in C.I. engines, therefore good S.I. fuels are poor C.I. fuels. In terms of fuel rating, diesel oil high cetane number (40-60) and low Octane numbr (30) and Petrol has high Octane number (80-90) and low Cetane number (20).

4.9 COMBUSTION CHAMBERS FOR C.I. ENGINES

The most important function of the CI engine combustion chamber is to provide proper mixing of fuel and air in a short time. For this purpose, an organised air movement, called air swirl, is provided to produce high relative velocity between the fuel droplets and air.

There are three basic methods of generating swirl in a CI engine combustion chamber.

1. By directing the flow of the air during its entry to the cylinder, known as induction swirl. This method is used in open combustion chambers shown in Fig. 4.15.

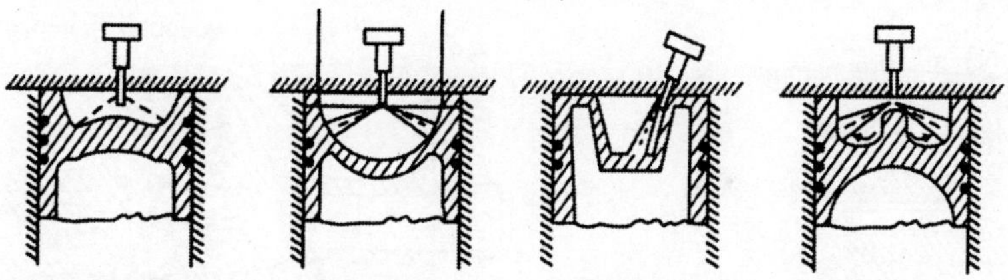

Fig. 4.15 Open combustion chambers.

2. By enforcing the air through a tangential passage into a separate swirl chamber during the compression stroke, known as compression swirl. This method is used in swirl chambers. This type of chamber is shown in Fig. 4.16.

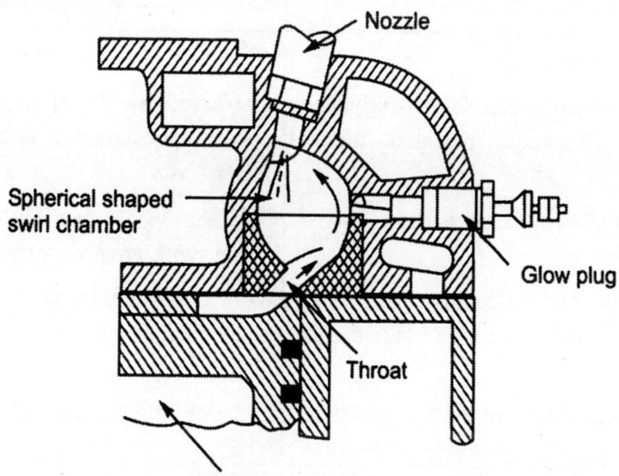

Fig. 4.16 Ricardo Swirl chamber comet, Mark II.

3. By use of the initial pressure rise due to partial combustion to create swirl turbulence, known as combustion induced swirl. This method is used in pre-combustion chambers and air-cell chambers. These type of chambers are shown in Fig. 4.17 (a) and (b).

The comparison of induction and compression swirl is given in Table 4.6 and comparison of combustion characteristics in Table 4.7.

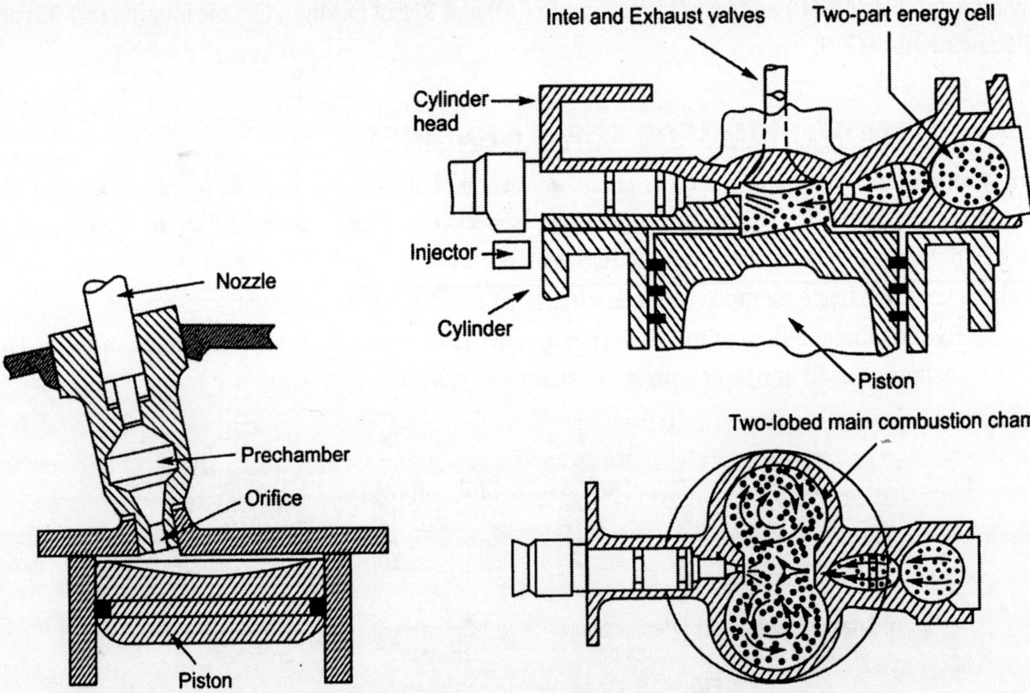

(a) Precombustion chamber

(b) Lanova air-cell combustion chamber

Fig. 4.17 Combustion induced swirl chambers.

Table 4.6 : Comparison of induction and compression swirl.

Induction Swirl	*Compression Swirl*
Advantages	**Disadvantages**
1. High excess air (low temperature), low turbulence (less heat loss). Therefore indicated thermal efficiency high.	Less excess air, lower indicated efficiency 5 to 8% more fuel consumption. Decreased exhaust valve life.
2. Due to low intensity of swirl easier starting.	Cold starting trouble due to high heat loss due to strong swirl, greater s/v ratio.
3. No additional work for producing swirl. High mechanical efficiency and hence high brake thermal efficiency.	Work absorbed in producing compression swirl, mechanical efficiency lower.
4. Used with low speeds. Therefore low quality of fuel can be used.	Cylinder more expensive in construction.
Disadvantages	**Advantages**
1. Weak swirl, multiorifice nozzle, high injection pressure, clogging of holes, high maintenance.	Single injector, pintle type (self cleaning), less maintenance.
2. Influences minimum quantity of fuel. Complication at high loads and idling.	Large valves, higher volumetric efficiency.
3. Shrouded valves, smaller valves, low volumetric efficiency.	Greater air utilization due to strong swirl. Smaller (cheaper) engine.

4. Weak swirl, low air utilisation (60%), lower mep, large size (costly) engine.	Swirl proportional to speed, suitable for variable speed operation.
5. Swirl not proportional to speed. Efficiency not maintained in a variable speed engine.	Smooth engine operation.

Table 4.7 : Comparison of combustion chamber characteristics (Mainly for 4-Stroke unsupercharged high speed engines)

Sl. No.	*Characteristics*	*Open chamber*	*M.A.N. 'M'-chamber*	*Swirl Chamber*	*Pre-combus combustion*	*Air-cell combustion*
1.	Speed capability, rpm	3500	3500	4000	3000	3500
2.	bmep, bar	6.6	6.9	8.3	6.3	6.3
3.	Rate of pressure rise, bar/degree crank angle	6.6	6.9	8.3	6.3	6.3
4.	Compression ratio	15-16	16-17	17-19	18	13-16
5.	Injection pressure, bar	560	280	350	140	280
6.	Nozzle holes	6-8	1	1	1	1
7.	Pumping loss	Low	Medium	High	High	High
8.	Volumetric efficiency	Low	Lowest	High	High	High
9.	Coolant load	Low	Medium	Highest	High	High
10.	Air motion	Medium	High	Highest	High	High
11.	Combustion noise	High	Lowest	Low	Low	Low
12.	Cold starting	Excellent	Fair	← Needs help →		

■ **Example 4.1** Explain clearly the terms pre-ignition, detonation and diesel knock.

Solution : Pre-ignition : After ignition, once a stable flame nucleus is formed, the flame will propagate from molecule to molecule. When the flame reaches the far end of the combustion chamber the peak pressure is normally attained. The end mixture experiences rapid compression first by the piston during the compression stroke and later by the advancing flame front. Thus the end charge experience a very high pressure and temperature. If the temperature exceeds the self-ignition temperature of the mixture, pre-ignition or auto-ignition takes place.

Detonation : It is a term used to signify an unusual sharp metallic sound which is clearly audible under abnormal conditions of engine operation.

Diesel Knock : If the delay period is long a large amount of fuel will be injected and accumulated in the combustion chamber of the engine. The auto-ignition of this large amount of fuel may cause high rate of pressure rise and high maximum pressure which may cause knocking in diesel engines, usually called diesel knock.

■ **Example 4.2** 'Factors which increase detonation in S.I. engine tends to reduce knocking in C.I. engines', discuss the validity of the statement.

Solution : The kncoking in SI engines and CI engines is fundamentally due to the auto-ignition lag of the fuel-air mixture. There are following differences in the kncoking phenomenon of SI and CI engines :

1. In the SI engine the detonation occurs near the end of combustion whereas in CI engine detonation occurs near the beginning of combustion, as shown is Fig. 4.18.

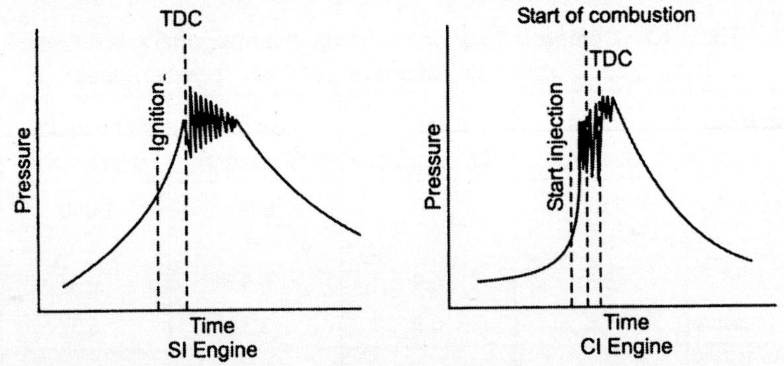

Fig. 4.18 Diagrams illustrating knocking combustion in SI and CI engines.

2. The detonation in SI engine is of a homogeneous charge causing very high rate of pressure rise and very high maximum pressure. In CI engine the rate of pressure rise is normally lower than that in the SI engine.
3. There is no question of pre-ignition in a CI engine as the fuel is injected in the cylinder at the end of the compression stroke.
4. In the SI engine it is relatively easy to distinguish between kncoking and non-kncoking operation as the human ear can easily find the distinction. But in the CI engine there is no definite distinction between normal and kncoking combustion.

The comparison of kncoking between SI and CI engines is given in the Table 4.8.

Table 4.8 : Factors tending to reduce kncoking in SI and CI engines.

Sl. No.	*Factors*	*SI Engine*	*CI Engine*
1.	Self-ignition temperature of fuel	High	Low
2.	Time lag or delay period for fuel	Long	Short
3.	Compression ratio	Low	High
4.	Inlet temperature	Low	High
5.	Inlet pressure	Low	High
6.	Combustion chamber wall temperature	Low	High
7.	Speed	High	Low
8.	Cylinder size	Small	Large

Hence the statement is valid.

4.10 RATING OF S.I. ENGINE FUELS-OCTANE NUMBER

The tendency to detonate depends upon the composition of fuel as mentioned earlier. Fuels differ widely in their ability to resist knock. the rating of a particular fuel is done by comparing its performance with that of a standard reference fuel which is usually a combination of *iso*-octane and *n*-heptane.

Iso-Octane offers great resistance to detonations therefore it is arbitrarily assigned a rating of 100 Octane number, *n*-heptane, a straight chain paraffin, on the other hand, deontates very rapidly therefore it is assigned a rating of zero-Octane number. It is possible to match the kncoking tendency of all commonly used fuels by blending these two reference fuels in different proportions.

The percentage of *iso*-Octane by volume in a mixture of *iso*-octane and *n*-heptane, which exactly matches the kncoking intensity of a given fuel, in a standard engine under given standard operating conditions, is termed as 'Octane number' of the fuel. An octane number of 80 means that the fuel has the same kncoking tendency as a mixture of 80% *iso*-octane and 20% *n*-heptane by volume.

It has already been mentioned that the kncoking tendency of an engine increases with increasing compression ratio. The higher the octane number of the fuel, the greater will be its resistance to knock and higher will be the compression ratio which may be used without kncoking. The power output and specific fuel consumption are functions of the compression ratio and therefore these can be considered as a function of the Octane number of the fuel. This fact indicates an importance of the octane number rating of fuels for S.I. engines. Higher Octane number fuels increase the efficiency of the engine as high compression ratio can be used with higher Octane number fuels.

The kncoking tendency of the fuels used in aero-engines is more as they are supercharged engines therefore it is necessary to use the fuels having highest octane number (largest resistance to knock). The fuels superior in anti-knock qualities to *iso*-octane (having Octane number greater than 100) are required. The addition of TEL (Tetra-Ethyl-Lead) to gasoline produces marked effect in reducing the knock tendency of the fuel. The addition of various amounts of one of these compounds (commonly used is tetraethyl lead) to *iso*-octane produces fuels of greater anti-knock quality than *iso*-octane alone. The anti-knock quality of fuels above 100 Octane number in U.S.A. is measured in terms of c.c. (cu-cm) of tetraethyl lead per U.S. gallon of iso-octane. A fuel having an Octane number of 110 means that the fuel has the same tendency to knock as a mixture of 10 c.c. of tetraethyl lead in 1 U.S. gallon of *iso*-octane. The importance of such fuels (commonly used in aero-engine) has diminished with an introduction of turbojet turbines in aircraft.

4.10.1 Anti-knock Agents

The knock resistance tendency of a fuel can be increased by adding anti-knock agents. The anti-knock agents are substances which decrease the rate of preflame reaction by delaying the auto-ignition of the end mixture in the engine until the flame generated by spark plug passes through the end mixture.

Among all, TEL [$Pb(C_2H_5)_4$]is most powerful antiknock agent among all known. It has helped to improve the efficiency of the engine and to increase the specific output of S.I. engines. Its use will not improve the performance of the engine which is not knocking unless the spark is advanced, the compression ratio is increased or a higher inlet pressure is used to take the advantage of an increase in Octane number. Advancing the spark will improve the performance of the engine only if optimum spark for a non-kncoking fuel is not used already.

Some other anti-knock agents with their relative effectiveness are lised below in Table 4.9.

Table 4.9 : Anti-knock agents

Compound	*Chemical Symbol*	*Weight for given effect (gm)*	*Relative weight*
Tetraethyl lead	$Pb(C_2H_5)_4$	0.0295	1
Aniline	$C_6H_5NH_2$	1.00	34

Ethyl iodide	C_2H_5I	1.55	53
Ethyl alcohol	C_2H_5OH	4.75	161
Xylene	$(C_6H_4CH_3)_2$	8.00	271
Toluene	$C_6H_5CH_3$	8.8	298
Benzene	C_6H_6	9.8	332

Knocking essentially results in an incomplete and nonuniform combustion of the fuel. It is caused by the sudden and spontaneous ignition of the unburned mixture, ahead of the flame front advancing through the combustion chamber. Partial oxidation reactions of the hydrocarbons in this mixture produce reactive intermediates that lead to such spontaneous ignition and overheating of the engine. This overheating can result in the wearing of engine parts. Antiknock agents react with these intermediates and reduce their ability to propagate the oxidation reaction that causes kncoking. Octane number of the gasoline is a measure of the effectiveness of the antiknock agent it contains. Higher this number lesser is the knock. Table 4.10 shows some of these agents and their relative effectiveness.

Table 4.10 : Relative effectiveness of antiknocks

Compound	*Relative effectiveness*
Tetraethyllead	100
Methylcyclopentadienyl	65
Manganese tricarbonyl ironcarbonyl	43
Copper methylaminomethylenecetate	40
Nickelcarbonyl	30
Triethylbismuth	20
Tetraethyltin	3
N-Methylaniline-Ethyliodide	11

Gasoline composition, methods of test and antiknock concentrations affect both the absolute and relative effectiveness of different compounds, even of the same element. Moreover, some compounds can function as an antiknock under some conditions but promote knock under other. Therefore, the comparison of active elements in various compounds is representative but subject to considerable variation.

The effectiveness of an element can vary greatly depending on the specific compound considered. Some aromatic amines are not antiknocks while others are twice as effective as N-methylamine (nitrogen in other forms is not effective). Most of the metallic compounds possessing the antiknock property have metal to carbon bonds but exceptions exist. In the case of organometallics, it is believed that the products released by the thermal decomposition of the compound at an appropriate stage of the precombustion process, are the active antiknock species.

Tetraethyllead, abbreviated as TEL is the most widely used (also known longest) of all antiknocks. A hydrogen atom in ethane is substituted by the heavy metal atom of lead in this organometallic compound. The average TEL usage, at present, in motor gasoline (2.0 gms lead per gallon in regular and 2.7 gms in premium) usually takes up the octane number by 6 to 10. Tetramethyl lead, a mixture of tetra methyl and ethyl leads, and the intermediate lead alkyls produced by these two are also commercially used as antiknocks. These higher volatile alkyls, offer advantages of better vaporisation and are most effective in some multicylinder engines. Concentrated and alkyls are hazardous to handle. Therefore, these antiknock agents are blended with gasoline in refineries under same conditions; the resulting leaded gasoline can be handled, as a motor fuel, safely. The use of leaded gasoline, however is not a perfect solution to the problem. It leads to the emission of lead into the atmosphere which is known to be hazardous.

4.10.2 Performance Number

The detonation tendency is also measured in terms of performance number. The performance number (PN) is defined as

$$\text{PN} = \frac{(KLIMEP)_{\text{test-fuel}}}{(KLIMEP)_{\text{Iso-octane}}}$$

where KLIMEP represents conception of knock limited indicated mean effective pressure.

The performance number is obtained for a specified engine, under specified set of conditions by varying the inlet pressure.

The PN method of rating enables to develop a scale beyond 100 Octane number. There is a relationship between performance number and Octane number as shown in Fig. 4.19.

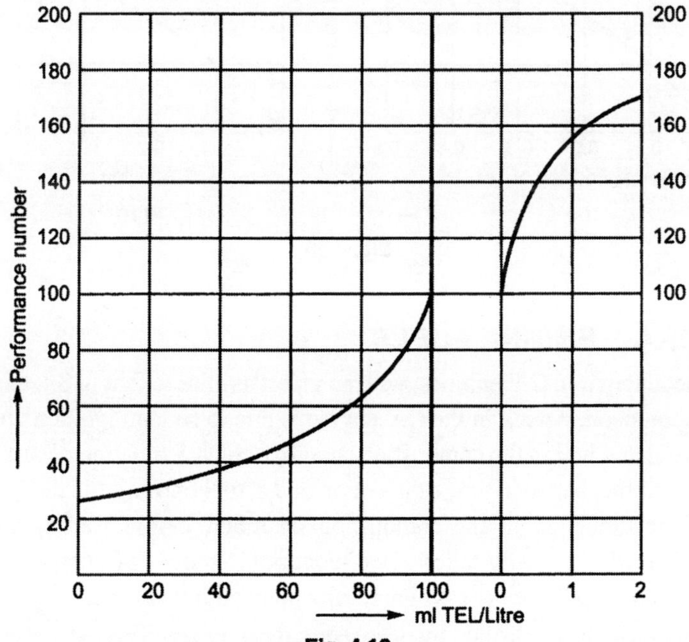

Fig. 4.19

It has already been mentioned that by adding TEL to *iso*-octane increases knock resistance tendency. Fig. 4.19 also shows the quantity of TEL by volume in milli-litre per litre of gasoline on X-axis for Octane numbers exceeding 100 and performance number on y-axis.

4.11 HIGHEST USEFUL COMPRESSION RATIO (HUCR)

In the development of S.I. engines, it was of great interest to use higher and higher compression ratio as the efficiency and power output of the engine increase with an increase in compression ratio. The maximum compression ratio of any S.I. engine is limited by its tendency to knock. Therefore, previously the rating of S.I. engine fuels was based on highest useful compression ratio that could be employed in a given engine under a given set of conditions. To find the highest compression ratio of a fuel, the compression ratio of the engine is increased (using a variable compression ratio engine) till knocking first become audible with certain temperature conditions and with mixture strength and ignition when both adjusted to give highest efficiency. The HUCR of few fuels determined by using Richardo E6 engine are listed below in Table 4.11.

Table 4.11 : HUCR for Various Fuels

Fuel	*Toluence*	*Benzene*	*Iso-octane*	*Cyclo-hextane*	*M-heptane*
HUCR	15	14.6	10.96	8.20	3.75

Through the measurement of HUCR of various fuels, a measure of their anti-knock value can be obtained.

The effect of anti-knock agents on HUCR is shown in Fig. 4.20.

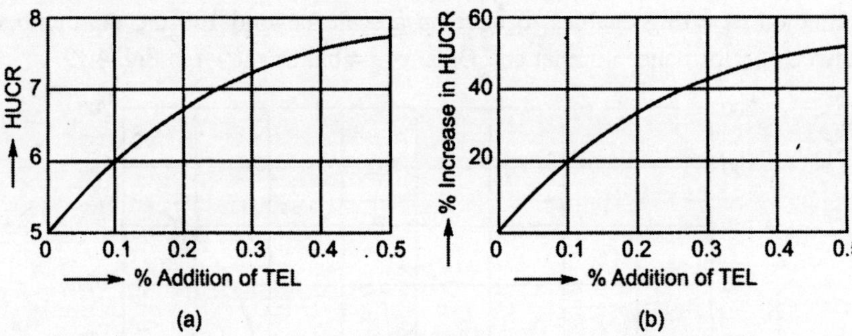

Fig. 4.20

4.12 RATING OF C.I. ENGINE FUELS

Knocking is also encountered in C.I. engines, with an effect similar to that in S.I. engines, although it is due to different phenomenon. Knock in the C.I. engine is due to sudden ignition and abnormally rapid combustion of accumulated fuel in the combustion chamber. Such a situation occurs because of ignition lag in the combustion of fuel between time of injection and actual burning. As the ignition lag increases, the amount of fuel accumulated in the combustion chamber, before starting of combustion, also increases. When combustion actually takes place, abnormal amount of energy is suddenly released, causing an excessive rate of pressure rise which results in an audible knock.

C.I. engine knock can be controlled by decreasing ignition lag. The shorter in an ignition lag, the less tendency to knock. Good C.I. engine fuel, therefore, will ignite more readily, *i.e.*, it will have short ignition delay. Furthermore, ignition lag affects the starting, warm up, and production of exhaust smoke in C.I. engines.

The property of ignition lag is generally measured in terms of *cetane number.* Cetane, a straight chain paraffin with good ignition quality, is arbitrarily assigned a rating of 100 Cetane Number (CN). It is mixed with alpha-methyl naphthalene, a hydrocarbon with poor ignition quality, which is assigned zero C.N. The mixture is matched with a fuel under test in standard (C.F.R.) engine running under prescribed conditions. The C.N of the fuel is then defined as the percent by volume of cetane in a mixture of cetane and alpha-methyl-napthalene that produces the same ignition lag as the fuel being tested, in the same engine and under the same operating conditions.

If test fuel is having a Cetane Number of 40 means a mixture containing 40% Cetane and 60% α-methyl napthalene by volume gives the same ignition delay as tested fuel.

The straight chain paraffins are the most suitable fuels for C.I. engines.

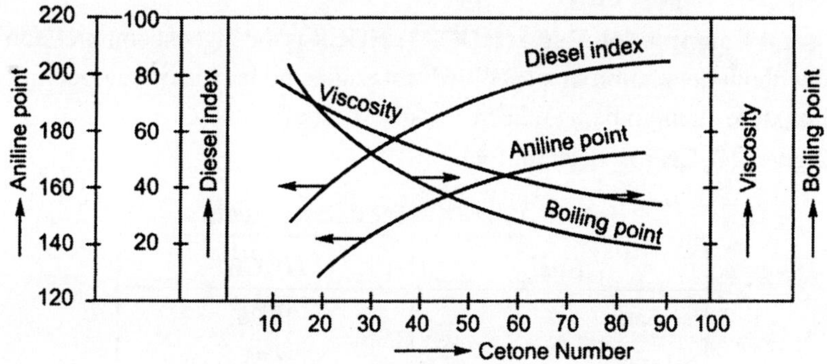

Fig. 4.21

The graph shown in Fig. 4.21 shows the relationship of the other propeties of the fuel with CN which are also responsible for the knocking of the engine.

4.13 I.C. ENGINE FUELS

The desirable properties of I.C. engine fuels are :

1. High energy density.
2. Good combustion qualities.
3. High thermal stability.
4. Low deposit forming tendencies.
5. Compatibility with the engine hardware.
6. Good fire safety.
7. Low toxicity.
8. Low pollution.
9. Easy transferability on board vehicle storage.

4.13.1 Fuels for Spark-Ignition Engines

(a) Gasoline

It is a mixture of various hydrocarbons such as paraffins, olefins, napthalenes, and aromatics. The requirements of an ideal gasoline are :

1. It should mix readily with air and afford uniform manifold distribution, *i.e.*, it should easily vaporise.
2. It must be knock resistant.
3. It should not pre-ignite easily.
4. It should not tend to decrease the volumetric efficiency of engine.
5. It should be easy to handle.

6. It should be cheap and easily available.
7. It should not corrade engine parts.
8. Its calorific value should be high.
9. It shold not form gum and varnish.

(b) Knock Rating of SI Engine Fuels

Highest Useful Compression Ratio (HUCR) : HUCR is the highest compression ratio at which a fuel can be used without detonation in a specified test engine under specified operating conditions and the ignition and mixture strength being adjusted to give best efficiency.

Table 4.12 gives HUCR for different fuels.

Table 4.12 : HUCR for different fuels

Fuel	*HUCR*
Iso-octane	10.96
N-heptane	3.75
Toluene	15.00
Cyclohexane	8.20

(c) Rating of SI Engine Fuels

The rating of fuel is done by comparing its performance with that of a standard reference fuel which is usuially a combination of iso-octane and *n*-heptane. Iso-Octane offers great resistance to deformations and is arbitrarily assigned a rating of 100-Octane number. On the other hand, *n*-heptane detonates very rapidly, and is assigned a rating of zero-octane number.

(d) Octane Number

It is the percentage of iso-octane by volume in a mixture is iso-octane and *n*-heptane, which exactly matches the knocking intensity of a given fuel, in a standard engine under given standard operating conditions.

The octane scale is extended beyond 100 by adding tetraethyl-lead (TEL) to *iso*-octane.

4.13.2 Fuels for C.I. Engines

Diesel fuels are used for C.I. engines. These are petroleum fractions that lies between kerosene and lubricating oils.

The main desirable characteristics of diesel fuels are :

1. Cleanliness—carbon residue, contamination, sulphur, etc.
2. Ignition quality—cetane number.
3. Fluditiy—viscosity, pour point, etc.
4. Volatility—flash point, carbon residue.

(a) Rating of C.I. Engine Fuels

The rating of a diesel fuel is measured in terms of cetane number. Cetane, a straight chain paraffin with good ignition quality is arbitrarily assigned a rating of 100-cetane number. It is mixed with a-methyl napthalene, a hydrocarbon with poor ignition quality, which is assigned zero-cetane number.

(b) Cetane Number

This is defined as the percentage of volume of Cetane in a mixture of Cetane and a-methyl napthalene that produce the same ignition lag as the fuel being tested, in the samc engine and under the same operating conditions.

4.14 SPECIAL COMBUSTION CHAMBERS

A few very special types of combustion chambers are developed and used on a particular automobiles.

1. Lanova or Air-cell Combustion Chamber

The Fig. 4.22 shows the arrangement of this chamber. This chamber consists of a small pre-combustion chamber into which the fuel is sprayed from a single hole nozzle. The walls of the chamber run hot so that the injected feel evaporates and starts burning instantaneously. And a high force created forces the burning fuel into the main combustion chamber with a very high velocity. The burning of fuel forms two vortices as shown in figure which further improves the mixing and burning.

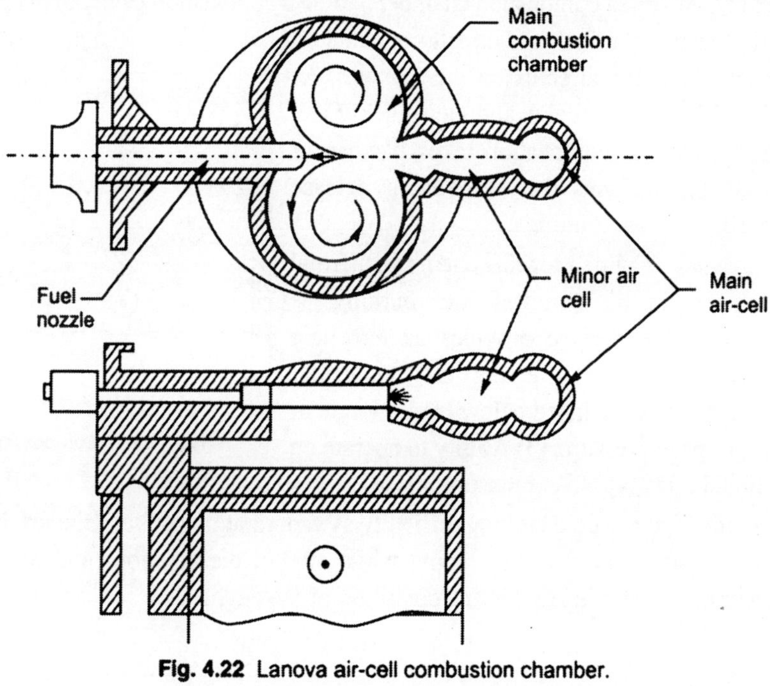

Fig. 4.22 Lanova air-cell combustion chamber.

The air-cell (pre-combustion chamber) has two parts as major and minor which are separated from each other and from the main combustion chamber by narrow orifices. The first burning starts in the minor cell and pushes the burning air in the main chamber. As the pressure in the minor cell falls because of discharging the gases, the major cell starts pushing the burning gases into the main combustion chamber. This step motion of the gases creates high swirling motion in the main combustion chamber and reduces delay period.

This design was very popular in America sometimes before. This is not suitable for variable speed engine as combustion induced swirl has no relationship to the speed of the engine. The advantages claimed are listed below.

1. The rate of pressure rise is low as mixing of air and fuel is minimised after starting the combustion.
2. Easy starting can be obtained by cutting off the cell from the main space by a valve which increases the compression ratio.
3. High power output is possible without excessive high pressure in the main combustion chamber.
4. A very good mixing of fuel and air takes place in the III-period, and avoids late burning and ensures complete combustion.
5. The engine runs smoothly as the maximum pressure in the main chamber is fairly low.

The only disadvantages is low thermal efficiency and high *bsfc*.

2. M-Combustion Chamber

A German Dr. Meurer developed a M-process engine after 20 years of research which ran without conventional diesel noise.

It is a special type of open combustion chamber having a combustion cavity in the piston as shown in Fig. 4.23. In this design, the fuel was directly injected on the surface of the cavity in the direction of the air-swirl. As the cavity remains very hot, the fuel evaporates quickly and mixed with swirling air and combustion takes place near homogeneous A : F mixture. Thus, the rate of energy release is equal to the rate of fuel evaporation.

The main advantage claimed for this chamber, thermal deposition, which occurs in atomized fuel burning, is prevented because the hot surface provides an excellent means of evaporation without affecting fuel molecular structure. A few advantages are listed below (1) Low rate of pressure rise (2) Low peak pressure, (3) Ability to operate on a wide range of liquid fuels (4) No combustion noise.

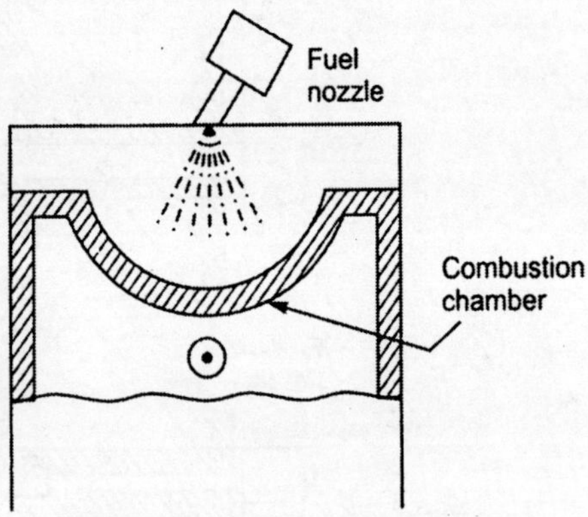

Fig. 4.23 M-combustion chamber.

The main disadvantage is cold starting difficulty as vaporization of fuel depends upon the surface temperature of the combustion chamber. Some white stroke, diesel odour and high hydro-carbon emission may occur during starting and idling condition of the engine.

SOLVED QUESTIONS FROM UNIVERSITY EXAMINATIONS

■ **Example 4.3**

(a) 'Factors tending to increase the detonation in S.I. Engines tend to reduce the knock in C.I. engine.' Discuss the validity of the above statement in the light of differences in the nature of two phenomenon and indicate the methods used to reduce the knock in C.I. engine.

(b) Explain the stages of combustion in a C.I. engine. [KU, Dec., 2005]

Solution : (a) Refer to Example 4.2.

(b) Refer to art. 4.8.

■ **Example 4.4**

(a) Define the following :

Octane rating of fuels, Ignition lag, Velocity of flame propagation, Pre-ignition.

(b) What are the theories of detonation and explain with graph the Auto-Ignition theory.

[KU, Dec., 2006]

Solution :

(a) Octane rating of fuels : Refer to art. 4.10.

Ignition lag : refer to art. 4.3.

Velocity of flame propagation : Refer to art. 4.4.

Pre-ignition : Refer to art. 4.2.

(b) Refer to art. 4.8.

■ **Example 4.5** What is 'M' combustion chamber? What advantages are claimed for this design of combustion chamber? [KU, Dec., 2007]

Solution : Refer to art. 4.14.

■ **Example 4.6** (a) What is detonation? Explain.

(b) Explain Octane raing of fuels. [KU, Sept., 2009]

Solution : (a) Refer to art. 4.8.

(b) Refer to art. 4.10.

■ **Example 4.7** Define the following terms : Octane rating of fuel, Pre-ignition, delay period, Variables affecting delay period. Knock in C.I. engines, Cetane rating. [K.U., Dec., 2007]

Solution :

Octane rating of fuel : Refer to art. 4.10.

Pre-ignition : Refer to art. 4.2.

Delay period : Refer to art. 4.3.

Variables affecting delay period : Refer to art. 4.3.1.

Knock in C.I. engines : Refer to art. 4.7.4.

Cetane rating : Refer to art. 4.12.

■ **Example 4.8** (a) What is Detonation? Explain.

(b) Explain the phenomenon of Pre-ignition [KU, Dec., 2009]

Solution : (a) Refer to art. 4.8.

(b) Refer to art. 4.2.

■ **Example 4.9** (a) Explain different stages of combustion in SI engine.

(b) Write a note on Octane rating. [KU, Dec., 2010]

Solution : (a) Refer to art. 4.2.

(b) Refer to art. 4.10.

■ **Example 4.10** With the help of neat sketch, explain the stages of combustion in C.I. Engines.

[MDU, Dec., 2007]

Solution : Refer to art. 4.7.

■ **Example 4.11** Define the following terms :

(i) Stages of combustion in C.I. Engine.

(ii) Types of injection system in C.I. Engine.

(iii) Various theories of detonation. [MDU, Dec., 2007]

Solution : (i) Refer to art. 4.7.

(ii) Refer to art. 3.3.

(iii) Refer to art. 4.8.

■ **Example 4.12** Define the following terms :

(i) Octane rating of fuel.

(ii) Theories of detonation.

(iii) Knock in C.I. engine. [MDU, May, 2008]

Solution : (i) Refer to art. 4.10.

(ii) Refer to art. 4.8.

(iii) Refer to art. 4.7.4.

■ **Example 4.13** Explain the effect of engine variables on detonation. [MDU, May, 2008]

Solution : Refer to art. 4.8.1.

■ **Example 4.14** State and explain various stages of combustion in S.I. engines through p-θ diagram for ideal as well as actual engines. [MDU, Dec. 2008]

Solution : Refer to art. 4.2.

■ **Example 4.15** How do we classify combustion in S.I. Engines? [MDU, Dec. 2008]

Solution : Refer to art. 4.2.

■ **Example 4.16** What is delay period? State and explain the variables affecting delay period. [MDU, May, 2009]

Solution : Refer to art. 4.3.

■ **Example 4.17** Name various theories of detonation. Explain pre-ignition through neat sketch. [MDU, May, 2009]

Solution : Theories of detonation : Refer to art. 4.8.

Pre-ignition : Refer to art. 4.2.

■ **Example 4.18** Briefly explain the stages of combustion in S.I. engine, eleborating the flame-front propagation. [MDU, Dec., 2009]

Solution : Refer to art. 4.2 and art. 4.4.

■ **Example 4.19** Explain the phenomenon of knock in C.I. engine and compare with S.I. engine knock. [MDU, Dec., 2009]

Solution : Refer to art. 4.7.4.

■ **Example 4.20** (a) Explain the effect of various engine variables on SI engine knock.

(b) Name various theories of detonation. Explain the pre-ignition with neat sketch.

[MDU, Dec., 2010]

Solution : (a) Refer to art. 4.5.1.

(b) Refer to art. 4.17.

■ **Example 4.21** (a) Explain the stages of Combustion in a diesel engine.

(b) What is Detonation? Explain effects of engine variable on detonation. [KU, Dec., 2008]

Solution : (a) Refer to art. 4.7.

(b) Refer to art. 4.8.

SHORT ANSWER TYPE QUESTIONS

1. Define combustion phenomenon in I.C. engines.
2. What are the stages of combustion in S.I. engines?
3. What is ignition lag?
4. List four factors affecting delay period.
5. List four factors which affect the velocity of flame propagation.
6. What is the phenomenon of knocking in S.I. engines?
7. List four factors which reduce knocking in S.I. engines.
8. What are the effects of knocking?
9. Write four design principles for combustion chambers.
10. What are the stages of combustion in C.I. engines?
11. Write four factors which reduce delay period in C.I. engines.
12. What is diesel knock?
13. What is detonation in C.I. engines?
14. List four factors which affect detonation in C.I. engines.
15. Give the names of few anti-knock agents.
16. Define performance number.
17. Define HUCR.
18. List four desirable properties of I.C. engine fuels.
19. Define Octane number and Cetane number.
20. What do you understand by rating of fuels?

EXERCISES

1. Explain the stages of combustion in S.I. engines with the help of p-θ diagram.
2. What do you understand by ignition lag? Discuss the effects of different factors which affect the ignition lag.
3. Explain the phenomenon of velocity of flame propagation in S.I. engine combustion chamber.
4. Describe the phenomenon of knocking and discuss the various factors affecting it.
5. How kncoking can be suppressed?
6. What are the requirements of a good combustion chamber for S.I. engine?
7. Draw neat layout of various types of combustion chambers for s.I. engine.

8. What are the effects of kncoking on engine? How kncoking can be controlled?
9. How S.I. engines are rated?
10. Define HUCR and explain its importance.
11. Explain the phenomenon of pre-ignition in S.I. engines.
12. Explain the combustion stages of C.I. engines.
13. What do you understand by ignition delay? Explain the phenomenon of ignition delay.
14. What are the factors which affect the delay period in C.I. engines?
15. Explain the phenomenon of diesel knock.
16. What are the different types of combustion chambers used in C.I. engines?
17. What are the requirements of combustion chambers for C.I. engines?
18. How C.I. engine fuels are rated?
19. Explain the phenomenon of detonation in C.I. engines.
20. What are the requirements for I.C. engine fuels?

❑❑❑

Chapter 5

Lubrication and Cooling Systems

Inside This Chapter

5.1 INTRODUCTION

All moving parts of an internal combustion engine have relative motion and rub against each other. The lubrication is required to reduce this rubbing action and increases the life of engine. The purpose of lubrication of engine is to reduce the rubbing action between different machine parts having relative motion and remove the heat generated inside the cylinder. Due to friction between rubbing parts and combustion of fuel in the cylinder the temperature of various parts, cylinder and piston may rise enormously if the engine is not cooled. The lubrication system will also be seriously affected leading to damage to cylinder and piston. Therefore, it is essential to provide lubrication and cooling to the various engine parts.

5.2 EFFECT OF ENGINE VARIABLES ON FRICTION

There are many geometrical parameters and physical parameters which are responsible for the frictional loss. Those parameters are considered in this article.

1. **Stroke to Bore Ratio (L/D) :** Lower (L/D) ratio tends to decrease IMEP. Its lower value reduces the friction losses as the surface area (A_s) decreases with decreasing L/D ratio with the same value of the stroke volume as shown in Fig. 5.1.

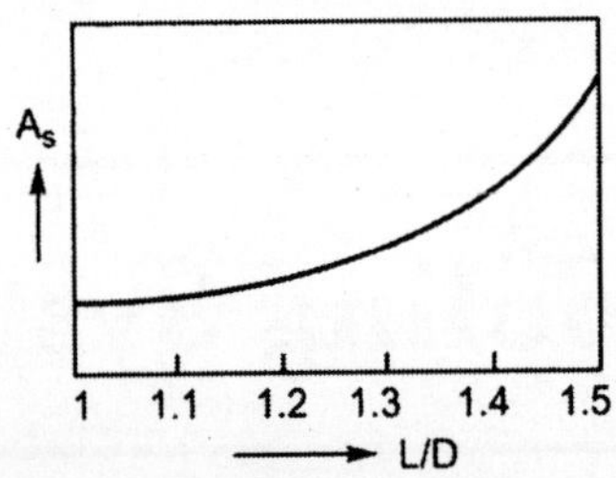

Fig. 5.1

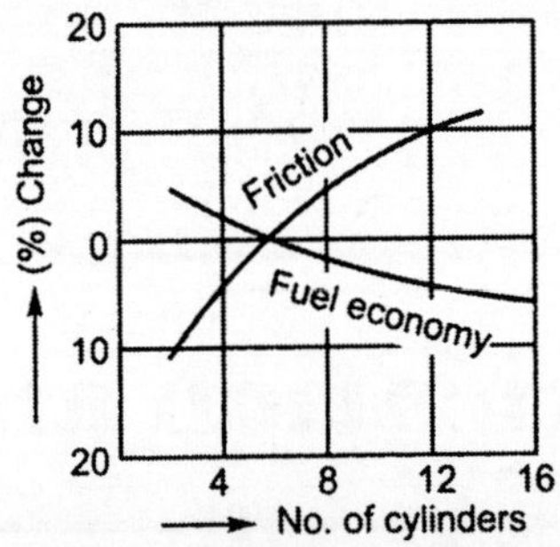

Fig. 5.2

2. **Cylinder Size and Number of Cylinders :** A smaller number of larger cylinders are preferred as the fuel economy is higher. This is because the proportion between piston area and its circumference area is reduced. Fig. 5.2 shows the effect of engine cylinders on the variation of friction for the same piston displacement. It is obvious from the figure that 4 and 6-cylinder engines are more efficient than 8-cylinder engine as fuel economy is concerned.
3. **Compression Ratio :** With an increase in compression ratio, IMEP increases as well as MEP_f by friction also increases but with a lower rate, so the mechanical efficiency of the engine increases with increasing compression ratio, as shown in Fig. 5.3.

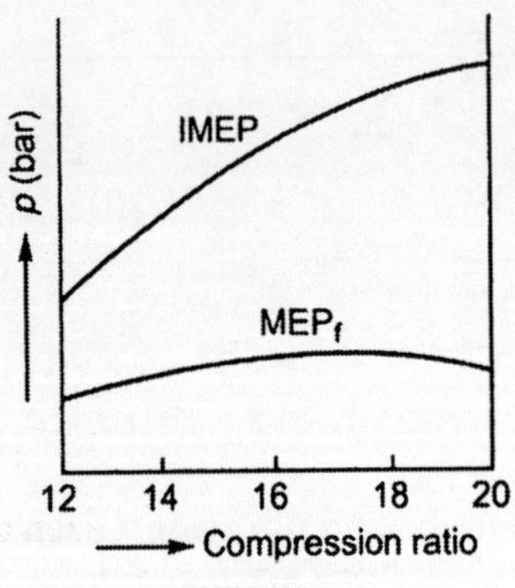

Fig. 5.3

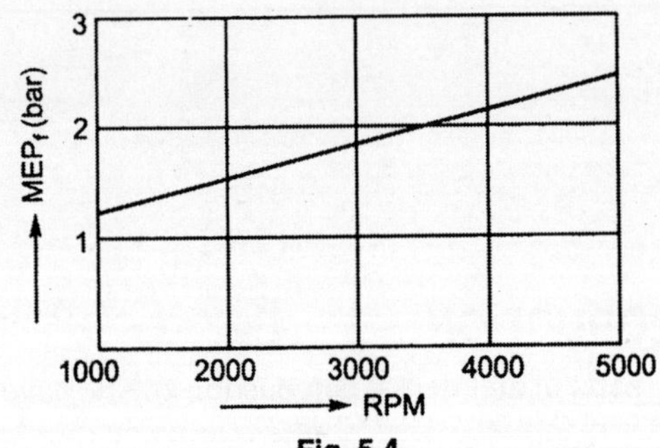

Fig. 5.4

4. **Engine Speed :** The friction loss increases with an increase in speed of the engine as rubbing between cylinder and piston occurs more times as engine speed increases. The best method to improve mechanical efficiency with higher speed engine is to increase the number of cylinders. The variation of frictional mean effective pressure (MEP_f) varies linearly with RPM as shown in Fig. 5.4.
5. **Engine Load :** As the load on the engine increases, the IMEP also increases and friction loss also increases. However, this increase in friction loss is compensated by decrease in viscosity of the lubricating oil due to higher temperature resulting from increased load.
6. **Cooling Water Temperature :** The rise in cooling water temperature reduces the frictional loss as the viscosity of oil at higher temperature is lower which reduces the friction loss.

The starting friction loss is higher as the water oil will be at the same temperature and oil viscosity is higher. This also adds rapid engine wear. The effect of cooling temperature on friction loss is shown in Fig. 5.5.

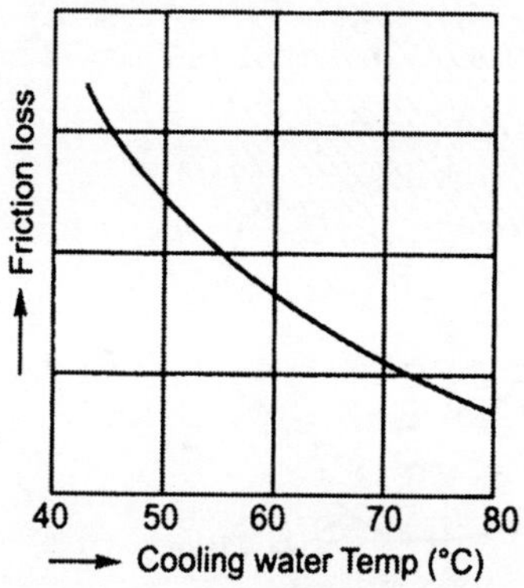

Fig. 5.5

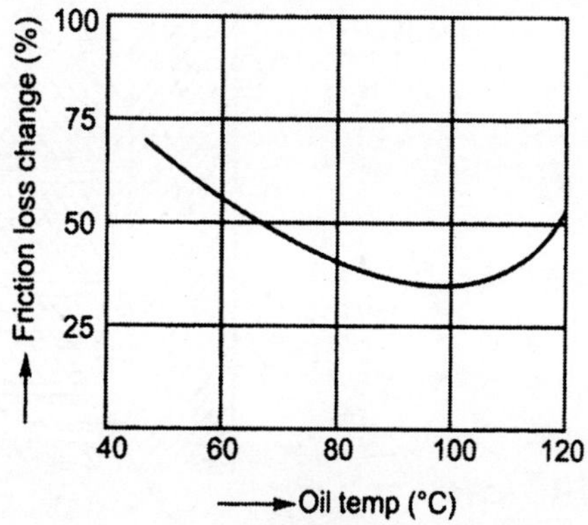

Fig. 5.6

7. **Oil Viscosity :** Higher is oil viscosity, the higher will be the pressure losses. As the oil temperature increases, the viscosity decreases and frictional loss also upto certain temperature of the oil as shown in Fig. 5.6. If the temperature goes higher than 100°C, again friction loss increases as local oil film is destroyed and results in metal to metal contact.
8. **Number of Piston Rings :** The effect of number of piston rings on friction is not significant as the selection depends on the size of the engine, lightness required and material used for rings. However, the effect is shown in Fig. 5.7 and it is obvious that generally 3-rings provide best fuel economy.

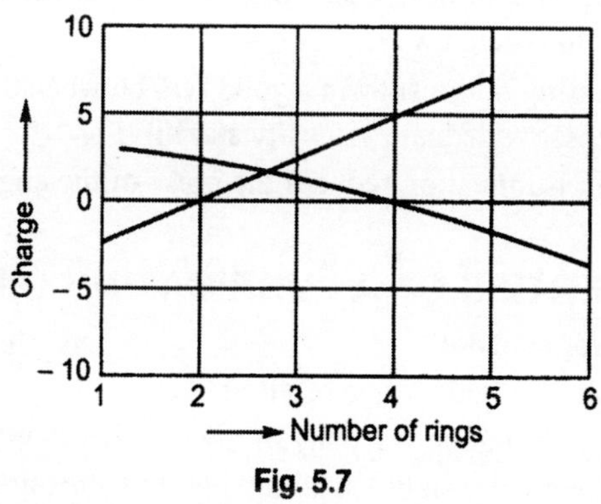

Fig. 5.7

The friction forces by the rings occur due to ring tension and due to gas pressure force behind the rings.

Fig. 5.8 shows the cylinder gas pressure behind the top ring. Because of ring tension, the ring pressure against the cylinder valve results in frictional losses. In addition to ring tension, the gas pressure behind the ring also causes friction loss. The pressure acting on the top piston ring is as high as gas pressure on the piston but it is much lower for other piston rings. For oil ring, no gas pressure acts on it.

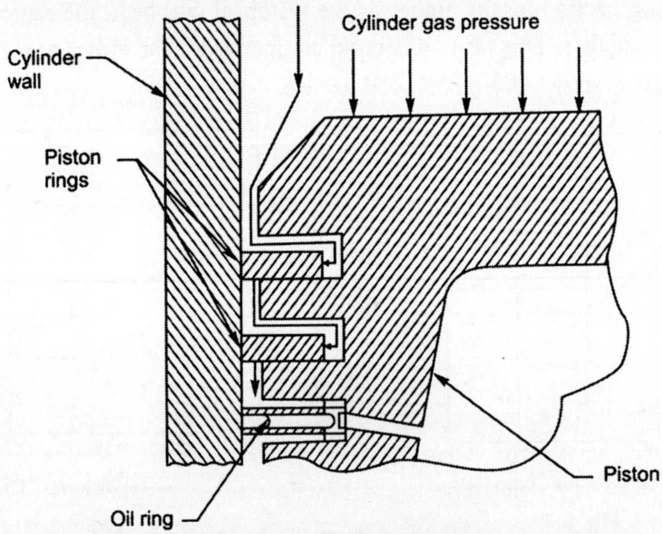

Fig. 5.8 Piston rings arrangement.

5.3 FUNCTIONS OF LUBRICATING SYSTEM

The important functions of a lubricating system are the following :

1. **Lubrication :** The main function of the lubricating system is to keep the moving parts sliding freely past each other and, thus, reduce the engine friction and wear.
2. **Cooling :** To keep the surfaces cool by taking away a part of their heat through the oil passing over them. This requires oil to posses good oxidation stability.
3. **Cleaning :** To keep the bearings and piston rings clean of the products of wear and combustion by washing them away.
4. **Sealing :** The lubricating oil must form a good seal between piston rings and cylinder walls. Therefore, it must posses adequate viscosity stability.
5. **Reduction of noise :** Lubrication reduces the noise of the engine and other sliding parts.

5.4 DESIRABLE PROPERTIES OF A LUBRICATING OIL

The desirable properties of a lubricant are :

1. It should be available in a wide range of viscosities.
2. There should be little change in viscosity of the oil with change in temperature.
3. The oil should be chemically stable at all temperature encountered in its application.
4. The oil should have sufficient specific heat to carry away frictional heat without abnormal rise in temperature.
5. It should be commercially available at a reasonable cost.

5.4.1 Properties of Lubricating Oils

Some of the important properties of lubricating oils are discussed below :

1. **Viscosity :** The viscosity determines the ability of the lubricant to lubricate under hydrodynamic state. It is the internal resistance to flow of liquids which determines the

ability of lubricating oil to support a load on fluid film. It is measured in Pa.s in SI system of units. The viscosity of oils decreases with rise in temperature. The variation in viscosity with temperature is called viscosity index (VI).

2. **Pour point :** It is the temperature where the oil will stop flowing under certain conditions. In hilly areas the power point of oil should be high.
3. **Flash point and fire point :** The flash point is the temperature at which the vapours of oil will just catch fire. The fire point is the temperature at which the oil vapuors will continue to burn. The flash point of oil should be high.
4. **Oiliness :** It is the property of oil to stick to the metal surface to provide a very thin layer of lubricant. Higher oiliness is most disirable property of a good lubricating oil.
5. **Oxidation stability :** It is the resistance to oxidation in the presence of oxygen at high temperature. Oils with higher carbon content form deposits on piston rings and cylinder. Oxidation inhibitors (amine and phenol) are added to oil to improve the oxidation stability.
6. **Acidity and neutralisation :** The lubricating oils should be neutral in reaction.
7. **Cleanliness :** The presence of water promotes corrosion of metal surfaces and insoluble foreign matter cause great wear of engine parts.
8. **Colour :** It is an indication of degree of oil refinement.
9. **Specific gravity :** It is an indication of weight and value and has no relation to lubricating property.

5.5 GRADING OF LUBRICATING OILS

The lubricating oils are classified according to their viscosity. The SAE (Society of Automotive Engineers, USA) has assigned the numbers to different oils and they are commonly used for grading the oil. SAE has assigned a number to oil whose viscosity at given temperature falls in certain range. There are two temperatures used as reference for assigning the number to oils – 18°C and 99°C SAE, 5W, 10W, 20W grades are defined in terms of viscosity at – 18°C and are the oils which render starting of engine in cold weather easy. While SAE, 20, 30, 40 and 50 grades are defined in terms of viscosity at 99°C, are the oils which work satisfactory in normal and hot weather conditions.

The above-mentioned numbers are nearly used for classification of oils according to viscosity and do not indicate the quality of oils since these numbers do not consider the factors like stability, oiliness etc. The grading system of lubricating oils is given in Table 5.1.

Table 5.1 : SAE grading system for lubricating oils

SAE Number	*Viscosity Units*	*Viscosity Range*			
		At – 18°C		*At 99°C*	
		Min.	*Max.*	*Min.*	*Max.*
5W	(a) Centipoise (CP)*	—	1200	—	—
	(b) SUS	—	6000	—	—
10W	(a) Centipoise	1200	2400	—	—
	(b) SUS	6000	12000	—	—

20W	(a) Centipiose	2400	9600	—	—
	(b) SUS	12000	48000	—	—
20	(a) Centistrokes	—	—	5.7	9.6
	(b) SUS	—	—	45	56
30	(a) Centistrokes	—	—	9.6	12.9
	(b) SUS	—	—	58	70
40	Centi-strokes	—	—	12.9	16.8
	(b) SUS	—	—	70	85
50	Centistokes	—	—	16.8	22.7
	SUS	—	—	85	110

* 1 CP = 1 mPa.s

5.5.1 Multigrade Oils

With the advent of additives such as viscosity index improvers, it is possible to develop an oil with more than one viscosity at different temperatures. Thus an oil may be in SAE-30 grade at 99°C and in the SAE-10 W grade at – 18°C oils of this type are known as multigrade oils. Other possible grades are 5W/20, 20W/20, 20W/40.

The multigrade oils have many advantages as listed below :

1. There is no necessity to change the oils as per the ambient temperature.
2. It extends the battery life due to ease of cranking at low ambient temperature.
3. It gives long life to the engine and required viscosity is maintained under different operating temperatures.
4. They provide easy starting and short warming up period and reduced oil consumption of low viscosity oil with low oil consumption of high viscosity oil. This reduces the oil consumption by 1 to 3% due to better lubrication and reduced viscous drag.
5. Most multigrade oils retard the build-up of carbon in the combustion chamber and prolong the mileage between the decarbonization.
6. Multigrade oils have excellent thermal and oxidation stability. They protect the engine parts against rust, corrosion and wear and are used in petrol and diesel engines operating over wide range of temperatures throughout the year.

5.5.2 Service-Grading of Oils

The SAE grading are based only on viscosity and they do not have any relationship to the quality of oil. It is already discussed that different additive influence their operating characteristics under different operating conditions. Therefore, API (American Petroleum Institute) adopted a system which divides the oil into three classes as : Regular type. Premium type and Heavy Duty type depending upon the properties of oil and operating conditions. Straight mineral oils fall under Regular class, oils containing oxidation inhibitors fall under Premium class and oils containing oxidation inhibitions with detergent additive fall under Heavy Duty type.

API further classified the oils used in petrol and diesel engines as the fuel used in both engines are totally different and pressure and temperature conditions differ significantly.

For petrol engines, 5-grades are given as SA, SB, SC, SD and SE (S-stands for spark-ingition and 4 grades are given for diesel engines as CA, CB, CC and CD (C-stands for compression ignition). The grade A stands for light duty and severity of the service increases towards the rating E which is severe duty. Table 5.2 gives the details of API specification.

Table 5.2 : API-Classification for spark ignition engines (S-Series)

Classification	*Description*
SA	Mineral oil, may contain antifoam and/or pour point additive.
SB	Mineral oil containing additive for oxidation stability.
SC	Meets all specification mentioned by Automotive manufactures (1967)
SD	Meets all specifications mentioned by Automotive manufactures (1997)
SE	Meets all specifications mentioned by Automotive manufactures in 1974
API-Classification for Compression Ignition Engines (C-Series)	
Classification	*Description*
CA	Naturally aspirated diesel engine using low sulphur content fuel.
CB	Naturally aspirated diesel engine using high sulphur content fuel.
CC	Super-charged diesel engines (MIL-I-2140 type)
CD	For super-charge diesel engine (Caterpillar series)

The viscosity of oils for different applications are listed in Table 5.3

Table 5.3 : Applications of lubricating oils

Application	*Viscosity in Centi Strokes At 25° C*	*Primary Function*
Engines Oils		
SAE 10W	60 – 90	Lubricate piston rings,
SAE 20	90 – 180	cylinder, valve gear,
SAE 80	180 – 280	Bearing, cool piston;
SAE 40	280 – 450	prevent deposition on
SAE 50	450 – 800	metal surfaces.
Gear Oils		
SAE 80	100 – 400	Prevent metal contact
SAE 90	400 – 1000	and wear of spur gear,
SAE 40	1000 – 2200	hypoid gear, worm gear;
Aviation Engine oils	220 – 700	cool gear cases, serves engine oil
Hydraulic brake fluid	35	Transmit power
Refrigerator oils	30 – 260	Lubricate compressor pump
Steam turbine oil	25 – 300	Lubricate and cool reduction gearing.
Steam cylinder oils	1500 – 3300	Lubricate in presence of steam at high temperatures

5.5.3 New Lubricants for Cheaper Maintenance

A west German company has developed a range of oils greases, pastes and sprays for the solution of lubrication problems for industrial customers. These lubricants function differently from conventional additive-bearing oils including EP (Extreme Pressure) lubricants and even lubricants with dispersed solids like Molybdenum di-sulphide (MoS_2) which are effective through development/formation of protecting films sandwiched between the mating surfaces.

In the new German technology, oil soluble organo-metallic substances of synergetic action are used which create an eutectic state at point of high working pressure and high force density such as in areas of maximum unevenness. This produces surfaces where, compared with the conventional additive-treated lubricants or lubricants with dispersed MoS_2, wear during running-in is claimed to be reduced by a factor of 10^{-2} to 10^{-3}. The eutectic action stops once residual peak-to-valley height of the surface drops to 10^{-3} to 10^{-4} mm. As a result, the actual area of contact increases substantially over conventional and solid dispersed lubricants and the surface is fully protected with much reduced pitting and scoring.

The benefits of this technology are claimed to be many; virtually no running-in, almost immediate running at full load, minimum wear, increased life of lubricant, reduced running noise and minimized precision mechanical surface finishing.

5.5.4 Two-stroke Engine Oils

Two-stroke cycle gasoline engines are used widely in different applications, mainly because of their low cost, high specific power, output using minimum number of moving parts. Some of the main applications are in outboard motors, snowmobiles, auto vehicles, lawn movers, motorcycles and scooters. In India, these two-stroke engine are very widely used in personal transport sector—motor cycles, mopeds and scooters. Of late, the number of motorcycles and scooters in India has rapidly grown.

There are two types of systems for use of two-stroke engine oils. One is called the Petroil system and the other, direct injection system. In the Petroil system, mixing of lubricating oil with petrol in certain proportion is recommended by the manufacturers, whereas in direct injection system, a separate tank is kept for lubricating oil and the quantity of lubricating oil is injected depending upon the speed and load on the engine.

Oil companies have been marketing two types of two-stroke engine oils, commonly known as 2T oils meeting API TA specifications and Super 2T oils meeting API TC specifications. Both are self-mixing oils.

The 2T oil meeting TA specification is designed to minimize spark plug fouling, deposits on pistons, combustion chamber and exhaust ports and to give maximum protection against piston, cylinder wear and bearing corrosion. However, Super 2T oils meeting TC specification have added advantage compared to 2T oils meeting TA specification, as they contain additives which have low ash content resulting in the following advantages: reduced spark plug fouling; high temperature engine cleanliness; reduced piston ring sticking; reduced combusion chamber deposits; reduced exhaust port block; reduction in silencer deposits; ability to prevent bearing rusting at lean oil to fuel ratio; and overall fuel economy and better performance of the engines.

It is because of these advantages that 2T oils conforming to TC specification can be used at lower dosage when compared with oils meeting TA specification. This results in substantial reduction of oil consumption. However, it is pertinent to note here that all over the world, environmental concern has forced manufacturers to change their recommendations to lean fuel to oil ratios and variable oil injection system. By using the oils conforming to TC specification, not only can the country save on precious lubricating oils, but also reduce smoke and hydrocarbon emission.

Keeping the above advantages in mind, *i.e.*, conservation of lube oil and environmental protection, the Indian government has decided to withdraw 2T oils conforming to TA specification and market only 2T oils conforming to TC specification.

In a four-stroke engine, the oil remains in circulation and therefore contains certain additives like anti-oxidant additives, added to increase the life of oil. In two-wheelers, the oil is either mixed with petrol (pre-mixed) or it is injected in the petrol through an injector system. In both these cases, the oils going for lubrication of a two-wheeler engine is supposed to burn completely without leaving any ash. If an ordinary engine oil is used in this application, the additives will leave behind white ash which will get deposited on the spark plug (known as spark plug fouling), resulting in a weakening of the spark. Due to lesser intensity of the spark, the fuel will not burn completely and carbon deposits will start building on the spark plug. Ultimately, the spark will become so weak that the engine will not start. At this stage, the spark plug will need cleaning and will give an impression to the user that non-starting is only due to carbon deposits, whereas the main problem is probably because of the use of a wrong lubricant, *i.e.*, engine oil meant for four-stroke engine.

5.6 GREASE LUBRICATION

A heavily loaded, low speed and large clearance bearings, thin oil is not suitable due to high fluidity.

Greases

Greases are semi-solid materials manufactured by dispersing a thickening agent in lubricating fluid. The lubricating fluid is generally mineral oil and the thickening agent is usually a metallic soap. The soap despersed in the oils forms a pattern of minute cris-crossing fibres which trap the oil and prevent its flow. The mineral components, to be precise, provides the lubrication while the soap component gives the body to the greases. The properties of greases mainly depend on the type of soap and lubricating fluids used. Some of the common soap-base greases are :

- **Calcium soap greases :** They are good general purpose greases and can be used at temperature upto about 50°C. They have good water resistance.
- **Sodium soap greases :** Can be used at higher temperatures than calcium greases, but they are water soluble that is, water breaks down their soap structure.
- **Lithium soap greases :** Are high temperature (upto 180°C) multipurpose greases with good water resistance.
- **Aluminium soap greases :** Have a smooth texture and are water resistant. In addition, they have good adhesive properties.
- **Non-soap greases :** In addition to ordinary soap-base greases, there is another category known as non-soap greases. They are manufactured by dispersing a thickening agent (for example, a specially processed clay or silica gel (in mineral oil or synthetic fluid. Clay gelled greases can be used at higher temperature than soap-based greases, specially when made with synthetic fluids.

Additives

Additives are used in greases in the same way as in lubricating oils to enhance certain properties for certain applications. The main types of additives used are :

- **Anti-oxidant :** Are used to increase oxidation stability.
- **Anti-rust :** Additives are used to improve protection against rusting.
- **Extreme pressure solid additives :** They are used to increase load bearing capacity, molybdenum disulphide is sometimess used to reinforce greases which have to lubricate sliding surfaces for prolonged periods without renewals.

Main control properties

- **Penetration or Consistency value :** Is a helpful guide to the resistance of the grease to permanent change of shape. Penetration is measured by NLGI (National Lubrication Grease Institute) No. Range varying from 000 (very very soft) to 6 (block greases).
- **Drop point :** Drop point is an indication of the temperature that the grease can resist before melting.

Main applications

A grease will stay in place in bearing or in a housing with comparatively elementary mechanical sealing arrangements, and on this account grease is often preferred as an alternative to oil in certain applications, particularly those which are not so easily accessible. On the other hand, the viscosity of a grease at low or moderate rates of shear remains much higher than that of lubricating oils in general, and although this is an advantage in some ways it often imposes limits on the operating speeds at which greases can be used, owing to power losses and heating due to shear or churning at excessive speeds. Moreover, greases are in comparison with oils, inefficient coolants. The maximum operating temperatures at which greases can be used are also limited, in comparison with those applicable to lubricating oils, because of breakdown of greases under heat, comparison with those applicable to to lubricating oils, because of breakdown of greases under heat, although there is considerable variation amongst greases in their thermal stability.

Consistency classification of greases NLGI ASTM worked penetration No. At 25°C.

000	445-475
00	400-430
0	355-385
1	310-340
2	265-295
3	220-250
4	175-205
5	130-160
6	85-115

5.7 LUBRICATING SYSTEMS

There are two types of lubricating systems :

1. Splash lubricating system, and
2. Pressure feed lubricating system

1. Splash Lubricating System

The arrangement of splash lubricating system is shown in Fig. 5.9. This method is generally used for a vertical engine with a closed crankcase. The sump is located at the bottom of the crankcase. When the engine crankshaft rotates, the big end of the connecting rod splashes oil by centrifugal action. The connecting rod big end has a hollow pipe called a scoop which is fitted to the bearing cap and pointed towards the direction of rotation of the crankshaft. The lubricating oil passing through the scoop lubricate the big end bearing and gudgeon pin bearing. All other parts are lubricated by the splash. Excess oil is collected in the troughs and are provided with overflows and collected in the main sump. The level of the oil in the trough is maintained constant. The dripping from the cylinders is also collected in the sump. The oil from the sump is reticulated with the help of a pump.

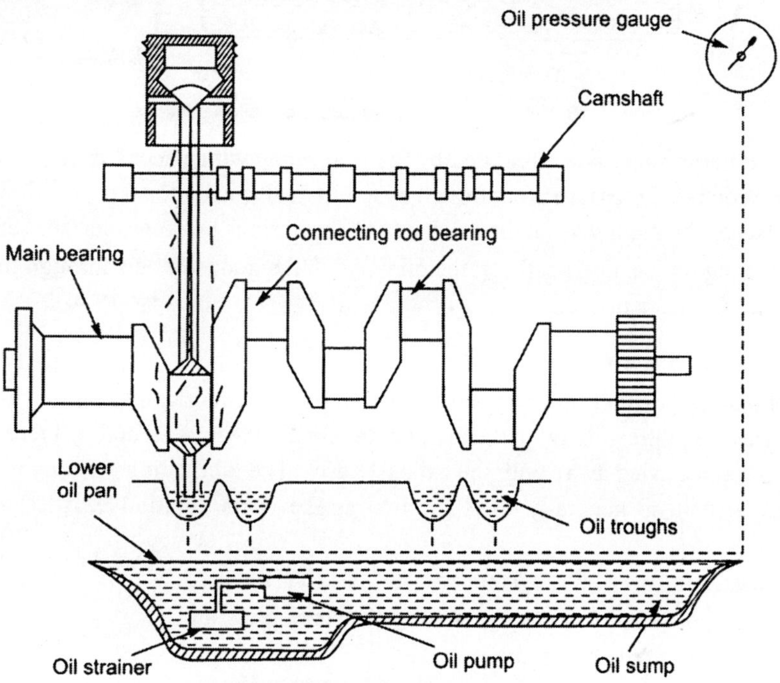

Fig. 5.9 Splash lubrication system.

The limitations of this system are : (i) Inability to regulate the quantity of oil splashed against the cylinder wall. (ii) Inability to keep the oil from getting past the piston head into the combustion chamber, burning with the fuel and passing out with exhaust gases.

2. Pressure Feed Lubricating System

This system is shown in Fig. 5.10. Such a system supplies oil under pressure directly to the connecting rod bearing, crankshaft bearings, valve gear and to the camshaft drive. Indirect supplies reach the cylinder walls, gudgeon pin, the distributor and pump drives.

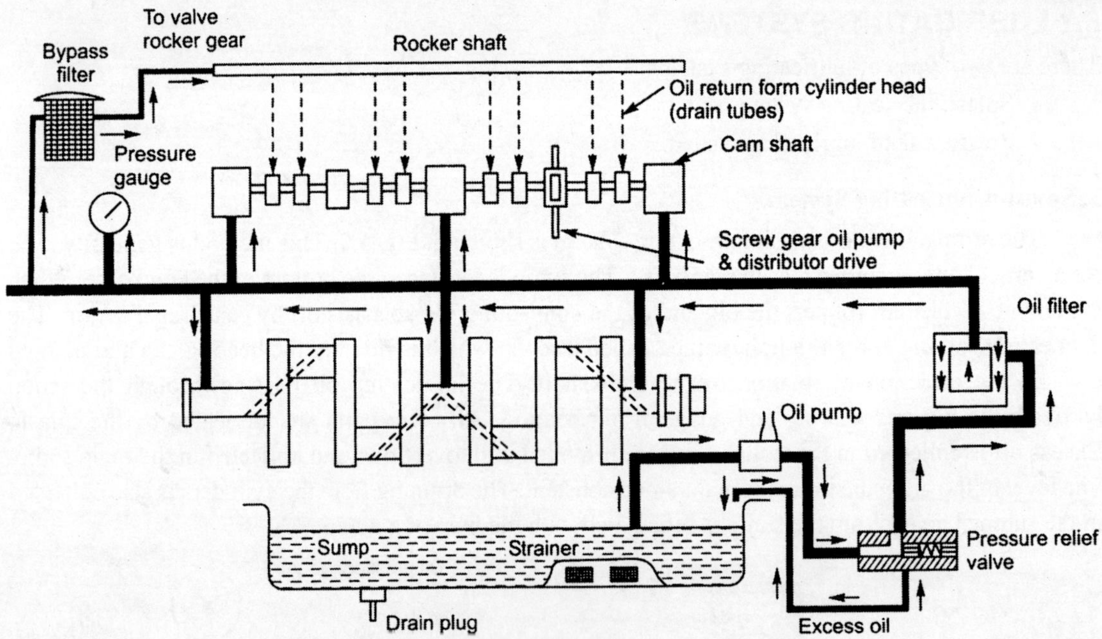

Fig. 5.10 Wet sump high pressure lubrication system.

Oil is carried in the sump and circulated by the gear pump which sucks from the sump through a strainer. The pump delivery pressure is controlled by a relief valve and the oil passes through a very fine filter before it reaches the main distributor gallery from the various bearings, surfaces and gears.

After lubricating the big end bearings, the oil is fed to the gudgeon pins through the oilway in the connecting rod and further squirted into the cylinder wall.

3. Charge Lubricating System

It does not require oil-filter and oil sump. In this system, the lubricating oil is pre-mixed with the petrol which carries the lubricating oil in the cylinder to lubricate the piston and cylinder. Most of the oil burns with the fuel and carried away with the exhaust gases. The lubricating oil cannot be recovered in this system. The system is generally used for two stroke spark ignition engines of scooters and motorcycles.

The main disadvantages of this system are :

(i) Carbon deposits due to burning of oil on the spark plug. and

(ii) Non-recovery of the oil used.

The advantages of this system are :

(i) It does not required separate lubricating system and is most economical.

(ii) There is no risk of failure of lubrication system.

(iii) The lubricating oil supplied is regulated at various loads and speeds by the increased fuel flow.

5.7.1 Lubricating System for I.C. Engines

Various lubricating systems used for internal combustion engines may be classified as :

1. Mist lubrication system

2. Wet sump lubrication system
3. Dry sump lubrication system.

1. Mist Lubrication System

This system is used for 2-stroke cycle engines. Most of these engines are crank charged, *i.e.,* they employ crankcase compression and, thus, are not suitable for crankcase lubrication.

Such engines are lubricated by adding 2 to 3 per cent lubricating oil in the fuel tank. The oil and the fuel mixture is inducted through the cerburettor. The gasoline is vaporised; and the oil, in its form of mist, goes via crankcase into the cylinder. The oil which impinges on the crankcase walls lubricates the main and connecting and bearing, and the rest of the oil which passes on to the cylinder during charging and scavenging periods lubricates the piston, piston rings and the cylinder.

The 2-stroke engine is very sensitive to particular oil and fuel combination. The composition of fuels and lubricants used influence the exhaust smoke, internal corrosion, bearing life, ring and cylinder bore wear, ring sticking, exhaust and combustion chamber deposits, and one of the most irritating and difficult problem of spark plug fouling and whishering. Therefore, specially formulated ashless oils are used for 2-stroke engines.

The fuel/oil ratio used is also important for good performance. A fuel/oil ratio of 40 to 50 : 1 is optimum. Higher ratios increase the rate of wear and lower ratios result in spark plug fouling.

The main advantage of this system is simplicity and low cost because no oil pump, filter, etc., are required. However, this simplicity is at the cost of many troubles some of which are enumerated below :

1. Some of the lubricating oil invariably in combustion chamber. This heavy oil when burned, and still worse, when partially burned in combustion chamber leads to heavy exhaust emissions and formation of heavy deposits on piston crown, ring grooves and exhaust port which interferes with the efficient engine operation.
2. One of the main functions of the lubricating oil is protection of anti-friction bearings, etc., against corrosion. Since the oil comes in close contact with acidic vapours produced during the combustion process, it rapidly loses its anti-corrosion properties resulting in corrosion damage of bearings.
3. For effective lubrication, oil and the fuel must be thoroughly mixed. This requires either separate mixing prior to use or use of some additive to give the oil good mixing characteristics.
4. One important limitation of this system is oil starvation of the working parts especially when the throttle is closed on a descent on a long hill. A closed throttle means no fuel, and, hence, no oil. The prolonged absence of oil so produced may result in overheating and piston seizure. This oil starvation can be controlled if the driver while descending on a hill periodically releases the throttle to replenish for the complete absence of oil.
5. Due to high exhaust temperature and less efficient scavenging the crankcase oil is diluted. In addition, some lubricating oil also burns in combustion chamber. This results in about 5 to 15 per cent high lubrication consumption for two-stroke engines as compared to four-stroke engines of similar size.
6. Since there is no control over the lubricating oil, once introduced with fuel, most of the two-stroke engines are over-oiled most of the time.

Some manufacturers use a separate oil injection pump to inject the oil directly into the carburettor and the amount of the oil is regulated according to engine load and speed. This system completely avoids the oils starvation problem discussed above. The oil consumption is also reduced. This results in less deposits and less spark plug fouling problems. Since in this system the main bearings are excluded from the crankcase and instead receive lubricating oil from a separate pump, the corrosion damage of bearings is also eliminated.

2. Wet Sump Lubricating System

The wet sump lubricating system is shown in Fig. 5.11. In this system, the sump is always full of oil. Oil is drawn from the sump by an oil pump through an oil strainer. A pressure relief valve is provided which automatically maintains the delivery pressure constant. If the pressure exceeds than the predetermined pressure, the valve opens and allows some of the oil to return to the sump and relieves the oil pressure in the system. The oil from the pump goes to the bearings and part of it passes through the filter which removes solid particles from the oil. As all the oil is not passed through the filter, the system is known as by-pass filtering system. The advantage of this system is that a clogged filter will not restrict the flow of oil to the engine.

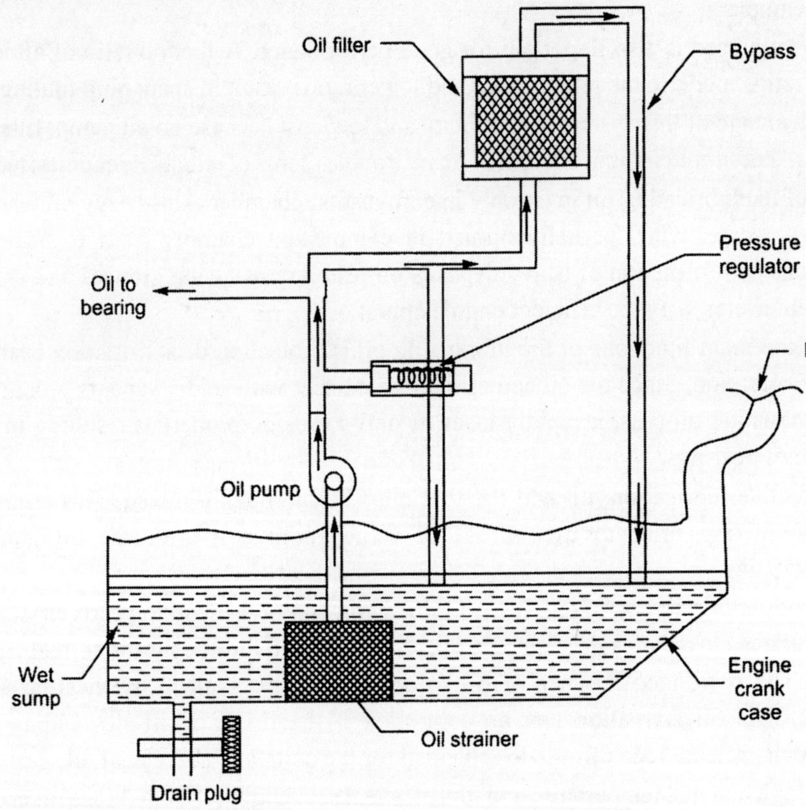

Fig. 5.11 Wet sump lubricating system.

3. Dry Sump Lubricating System

In the dry sump lubricating system shown in Fig. 5.12, the sump does not contain oil and all the oil required for lubrication remains in circulation. High speed racing cars and military jeeps use this type of system.

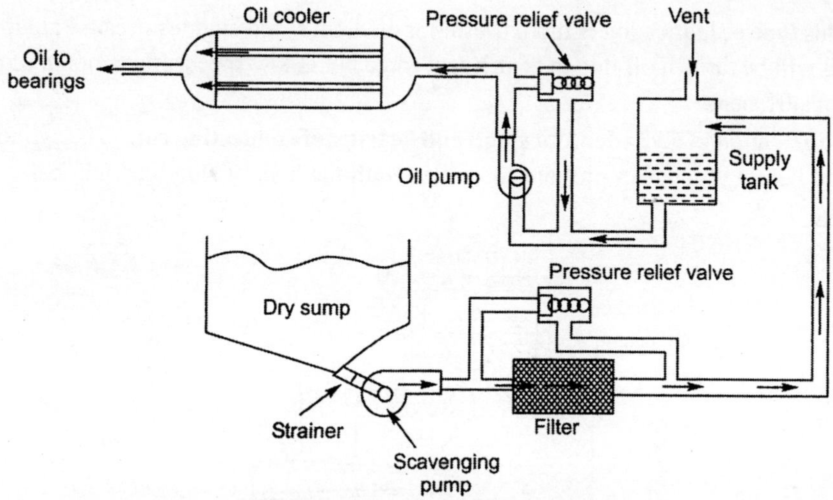

Fig. 5.12 Dry sump lubricating system.

An auxuliary tank is used to supply the oil to the main bearings with the help of the pump. The oil returns back to the tray and then returned to the auxiliary tank by scavenging pump. If the filter gets clogged, the pressure relief valve opens permitting oil to flow by passing the filter and reaches the supply tank. The oil is then recirculated to the bearings from the supply tank. A separate oil cooler is used to cool the oil.

5.8 LUBRICATION OF DIFFERENT ENGINE PARTS

1. **Lubrication of Main Bearings :** The main bearings are lubricated satisfactorily with the help of ring or chain type feeder. The arrangement of the system is shown in Fig. 5.13.

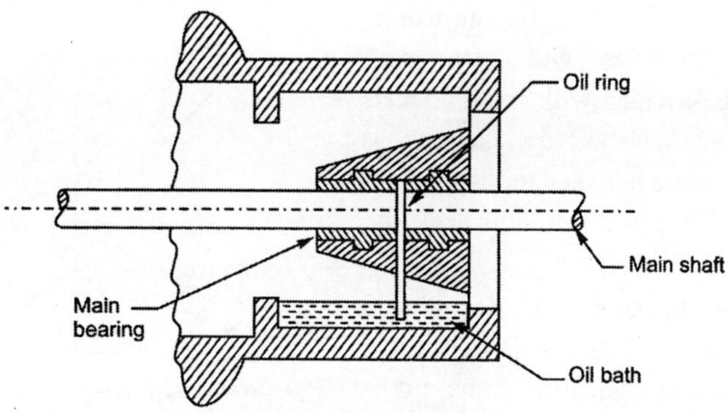

Fig. 5.13 Ring lubrication.

The oil ring rests on the main shaft where a small portion of the main bearing shell has been cut away as shown in the figure. The lower end of the oil ring is allowed to submerge in the oil bath as shown in the figure. The oil ring rotates with the main shaft and carries the oil from the oil bath to the bearing and it is distributed to the bearing through the oil groove. The surplus oil flows to the ends of the bearing and drops back into the oil bath. Chains or more rings are provided instead of single ring to carry more oil.

This type of lubrication is more useful for medium speed engines because at high speeds the oil will be thrown off due to centrifugal force but at low speeds, the amount of oil carried is not sufficient.

2. **Lubrication of Cylinder and small end bearing of connecting rod :** The cylinder and small end bearing (gudgeon pin) are lubricated with the help of side feed lubricator as shown in Fig. 5.14.

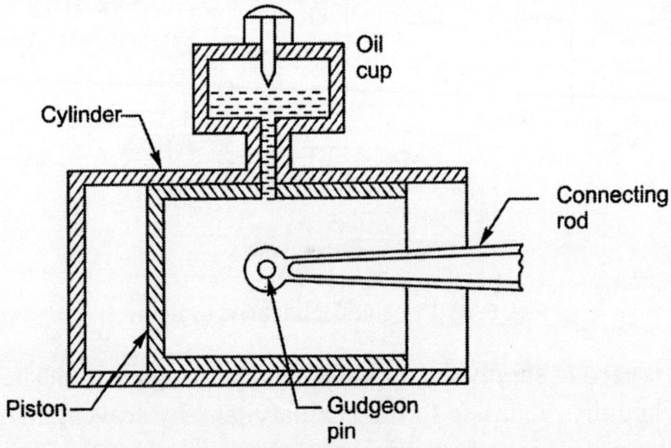

Fig. 5.14 Side feed lubricator.

The oil is supplied to the surface of piston through the hole provided in the cylinder wall. The oil falling on the surface of the piston is spread over the surface of the cylinder walls.

3. **Lubrication of crank and Gudgeon Pin :** The scoops are provided to the caps of big-end-bearing of the connecting rods as shown in Fig. 5.15. These scoops dip into the splash toughs when the pistons are at B.D.C. and splashes the oil in all directions with the rotation of cranks. A thin mist of splashed oil settles on the surface of the cylinder and on the other parts which are to be lubricated. The main bearings and cam shaft bearings are provided with small pockets and the oil splashed by scoops is collected in those pockets and is used for lubrication. The surplus oil falls back into the oil sump.

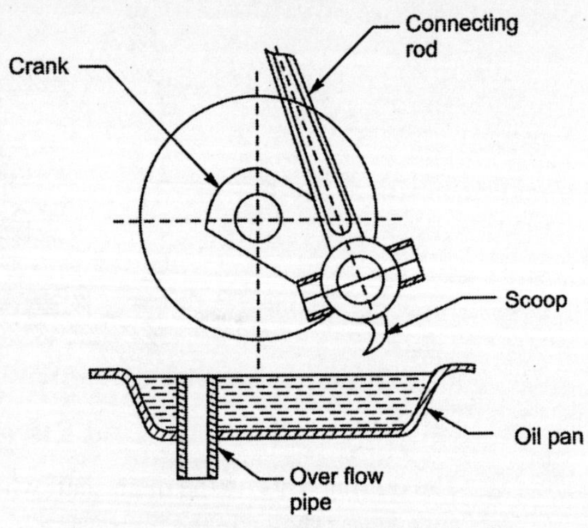

Fig. 5.15 Scoope used in splash lubrication.

Many times, a through hole is provided in the connecting rod and it is connected with the scoop, therefore some oil is supplied to the crank pin and through the hole to the gudgeon pin also. The oil level in the splash tough is maintained with the help of oil pump as shown in the figure. The arrangement of scoop with connecting rod is shown in Fig. 5.21.

4. **Lubrication of big end (crank pin) and small end (Gudgeon pin) of connecting rod :** The crank pin and gudgeon pin are lubricated by ring as shown in Fig. 5.16.

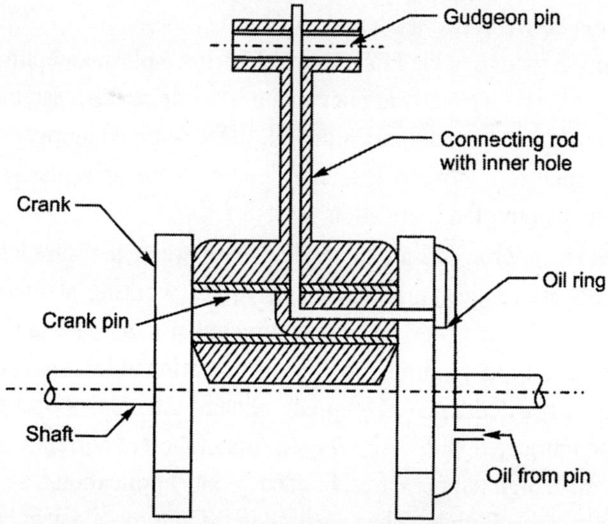

Fig. 5.16 Lubrication system for big end of connecting rod.

The arrangement of the system used is shown in Fig. 5.16. In this system, the oil is supplied to the ring from the oil feed pump which is fitted concentric on the crank shaft. The centrifugal force is given to the oil as the oil ring rotates with the crank shaft and the lubricating oil is forced to the crank pin and to the gudgeon pin through the holes provided in the crank pin and connecting rod as shown in the figure. This system of lubrication is known as Ring lubrication.

5.9 LUBRICATING METHODS FOR BEARINGS

The various methods of lubrication methods are given in Table 5.4.

Table 5.4 : Methods of liquid lubricant supply for plain bearings

Method of supply	*Main characteristics*	*Applications*
1. Hand oiling	Non-automatic, irregular, low initial cost, high maintenance cost.	Low speed, cheap journal bearings.
2. Drip and wick feed	Non-automatic, adjustable, moderately efficient, cheap	Journals in some machine tools, axles
3. Ring and collar feed	Automatic, reliable, efficient, fairly cheap, mainly horizontal bearings.	Journals in pumps, blowers, large electric motors.
4. Bath and splash lubrication	Automatic, reliable, efficient, oil-tight housing required, high initial cost.	Thrust bearings, bath only, engines process machinery, general appliances.
5. Pressure fed	Automatic, positive and adjustable, reliable and efficient, high initial cost.	High speed and heavily loaded journal and thrust bearings in machine tools, engines and compressors.

5.10 LUBRICATION IN GEAR TRANSMISSION SYSTEM

For the satisfactory performance and durability of the gears, proper lubrication of the gear teeth is essential. The main function of the lubricant is to separate the tooth surfaces and remove heat from the gears. The relative motion between two teeth in contact is a combination of rolling and sliding motion. This combination of motion increases the lubrication problems. Splash and pumped lubrication are the two methods by which lubricant is usually applied to a gear system. In splash lubrication, the gears are enclosed in a box and dipped in a bath of lubricating oil. In the pumped lubrication system, a jet of oil is directed towards the meshing teeth. For low velocities, under 15 m/s, the splash system is used, whereas for velocities above 25 m/s, pump fed lubrication is preferred.

Gears are lubricated by grease, straight mineral oils or extreme pressure lubricants. Grease is used as a lubricant for hand-operated gear machanisms or drives working at low speeds and low loads working intermittently. For enclosed gear drives operating under high loads and normal speeds, straight mineral oils provide satisfactory lubrication. Small gears operating at low speeds (below 1200 rpm) and lightly loaded should be lubricated by SAE 80 grade mineral oils. Gears operating at low speeds and heavily loaded should be lubricated with SAE 90 grade mineral oil at normal ambient temperatures and SAE 140 grade mineral oil at high temperatures. For heavy duty applications, EP lubricants are used. EP lubricants are mineral oils with additives. They are used in the automobile gear box and heavy industrial gear boxes. Synthetic oils should be used for aircraft turbine engines.

5.11 LUBRICANTS

Both mineral and synthesis oils are used as lubricants. Mineral oils are basically hydrocarbons, which may be classified as follows :

- **Paraffins :** They contains significant amounts of waxy hydrocarbons but little or no asphaltic matter.
- **Naphatheline :** It contains asphaltic matter in least volatile fractions, but little or no wax.
- **Mixed base :** It contains both waxy and asphaltic materials

Plain mineral oils are used in many units and systems for the lubrication of bearings where their oxidation ability, operating temperature range, ability to prevent wear, etc. are adequate. Special chemicals or additives are added to many oils to improve their properties.

Synthesis oils, most commonly used are : di-ester, silicones, polyglycol, chlorinated diphenyl, polyphenyl ether, fluorocarbon and silicate ester, etc.

A grease may be defined as a solid to semi-fluid lubricant consisting of a dispersion of a thickening agent in a lubricating fluid. The thickening agent may consist of a soap, a clay or a dyestuff. The lubricating fluid is usually a mineral oil, a diester or a silicone. They should be used for operating speeds below 1 to 2 m/s.

The solid fluids are used when fluids are undesirable and ineffective. The solid lubricants may be classified as follows :

Lamellar solids	: MoS_2, WS_2, Graphite, TaS_2 and CaF_2.
Polymers	: FEP, PTFCE, PTFE, Nylons, Acetals and Polyurethanes
Soft metals	: Pb, Au, Ag, Sn, In
Oxides	: MoO_3, PbO/SiO_2, B_2O_3/PbS
Salts	: Na_2WO_4, $PbMoO_4$

5.12 NECESSITY OF I.C. ENGINE COOLING

The cooling is provided to avoid the bad effects of overheating as listed below :

1. The high temperature reduces the strength of the piston and piston rings and uneven expansion of cylinder and piston may cause the seizure of the piston.
2. The high temperature may cause the decomposition of the lubricating oil and lubrication between the cylinder wall and piston and may break down resulting in a scuffing of the piston.
3. If the temperature around the valve exceeds 250°C, the over-heating of the valves may cause the scuff of the valve guides due to lubrication break down. Further increase in temperature may cause the burning of valves and valve seats.
4. The tendency of the detonation increases with an increase in temperature of the cylinder body.
5. The pre-ignition of the charge is possible in spark ignition engines if the ignition parts initially are at high temperature.

5.13 TYPES OF COOLING SYSTEMS

There are mainly two types of cooling system as per the coolant used.

1. Air Cooling System

In this system, air is used as a cooling medium and it is used for small capacity engines. Earlier it is used for big capacity air-craft engines as the water cooling was not practically possible as the weight of water cooling system is very high compared with air-cooling system.

The heat transfer coefficient for air-cooling is very low as mentioned earlier. This heat transfer coefficient can be increased further by using the forced flow of air over the engine surface as done in aero-engines. The heat transfer coefficient with air-cooling with forced circulation is also considerably lower (50 W/m^2-K) compared with water cooling system. The other method of increasing the heat transfer rate from the cylinder surface is to increase the surface area by providing the fins. The use of fins

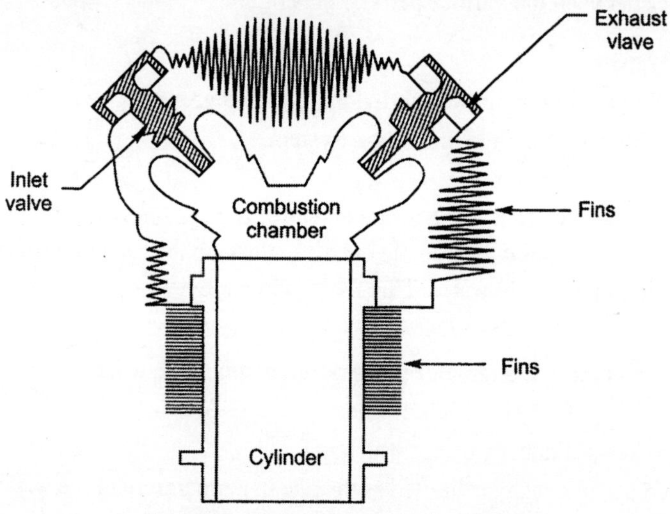

Fig. 5.17 Air cooling of engine cylinder.

increases the heat transfer surface by 5 to 10 times of its original value. The air-cooled systems, the forced circulation of air with increased surface area by using the fins, is commonly used in practice for aero-engines and motor cycle engines.

The sectional view of an engine-cylinder with fins is shown in Fig. 5.17. More fins are used near the exhaust valve and cylinder head where the possibility of occurrence of maximum temperature exists.

The cooling fins are either cast integral with the cylinder and cylinder head or they are fixed to the cylinder block separately. The number of the fins used are 2 to 3 in case of cast fins and 4 to 5 in case of machined fins per centrimetre. The height of fin depends on the type of material and manufacturing process used. Generally the height of fin used lies between 2 cm to 5 cm and fin spacing is limited to 2.5 mm.

Advantages of air cooling :

(i) Water jacket, radiator and water pump are not required. This reduces to size of the engine and its weight.
(ii) Simpler engine design.
(iii) Less sensitive to climatic conditions engine.
(iv) Due to small thermal losses the specific fuel consumption is low.
(v) Better warm-up performance results in low wear of cylinder.
(vi) Sustained engine performance due to reduced carbon deposit on combustion chamber.
(vii) Easier control of cooling systems compared to water cooled engines.
(viii) An air-cooled engine can take up some degree of damage.

Disadvantages of air cooling :

(i) Air fan creates noise.
(ii) The volumetric efficiency is lower due to higher cylinder head temperature.
(iii) High specified output engines can not be air-cooled due to the complex nature of the fins that are required.
(iv) Air cooling results in higher engine temperature. This necessitates the provision of bigger clearances between the various parts of the engine.

2. Water Cooling System

The water cooling system is further subdivided into two groups :

1. Thermo-siphon or Natural Circulation System.
2. Forced circulation system.

1. **Thermo-siphon System :** The arrangement of the system is shown in Fig. 5.18. The force (pressure head) required to circulate the water through the system is the difference in pressure head due to hot and cold water. This force is given by

$$F = h(\omega_c - \omega_h) = h \cdot \Delta\omega.$$

where h = Height of the radiator tubes through which the water is circulated.

ω_c = Weight density of cold water.

ω_h = Weight density of hot water.

The difference in density is limited as the rise in temperature of the water passing through the engine is limited. The rate of circulated is less as the force causing the flow of water is

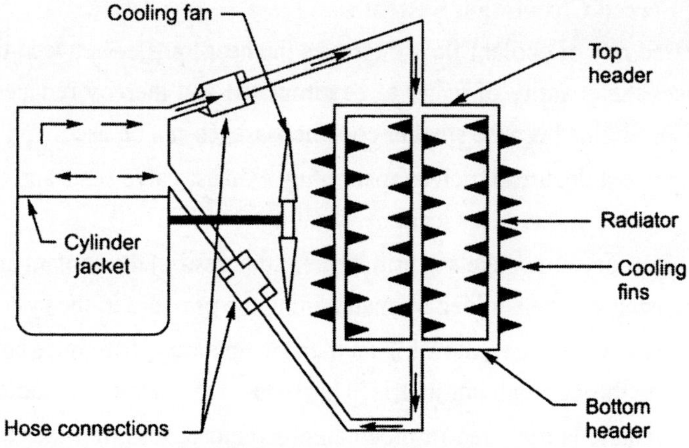

Fig. 5.18 Thermo siphon system.

limited. This cannot be used for heavy duty engines as the heat carrying capacity of this system is limited.

The water as passed through the radiator is cooled by the flow of air passed over the radiator tubes by the cooling fans. The cooled water rises to the cylinder jacket; takes the heat from the cylinder wall and then it enters into the radiator from the top header and comes down. It is cooled as it is passed through the radiator tubes.

The limitations of this system are listed below :

1. The engine should be placed as low as possible in relation to the radiator as the force causing the flow is limited by the temperature difference of hot and cold water.
2. The water level in the system should not fall below the level of the delivery pipe; otherwise the circulation of water in the system will stop. This causes the boiling of water and formation of steam resulting in further loss of water which may damage the engine in a short period.
3. The use of this system is recommended for small capacity engines.
4. The main drawback of this system is that, the cooling depends only on the temperature and is independent of engine speed. The rate of circulation is slow.

Because of many limitations this system is rarely used presently.

2. **Forced Pump System :** In thermo-siphon system, the height of the radiator should be kept above the engine as the flow of water takes place by natural circulation.

To avoid this limitation, a pump is incorporated in the system for water circulation. The arrangement of the system is shown in Fig. 5.19.

The pump is driven from the engine shaft with the help of the belt.

The major advantage of this system is effective and positive cooling of all the parts of the engine. It can easily take the overload as the engine speed increases, water circulation is also increased and same effective cooling can be maintained by this system.

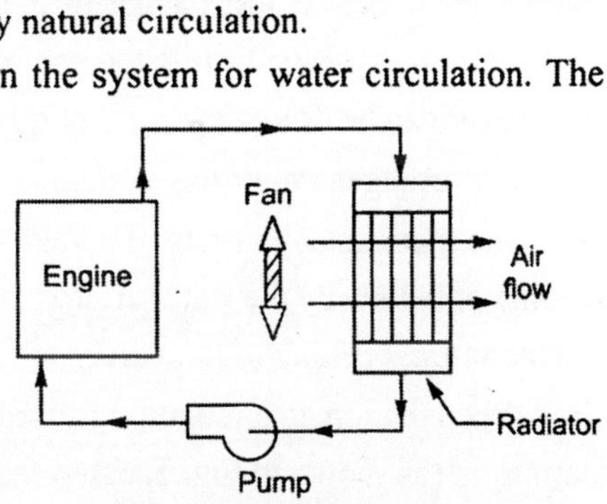

Fig. 5.19 Forced pump system.

Advantages of Forced Circulation System :

1. The increased rate of coolant flow improves the heat transfer between the gases and water. This reduces the quantity of water to be circulated and thereby reduces both the bulk and weight of the engine because smaller coolant passages can be used.
2. The high temperature areas such as spark plug; exhaust valve seats and exhaust ports can be effectively cooled and reduces local overheating.
3. The formation of steam pockets in critical areas that restrict the coolant circulation and hence the heat transfer, can be avoided by maintaining the pressure in the system.
4. Increased coolant velocities allow higher mean temperature difference between the water and the air passing through radiator matrix. This reduces the size of the radiator.
5. Where the vehicle is operated in mountainous regions, loss of water through boiling may occur is thermo-siphon system is used because the boiling temperature of water decreases with the decrease in atmospheric pressure at high altitudes.

Some limitations of this system are listed below :

1. There is possibility of overcooling as the cooling is independent of temperature.
2. There is possibility of overheating the engine when the automobile is climbing a hill where the fuel burning rate is increased and water flow rate is reduced as part of the head is lost to overcome the inclination of the road.
3. As soon as engine stops, the cooling also stops. This is undesirable because the cooling should continue for some time till the temperatures are brought to normal values.

3. Cooling with Thermostatic Regulator

An important feature of the cooling system of a present day cars is inclusion of a thermostat. This is generally fitted in the upper radiator hose and its purpose is to prevent the circulation of water through radiator until the cylinder block and head have reached a temp of about 75°C. If the water circulation starts immediately as the engine has been started (as in the absence of thermostat), the engine cannot run efficiently centi all the water in the cooling system has been raised to a temperature near the boiling point (90°). By prevents, circulation before the engine is warm, the engine gets warm much more quickly. In consequence, there is less danger of the oil being washed away from the cylinder walls by wet petrol drawn into the cylinders while the engine is cold. Less petrol is wasted while the engine is warming up and the car can be driven away more quickly after the engine has been started.

The arrangement of this system is shown in Fig. 5.20.

The thermostat consists of a valve attached to a bellows containing volatile liquid such as ether. Heating of bellow by the water around them causes vaporization of the liquid, expansion of the bellows, and therefore, opening of the valve.

When the forced circulation is used, it is necessary to use a bypass pipe from the engine side of the thermostat (as shown in Fig. 5.20) to the lower water hose so that the circulation pressure of the water cannot force to open the valve before the correct temperature has been reached.

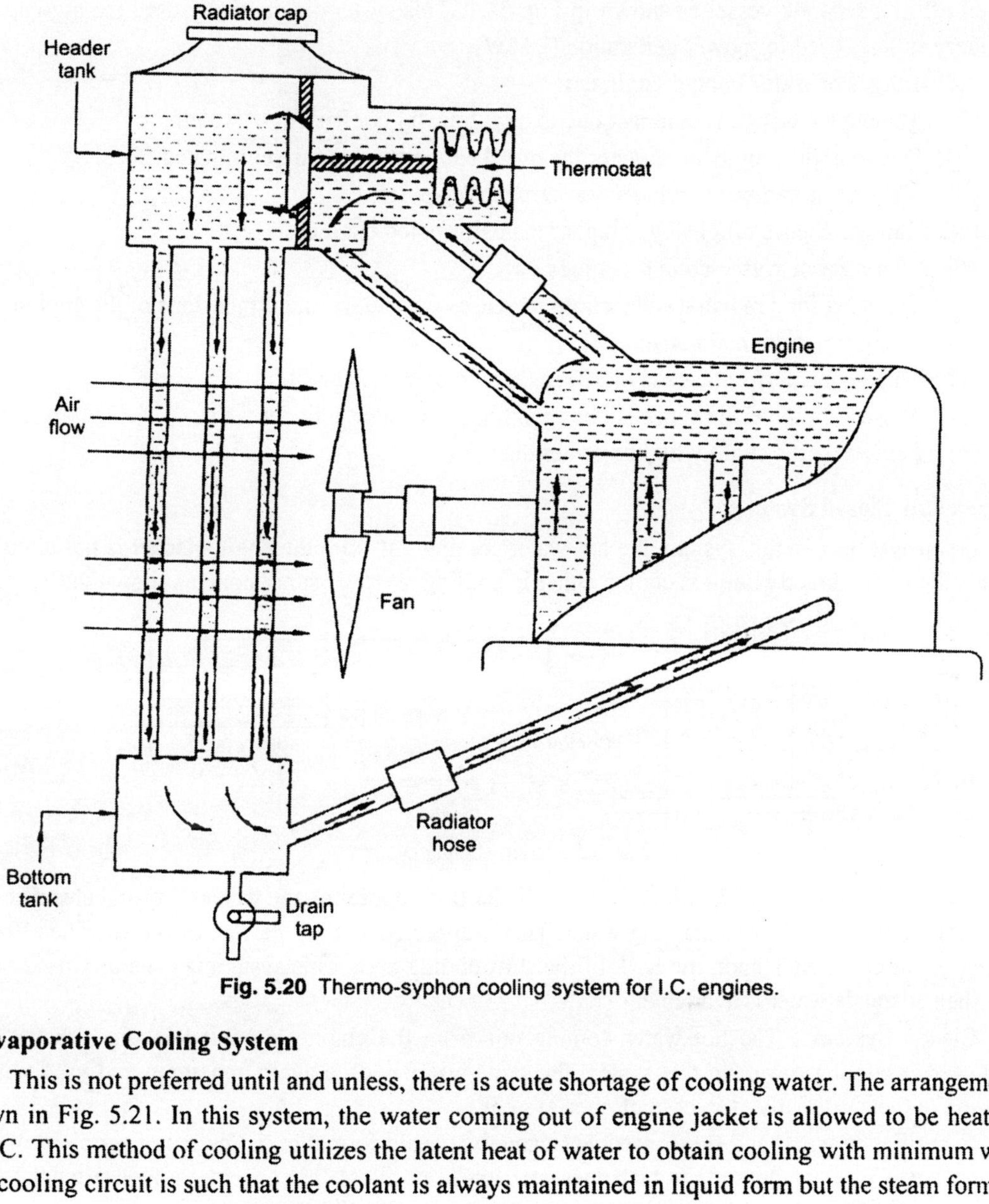

Fig. 5.20 Thermo-syphon cooling system for I.C. engines.

4. Evaporative Cooling System

This is not preferred until and unless, there is acute shortage of cooling water. The arrangement is shown in Fig. 5.21. In this system, the water coming out of engine jacket is allowed to be heated to 100°C. This method of cooling utilizes the latent heat of water to obtain cooling with minimum water. The cooling circuit is such that the coolant is always maintained in liquid form but the steam formed is

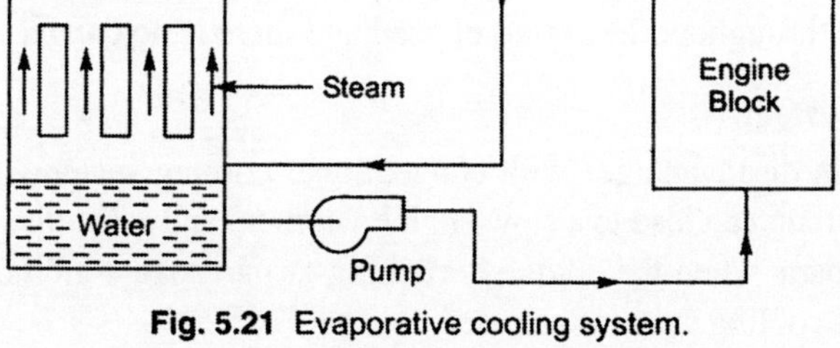

Fig. 5.21 Evaporative cooling system.

flashed off in a separate vessel as shown in Fig. 5.21. This system is generally used for big-capacity stationary engines used for power generation (2 MW).

Advantages of water cooled engines :

(i) The engine design is compact due to high transfer coefficient of water.

(ii) Due to high latent heat of water the over heating troubles are eliminated.

(iii) The engine can be installed any where in the vehicle.

(iv) The volumetric efficiency is higher than air-cooled engine.

Disadvantages of water-cooled engines :

(i) The need for a radiator and a pump increases the weight and dimensions of the engine.

(ii) It requires more maintenance.

(iii) The engine performance is more sensitive to climatic conditions.

(iv) The warm-up performance is poor resulting in greater cylinder wear.

(v) Pump absorbs slightly higher power than fan.

5. Open and Closed System

Open system : In this system, the hot water coming out from the engine jacket is not circulated through the pump directly but it is cooled either in cooling tower or spray pond as shown in Fig. 5.22.

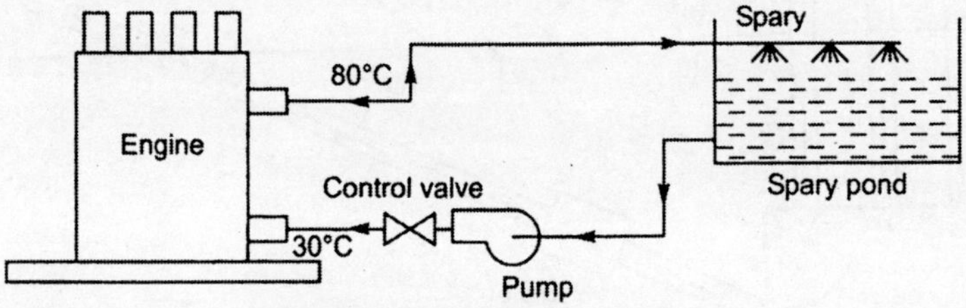

Fig. 5.22 Open cooling system.

With the help of spray, the water is cooled by partly evaporating the water (1%) and absorbing the latent heat required from the remaining water. This evaporated water is carried by the air. The effective cooling of water depends upon the R.H. of the surrounding area. This system is generally used where more than adequate water is available.

Closed System : The hot water coming out from the engine is cooled in the radiator, heat exchanger or cooling tower. In this system the same water is used again and again and used without exposing to the air directly as in case of open system.

Impurities in water and the consequent danger of scale formation is the major problem in open system. Due to water evaporation (which does not carry impurities), the concentration of salt in the water after every pass increases and danger of scaling also increases.

In closed system, the hot water cooling methods are shown in Fig. 5.23.

In closed system, the same water remains in the system and recooled again and again. If it is pure at the start, it remains pure throughout the period of used and there is no danger of scaling.

6. Advanced Cooling System

There are two main disadvantages with conventional cooling systems. Firstly, large volume of coolant in the primary circuit can lead to a slow engine warm up and secondly, the cooling system tends to overcool the engine parts when the engine is running at part load conditions. The overcome these difficulties, an advanced cooling systems are used.

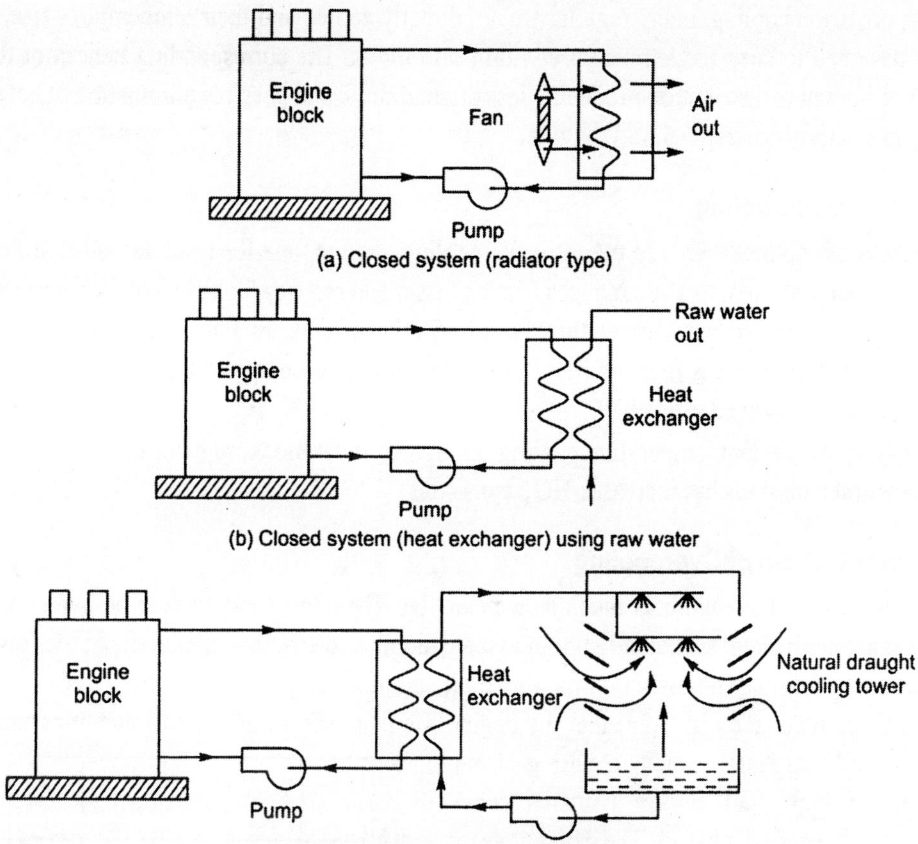

Fig. 5.23 Closed cooling systems.

The main purpose of these systems to maintain more uniform temperature and less sensitive to the load and speed of the engine. The friction loss between piston and cylinder is reduced if the liner temperature is raised. The significance of reduction in friction loss increases as load is reduced because this leads to reduction in fuel consumption. Rise in liner temperature may increase the emission of NO_x. This can be reduced by either retarding the ignition timing or using exhaust gas recirculation (EGR). Both these remedies impose a fuel consumption penalty and fuel saving gained by reducing the frictional loss is lost. The NO_x emission is problematic in diesel engine at full load conditions.

With S.I. engines, another problem that can occur at full load is the reduction in the knock margin caused by higher cylinder temperatures. This also can be remedied by EGR or retarding the ignition timing.

A few methods used for this purpose are described in short.

5.13.1 Precision Cooling

It is developed for cooling diesel engines and to provide cooling only where it is needed; at a rate proportional to the local heat flux.

A precision cooled system has small local passages which are used to cool critical regions such as injector nozzle, exhaust valve region and valve guides, with much space in the head filled with air.

With precision cooling, many regions are not directly cooled and their temperatures rise, but the system is designed to keep the temperature within safe limits. The corresponding benefit of this is, a more even temperature distribution producing less thermal strain and possible elimination of hot spots to allow higher compression ratios in S.I. engine.

5.13.2 Dual Circuit Cooling

In this process of cooling, there are two separate cooling circuits, one for head and other for cylinder block. The higher block temperature reduces friction losses and reduces fuel consumption particularly at part loads. The lower coolant temperature in cylinder head tends to reduce the indicated thermal efficiency but risk of knock is reduced in S.I. engines and higher compression ratio can be used. The overall effect is an improved efficiency.

Increasing the coolant temperature passing through cylinder block, reduces the fuel consumption and hydrocarbon emissions but increases NO_x emissions.

5.13.3 Disadvantages of Overcooling

Should we cool the engine as much as we can? No The engine must never be overcooled. The engine must always be kept sufficiently hot to assure smooth and efficient operation. At too low engine temperatures starting may be difficult and above all, the low temperature corrosion assumes such a significant magnitude (see Fig. 5.24) that the engine life is greatly reduced. At low temperatures, the sulphurous and sulphuric acids resulting from combustion of fuel (fuel always contains some sulphur) attack the cylinder barrel. The dew points of these acids vary with pressure and hence the critical temperature, at which corrosion assumes significant properties, varies along the cylinder barrel. To avoid condensation of acids the coolant temperature should be greater than 70°C. Thus, the cooling system should not only cool but must also keep the cylinder liner temperature above a minimum level to avoid corrosion and ensure good warm up performance of the engine.

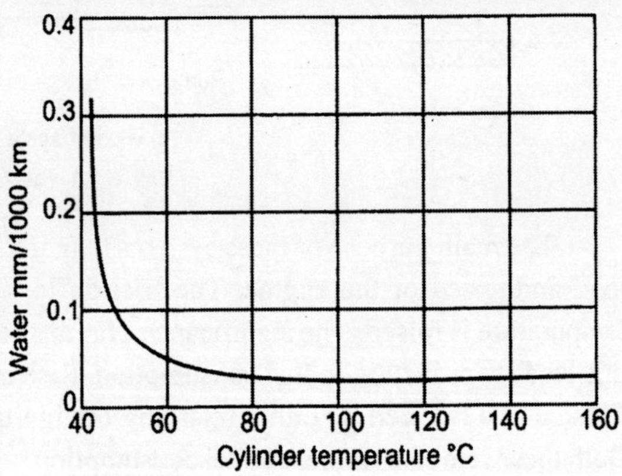

Fig. 5.24 Effect of temperature on corrosion.

5.14 ENGINE RADIATORS

The function of the radiator is to cool hot water coming out of engine with the help of air. In the radiator, the heat transfer from hot water to the air takes place by conduction and convection.

The cooling effect in the radiator is enhanced by providing the fins over the surface of tubes carrying the hot water to be cooled. The fins are also so arranged that some turbulence is generated in air passages which further helps to increase heat transfer coefficient to the air side.

A commonly used cross-flow type radiator is shown in Fig. 5.25. The fins are not shown in figure.

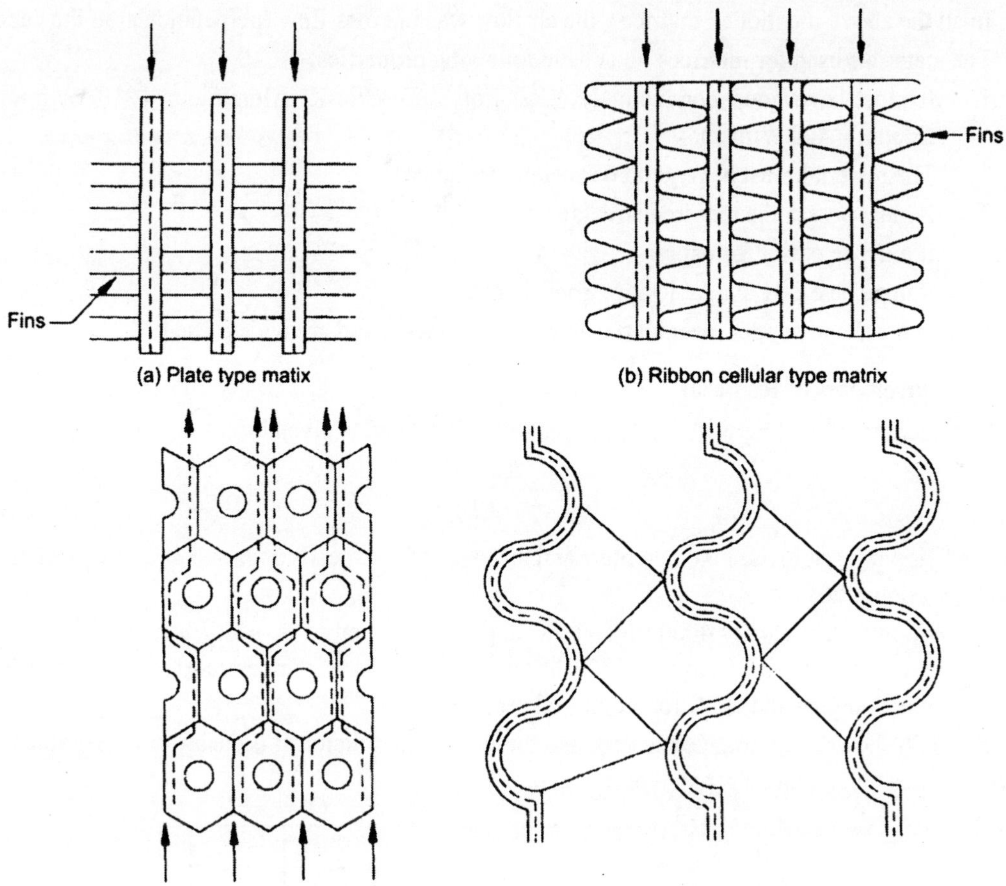

(a) Plate type matix (b) Ribbon cellular type matrix

(c) Corrugated fin matrix (d) Honey comb matrix

Fig. 5.25 Different types of radiator matrics.

Radiator Matrix

Figure 14.21 shown 4-different types of matrices (fins) commonly used in practice.

In the tube and fin type shown in Fig. 5.25 (a), a series of long tubes extending from top to bottom of the radiator are surrounded by simple straight metallic fins. The matrix has greater structural strength.

The improvement over the tube and fin type is tube and corrugated fin type matrix as shown in Fig. 5.25 (b). The water tubes are made of oval shape section and zig-zag copper ribbons are used to provide secondary heat transfer areas and air turbulence. This combines the good heat transfer characteristics of film type matrix and structural strength of the tube and fin type matrix.

Figure 5.25 (c) shows the film type matrix which is also known as ribbon-cellular matrix. It consists of a pair of thin metal ribbons soldered together along their edges so as to form a water-way running from header tank to the bottom tank of the radiator. The zig-zag copper ribbons forms an air passage. In this case, metal surface area is significantly increased and higher turbulence is also created.

Earlier radiators used honey-comb matrix as shown in Fig. 5.25 (d). The hexagons were packed in contact and bound by solder. This provided a continuous water passage between the circular parts of the tube. The cooling air was passed through the circular tubes. This type of matrix is rarely used in present types of radiators.

In all the above-mentioned matrices, the air flow was in cross-flow (perpendicular to the paper). The materials used for matrices must have following properties.

1. It must have good conductively (Cast iron = 60 W/m-K, Aluminium = 210 W/m-K and Copper = 400 W/m-k).
2. The material must have good corrosion resistance.
3. It must posses the required strength.
4. It must be easily formable.
5. It must be easily available at reasonable cost.

Yellow brass and copper meet all the above requirements and are widely used.

Water Requirements of Radiator

The quantity of cooling water (Q) required to be circulated is given by

$$Q = \frac{K \times BP}{(\Delta T)}$$

where (ΔT) is permissible rise in temperature and the value of K depends upon specific fuel consumption and compression ratio.

The cooling water carries 4000 kJ/kWh for large engines and 550C kJ/kWh for small engines. If the rise in temperature is limited to 10–15°C, the outlet cooling water temperature is limited to 60°C for small or medium engine and 80°C for automobile engines.

Fig. 5.26 shows the amount of air required for air-cooled radiator. It is clear from the graph, petrol engine requires much more air than diesel engine.

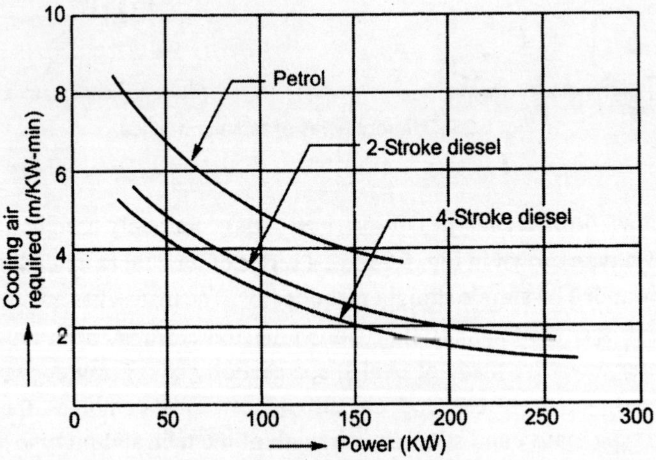

Fig. 5.26

The power required for cooling the engine is given by

$$P_{\text{cooling}} = \frac{A_e (IP)^{2.25}}{(A_f \cdot \Delta T_f)^{5.75} \rho_a^2}$$

where A_f and ΔT_f are the fin area and temperature difference between the fin average surface temperature and air temperature and ρ_a is air density. A_e is effective area.

The above equation is developed with following assumptions.

1. A : F ratio supplied to engine is constant.
2. The temperature change in fin is assumed small.
3. The density and temperature of the cooling media are assumed to be unaffected by its flow through the radiator fin.

Fans

As mentioned earlier, an engine driven fan is mounted behind the radiator to increase the air flow over the radiator tubes. This increases the air-side heat transfer coefficient and ultimately overall heat transfer coefficient.

Radiator fan is usually made of pressed steel blades, four or six in number depending upon the size of the radiator (capacity of engine). The power consumed by the fan depends upon the amount of air handled, diameter of the fan; its shape and speed of the engine.

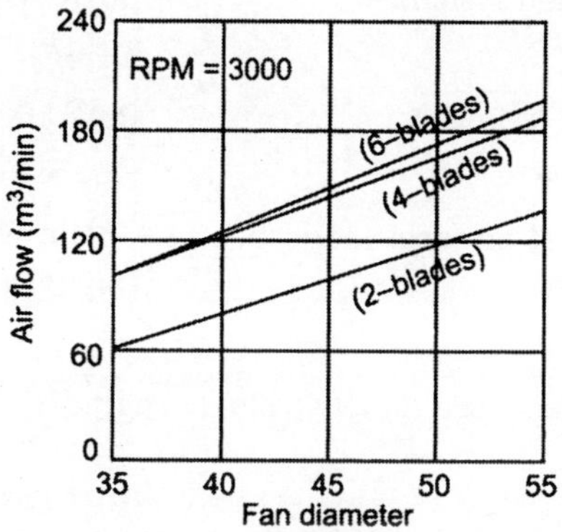

Fig. 5.27

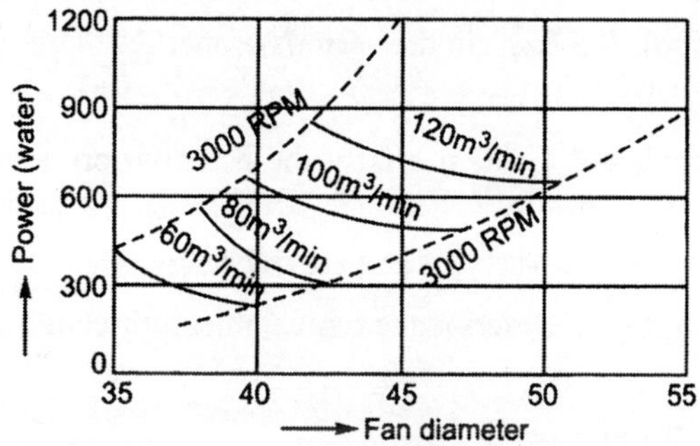

Fig. 5.28

Fig. 5.27 shows the effect of speed and diameter of fan on the quantity of air flow. Fig. 5.28 shows the effect of fan diameter and air flow on the power consumed by the fan. For the given flow rate of air, the power required decreases slightly as the fan diameter increases, while for the same RPM, the power required increases with the fan diameter.

5.15 COOLING OF EXHAUST VALVE

In the liquid cooled engines the spaces around the exhaust valve seating should be sufficiently large for the coolant to flow. Sometimes, the water circulation provides the jets of water to be directed on the valve seating metal.

The efficient valve cooling requires methods of conducting the heat from the valve head to the cylinder, improved guides and metal, also to the lower part of the valve stem operating under cooler air conditions.

In order to enhance the conductivity of the exhaust valve, the practice of making the valve stem and even the heads hollow is becoming popular. The hollow stem is filled with sodium which melts at 97°C and boils at 880°C. Thus at valve operating temperatures, the valve is filled with conducting material. Such a valve is shown is Fig. 5.29. With a tapered plug seal driven into the hollow stem. The end of the stem is provided with Stellite button, fused on the seat to prolong the wearing life of stem and end.

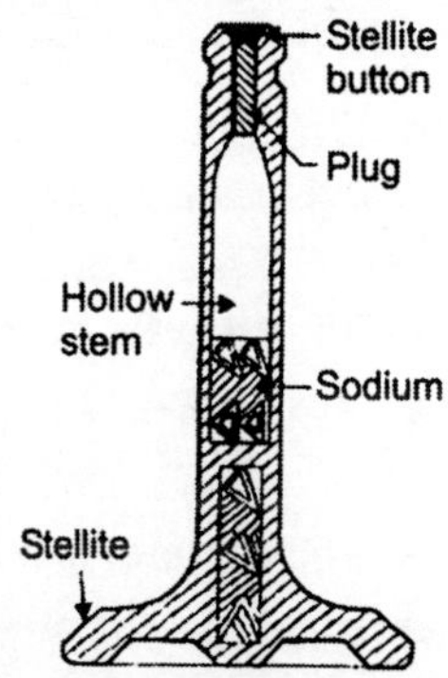

Fig. 5.29 Cooling of exhaust valve.

SOLVED QUESTIONS FROM UNIVERSITY EXAMINATIONS

■ **Example 5.1** What are the various types of lubrication systems ? With the help of a neat sketch explain the wet sump lubrication system. [MDU, Dec., 2007]

Solution : Refer to art. 5.7.1.

■ **Example 5.2** With the help of neat sketch explain the wet sump and dry sump system used for lubrication. [MDU, May, 2008]

Solution : Refer to art. 5.7.1.

■ **Example 5.3** Name various types of lubrication systems. Explain mist lubrication system. Compare mist lubrication with other lubrication systems. [MDU, Dec. 2008]

Solution : Refer to art. 5.7.1.

■ **Example 5.4** State essential requirements of a lubrication system. [MDU, Dec. 2008]

Solution : Refer to art. 5.3.

■ **Example 5.5** Explain the various properties of an ideal lubricant. [MDU,Dec., 2009]

Solution : Refer to art. 5.4

■ **Example 5.6** Explain briefly the essential properties of an ideal lubricant and explain wet sump lubrication briefly. [MDU Dec., 2010]

Solution : Refer to art. 5.4 and art. 5.7.1.

■ **Example 5.7** Describe the typical mist lubrication system used for a two stroke engine. [KU, Dec., 2005]

Solution : Refer to art. 5.7.6.

■ **Example 5.8** (a) What are the functions of lubricating system ? Explain.

(b) What is the necessity of engine cooling ? Explain. Also explain the disadvantages of overcooling. [KU, Sept., 2009]

Solution : (a) Refer to art. 5.3.

(b) Refer to art. 5.12.

■ **Example 5.9** (a) What are the functions of Lubricating system ?

(b) Explain Wet Sump Lubricating system. [KU, Dec., 2010]

Solution : (a) Refer to art. 5.3.

(b) Refer to art. 5.7.1.

■ **Example 5.10** What is the necessity of engine cooling and what are the disadvantages of overheating ? What is the difference between air cooling and water cooling system ? [MDU, Dec., 2007]

Solution : Refer to art. 5.12 and art. 5.13.

■ **Example 5.11** Explain the properties of a lubrication oil. What is the SAE rating of lubricants ? [MDU, May, 2008]

Solution : Refer to art. 5.4 and art. 5.5.

■ **Example 5.12** What is the necessity of engine cooling ? State the effect of overcooling. Compare air and water cooling. [MDU, May, 2009]

Solution : Refer to art. 5.12 and art. 5.13.3.

■ **Example 5.13** Explain thermostat cooling through a neat sketch. What will happen if thermostat is not working in hot weather ? [MDU, May, 2009]

Solution : Refer to art. 5.13.

■ **Example 5.14** Explain the following with neat sketches :

(a) Evaporative cooling system.

(b) Pressure cooling system. [MDU, Dec., 2009]

Solution : (a) Refer to art. 5.13, Sl. No. 4.

(b) Refer to art. 5.13, Sl. No. 2.

■ **Example 5.15** (a) What is effect the following on engine friction : (i) Stroke/Bore ratio, (ii) CR, (iii) Oil viscosity, (iv) Engine load, and (v) Cooling water temperature.

(b) Explain with neat sketch the `Thermo-syphon cooling.' [KU, Dec., 2008]

Solution : (a) Refer to art. 5.2.

(b) Refer to art. 5.13.

Example 5.16 (a) Explain the necessity of Cooling system.

(b) What are the disadvantages of Over cooling ? Explain. [K.U. Dec., 2009]

Solution : (a) Refer to art. 5.12.

(b) Refer to art. 5.13.3.

■ **Example 5.17** Explain the two types of cooling systems and compare them. [MDU, Dec., 2010]

Solution : Refer to art. 5.13.

■ **Example 5.18** (a) Describe with the help of a suitable sketch the following :

(i) Wet Sump Lubrication System

(ii) Dry Sump Lubrication System.

(b) What are the various types of cooling system used in I.C. Engines ? With the help of neat sketch explain the function and working of a radiator. [KU, Dec., 2006]

Solution : (a) Refer to art. 5.7.1.

(b) Refer to art. 5.13.

■ **Example 5.19** (a) Discuss with the help of suitable sketch the following :

(i) Wet sump lubrication system.

(ii) Dry sump lubrication system.

(b) What is the function of a radiator ? Explain the construction of a typical radiator. [KU, Dec., 2007]

Solution : (a) Refer to art. 5.7.1.

(b) Refer to art. 5.14.

SHORT ANSWER TYPE QUESTIONS

1. List the engine variables which affect friction in I.C. engines.
2. What are the functions of a lubricating system?
3. Define : viscosity, pour point, flash point and oiliness.
4. Expand S.A.E. and A.P.I.
5. What are multigrade oils?

6. What are Extreme Pressure (EP) lubricants?
7. What are 2T oils?
8. What are greases and where they are used?
9. What are the lubricating systems for I.C. engines?
10. What is the necessity of cooling system for an I.C. engine?
11. Name the cooling systems generally used for I.C. engines.
12. What is overcooling? What are its disadvantages?

EXERCISES

1. Why lubrication and cooling of I.C. engine components necessary ? Explain.
2. What are the effects of engine variables on friction ?
3. What are the functions of lubricating system ?
4. What are the desirable properties of a lubricating oil ?
5. What are the properties of lubricating oils ?
6. Explain the SAE grading system of lubricating oils.
7. What are multigrade oils ?
8. Differentiate between splash and pressure lubrication system and describe their operation with the help of neat diagrams.
9. What is meant by dry and wet sump lubrication? Where dry sump is preferred and why?
10. Discuss the two-stroke engine oils.
11. Explain the lubrication of the following parts :
    (a) Crank pin and gudgeon pin (b) Main bearing
    (c) Cylinder and piston (d) Exhaust valve.
12. Discuss the lubricating systems.
13. What is the necessity of I.C. engine cooling ?
14. Discuss the types of cooling systems for I.C. engine.
15. What are the effects of under cooling and over cooling of an engine ?
16. Draw a neat diagram of thermostatic controlled water cooling system and explain its working.
17. Discuss the merits and demerits of water cooling with air cooling system.
18. Define the function of a radiator. Discuss different types of matrices used with these radiators.
19. What is the function of fan in radiator system ? Discuss its performance with change of speed and load on the engine.
20. What do you understand by evaporative cooling ? When it is preferred ?
21. What is the difference between open system and close system cooling used for I.C. engines ? Discuss their relative merits and demerits.
22. Discuss the relative maerits and demerits of air cooling system over water cooling system.

❑❑❑

Chapter 6

Engine Testing and Performance

Inside This Chapter

6.1 PERFORMANCE PARAMETERS

The engine performance is an indication of the degree of success with which the conversion of chemical energy contained in the fuel is done into useful mechanical work. The degree of success is compared on the basis of following parameters :

(i) Specific fuel consumption (SFC).

(ii) Brake mean effective pressure (BMEP)

(iii) Specific power output (SP)

(iv) Specific weight (SW)

(v) Exhaust stroke and other emissions.

However, in the evaluation of engine performance, the following basic performance parameters are chosen :

1. Power and mechanical efficiency.
2. Mean effective pressure and torque.
3. Specific output.
4. Fuel-air ratio.
5. Volumetric efficiency.

6. Specific fuel consumption.
7. Thermal efficiency and heat balance.
8. Exhaust smoke and other emissions.
9. Specific weight.

Indicated power, $$\text{IP} = \frac{p_{im}\,\text{LAN}}{60 \times 10^3 \times n}\ \text{kW} \qquad \text{...(6.1)}$$

where p_{im} = indicated mean effective pressure, Pa

$$= \frac{\text{area of indicator diagram, mm}^2}{\text{length of indicator diagram in mm} \times \text{spring constant (N/mm)}}$$

L = Length of stroke, m

A = Area of stroke, m^2

N = rpm

n = number of strokes per revolution

= 2 for four-stroke engine

= 1 for two-stroke engine

Brake power, $$\text{BP} = \frac{p_{bm}\,\text{LAN}}{60 \times 10^3 \times n}\ \text{kW} \qquad \text{...(6.2)}$$

where p_{bm} = brake mean effective pressure, Pa

Frictional power, $$\text{FP} = \text{IP} - \text{BP} \qquad \text{...(6.3)}$$

Mechanical efficiency, $$\eta_{mech} = \frac{\text{BP}}{\text{IP}} = \frac{p_{bm}}{p_{im}} = \frac{\text{BP}}{\text{BP} + \text{FP}} \qquad \text{...(6.4)}$$

Torque, $$T = \frac{p_{bm}AL}{2\pi n} \qquad \text{...(6.5)}$$

Specific output : It is defined as the brake output per unit of piston displacement.

$$\text{Specific output} = \frac{\text{BP}}{\text{AL}} \qquad \text{...(6.6)}$$

Volumetric efficiency,

$$\eta_v = \frac{\text{Actual volume inhaled during suction stroke at NTP}}{\text{Swept volume of the piston}}$$

$$= v_a / v_s \qquad \text{...(6.7)}$$

Brake specific fuel consumption,

$$\text{BSFC} = \frac{60 \times m_f}{\text{BP} \times t}\ \text{kg/(BP.h)} \qquad \text{...(6.8)}$$

where m_f = mass of fuel consumed, kg

t = time in minutes

Indicated specific fuel consumption,

$$\text{ISFC} = \frac{60 \times m_f}{\text{IP} \times t}\ \text{kg/(IP.h)} \qquad \text{...(6.9)}$$

Brake thermal efficiency, $$\eta_{bt} = \frac{BP \times 60}{m_f \times LCV} \quad ...(6.10)$$

where LCV = lower calorific value of fuel, kJ/kg

Indicated thermal efficiency, $$\eta_{it} = \frac{IP \times 60}{m_f \times LCV} \quad ...(6.11)$$

Relative efficiency, $$\eta_r = \frac{\eta_{bt}}{\eta_a} \text{ or } \frac{\eta_{it}}{\eta_a} \quad ...(6.12)$$

where η_a = air standard efficiency

Fuel-air ratio, $$F/A = \frac{\text{Mass of fuel}}{\text{Mass of air}} \quad ...(6.13)$$

Air-fuel ratio, $$A/F = \frac{\text{Mass of air}}{\text{Mass of fuel}} \quad ...(6.14)$$

Relative fuel-air ratio, $$F_R = \frac{\text{Actual fuel-air ratio}}{\text{Stoichiometic fuel-air ratio}} \quad ...(6.15)$$

Equivalence ratio, $$\phi = \frac{\text{Stoichiometic A/F ratio}}{\text{Actual A/F ratio}} \quad ...(6.16)$$

$$\text{Specific weight} = \frac{\text{Weight of the engine}}{\text{Brake power, BP}} \quad ...(6.17)$$

Exhaust smoke and other emissions : Oxides of nitrogen, unburned hydrocarbons, etc.

6.2 BASIC ENGINE MEASUREMENTS

The basic measurements required to be carried out to evaluate the performance of an engine are :

1. Speed
2. Fuel consumption
3. Air consumption
4. Smoke density
5. Brake power
6. Indicated power and friction power
7. Heat loss to cooling water
8. Heat going to exhaust
9. Exhaust gas analysis.

1. Speed Measurement

The angular speed of engine (crankshfat) may be measured by a tachometer. Tachometers may be classified into two categories :

(i) Mechanical tachometers, and

(ii) Electrical tachometers.

They can further be classified as contact type and non-contact type.

Contact type tachometers are :

(a) Mechanical type : Revolution counter and timer, slipping clutch tachometers, centrifugal free tachometers.

(b) Electrical type : Drag cup tachometer, electromagnetic tachogenetator. Non-contact type tachometers are : Photoelectric tachometer, variable reluctance tachometer, stroboscope, capacitance pick up tachometer.

2. Fuel Consumption Measurement

The fuel consumption of an engine is measured by determining the volume flow in a given time interval and multiplying by the specific gravity of the fuel. Another method is to measure the time required for consumption of a given mass of fuel. Accurate methods for fuel measurement are :

(i) Two glass-chambers method.

(ii) Rotameter, and

(iii) Positive displacement metre.

3. Measurement of Air Consumption

Accurate measurements for air consumption can be carried out by the following methods :

(i) Orifice chamber method

(ii) Viscous-flow air meter

4. Measurement of Exhaust Smoke

A smoke meter may be used, which is a basically sort density measuring device.

5. Measurement of Heat Carried Away by Exhaust Gases

An exhaust gas calorimeter is commonly used for the measurement of heat carried by exhaust gases.

Mass of air supplied per *hg* of fuel,

$$m_a = \frac{N \times C}{33(C_1 + C_2)} \qquad ...(6.18)$$

Heat carried away by exhaust gas per kg of fuel,

$$Q_g = (m_a + 1)c_{pg}(T_{ge} - T_a) \text{ kJ/kg of fuel} \qquad ...(6.19)$$

where $(m_a + 1)$ = mass of exhaust gases formed per kg of fuel suppled.

C_{pg} = specific heat of exhaust gases, kJ/kg.k

T_{ge} = temperature of exhaust gases coming out from engine, °C

T_a = temperature of ambient air, °C.

6. Measurement of Heat Carried Away by Cooling Water

The heat carried away by cooling water is generally measured by measuring the water flow rate through the cooling jacket and the rise is temperature of water during flow through the engine. The inlet and outlet temperatures of water are measured by thermometers inserted in the pockets provided at inlet and outlet from the engine. The quantity of water flowing is measured by collecting the water in a bucket for a specified period.

Heat carried away by cooling water,

$$Q_w = c_{pw}m_w(T_{w0} - T_{w_i}) \text{ kJ/min} \qquad ...(6.20)$$

where m_w = mass of water collected/min

c_{pw} = specific heat of water, kJ/kg °C

T_{wi} = inlet temperature of water, °C

T_{w0} = outlet temperature of water, °C

6.3 HEAT BALANCE SHEET

A heat balance sheet is an account of heat supplied and heat utilised during engine trial.

(a) Heat supplied to the engine,

$$Q_s = \dot{m}_f \times \text{L.C.V.} \quad \ldots(1)$$

where $\dot{m}_f$ = mass of fuel supplied per hours.

L.C.V. lower calorific value of fuel, kJ/kg

(b) Heat utilised in various ways :

(i) Heat equivalent of *BP*,

$$Q_1 = 3600 \times kw \text{ kJ/h} \quad \ldots(2)$$

(ii) Heat carried away by cooling water,

$$Q_2 = c_{pw} \cdot m_w (T_{w0} - T_{w_i}) \text{ kJ/h} \quad \ldots(3)$$

(iii) Heat carried away by exhaust gases,

$$Q_3 = c_{pg} \cdot m_g (T_{ge} - T_a) \text{ kJ/h} \quad \ldots(4)$$

(iv) Heat unaccounted for,

$$Q_4 = Q_s - (Q_1 + Q_2 + Q_3) \quad \ldots(5)$$

6.4 WILLAN'S LINE METHOD

It is a method to determine the frictional power of an engine. In this method, gross fuel consumption *v*'s brake power at a constant speed is plotted and the graph is extrapolated back to zero fuel consumption as shown in Fig. 6.1. The point where this graph cuts the BP axis is an indication of the friction power of the engine at that speed. The test is applicable to only compression ignition engines.

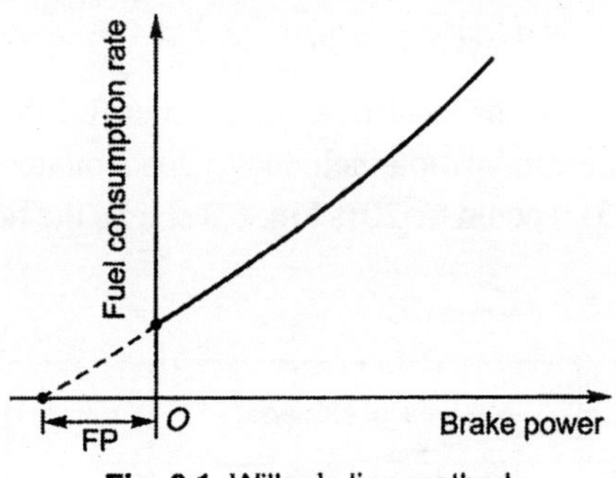

Fig. 6.1 Willan's line method.

6.5 MORSE TEST

This is a method to determine the indicated power of a cylinder is a multi-cylinder engine. In this method, the engine is first run at the required speed and the output is measured. Then one cylinder is cut out by short circuiting the spark plug or by disconnecting the injector as the case may be. The output is measured by keeping the speed constant at its original value. The difference in the output is a measure of the indicated power of the cutout cylinder. Thus for each cylinder the IP is obtain and is added together to find the total IP the engine.

IP of *n* cylinder, $\quad \text{IP}_n = \text{BP}_n + \text{FP}_n$

IP of $(n-1)$ cylinders, $\quad \text{IP}_{n-1} = \text{BP}_{n-1} + \text{FP}_n$

IP of *n*th cylinder, $(\text{IP})_{nth} = \text{BP}_n - \text{BP}_{n-1} = \text{IP}_n - \text{IP}_{n-1}$

Total IP of engine, $\text{IP}_n = \Sigma(\text{IP})_{nth}$

Frictional power of the engine, $\text{FP} = \text{IP}_n - \text{BP}_n$ ...(6.21)

6.6 PERFORMANCE OF SI ENGINES

The performance of an engine is generally given by heat balance-sheet. The main components of the heat balance are (i) heat equivalent to the effective (brake) work of the engine, (ii) heat rejected to the cooling medium, (iii) heat carried away from the engine with the exhaust gases, (iv) unaccounted losses. The unaccounted losses include the radiation losses from the various parts of the engine and heat loss due to incomplete combustion. The friction loss is not shown as a separate item to the heat balance sheet as the friction loss ultimately reappears as heat in cooling water, exhaust and radiation.

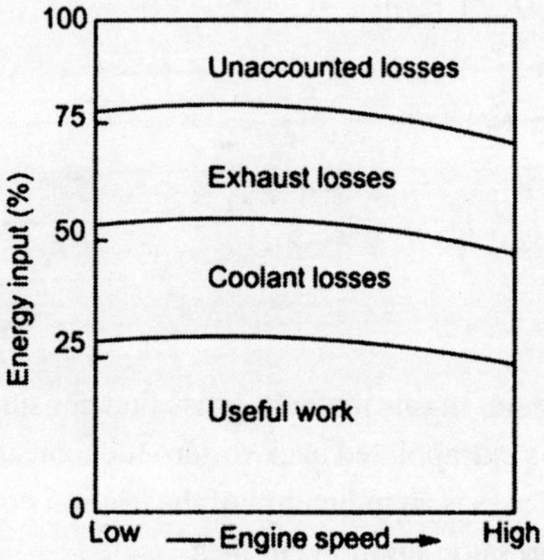

Fig. 6.2 Heat balance diagram for a typical SI engine.

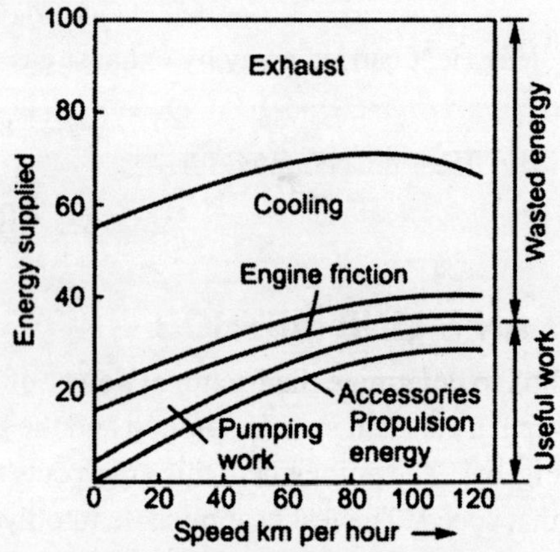

Fig. 6.3 Uncontrolled otto engine.

Fig. 6.2 shows the heat balance for a petrol engine run at full throttle over its speed range. In SI engines the loss due to incomplete combustion included in unaccounted losses can be rather high. For a rich mixture (A/F ratio 12.5 to 13) it could be 20% Fig. 6.3 shows the heat balance of uncontrolled Otto engine at different loads.

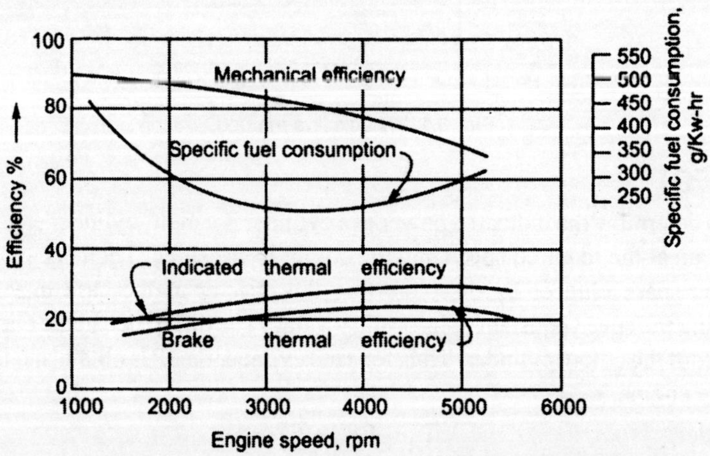

Fig. 6.4 Efficiency and specific fuel consumption vs. speed for a petrol engine at full throttle.

Fig. 6.4 shows the brake thermal efficiency, indicated thermal efficiency, mechanical efficiency and specific fuel consumption for the above SI engine.

Fig. 6.5 shows the IP, BP, FP (by difference) brake torque, brake mean effective pressure and brake specific fuel consumption of a high compression ratio (9) automotive SI engine at full or wide open throttle (W.O.T.)

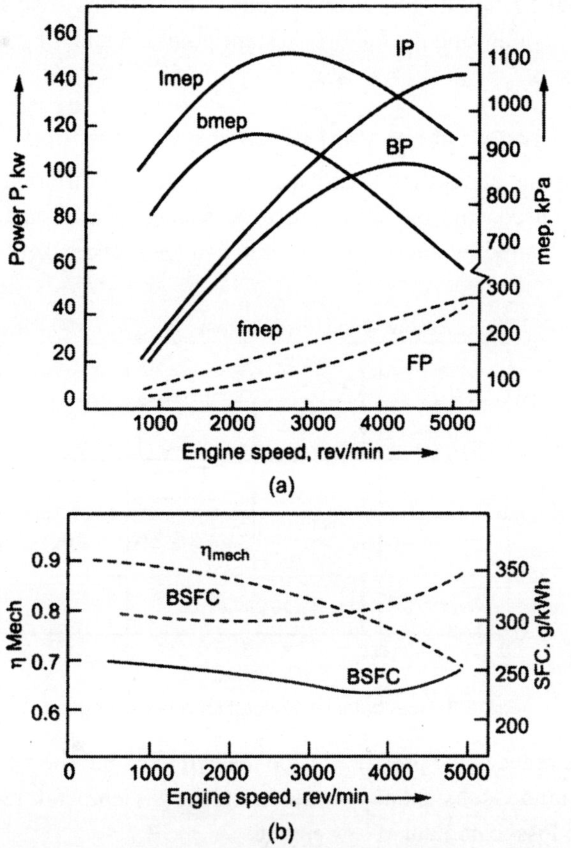

Fig. 6.5 Variable speed test of automotive SI engine at full throttle (CR = 9).

The following conclusion can be drawn from the above figures :

1. At full throttle the brake thermal efficiency at various speeds varies from 20 to 27 per cent, maximum efficiency being at the middle speed range.
2. The percentage heat rejected to coolant is more at lower speed (>> 35 percent) and reduces at higher speeds (>> 25 per cent). Considerably more heat is carried by exhaust at higher speeds.
3. Torque and mean effective pressure do not strongly depend on the speed of the engine, but depend on the volumetric efficiency and friction losses. Maximum torque position corresponds with the maximum air charge or maximum volumetric efficiency position.

   Torque and mep curves peak at about half that of the horse-power.

   Note : If size (displacement) of the engine were to be double, torque would also double, but mean effective pressure (mep) is a 'specific' torque – a variable independent of the size of the engine.

4. High horsepower arises from the high speed. In the speed range before the maximum horsepower is obtained, doubling the speed doubles the power.
5. At low engine speed the friction horsepower is relatively low and b.p. is nearly as large is i.p. As engine speed increases, however, F.P. increases at continuously greater rate and therefore b.p. reaches a peak and starts reducing even through i.p. is increasing. As engine speeds above the usual operating range. F.P. increases very rapidly. Also at these higher speeds I.P. will reach a maximum and the fall off. At some point I.P. and F.P. will be equal, and B.P. will then drop to zero.

6.6.1 Performance of SI engine at constant speed and variable load

The performance of SI engine at constant speed and variable load is different from the performance at full throttle and variable speed, Fig. 6.6 shows the heat balance of a SI engine at constant speed and variable load. The load is varied by altering the throttle and the speed is kept constant by resetting the dynamometer.

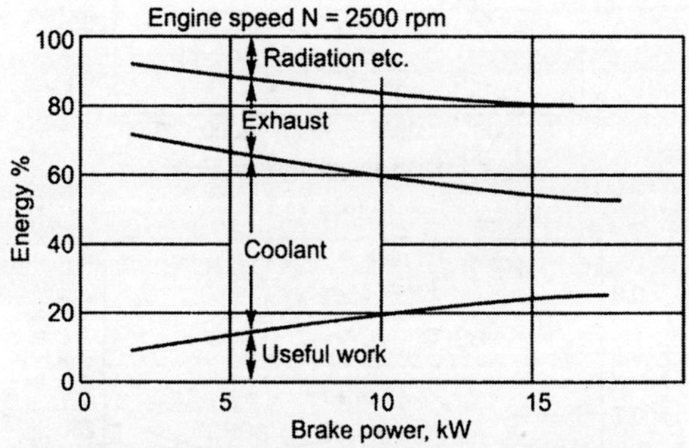

Fig. 6.6 Heat balance vs. load for a petrol engine.

Closing the throttle reduces the pressure inside the cylinders but the temperature is affected very little because the air/fuel ratio is substantially constant, and the gas temperatures throughout the cycle are high. This results in high loss to coolant at low engine.

At low loads the efficiency is about 10 percent, rising to about 25 percent at full load.The loss to coolant is about 60 percent at low loads and 30 percent at full load. The exhaust temperature rises very slowly with load and as mass flow rate of exhaust gas is reduced because the mass flow rate of fuel into the engine is reduced, the percentage loss to exhaust remains nearly constant (about 21% at low loads to 24% at full load). Percentage loss to radiation increases from about 7% at low loads to 20% at full load.

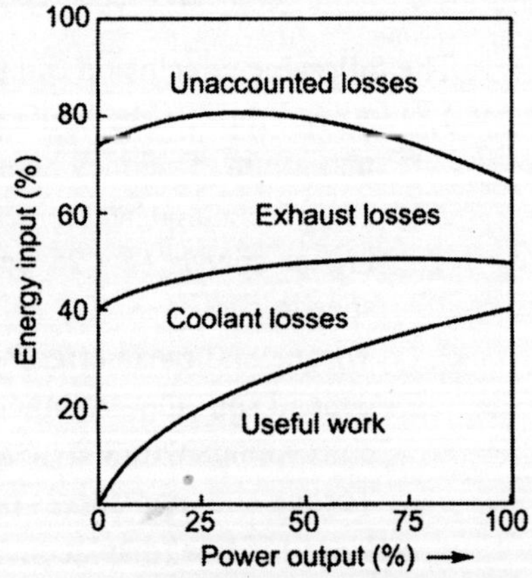

Fig. 6.7 Heat balance diagram for a typical CI engine.

6.7 PERFORMANCE OF CI ENGINES

The performance of a CI engine at constant speed variable load is shown in Fig. 6.7. As the efficiency of the CI engine is more than the SI engine the total losses are less. The coolant loss is more at

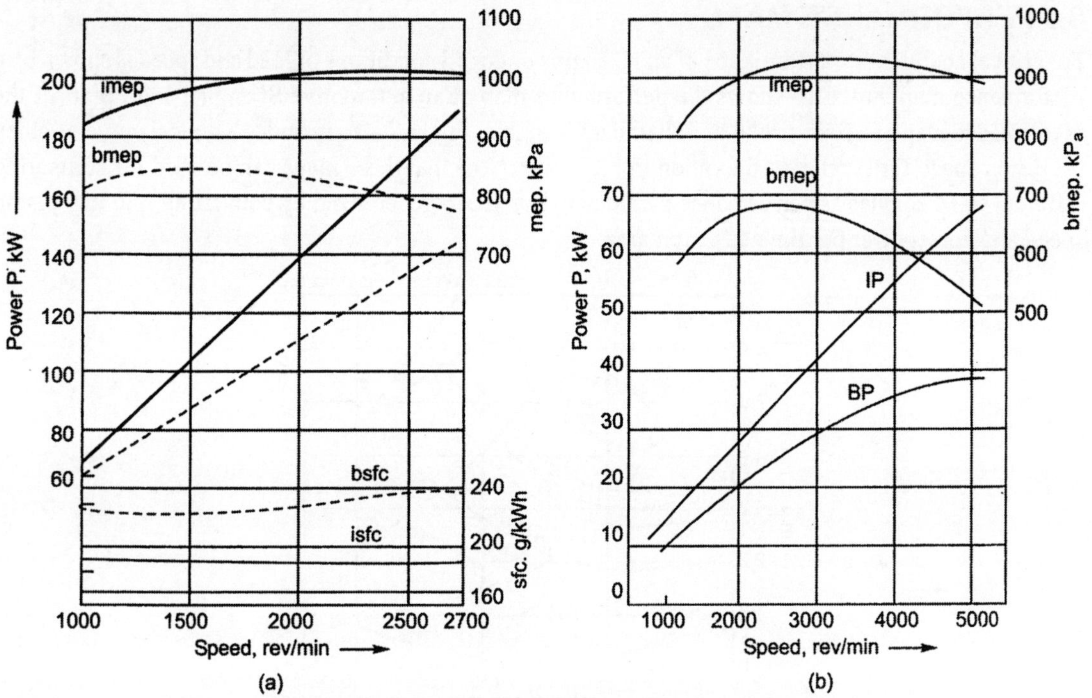

Fig. 6.8 IP, BP, imep, bmep and SFC for a CI engine.

low loads and radiation, etc. losses are more at high loads. The bmep. B.P. and torque directly increases with load, as shown in Fig. 6.8. Unlike the SI engine b.p. and bmep are continuously rising curves and are limited only by the smoke. The exhaust temperature is also nearly proportional to the load. The lowest brake specific fuel consumption and hence the maximum efficiency occurs at about 80 per cent of the full load.

Fig. 6.9 shows the performance curves of a variable speed GM 7850 cc four cycle V-6 Toro-flow diesel engine. The maximum torque value is at about 70 per cent of maximum speed compared to about 50 per cent in the SI engine. Also, the bsfc is low through most of the speed range for the diesel engine and is better than the SI engine.

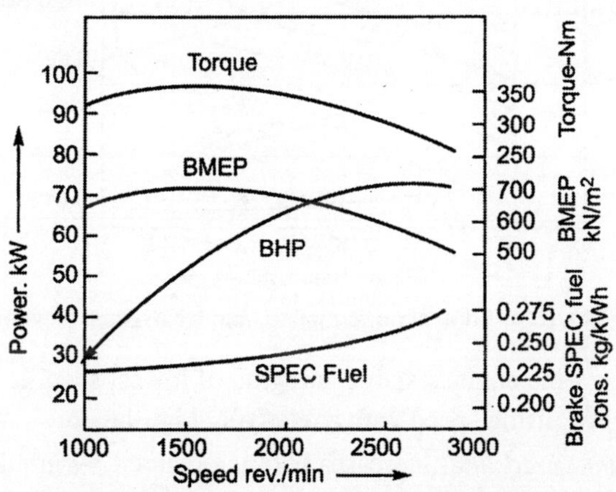

Fig. 6.9 Performance curves of a diesel engine.

6.8 PERFORMANCE MAPS

For critical analysis the performance of an IC engine under all conditions of load and speed is shown by a performance map. Fig. 6.10 shows the performance map of an automotive SI engine and Fig. 6.11 the performance map of a four-stroke pre-chamber CI engine. Fig. 6.7 also includes a typical curve of bmep vs. piston speed for level road operation in high gear. Note that these maps can be used for comparing different sized engines as performance parameters have been generalized by converting rpm into piston speed and horsepower per time of piston area.

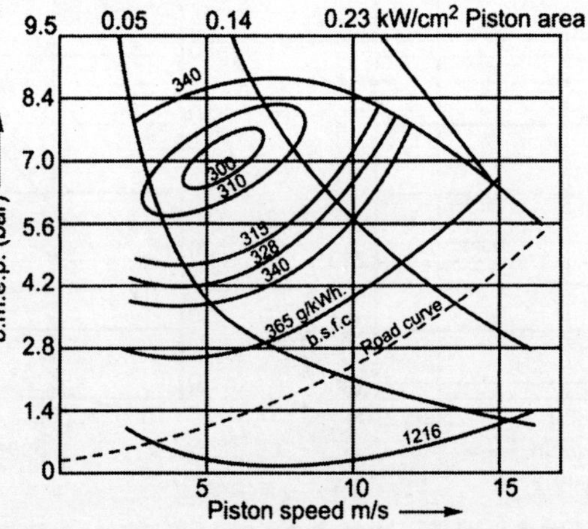

Fig. 6.10 Form of performance map for a petrol engine.

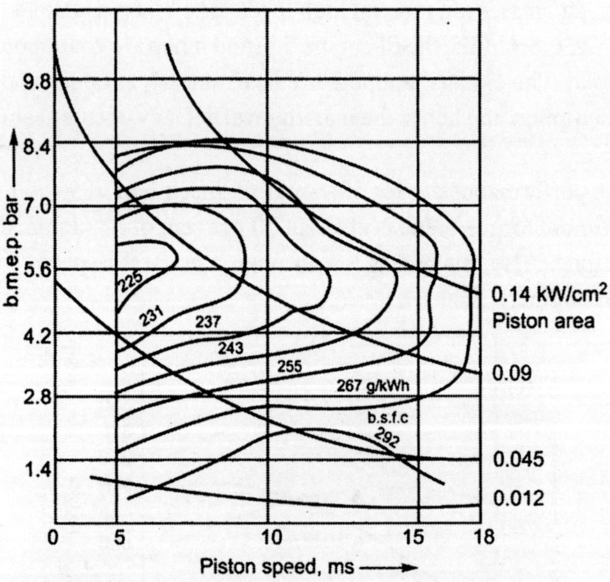

Fig. 6.11 Form of performance map for a diesel engine.

Generally speaking, all the engines shown a region of lowest specific fuel consumption (highest efficiency) at a relatively low piston speed with a relatively high bmep.

SI Engine : *Constant speed line*. Increases bsfc is obtained by moving upward along constant speed line, because of mixture enrichment at high load which more than offsets increase in mechanical efficiency. Moving to lower bmeps, the bsfc increases because of the reduced mechanical efficiency (imep decreases while fmep remains constant).

Constant bmep line. Moving from the region of highest efficiency along a line of constant bmep, the bsfc increases due to increased friction at higher piston speeds. Moving to the left towards lower piston speed, although friction mep decreases, indicated efficiency falls off owing to poor fuel distribution and increased relative heat losses.

CI Engine : In the CI engine the bsfc increases at high loads owing to the increased fuel waste (smoke) associated with high fuel-air ratios. At lower load bsfc increases due to decrease in mechanical efficiency (same in the SI engine).

As the speed is reduced from the point of best economy along a line of constant bmep, the product of mechanical and indicated thermal efficiency appears to remain about constant down to the lowest operating speed. The reduction in fmep with speed is apparently balanced by a reduction is indicated thermal efficiency due to poor spray characteristics at very low speeds.

An interesting feature of performance curves that they show that power at maximum economy is about half of the maximum power.

■ **Example 6.1** A 16-cylinder diesel engine has a power output of 800 kW at 900 revolutions per minute. The engine works on the four stroke cycle and has a fuel consumption of 0.238 kg/kW.hr. The pressure in the cylinder at the beginning of injection is 32.4 bar and the maximum cylinder pressure is 55 bar. The injector is set at 214 bar and maximum pressure at the injector is around 0.6. The specific gravity of the fuel is 0.86. Calculate the orifice area required per injector, if the injection takes place over 10 degree crank angle. [IES, 2001]

Solution : Number of cylinder, $n = 16$

Power output, $P_t = 800$ kW

$$\text{Power per cylinder} = \frac{P_t}{n} = \frac{800}{16} = 50 \text{ kW}$$

$$\text{Fuel consumption per cylinder} = 0.238 \times 50$$

$$= 1.19 \text{ kg/h} \text{ or } 3.3056 \times 10^{-3} \text{ kg/s}$$

$$\text{Fuel to be injected per cycle} = \frac{3.3056 \times 10^{-3}}{\left(\frac{900}{60}\right)\left(\frac{1}{2}\right)} = 4.0747 \times 10^{-4} \text{ kg}$$

$$\text{Fuel injection time} = \left(\frac{10}{360}\right)\left(\frac{60}{900}\right) = 1.852 \times 10^{-3} \text{ s}$$

$$\text{Fuel mass rate per second, } \dot{m}_f = \frac{4.40747 \times 10^{-4}}{1.852 \times 10^{-3}} = 0.238 \text{ kg/s}$$

Pressure difference in the beginning, $\Delta p_i = 214 - 32.4 = 181.6$ bar

Pressure difference at the end,

$$\Delta p_e = 600 - 55 = 545 \text{ bar}$$

$$\text{Average pressure difference, } \Delta p_a = \frac{\Delta p_i + \Delta p_e}{2} = \frac{181.6 + 545}{2} = 363.3 \text{ bar}$$

Now $$\dot{m}_f = C_d A_i [2\rho_f \Delta p_a]^{1/2}$$

$$0.238 = 0.6 \times A_i [2 \times 0.86 \times 10^3 \times 363.3 \times 10^5]^{1/2}$$

$$A_i = 1.587 \times 10^{-6} \text{ m}^2 \text{ or } 1.587 \text{ mm}^2$$

■ **Example 6.2** A six cylinder, four stroke spark-ignition of 10 cm × 2 cm (bore *x* stroke) with a compression ratio of 6 is tested at 4800 rpm on a dynamometer of arm 55 cm. During a 10 minutes test, the dynamometer reads 45 kg and the engine consumed 45 kg of petrol of calorific value 45 MJ/kg. The carburetor receives the air at 9°C and a bar at the rate of 10 kg/min. Calculate :

(i) the brake power,
(ii) the brake mean effective pressure,
(iii) the brake specific fuel consumption,
(iv) the brake specific air consumption,
(v) the brake thermal efficiency,
(vi) the air-fuel-ratio.

Solution :

(i) Brake power, $$\text{BP} = \frac{2\pi NT}{60 \times 10^3}$$

$$T = 45 \times +9.81 \times 55 \times 10^{-2} = 242.79 \text{ N-m}$$

$$\text{BP} = \frac{2\pi \times 4800 \times 242.79}{60 \times 10^3} = 122 \text{ kW}$$

(ii) $$\text{BP} = p_{bm} [(\text{LA})N / i] \times \text{Number of cylinders}$$

$$122 \times 10^3 = p_{bm} \times \left(\frac{\pi \times 0.1 \times 0.12}{4} \right) \times \frac{4800 \times 6}{2 \times 60}$$

$$p_{bm} = 5.39 \text{ bar}$$

(iii) $$\dot{m}_f = \frac{5}{10} \text{ kg/min or 30 kg/h}$$

$$\text{BSFC} = \frac{30}{122} = 0.45 \text{ kg/kW.h}$$

(iv) $$\dot{m}_a = 10 \text{ kg/min or 600 kg/h}$$

$$\text{BSFC} = \frac{600}{122} = 4.918 \text{ kg/kw.h}$$

(v) $$\eta_a = \frac{\text{BP}}{\dot{m}_f \times \text{CV}} = \frac{122 \times 10^3}{\frac{30 \times 45 \times 10^6}{3600}} = 0.325 \text{ or } 32.5\%$$

(vi) $$\text{A/F ratio} = \frac{\dot{m}_a}{\dot{m}_f} = \frac{600}{30} = 20:1$$

■ **Example 6.3** A 6-cylinder, four stroke cycle, 10 cm × 12.5 cm stroke, diesel engine develops 50 kW at 1000 rpm. The various efficiencies are : mechanical 76%; volumetric 80% under room conditions; indicated relative 88%; theoretical thermal 52%. The lower calorific value of the liquid fuel is 450 kJ/kg. Compute (a) b.m.e.p; (b) air-fuel ratio; (c) specific fuel consumption. Assume air density as 0.12 kg/m^3. Under room conditions.

Solution :

(a) Power developed per cylinder $= \frac{50}{6}$ kW

$$BP = \frac{p_{bm}\ LAN}{60 \times 10^3 \times 2}$$

$$p_{bm} = \frac{50 \times 60 \times 10^3 \times 2}{6 \times 12.5 \times 10^{-2} \times \frac{\pi}{4}(0.1)^2 \times 1000} = 10.186 \text{ bar}$$

(b)
$$IP = \frac{BP}{\eta_{mech}} = \frac{50}{0.76} = 65.79 \text{ kW}$$

$$\eta_{ith} = \eta_{ir} \times (\eta_{th})_{theo} = 0.88 \times 0.52 = 0.4576$$

$$= \frac{IP}{\dot{m}_f \times LCV} = \frac{65.79}{\dot{m}_f \times 45000}$$

$$\dot{m}_f = \frac{65.79}{0.4576 \times 45000}$$

$$= 3.195 \times 10^{-3} \text{ kg/s}$$

Swept volume, $V_s = \frac{\pi}{4}(0.1)^2 \times 0.125 = 9.8175 \times 10^{-4} \text{ m}^3$

Actual air intake per cylinder,

$$V_a = V_s\ \eta_v = 9.8175 \times 10^{-4} \times 0.8 = 7.854 \times 10^{-1} \text{ m}^3$$

Actual air taken in per second per cylinder $= 7.854 \times 10^{-4} \times \frac{1000}{60 \times 2} = 6.545 \times 10^{-3} \text{ m}^3/\text{s}$

Mass of air taken in per second per cylinder

$$= 6.545 \times 10^{-3} \times 0.12 = 7.854 \times 10^{-4} \text{ kg/s}$$

Air-fuel ratio $= \frac{7.854 \times 10^{-4} \times 6}{3.195 \times 10^{-3}} = 1.475$

(c) Specific fuel consumption $= \frac{3.195 \times 10^{-3} \times 3600}{50} = 0.23$ kg/(BP.h)

■ **Example 6.4** During the trial of a single oil engine, cylinder diameter 20 cm, stroke 28 cm, working on the two-stroke cycle and firing every cycle, the following observations were made :

Duration of trial	= 1 hour
Total fuel used	= 4.22 kg

Calorific value of fuel	= 44670 kJ/kg
Proportion of hydrogen in fuel	= 15%
Total number of revolutions	= 21000
Mean effective pressure	= 2.74 bar
Net brake load applied to a drum of 100 cm diameter	= 600 N
Total mass of cooling water circulated	= 495 kg
Temperature of cooling water : inlet 13°C, outlet 38°C	
Air used	= 135 kg
Temperature of air in test room	= 20°C
Temperature of exhaust gases	= 370°C
Assume c_p of gases	= 1.005 kJ/kg.K
c_p of steam	= 2.093 kJ/kg.K at atmospheric pressure

Calculate the thermal efficiency and draw up the heat balance sheet. [IES, 1997]

Solution :

$$N = \frac{21000}{60} = 350 \text{ RPM}$$

Torque, $T = 600 \times 0.5 = 300 \text{ N.m}$

Brake power, $\text{BP} = \dfrac{2\pi NT}{60 \times 10^3} = \dfrac{2\pi \times 350 \times 300}{60 \times 10^3} = 10.996 \text{ kW}$

Indicated power,

$$\text{IP} = \frac{p_m \text{ LAN}}{60 \times 10^3} = \frac{2.74 \times 10^5 \times 0.28 \times \pi \times (0.2)^2 \times 350}{4 \times 60 \times 10^3} = 14.06 \text{ kW}$$

Thermal efficiency, $\eta_{th} = \dfrac{\text{IP}}{\dot{m}_f \times \text{LCV}} = \dfrac{14.06 \times 3600}{4.22 \times 44670} = 0.2685 \text{ or } 26.85\%$

Swept volume, $V_s = \dfrac{\pi}{4} D^2 \text{LN} = \dfrac{\pi}{4} (0.2)^2 \times 0.28 \times 350 = 3.08 \text{ m}^3/\text{min}$

Heat input $= \dfrac{4.22}{60} \times 44670 = 3141.79 \text{ kJ/min}$

Heat equivalent of BP $= 10.996 \times 60 = 659.76 \text{ kJ/min}$

Heat lost to cooling water $= \dot{m}_w \, c_w \cdot \Delta t = \dfrac{495}{60} \times 4.18\,(38 - 13) = 862.125 \text{ kg/min}$

Heat carried away by exhaust gases :

$$\text{Mass of flue gases} = \dot{m}_a + \dot{m}_f = \frac{135 + 4.22}{60} = 2.32 \text{ kg/min}$$

Steam in exhaust gases $\quad \dot{m}_s = 9 \times 0.15 \times \dfrac{4.22}{60} = 0.095 \text{ kg/min}$

Mass of dry exhaust gases $\dot{m}_g = 2.32 - 0.095 = 2.225 \text{ kg/min}$

Heat in steam in exhaust gases

$$= \dot{m}_s[100 - 20] + h_{fg} + c_p \text{ steam } (370 - 100)$$

$$= 0.095\,[(100 - 20) + 2256.9 + 2.093\,(370 - 100)] = 275.7 \text{ kJ/min}$$

Heat in dry exhaust gases $= \dot{m}_g \times c_{pg}(370 - 100)$

$$= 2.225 \times 1.005 \times 270 = 603.75 \text{ kJ/min}$$

Unaccounted heat loss $= 3141.79 - [659.76 + 603.75 + 275.7 + 862.125]$

$$= 740.455 \text{ kJ/min}$$

Heat balance sheet :

Heat Input	***kJ/min***	***Heat expenditure***	***kJ/min***	**%**
Heat supplied by fuel	3141.79	1. Heat equivalent to BP	659.79	21.00
		2. Heat in cooling water	862.125	27.44
		3. Heat in steam in exhaust gases	275.7	8.77
		4. Heat in dry exhaust gases	603.75	19.22
		5. Unaccounted heat	704.455	23.57
Total	3141.79		3141.79	100.00

■ **Example 6.5** A sharp edged circular orifice of diameter 3.8 cm and coefficient of discharge as 0.6 is used to measure air-consumption of a four stroke petrol engine. Pressure drop through the orifice is 145 mm of water and barometer reads 75.5 cm of Hg. The compression ratio of the engine is 6 and the piston displacement volume is 2000 cm^3.

The temperature of air is taken to be 26°C. At 2600 rpm, the engine brake power recorded is 29.5 kW. The fuel consumption is 0.14 kg/min and the calorific value of fuel used is 43960 kJ/kg.

Calculate the following :

(i) Volumetric efficiency

(ii) Air fuel ratio

(iii) Brake mean effective pressure

(iv) Brake thermal efficiency

(v) Relative efficiency.

Solution :

$$d = 3.8 \text{ cm}, C_d = 0.6, h_w = 14.5 \text{ cm of water}$$

$$p_a = 75.5 \text{ cm of Hg},$$

$$r = 6, V_s = 2000 \text{ cm}^3$$

$$N = 2600 \text{ rpm, BP} = 29.5 \text{ kW}$$

$$\dot{m}_f = 0.14 \text{ kg/min, CV} = 43960 \text{ kJ/kg}$$

Density of air at atmospheric conditions,

$$\rho_a = \frac{m}{V} = \frac{p}{RT} = \frac{1 \times 10^5}{287 \times 299} = 1.1653 \text{ kg/m}^3$$

$$\text{Head in } m \text{ of air} = \left(\frac{h_w}{100}\right)\left(\frac{\rho_w}{\rho_a}\right) = \frac{14.5 \times 1000}{100 \times 1.1653} = 124.43 \text{ m of air}$$

Velocity of air passing through orifice,

$$c_a = [2gH]^{0.5} = [2 \times 9.81 \times 424.43]^{0.5} = 49.41 \text{ m/s}$$

Volume of air passing through orifice,

$$V_a = A_0 c_a C_d = \frac{\pi}{4}\cdot(3.8)^2 \times 10^4 \times 49.41 \times 0.6$$

$$= 0.033622 \text{ m}^3/\text{s or } 2.0173 \text{ m}^3/\text{min}$$

(i) Volumetric efficiency

$$\frac{V_a}{V_s} = \frac{2.0173}{2000 \times 10^{-6} \times 260} = 0.3879 \text{ or } 38.79\%$$

(ii) Mass flow rate of air,

$$\dot{m}_a = V_a \rho_a = 2.0173 \times 1.1653 = 2.35 \text{ kg/min}$$

$$\text{A/F ratio} = \frac{\dot{m}_a}{\dot{m}_f} = \frac{2.35}{0.14} = 16.79:1$$

(iii) $$\text{BMEP} = \frac{\text{BP}}{V_s} = \frac{29.5 \times 10^3}{\dfrac{2000 \times 10^{-6} \times 2600}{60}} = 0.3404 \text{ MPa}$$

(iv) $$\eta_{bt} = \frac{\text{BP}}{m_f \times \text{CV}} = \frac{29.5 \times 3600}{0.14 \times 60 \times 43960} = 0.2895 \text{ or } 28.95\%$$

(v) Air standard efficiency, $\eta_a = 1 - \frac{1}{r^{\gamma-1}} = 1 - \frac{1}{6^{0.4}}$

$$= 0.5116 \text{ or } 51.16\%$$

(vi) Relative efficiency $= \frac{\eta_{bt}}{\eta_a} = \frac{28.85}{51.16} = 0.5639$ or 56.39%

■ **Example 6.6** The air flow to a four cylinder, four-stroke petrol engine is measured by means of a 7.5 cm sharp-edged orifice, $C_d = 0.6$. During a test on the engine, following data was recorded.

Bore = 12 cm, stroke = 14 cm, engine speed = 2200 rpm, brake power = 35 kW, fuel consumption = 10 kg/h, LCV = 42000 kJ/kg, pressure drop across the orifice = 4 cm of water. Atmospheric temperature = 20°C, and atmospheric pressure = 1.01325 bar.

Calculate :

(a) Thermal efficiency on BP basis,

(b) BMEP, and

(c) Volumetric efficiency based on free air conditions.

Solution :

(a) Brake thermal efficiency,

$$\eta_{bth} = \frac{\text{BP} \times 3600}{\dot{m}_f \times \text{LCV}} = \frac{35 \times 3600}{10 \times 24000} = 0.3 \text{ or } 30\%$$

(b) Brake mean effective pressure,

$$p_{bm} = \frac{\text{BP}}{\text{LAN}} = \frac{(35/4) \times 60 \times 10^3}{0.14 \times \frac{\pi}{4}(0.12)^2 \times 2200} = 1.507 \text{ bar}$$

(c) Air inhaled/s,

$$\dot{V}_a = C_d A_o \sqrt{2\Delta p \cdot \rho_a}$$

$$\Delta p = \rho g h = 1000 \times 9.81 \times 4 \times 10^{-2} = 392.4 \text{ N/m}^2$$

$$\rho_a = \frac{p}{RT} = \frac{1.01325 \times 10^5}{287 \times 293} = 1.205 \text{ kg/m}^3$$

$$\dot{V}_a = 0.6 \times \frac{\pi}{4}(7.5 \times 10^{-2})^2 \sqrt{2 \times 392.4 \times 1.205} = 0.0185 \text{ m}^3/\text{s}$$

$$\text{Swept volume/s } \dot{V}_s = i\frac{\pi}{4}D^2 L \times \frac{N}{2 \times 60} = 4 \times \frac{\pi}{4}(0.12)^2 \times 0.14 \times \frac{2200}{2 \times 60}$$

$$= 0.1161 \text{ m}^3/\text{s}$$

$$\text{Volumetric efficiency, } \eta_v = \frac{\dot{V}_a}{\dot{V}_s} = \frac{0.0815}{0.1161} = 0.702 \text{ or } 70.2\%$$

■ **Example 6.7** The air flow to a cylinder, four-stroke oil engine is measured by means of a 5 cm diameter orifice, having a coefficient of discharge of 0.6. During a test on the engine the following data was recorded :

Bore = 10 cm, stroke = 12 cm, engine speed = 1200 rpm, brake torque = 147 N-m, fuel consumption = 5.5 kg/h, calorific value of fuel = 43100 kJ/kg, head across orifice = 5.8 cm of water, ambient temperature and pressure = 20°C and 1.013 bar. Calculate (a) Brake thermal efficiency, (b) Brake mean effective pressure, and (c) Volumetric efficiency based on free air condition.

Solution :

$$\text{(a) BP} = \frac{2\pi NT}{10^3 \times 60} = \frac{2\pi \times 1200 \times 147}{10^3 \times 60} = 18.472 \text{ kW}$$

$$\text{Thermal efficiency} = \frac{\text{BP} \times 3600}{\dot{m}_f \times \text{LCV}} = \frac{18.472 \times 3600}{5.5 \times 43100} = 0.2805 \text{ or } 28.05\%$$

$$\text{(b)} \quad p_{bm} = \frac{\text{BP} \times n}{\text{LAN} \times i} = \frac{18.472 \times 2 \times 10^3 \times 60 \times 10^{-5}}{0.12 \times \frac{\pi}{4}(0.1)^2 \times 1200 \times 4} = 4.9 \text{ bar}$$

$$\text{(c)} \quad \text{Volume flow of air } \dot{V}_a = C_d A_0 \sqrt{2gh_w \frac{\rho_\omega}{\rho_\alpha}}$$

$$\rho_a = \frac{p}{RT} = \frac{1.013 \times 10^5}{287 \times 293} = 1.2 \text{ kg/m}^3$$

$$\dot{V}_a = 0.6 \times \frac{\pi}{4}(0.05)^2 \sqrt{2 \times 9.81 \times 5.8 \times 10^{-2} \times \frac{1000}{1.2}}$$

$$= 0.03628 \text{ m}^3/\text{s}$$

Swept volume,

$$\dot{V}_s = i \times \text{AL} \times \frac{N}{n \times 60} = 4 \times \frac{\pi}{4}(0.1)^2 \times 0.12 \times \frac{1200}{2 \times 60} = 0.0377 \text{ m}^3/\text{s}$$

Volumetric efficiency,

$$\eta_v = \frac{\dot{V}_a}{\dot{V}_s} = \frac{0.03628}{0.0377} = 0.9623 \text{ or } 96.23\%$$

■ **Example 6.8** A four-stroke cycle petrol engine has six single-acting cylinders of 7.5 cm bore and 9 cm stroke. The engine is coupled to a brake having a torque arm radius of 38 m. At 3300 rpm, with all cylinders operating the net brake load is 324 N. When each cylinder in turn is rendered in operative, the average net brake load produced at the same speed of the remaining five cylinders is 245 N. Estimate the indicated mean effective pressure of engine.

Solution : (a) BP when all cylinder are working

$$= 2\pi NT = 2\pi \times \frac{300}{60} = (324 \times 0.38) \times 10^{-3}$$

$$= 42.55 \text{ kW}$$

BP when each cylinder is cut-off in turn

$$= 2\pi \times \frac{3300}{60} \times (245 \times 0.38) \times 10^{-3} = 32.17 \text{ kW}$$

IP of cylinder cut off $= 42.55 - 32.17 = 10.38$ kW

Total IP of engine $= 6 \times 10.38 = 62.3$ kW

Indicated mean effective pressure

$$= \frac{\text{IP} \times n}{i \times \text{LAN}} = \frac{62.3 \times 10^3 \times 2 \times 60 \times 10^{-5}}{6 \times 0.09 \times \frac{\pi}{4}(0.0745)^2 \times 3300}$$

$$= 9.5 \text{ bar}$$

■ **Example 6.9** A six-cylinder petrol engine operates in the four-stroke cycle. The bore of each cylinder is 70 mm and the stroke 100 mm. The clearance volume per cylinder is 67 cm^3. At a speed of 3960 rpm the fuel consumption is 19.5 kg/h and the torque developed is 140 N-m. Calculate (a) BP, (b) BMEP, (c) the brake thermal efficiency of LCV of fuel is 44000 kJ/kg and (d) the relative efficiency on brake power basis. The engine works on the constant volume cycle. Assume $\gamma = 1.4$ for air.

Solution :

(a) $$\text{BP} = \frac{2\pi NT}{60 \times 10^3} = \frac{2\pi \times 3960 \times 140}{60 \times 10^3} = 58 \text{ kW}$$

(b) $$p_{bm} = \frac{\text{BP} \times n}{\text{LAN} \times i} = \frac{6 \times 58 \times 10^3 \times 2 \times 60 \times 10^{-5}}{0.01 \times \frac{\pi}{4}(0.07)^2 \times 3960 \times 6} = 7.61 \text{ bar}$$

(c) $$\eta_{bth} = \frac{\text{BP}}{m_f \times \text{LCV}} = \frac{58 \times 3600}{19.5 \times 44000} = 0.244 \text{ or } 24.4\%$$

(d) Swept volume per cylinder,

$$V_s = AL = \frac{\pi}{4}(0.07)^2 \times 0.1 = 3.848 10^{-4}\ m^3$$

Compression ratio, $$r = \frac{V_s + V_c}{V_c} = \frac{384.8 + 67}{67} = 6.743$$

Air standard efficiency, $$\eta_a = 1 - \frac{1}{r^{\gamma - 1}} = 1 - \frac{1}{(6.473)^{0.4}} = 0.534$$

Relative efficiency, $$\eta_r = \frac{\eta_{bth}}{\eta_a} = \frac{0.244}{0.534} = 0.457 \text{ or } 45.7\%$$

6.9 MEASUREMENT OF AIR-CONSUMPTION BY AIR-BOX METHOD

The arrangement to measure the consumption of air by air-box method is shown in Fig. 6.12. It consists of an air-tight box fitted with a sharp-edged orifice of known coefficient of discharge. Due to the suction of engine, there is a pressure depression in the box which causes the flow through orifice for obtaining a steady flow. The pressure difference causing the flow through the orifice is measured with the help of a water manometer.

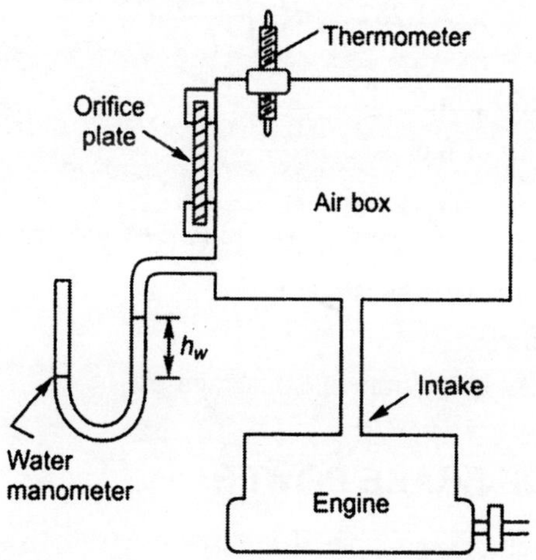

Fig. 6.12 Measurement of air by air box method.

Let

A_0 = area of orifice, m^2

h_w = head of water causing flow, cm

C_d = coefficient of discharge for orifice

d_0 = diameter of orifice, cm

ρ_a = density of air, kg/m^3

Head of air, $$H = \frac{h_w}{100} \times \frac{\rho_w}{\rho_a} = \frac{h_w}{100} \times \frac{1000}{\rho_a} = \frac{10\, h_w}{\rho_a} \text{ m of air}$$

Velocity of air through the orifice,

$$c_a = \sqrt{2gH} = \sqrt{2g \times \frac{10\,h_w}{\rho_a}} \text{ m/s}$$

Volume of air passing through the orifice,

$$V_a = A_0 c_a \cdot C_d = A_0 C_d \sqrt{\frac{2g \times 10\,h_w}{\rho_a}} = 14\,A_0 C_d \sqrt{\frac{h_w}{\rho_a}} \text{ m/s}$$

Mass flow rate of air passing through the orifice,

$$\dot{m}_a = V_a \cdot \rho_a = 14 \times \frac{\pi}{4} d_0^2 \times 10^{-4} \times C_d \times \rho_a \sqrt{\frac{h_w}{\rho_a}}$$

$$= 11 \times 10^{-4}\, C_d d_0^2 \sqrt{h_w \rho_a} \text{ kg/s}$$

or $$0.066\, C_\alpha d_0^2 \sqrt{h_w \rho_a} \text{ kg/min} \qquad \ldots(6.22)$$

Density of air, $$\rho_a = \frac{p_a}{RT_a}$$

$\therefore$ $$\dot{m}_a = 0.066\, C_d d_0^2 \sqrt{\frac{h_w \times p_a \times 10^5}{287\, T_a}}$$

$$= 1.232\, C_d d_0^2 \sqrt{\frac{h_w p_a}{T_a}} \qquad \ldots(6.23)$$

where h_w is in cm, p_a in bar and T_a in K

Mass of air supplied per kg of fuel used,

$$m_a/\text{kg of fuel} = \frac{N \times C}{33\,(C_1 + C_2)} \qquad \ldots(6.24)$$

where N = % of N_2 by volume in exhaust gases

C = % of carbon in fuel

C_1, C_2 = % of CO_2 and CO by volume in exhaust gases.

6.10 MEASUREMENT OF BRAKE POWER

(a) Rope Brake Dynamometer

A rope brake dynamometer is shown in Fig. 6.13. A rope is wound around the brake drum, whose one end is connected to the spring balance S suspended from overhead and the other end carries the load W. In this arrangement, the whole power developed by the engine is absorbed by the friction produced at the rim of the brake drum. The rim of the brake drum is generally water cooled to absorb the heat generated due to rubbing action of rope on rim.

Let

W = dead weight on the rope, N

S = spring pull, N

D = outer diameter of brake drum, m

d_r = diameter of rope, m

N = sped of engine, rpm

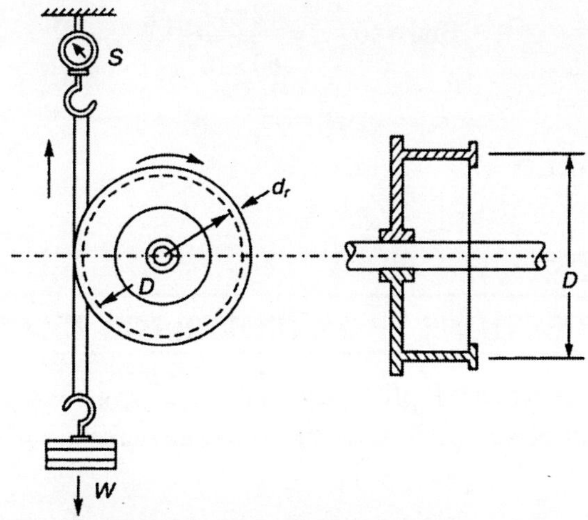

Fig. 6.13 Rope brake dynamometer.

Net load acting on brake drum = $(W - S)$ N

Effective radius of brake drum, $R = \frac{1}{2}(D + d_r)$

Frictional torque rating on brake drum,

$$T_f = (W - S) \times R \quad \text{N.m}$$

Power developed by engine $= \frac{2\pi N T_f}{60 \times 10^3}$ kW ...(6.25)

(b) Prony Brake

The arrangement of this braking system is shown in Fig. 6.14. It consists of brake shoes which are clamped on the rim of the brake drum by means of bolts. The pressure on the rim is adjusted with the help of nuts and springs. A load lever extends from top of the brake and a load carrier is attached to the end of the load lever. The weights kept on this load carrier is balanced by the reaction torque in the shoes. The load lever is kept horizontal to keep its length constant.

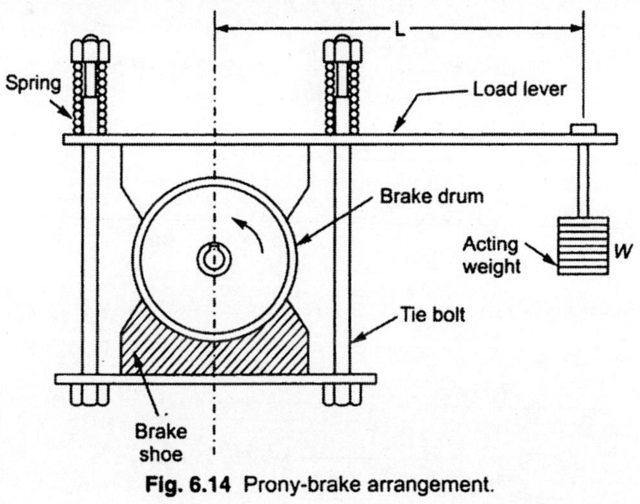

Fig. 6.14 Prony-brake arrangement.

$$\text{Brake power} = \frac{2\pi NT}{60 \times 10^3}\ \text{kW}$$

where $$T = WL \qquad \text{...(6.26)}$$

W = load on load carrier, N

L = distances from centre of shaft to load, m

SOLVED QUESTIONS FROM UNIVERSITY EXAMINATIONS

■ **Example 6.10** Define the volumetric efficiency for an I.C. Engine. What is the effect of volumetric efficiency on : (i) Engine power, (ii) Specific fuel consumption? [MDU, Dec., 2007]

Solution : Refer to Eq. (6.7).

■ **Example 6.11** A four cylinder engine running at 1200 rpm gave 18.6 kW brake power. The average torque when one cylinder was cut-off was 105 N.m. Determine the indicated thermal efficiency if the calorific value of the fuel is 42000 kJ/kg and the engine uses 0.34 kg of petrol per brake power hour. [MDU, Dec., 2007]

Solution : Given, $i = 4$, $N = 1200$ rpm, $(\text{BP})_n = 18.6$ kW, $T_3 = 105$ N.m, $CV = 42{,}000$ kJ/kg,

$$\text{BSFC} = 0.34\ \text{kg/B.P. h}$$

$$(\text{BP})_{n-1} = \frac{2\pi NT_3}{60 \times 10^3} = \frac{2\pi \times 1200 \times 10^5}{60 \times 10^3} = 13.2\ \text{kW}$$

$$(\text{BP})_n - (\text{BP})_{n-1} = (\text{IP})_n - (\text{IP})_{n-1} = (\text{IP})_1$$

$$\therefore \qquad (\text{IP})_1 = 18.6 - 13.2 = 5.4\ \text{kW}$$

Indicated power of 4 cylinders,

$$(\text{IP})_n = i \times (\text{IP})_1 = 4 \times 5.4 = 21.6\ \text{kW}$$

$$\eta_{\text{mech}} = \frac{(\text{BP})_n}{(\text{IP})_n} = \frac{18.6}{21.6} = 0.861$$

$$\text{ISFC} = \frac{\text{BSFC}}{\eta_{\text{mech}}} = \frac{0.34}{0.861} = 0.3948\ \text{kg/IP.h}$$

Fuel consumption, $$\dot{m}_f = \frac{\text{ISFC} \times (\text{IP})_n}{3600}$$

$$= \frac{0.3948 \times 21.6}{3600} = 2.3688 \times 10^{-3}\ \text{kg/s}$$

Indicated thermal efficiency,

$$(\eta_{\text{th}})_i = \frac{(\text{IP})_n}{\dot{m}_f \times CV}$$

$$= \frac{21.6}{2.3688 \times 10^{-3} \times 4200} = 0.2171 \text{ or } 21.71\%$$

■ **Example 6.12** An engine uses 6.5 kg of oil per hour of calorific value 30,000 kJ/kg. If the B.P. of the engine is 22 kW and mechanical efficiency 85%. Calculate (i) Indicated efficiency (ii) Brake thermal efficiency, and (iii) specific fuel consumption in kg/B.P/h. [MDU, May, 2008]

Solution : Given : $\dot{m}_f = 6.5$ kg/h, $CV = 30{,}000$ kJ/kg, $BP = 22$ kW, $\eta_{mech} = 0.85$

Indicated power, $$IP = \frac{BP}{\eta_{mech}} = \frac{22}{0.85} = 25.882 \text{ kW}$$

(i) Indicated thermal efficiency,

$$(\eta_{th})_i = \frac{IP \times 3600}{\dot{m}_f \times CV} = \frac{25.882 \times 3600}{6.5 \times 30{,}000} = 0.4778 \text{ or } 47.78\%$$

(ii) Brake thermal efficiency,

$$(\eta_{th})_b = \frac{BP \times 3600}{\dot{m}_f \times CV} = \frac{22 \times 3600}{6.5 \times 30{,}000} = 0.04061 \text{ or } 40.61\%$$

(iii) Specific fuel consumption,

$$\text{BSFC} = \frac{\dot{m}_f}{BP} = \frac{6.5}{22} = 0.295 \text{ kg/BP/h}$$

■ **Example 6.13** A four cylinder petrol engine operates on the four stroke cycle. The bore of each cylinder is 70 mm and the stroke 100 mm. The clearance volume per cylinder is 60 cm. At a speed of 4000 rpm, the fuel consumption is 20 kg/h and the torque developed is 150 N-m. Calculate : (a) The brake power (b) The brake mean effective pressure (c) The brake thermal efficiency if calorific value of the fuel is 44,000 kJ.kg. (d) The relative efficiency on brake power basis assuming that the engine works on constant volume cycle. [MDU, Dec., 2008]

Solution : Given : $i = 4, n = 2, d = 70$ mm, $L = 100$ mm, $V_c = 60$ cm^3

$$N = 4000 \text{ rpm}, \dot{m}_f = 20 \text{ kg/h}, T = 150 \text{ N.m}, CV = 44{,}000 \text{ kJ/kg}$$

Swept volume, $$V_s = \frac{\pi}{4} d^2 L = \frac{\pi}{4} (0.070)^2 \times 0.1 = 3.848 \times 10^{-4} \text{ m}^3$$

(a) Brake power,

$$BP = \frac{2\pi NT}{60 \times 10^3} = \frac{2\pi \times 4000 \times 150}{60 \times 10^3} = 62.832 \text{ kW}$$

(b) $$\text{BMEP} = \frac{10^3 \times BP \times 60}{V_s \times \frac{N}{n} \times i} = \frac{62.832 \times 10^3 \times 60}{3.848 \times 10^{-4} \times \frac{4000}{2} \times 4} = 12245 \times 10^6 \text{ N/m}^2$$

(c) Brake thermal efficiency,

$$(\eta_{th})_b = \frac{BP \times 3600}{\dot{m}_f \times CV} = \frac{62.832 \times 3600}{20 \times 44{,}000} = 0.2570 \text{ or } 25.7\%$$

(d) Relative efficiency, $$\eta_r = \frac{(\eta_{th})_b}{\eta_a}$$

For a constant volume cycle, $\eta_a = 1 - \dfrac{1}{r^{\gamma-1}}$

Compression ratio, $r = \dfrac{V_1}{V_c} = \dfrac{V_c + V_s}{V_c} = 1 + \dfrac{V_s}{V_c}$

$$= 1 + \frac{3.848 \times 10^{-4}}{60 \times 10^{-6}} = 7.41$$

$$\eta_a = 1 - \frac{1}{(7.41)^{0.4}} = 0.5513 \text{ or } 55.13\%$$

$$\eta_r = \frac{25.70}{55.13} = 0.4662 \text{ or } 46.22\%$$

■ **Example 6.14** A single cylinder 4-stroke oil engine works on diesel cycle. The following readings correspond to full load conditions :

Area of indicator diagram	$= 3\text{ cm}^2$
Length of the diagram	$= 5$ cm
Spring constant	$= 12\text{ bar/cm}^2\text{.cm}$
Speed of the engine	= 500 rpm
Load on the brake	= 400 N
Spring reading	= 50 N
Diameter of brake drum	= 120 cm
Fuel consumption	= 2.75 kg/h
Calorific value of fuel	= 42,000 kJ/kg
Diameter of cylinder	= 16 cm
Stroke of piston	= 20 cm

Find the following :

(a) Friction power

(b) Mechanical efficiency

(c) Brake thermal efficiency

(d) Brake mean effective pressure.

Solution : Given, $n = 2$, $A_i = 3\text{ cm}^2$, $L_i = 5$ cm, $C = 12\text{ bar/cm}^2\text{.cm}$, $N = 500$ rpm, $W = 400$ N, $S = 50$ N, $D_b = 120$ cm, $\dot{m}_f = 2.75$ kg/h, $CV = 42{,}000$ kJ/kg, $d = 16$ cm, $L = 20$ cm

IMEP, $p_{im} = \dfrac{A_i C}{L_i} = \dfrac{3 \times 12}{5} = 72$ bar

$$\text{IP} = \frac{p_{im}\text{ LAN}}{n \times 60 \times 10^3} = 72 \times 10^5 \times 0.2 \times \frac{\pi}{4}(0.16)^2 \times \frac{500}{2} \times \frac{1}{60 \times 10^3} = 12.06 \text{ kW}$$

$$\text{BP} = (W - S)\frac{D_b}{2} \times \frac{2\pi N}{60 \times 10^3} = \frac{(400 - 50) \times 12 \times 2\pi \times 500}{60 \times 10^3 \times 2} = 10.99 \text{ kW}$$

(i) Frictional power,

$$FP = IP - BP = 12.06 - 10.99 = 1.07 \text{ kW}$$

(ii) $\eta_{mech} = \frac{BP}{IP} = \frac{10.99}{12.06} = 0.9122$ or 91.12%

(iii) $(\eta_{th})_b = \frac{BP \times 3600}{\dot{m}_f \times CV} = \frac{10.99 \times 3600}{2.75 \times 42{,}000} = 0.3425$ or 34.25%

(iv) BMEP,

$$p_{bm} = \frac{BP \times 10^3 \times 60 \times n}{\text{LAN}} = \frac{10.99 \times 10^3 \times 60 \times 2}{0.2 \times \frac{\pi}{4}(0.16)^2 \times 500} = 0.6559 \times 10^5 \text{ N/m}^2 \text{ or } 6.559 \text{ bar.}$$

■ **Example 6.15** The power output of a 6-cylinder, 4-stroke engine is absorbed by a water brake for which the law is WN/20000, where the brake load W is in N, and speed N in rpm. The air consumption is measured by an air-box with sharp edged orifice system. The following readings are obtained :

Orifice diameter	= 30 mm
Bore	= 100 mm
Stroke	= 120 mm
Speed	= 2400 rpm
Brake load	= 560 N
C/H ratio by mass	= 83/17
Coefficient of discharge	= 0.6
Ambient pressure	= 1 bar
Pressure drop across orifice	= 14.5 cm of Hg
Time taken for 100 cc of fuel consumption	= 20 s
Ambient temperature	= 27°C
Fuel density	= 831 kg/m^3

Calculate :

Brake power, the torque, the brake specific fuel consumption, the percentage of excess air and volumetric efficiency. [MDU, Dec., 2009]

Solution : $i = 6$, $N = 2$, $BP = \frac{WN}{20{,}000}$, W in N and N is rpm, $d_0 = 30$ mm,

$d = 100$ mm, $L = 120$ mm, $W = 560$ N, $C/H = 83/17$, $C_{d_0} = 0.6$, $p = 1$ bar,

$T_a = 273 + 27 = 300$ K, $h = 14.5$ cm of Hg, $Q_f = 100$ cc, $t = 20$ s, $\rho_f = 831$ kg/m^3,

Density of air, $$\rho_a = \frac{p_a}{RT_a} = \frac{1 \times 10^5}{287 \times 300} = 1.16 \text{ kg/m}^3$$

Volume flow rate of air, $$\dot{Q}_a = C_{d_0} A_0 \sqrt{2gh\left(\frac{\rho_{Hg}}{\rho_a}\right)}$$

$$= 0.6 \times \frac{\pi}{4} \times (0.03)^2 \sqrt{2 \times 9.81 \times 0.145\left(\frac{13.6}{1.16}\right) \times 10^3}$$

$$= 0.07746 \text{ m}^3/\text{s}$$

Swept volume rate, $\dot{V}_s = \frac{\pi}{4} d^2 L \times \left(\frac{N}{n \times 60}\right) \times i$

$$= \frac{\pi}{4}(0.1)^2 \times 0.12 \times \frac{2400}{2 \times 60} \times 6 = 0.11309 \text{ m}^3/\text{s}$$

Volumetric efficiency, $\eta_{vol} = \frac{\dot{Q}_a}{\dot{V}_a} = \frac{0.07746}{0.11309} = 0.6849 \text{ or } 68.49\%$

$$BP = \frac{WN}{2000} = \frac{560 \times 2400}{2400} = 67.2 \text{ kW}$$

$$\frac{BP}{i} = p_{bm} \times \frac{\text{AL } N}{n \times 60}$$

$$\frac{67.2}{6} = p_{bm} = \frac{\pi}{4}(0.1)^2 \times 0.12 \times \frac{2400}{2 \times 60}$$

BMEP, $p_{bm} = 594.178 \text{ kN/m}^2$ or 5.94178 bar

Also $\text{BP} = 2\pi NT/(60 \times 10^3)$

$$T = \frac{67.2 \times 10^3 \times 60}{2\pi \times 2400} = 267.38 \text{ N.m},$$

Fuel flow rate, $\dot{m}_f = \left(\frac{100}{193} \times 3600\right) \times \frac{1}{10^3} = 0.831 = 15.5 \text{ kg/h}$

Brake specific fuel consumption,

$$\text{BSFC} = \frac{\dot{m}_f}{BP} = \frac{15.5}{67.2} = 0.2306 \text{ kg/kW.h}$$

Composition of fuel : $C = 83\%$, $H_2 = 17\%$

Oxygen required per kg of fuel $= 0.83 \times \frac{32}{12} \times 0.17 \times \frac{16}{2} = 3.573$ kg/kg of fuel

Theoretical air required per kg of fuel $= \frac{3.573}{0.2306} = 15.494$ kg/g of fuel

Actual mass flow rate $= \dot{Q}_a \times \rho_a = 0.07746 \times 1.16 = 0.0898$ kg/s

$$A:F \text{ ratio} = \frac{\dot{m}_a}{\dot{m}_f} = \frac{0.0898 \times 3600}{15.5} = 20.87$$

$$\text{Percentage excess air} = \frac{\text{Actual air} - \text{theoretical air}}{\text{Theoretical air}} \times 100$$

$$= \frac{20.87 - 15.494}{15.494} \times 100 = 34.69\%$$

■ **Example 6.16** The following observations were recorded during a trial of a 4-stroke, single cylinder oil engine. Duration of trial is 30 minute : Oil-consumption is 4 litres. Calorific value of oil is 43 MJ/kg. Specific-gravity of fuel is 0.8, avg. area of the indicator diagram $= 8.5 \text{ cm}^2$. Length of indicator diagram is 8.5 cm, spring-constant $= 5.5$ bar/cm, brake load $= 150$ kg, spring balance reading is 20 kg. Effective brake-wheel diameter is 1.5 m, speed 200 rpm, cylinder-diameter is 30 cm, stroke = 45 cm, Jacket cooling-water = 10 kg/min. temperature rise is 36°C.

Calculate :

(i) Indicated power
(ii) Brake power
(iii) Mechanical efficiency
(iv) Brake specific fuel consumption in kg/kwh
(v) Indicated thermal efficiency. [MDU, Dec., 2010]

Solution : Given : $n = 2$, $t = 30$ min, $Q = 4$ litres, $CV = 43$ MJ/kg, $S_f = 0.8$, $A_i = 8.5\ \text{cm}^2$, $L_i = 8.5$ cm, $C = 5.5$ bar/cm, $W = 150 \times 9.81 = 1471.5$ N, $S = 20 \times 9.81 = 196.2$ N, $D_b = 1.5$ m, $N = 200$ rpm, $d = 30$ cm, $L = 45$ cm, $m_w = 10$ kg/min, $T_{w0} - T_{wi} = 36°\text{C}$

IMEP, $$p_{im} = \frac{A_i C}{L_i} = \frac{8.5 \times 5.5}{8.5} = 5.5 \text{ bar}$$

(i) $$\text{IP} = \frac{p_{im}\text{ LAN}}{60 \times 10^3 \times n} = 5.5 \times 10^5 \times 0.45 \times \frac{\pi}{4}(0.3)^2 \times \frac{200}{60 \times 10^3 \times 2} = 29.158 \text{ kW}$$

(ii) $$\text{BP} = (W - S)\frac{D_b}{2} \times \frac{2\pi N}{60 \times 10^3} = \frac{(1471.5 - 196.2) \times 1.5}{2} \times \frac{2\pi \times 200}{60 \times 10^3} = 20.032 \text{ kW}$$

(iii) $$\eta_{mech} = \frac{\text{BP}}{\text{IP}} = \frac{20.032}{29.158} = 0.6870 \text{ or } 68.7\%$$

(iv) $$\text{BSFC} = \frac{60 \times m_f}{\text{BP} \times t} \text{ kg/kWh}$$

$$\frac{m_f}{t} = \frac{Q \times S_f \times 10^3}{t} = \frac{4 \times 10^{-3} \times 0.8 \times 10^3}{30} = 0.1067 \text{ kg/min}$$

$$\text{BSFC} = \frac{60 \times 0.1067}{20.032} = 0.3195 \text{ kg/kWh}$$

(v) $$(\eta_{th})_i = \frac{\text{IP} \times 60}{m_f \times CV} = \frac{29.158 \times 60}{0.1067 \times 43 \times 10^3} = 0.3813 \text{ or } 38.13\%$$

■ Example 6.17

(a) Define the following : BHP, IHP, Mechanical Efficiency, Volumetric Efficiency, Specific fuel consumption, Thermal efficiency.

(b) Find the engine dimensions of a two-cylinder, two-stroke I.C. Engine from the following data, Engine Speed = 4000 rpm.

Volumetric efficiency	= 0.77
Mechanical efficiency	= 0.75
Fuel Consumption	= 7.3 kg/h
Air Fuel Ratio	= 18 : 1
Piston Speed	= 600 m/min.

Indirected mean effective pressure = 5 bar. Find also the brake power

Take $R = 287$ J/kg.K at STP. [KU, Dec., 2006]

Solution :

(a) Refer to Art. 6.1.

(b) Given : $N = 4000$ rpm, $n = 1$, $i = 2$, $\eta_{iwl} = 0.77$, $\eta_{mech} = 0.75$, $\dot{m}_f = 73$ kg/h, $A:F = 18:1$, $c = 600$ m/min,

$$p_{im} = 5 \text{ bar}, R = 287 \text{ J/kg.k at STP.}$$

Let d = diameter of piston, m

L = length of stroke, m

Swept volume by piston,

$$\dot{V}_s = i \times \frac{\pi}{4} d^2 \times c = 2\frac{\pi}{4} d^2 \times 600 = 942.48\, d^2 \text{ m}^3/\text{min}$$

$$\eta_{iwl} = \frac{\dot{V}_a}{\dot{V}_s}, \dot{V}_a = 0.77 \times 942.48\, d^2 = 725.7\, d^2 \text{ m}^2/\text{min}$$

Now $$\dot{m}_a = \dot{m}_f \times A:F \text{ ratio} = \frac{73}{60} \times 18 = 2.19 \text{ kg/min}$$

$$p\dot{V}_a = \dot{m}_a RT$$

$$\dot{V}_a = \frac{2.19 \times 287 \times 288}{1.013 \times 10^5} = 1.787 \text{ m}^3/\text{min}$$

$\therefore$ $$725.7 d^2 = 1.787$$

$$d = 0.0496 \text{ m or} \simeq 50 \text{ mm}$$

Also $$\dot{V}_a = i\frac{\pi}{4} d^2 \cdot \frac{LN}{60}$$

$$1.787 = 2\frac{\pi}{4}(0.05)^2 \times L \times 4000$$

$$L = 0.1137 \text{ m or} \simeq 114 \text{ mm}$$

$$\text{IP} = ip_{im} \frac{\text{LAN}}{60 \times 10^3} = 2 \times 5 \times 10^5 \times 0.114 \times \frac{\pi}{4}(0.05)^2 \times \frac{4000}{60 \times 10^3} = 19 \text{ kW}$$

$$\text{BP} = \text{IP} \times \eta_{mech} = 19 \times 0.75 = 14.25 \text{ kW}$$

■ **Example 6.18** A four stroke petrol engine 80 mm bore 100 mm stroke is tested at full throttle at constant speed. The fuel supply is fixed at 0.08 kg/min and the plugs of the four cylinders are successively short circuited without change of speed, brake torque being correspondingly adjusted. The brake power measurements are the following :

With all cylinder firing = 12.5 kW

With cylinder No. 1 cut-off = 9 kW

With cylinder No. 2 cut-off = 9.15 kW

With cylinder No. 3 cut-off = 9.2 kW

With cylinder No. 4 cut-off = 9.1 kW

Determine the IP of the engine under these conditions. Also determine the indicated thermal efficiency. Calorific value of the fuel is 44100 kJ/kg. Compare this efficiency with the air standard value, clearance volume of one cylinder is 70×10^3 mm^3. [KU, Dec., 2007]

Solution : Given : $n = 2$, $d = 80$ mm, $L = 100$ mm, $\dot{m}_f = 0.08$ kg/min, $(BP)_n = 12.5$ kW, $(BP)_1 = 9$ kW, $(BP)_2 = 9.15$ kW, e$(BP)_3 = 9.2$ kW, $(BP)_4 = 9.1$ kW, $CV = 44100$ kJ/kg

$$(IP)_1 = (BP)_n - (BP)_{n-1}$$

$$\therefore \quad (IP)_1 = 12.5 - 9 = 3.5 \text{ kW}$$

$$(IP)_2 = 12.5 - 9.15 = 3.35 \text{ kW}$$

$$(IP)_3 = 12.5 - 9.2 = 3.3 \text{ kW}$$

$$(IP)_4 = 12.5 - 9.1 = 3.4 \text{ kW}$$

$$\text{Indicated power} = \sum_{n=1}^{4} (IP)_n = 3.5 + 3.35 + 3.3 + 3.4 = 13.55 \text{ kW}$$

$$\text{Indicated thermal efficiency, } (\eta_{th})_i = \frac{60 \times IP}{\dot{m}_f \times CV} = \frac{60 \times 13.55}{0.08 \times 44100}$$

$$= 0.2304 \text{ or } 23.04\%$$

$$\text{Swept volume,} \quad V_s = \frac{\pi}{4} d^2 L = \frac{\pi}{4} (0.08)^2 \times 0.1 = 5.026 \times 10^{-4} \text{ m}^3$$

$$V_c = 70 \times 10^3 \text{ mm}^3 \text{ or } 70 \times 10^{-6} \text{ m}^3$$

$$\text{Compression ratio, } r = \frac{V_c + V_s}{V_c} = 1 + \frac{V_s}{V_c} = 1 + \frac{5.026 \times 10^{-4}}{70 \times 10^{-6}} = 8.18$$

Air standard efficiency of petrol engine,

$$\eta_a = 1 - \frac{1}{r^{\gamma - 1}} = 1 - \frac{1}{(8.18)^{0.4}} = 0.5686 \text{ or } 56.86\%$$

■ **Example 6.19** The following data relates to a four cylinder four stroke petrol engine :

Diameter of piston = 80 mm, length of stroke = 120 mm

Clearance volume = 100×10^3 mm^3

Fuel supply = 4.8 kg/hr. calorific value = 44100 kJ/kg

When the Morse test was performed on a single cylinder, the following data were obtained :

B.P. with all cylinders working	= 14.5 kW
B.P with cylinder 1 cut-off	= 9.8 kW
B.P. with cylinder 2 cut-off	= 10.3 kW
B.P. with cylinder 3 cutt-off	= 10.14 kW
B.P. with cylinder 4 cutt-off	= 10 kW.

Find the I.P. of the engine and also calculate indicated thermal efficiency. Brake thermal efficiency and relative efficiency. [KU, Jan., 2008]

Solution : Given, $i = 4, n = 2, d = 80$ mm, $L = 120$ mm, $V_c = 100 \times 10^3$ mm^3, $\dot{m}_f = 4.8$ kg/k, $CV = 44100$ kJ/kg, $(\text{BP})_n = 14.5$ kW, $(\text{BP})_1 = 9.8$ kW, $(\text{BP})_2 = 10.3$ kW, $(\text{BP})_3 = 10.14$ kW, $(\text{BP})_4 = 10$ kW

$$(\text{IP})_1 = (\text{BP})_n = (\text{BP})_{n-1}$$

$$\therefore \quad (\text{IP})_1 = 14.5 - 9.8 = 4.7 \text{ kW}$$

$$(\text{IP})_2 = 14.5 - 10.3 = 4.2 \text{ kW}$$

$$(\text{IP})_3 = 14.5 - 10.14 = 4.36 \text{ kW}$$

$$(\text{IP})_4 = 14.5 - 10 = 4.5 \text{ kW}$$

Indicated power,
$$\text{IP} = \sum_{n=1}^{4} (\text{IP})_n = 4.7 + 4.2 + 4.36 + 4.5 = 17.76 \text{ kW}$$

Indicated thermal efficiency,

$$(\eta_{\text{th}})_i = \frac{360 \times \text{IP}}{\dot{m}_f \times CV} = \frac{3600 \times 17.76}{4.8 \times 44100} = 0.3020 \text{ or } 30.2\%$$

Brake thermal efficiency,

$$(\eta_{\text{th}})_b = \frac{3600 \times \text{BP}}{\dot{m}_f \times CV} = \frac{3600 \times 14.5}{4.8 \times 44100} = 0.2466 \text{ or } 24.66\%$$

Swept volume,
$$V_s = \frac{\pi}{4} d^2 L = \frac{\pi}{4} (80)^2 \times 120 = 603.185 \times 10^3 \text{ mm}^3$$

Compression ratio,
$$r = 1 + \frac{V_s}{V_c} = 1 + \frac{603.185 \times 10^3}{100 \times 10^3} = 7.03$$

Air standard efficiency,

$$\eta_a = 1 - \frac{1}{r^{\gamma - 1}} = 1 - \frac{1}{(7.03)^{0.4}} = 0.5416 \text{ or } 54.16\%$$

Relative efficiency,
$$\eta_i = \frac{(\eta_{\text{th}})_i}{\eta_a} = \frac{30.2}{54.16} = 0.5576 \text{ or } 55.76\%$$

■ **Example 6.20** During a trial on a four cylinder petrol engine running at 3000 r.p.m. the brake load was 267 N when all cylinders were working. When each cylinder was cut out in turn and speed returned to 3000 r.p.m., the brake readings were 178 N, 187 N, 182 N and 182 N. For the brake $\text{b.p.} = \dfrac{W \times \text{r.p.m.}}{27300}$ where W is brake load in N. Determine :

(i) Friction power of the engine,

(ii) The indicated power, and

(iii) Mechanical efficiency. [KU, Dec., 2008]

Solution : Given : $i = 4$, $N = 3000$ rpm, $W = 267$ N, $W_1 = 178$ N, $W_2 = 187$ N, $W_3 = 182$ N, $W_4 = 182$ N

$$(\text{BP})_n = \frac{W \times N}{27300} \text{ kW} = \frac{267 \times 3000}{27300} = 29.34 \text{ kW}$$

$$(BP)_1 = \frac{W_1 N}{27300} = \frac{178 \times 3000}{27300} = 19.56 \text{ kW}$$

$$(BP)_2 = \frac{W_2 N}{27300} = \frac{187 \times 300}{27300} = 20.55 \text{ kW}$$

$$(BP)_3 = \frac{W_3 N}{27300} = \frac{182 \times 300}{27300} = 20 \text{ kW}$$

$$(BP)_4 = \frac{W_4 N}{27300} = \frac{182 \times 3000}{27300} = 20 \text{ kW}$$

$$(IP)_1 = (BP_n) - (BP)_{n-1} = 29.34 - 19.56 = 9.78 \text{ kW}$$

$$(IP)_2 = 29.34 - 20.55 = 9.79 \text{ kW}$$

$$(IP)_3 = 29.34 - 20 = 9.34 \text{ kW}$$

$$(IP)_4 = 29.34 - 20 = 9.34 \text{ kW}$$

$$IP = \sum_{n=1}^{4} (IP)_n = 9.78 + 8.79 + 9.34 + 9.34 = 37.25 \text{ kW}$$

(i) $FP = IP - BP = 37.25 - 29.34 = 9.41 \text{ kW}$

(ii) $IP = 37.25 \text{ kW}$

(iii) $\eta_{mech} = \frac{BP}{IP} = \frac{29.34}{37.25} = 0.7876 \text{ or } 78.76\%$

■ **Example 6.21** The following details were noted in a test on a 4-stroke, 4-cylinder engine, Diameter-100 mm, stroke-120 mm, speed of engine-1600 rpm, fuel consumption-0.2 kg/minute, calorific value of fuel-44000 kJ/kg, difference in tension on either side of brake pulley-40 kg, brake circumference-300 cm. If the mechanical efficiency is 80% calculate

(i) brake thermal efficiency
(ii) indicated thermal efficiency
(iii) indicated mean effective pressure
(iv) brake specific fuel consumption. [KU, Sept., 2009]

Solution : Given : $i = 4$, $n = 2$, $d = 100$ mm, $L = 120$ mm, $N = 1600$ rpm, $\dot{m}_f = 0.2$ kg/min, $CV = 44000$ kJ/kg, $T_1 - T_2 = 40$ kg, $\pi D_b = 300$ cm, $\eta_{mech} = 0.80$.

$$D_b = \frac{300 \times 10^{-2}}{\pi} = \frac{3}{\pi} \text{ m}$$

$$BP = (T_1 - T_2)\frac{D_b}{2} \times \frac{2\pi N}{60 \times 10^3} \text{ kW}$$

$$= 40 \times 9.81 \times \frac{3}{2\pi} \times \frac{2\pi \times 1600}{60 \times 10^3} = 31.392 \text{ kW}$$

$$IP = \frac{BP}{\eta_{mech}} = \frac{31.392}{0.8} = 39.24 \text{ kW}$$

(i) Brake thermal efficiency,

$$\eta_{bt} = \frac{60 \times BP}{\dot{m}_f \times CV} = \frac{60 \times 31.392}{0.2 \times 44000} = 0.2140 \text{ or } 21.4\%$$

(ii) Indicated thermal efficiency,

$$\eta_{it} = \frac{60 \times \text{IP}}{\dot{m}_f \times CV} = \frac{60 \times 3924}{02 \times 44000} = 02675 \text{ or } 26.25\%$$

(iii)
$$\text{IP} = i \times p_{im} \times \frac{\pi}{4} d^2 \times \frac{LN}{60 \times 10^3 \times n}$$

$$3924 = 4 \times p_{im} \times \frac{\pi}{4} (0.1)^2 \times 0.12 \times \frac{1600}{60 \times 10^3 \times 2}$$

$$p_{im} = 7.8 \text{ bar}$$

(iv) $\text{BSFC} = \frac{60 \times \dot{m}_f}{\text{BP}} = \frac{60 \times 0.2}{31392} = 03822 \text{ kg/kwh.}$

■ **Example 6.22** An eight cylinder, 4-stroke engine of 9 cm bore and 8 cm stroke with compression ratio of 7 is tested at 4500 rpm on a dynamometer which has 54 cm arm. During a ten minute test, the dynamometer scale beam reading was 42 kg and the engine consumed 4.4 kg of petrol with a calorific value of 44000 kJ/kg. Air at 27°C and 1 bar was supplied to the carburettor at the rate of 6 kg/minute. Find :

(a) the brake power,
(b) the brake mean effective pressure,
(c) the brake specific fuel consumption,
(d) the brake specific air consumption,
(e) the brake thermal efficiency,
(f) the volumetric efficiency, and
(g) the air-fuel ratio. [KU, Dec., 2009]

Solution : Given $i = 8, n = 2, d = 9$ cm, $L = 8$ cm, $r = 7$, $N = 4500$ rpm, $l = 54$ cm, $t = 10$ min, $W = 42$ kg, $m_f = 4.4$ kg, $CV = 44000$ kJ/kg, $T = 273 + 27 = 300$ K, $p = 1$ bar, $\dot{m}_a = 6$ kg/min

(a) $\text{BP} = \frac{2\pi N \times Wl}{60 \times 10^3} = \frac{2\pi \times 450 \times 42 \times 9.81 \times 0.54}{60 \times 10^3} = 104.846 \text{ kW}$

(b) BMEP :

$$\text{BP} = \frac{i \times p_{bm} \times \text{LAN}}{60 \times 10^3 \times n}$$

$$104.846 = \frac{8 \times p_{bm} \times 0.08 \times \frac{\pi}{4} (0.90)^2 \times 4500}{60 \times 10^3 \times 2}$$

$$p_{bm} = 6.867 \text{ bar}$$

(c) $\text{BSFC} = \frac{60 \times m_f}{\text{BP} \times t} = \frac{60 \times 4.4}{104.846 \times 10} = 0.2518 \text{ kg/kWh}$

(d) $\text{BSAC} = \frac{60 \times \dot{m}_a}{\text{BP}} = \frac{60 \times 6}{104.846} = 3.434 \text{ kg/kWh}$

(e) $\eta_{bt} = \frac{60 \times BP \times t}{m_f \times CV} = \frac{60 \times 104.846 \times 10}{4.4 \times 44000} = (3249 \text{ or } 32.49\%$

(f) $\dot{V}_s = i \times \frac{\pi}{4} d^2 \frac{LN}{n} = 8 \times \frac{\pi}{4}(0.08)^2 \times 0.09 \times \frac{4500}{2} = 8.143 \text{ m}^3/\text{min}$

$$p\dot{V}_a = \dot{m}_a RT$$

$$1 \times 10^5 \times \dot{V}_a = 6 \times 287 \times 300$$

$$\dot{V}_a = 5.166 \text{ m}^3/\text{min}$$

$$\eta_{vol} = \frac{\dot{V}_a}{\dot{V}_s} = \frac{5.166}{8.143} = 0.6344 \text{ or } 63.44\%$$

(g) $A : F = \frac{\dot{m}_a}{\dot{m}_f} = \frac{6 \times 10}{4.4} = 13.64$

■ **Example 6.23** A six cylinder, petrol engine operates on 4-stroke cycle. The bore of each cylinder is 80 mm and the stroke is 100 mm. The clearance volume per cylinder is 70 ml. At a speed of 4000 rpm the fuel consumption is 20 kg/hour and the torque developed is 150 N.m. Calculate (i) Brake power, (ii) Brake mean effective pressure, (iii) Brake thermal efficiency if the calorific value of fuel is 43000 kJ/kg and (iv) The relative efficiency on a brake power basis assuming the engine works on the constant volume cycle. $\gamma = 1.4$ for air. [KU, Dec., 2010]

Solution : Given : $i = 6$, $n = 2$, $d = 80$ mm, $L = 100$ mm, $V_c = 70$ ml, $N = 400$ rpm, $\dot{m}_f = 20$ kg/h, $T = 150$ N.m , $CV = 43000$ kJ/kg, $\gamma = 1.4$

(i) $$BP = \frac{2\pi NT}{60 \times 10^3} = \frac{2\pi \times 4000 \times 150}{60 \times 10^3} = 62.832 \text{ kW}$$

(ii) BMEP :

$$BP = \frac{i \times p_{bm} \times LAN}{60 \times 10^3 \times n}$$

$$62.832 = 6 \times b_{bm} \times 0.1 \times \frac{\pi}{4}(0.08)^2 \times \frac{100}{60 \times 10^3 \times 2}$$

$$p_{bm} = 6.251 \text{ bar}$$

(iii) $\eta_{bt} = \frac{3600 \times BP}{\dot{m}_f \times CV} = \frac{3600 \times 62.832}{20 \times 43000} = 0.2630 \text{ or } 26.3\%$

(iv) $V_s = \frac{\pi}{4} d^2 L = \frac{\pi}{4}(0.08)^2 \times 0.1 = 0.5026 \times 10^{-3} \text{ m}^3$

$V_c = 70 \times 10^{-6} \text{ m}^3$

Compression ratio, $r = 1 + \frac{V_s}{V_c} = 1 + \frac{0.5026 \times 10^{-3}}{70 \times 10^{-6}} = 8.18$

Air standard efficiency, $\eta_a = 1 - \frac{1}{r^{\gamma - 1}} = 1 - \frac{1}{(8.18)^{0.4}} = 0.5686 \text{ or } 56.86\%$

SHORT ANSWER TYPE QUESTIONS

1. Explain the method to measure the brake power of a small engine.
2. Describe the method to measure the heat lost in exhaust gases of an I.C. engine.
3. Explain the Morse test to measure the indicated power of a multi cylinder engine.
4. Describe the method commonly used in laboratory to measure the air supplied to an I.C. engine.
5. Derive the formula used for finding the mass of air supplied to an engine using an orifice meter and tank.
6. Draw the following curves for a single-cylinder, 4-stroke petrol and diesel engines :
   (a) BP *v*'s fuel consumption (b) BP *v*'s SFC
   (c) BP *v*'s brake thermal efficiency.
7. Discuss the various performance parameters of I.C. engines.
8. What are the effects of load on the following :
   (a) η_{vol} (b) η_{mech}
   (c) η_{bt} (d) BSFC of S.I. engines.
9. Define the following :
   (a) BP (b) IP (c) FP
   (d) BSFC (e) η_r
10. Define the following :
    (a) Equivalence ratio (b) ISFC (c) Specific weight (d) p_{imr}

EXERCISES

1. A single cylinder, 4-stroke cycle oil engine works on diesel cycle. The following data is available from a test :
   Area of indicator diagram = 3 cm^3, length of indicator diagram = 4 cm, Spring constant = 10 bar/cm^2-cm, Engine speed = 400 rpm, Load on the brake = 380 N, Spring reading = 50 N, Diameter of brake drum = 1.2 m, Fuel consumption = 2.8 kg/h, LCV = 42 MJ/kg, Cylinder diameter = 16 cm, Piston stroke = 20 cm.
   Calculate : (a) FP (b) η_{mech} (c) Brake thermal efficiency, and (d) BMEP.
   [Ans. 1.71 kW, 83%, 25%, 6.15 bar]
2. A 4-cylinder, 4-stroke petrol engine 6 cm bore and 9 cm stroke was tested at constant speed. The fuel supply was fixed to 0.13 kg/min and spark plugs of 4-cylinders were successively short-circuited without change of speed.
   The power measurements were as follows :
   With all cylinder firing = 16.25 kW, with first cylinder cut-off = 11.55 kW,
   With 2nd cylinder cut-off = 11.65 kW, with 3rd cylinder cut-off = 11.70 kW,
   With 4th cylinder cut-off = 11.50 kW.
   Determine :
   (a) IP of the engine
   (b) mechanical efficiency
   (c) Indicated thermal efficiency if LCV of fuel used is 42 kJ/kg, and
   (d) Relative efficiency on IP basis assuming clearance volume 65 cm^3. **[Ans.** 18.5 kW, 88%, 20.4%, 42%]

3. The following readings were taken during the test on a single cylinder, 4-stroke I.C. engine :
Speed of engine = 240 rpm, Orifice diameter of tank = 20 mm,
Pressure causing flow through the orifice = 10 cm of H_2O,
Ambient temperature = 30°C, Berometer reading = 760 mm of Hg.
Gas consumption = 3.8 m^3/h at 30°C and 10 cm of H_2O above atmosphere.
Find (a) The air consumption per hour, (b) A : F ratio by volume and weight.
Take $C_d = 0.7$ for air orifice, $R = 287$ J/kg.K for air and 280 J/kg.K for gas.

[Ans. 0.544 m^3/min, 85 : 1, 8 : 1]

4. The following data was obtained from a test on a single-cylinder, 4-stroke oil engine.
Cylinder bore = 15 cm, stroke = 25 cm, area of indicator diagram = 540 mm^2, length of indicator diagram = 50 mm, indicator spring rating = 1.2 mm for a pressure of 8.8 N/cm^2, engine speed = 400 rpm, brake torque = 225 N-m, fuel consumption = 3 kg/h, calorific value of fuel = 44200 kJ/kg, cooling water flow rate = 4 kg/min, cooling water temperature rise = 42°C, specific heat of water = 4.1868 kJ/kg.°C.
Calculate : (a) The mechanical efficiency, (b) the brake thermal efficiency, (c) the specific fuel consumption, and (d) draw heat balance sheet.

[Ans. 87.07%, 25.6%, 0.31 kg/kwh, $Q_s = 36.83$ kW, $Q_w = 11.72$ kW, $Q_{ge} = 15.68$ kW, $Q_b = 9.43$ kW]

5. The following particulars were obtained in a trial for one hour on a 4-stroke gas engine :
Speed = 16000 rpm, Missed cycles = 600 rpm, Net brake load = 1600 N, Brake circumference = 4 m, IMEP = 8 bar, Gas consumption = 2200 litres, *CV* of gas = 20 kJ/litre, $d = 25$ cm, $L = 40$ cm, $r = 6.5$.
Find (a) IP and BP (b) BSFC, (c) η_{bt} and η_r. **[Ans.** 32.3 kW, 28.4 kW, 773.4 litres/kwh, 23.3%, 44%]

6. An 8-cylinder, 4-stroke petrol engine of 80 mm bore and 90 mm stroke has a compression ratio of 7. It is tested at 4000 rpm on a dynamometer which has 54 cm arm. During a 10 minute test, the dynamometer reads 400 N and the engine consumes 4.75 kg of fuel. Air is supplied at 1 bar and 27°C at the rate of 6.5 kg/min. *CV* of fuel = 44 MJ/kg.
Determine : (a) BP (b) BMEP (c) η_{vol} (d) A : F ratio, and (e) η_{rel}

[Ans. 90.8 kW, 6 bar, 77.32%, 13.68 : 1, 48%]

□□□

Chapter 7

Air Pollution from I.C. Engines

Inside This Chapter

7.1 S.I. ENGINE-EMISSIONS

S.I. engine emissions are divided into three categories as exhaust emission, evaporative emission and crank case emission.

The major constituents which contribute to air pollution are CO, NO_x and HC coming from S.I. engine exhaust.

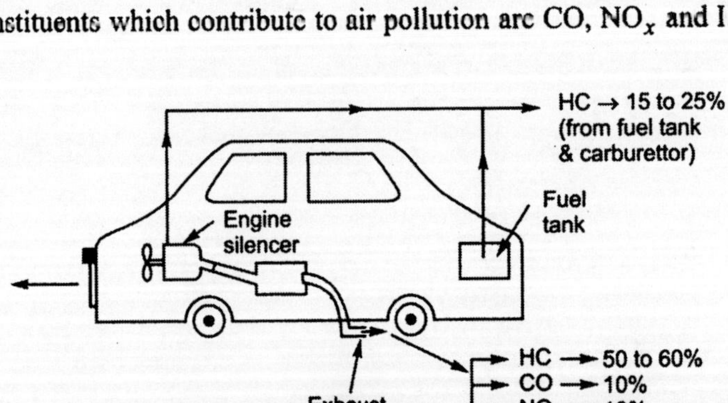

Fig. 7.1 Emissions from S.I. engines.

The percentages of different constituents coming out from the above three mentioned sources are shown in Fig. 7.1.

The relative amounts depends on engine design and operating conditions but are of order, NOx → 500 to 1000 ppm (20 g/kg of fuel), CO → 1 to 2% (200 g/kg of fuel) and HC → 3000 ppm (25 g/kg of fuel). Fuel evaporation from fuel tank and Carburettor exists even after engine shut down and these are unburned hydrocarbons. However in most modern engines, these non-exhaust unburned HC are effectively controlled by returning the blow by gases from the crank case to the engine intake system by venting the fuel tank and through a vapour absorbing carbon cannister which is purged as some of the engine intake air during normal engine operation.

The other constituents include SO_2 and lead compounds. The petrol rarely contains sulphur therefore, SO_2 is not a pollutant from S.I. engine exhaust. Petrol contains lead in small percentages but its effect is more serious on human health.

The processes by which pollutants form within the cylinder in conventional S.I. engine are illustrated qualitatively in Fig. 7.1. The figure shows the formation of pollutants during 4-strokes of the cycle. NO forms throughout the high temperature burned gases behind the flame through chemical reactions. NO formation rate increases with an increase in gas temperature. As the burned gases cool, during expansion stroke, the reactions involving NO freeze and leave NO concentrations far in excess of levels corresponding to equilibrium temerature at exhaust conditions.

CO also forms during combustion process with lean A : F mixtures, there is sufficient O_2 to burn all the carbon in the fuel to CO_2. But in high temperature products even with lean mixtures, there is sufficient CO in exhaust because of dissociation of CO_2. Later in expansion stroke, the CO oxidation process also freezes as the gas temperature falls.

The unburned hydrocarbon emission comes from different sources. During compression and combustion, the increasing cylinder pressure forces some of the gases in the cylinder into crevices connected to combustion chamber, the volumes between the piston rings and cylinder wall are the largest of these. Most of this gas entering into crevices is unburned air fuel mixture escaped from primary combustion zone. This is because, the flame cannot enter into these narrow crevices. This gas which leaves these crevices later in the expansion and exhaust processes, is one source of unburned hydro-carbons emissions. Another possible source is the combustion chamber walls. A quench layer containing unburned and partially burned A : F mixture is left at the wall when the flame dies as it approaches the wall. This unburned HC in this layer (0.1 mm) burns rapidly if the co.nbustion chamber walls are clean. The next source of HC is the thin layer of lubricating oil on cylinder wall, and piston which absorbs HC before and after combustion. A final source of HC in engines is incomplete combustion due to bulk quenching of the flame in that fraction of engine cycle where the combustion is specially slow. This unburned HC reinained near the cylinder wall are exhausted out during exhaust stroke as the piston pushes gases out.

7.1.1 Exhaust Emissions

Major exhaust emissions are :

(i) Unburnt hydrocarbons, (HC)
(ii) Oxides of carbon, (CO and CO_2)
(iii) Oxides of nitrogen, (NO and NO_2)
(iv) Oxides of sulphur, (SO_2 and SO_3)
(v) Particulates
(vi) Soot and smoke.

The various exhaust emissions from petrol engine are :

1. **Unburnt hydrocarbons (HC) :** The causes for the emissions of HC are :
   Incomplete combustion, crevice volumes and flow in crevices, leakage past the exhaust valve, valve overlap, deposits on walls and oil on combustion chamber walls.
   The reasons for incomplete combustion are : improper mixing of air and fuel, and flame quenching at the walls of the cylinder. Low load and idle conditions increase HC.
2. **Oxides of Carbon (CO and CO_2) :** Carbon monoxide is generated with a rich fuel-air ratio mixture. This happens during starting and accelerating under load. Poor mixing, local rich regions, and incomplete combustion are the sources of CO emissions.
3. **Oxides of Nitrogen (NO_x) :** NO_x are created mostly from nitrogen in the air and fuel blends. In addition to temperature, the formation of NO_x depends on pressure, A/F ratio, combustion duration and location of spark plug.
4. **Oxides of Sulphur (SO_x):** These are generated due to presence of sulphur in the fuel. SO_2 and SO_3 react with water to give rise to H_2SO_3 and H_2SO_4 which causes acid rain.
   The variation of emissions from a petrol engine with A/F ratio is shown in the Fig. 7.2.

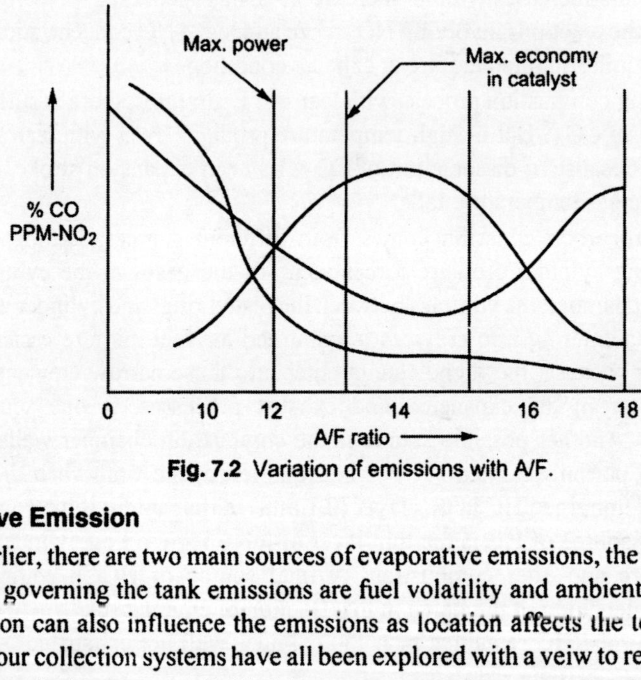

Fig. 7.2 Variation of emissions with A/F.

7.1.2 Evaporative Emission

As mentioned earlier, there are two main sources of evaporative emissions, the fuel tank and carburettor. The main factors governing the tank emissions are fuel volatility and ambient temperature but the tank design and location can also influence the emissions as location affects the temperature. Insulation of fuel tank and vapour collection systems have all been explored with a veiw to reduce the tank emissions.

Carburettor emission may be divided into two categories as running losses and parking losses. Although most internally vented carburettors have an external vent which open at idle throttle position. The existing pressure forces prevent outflow of vapours to the atmosphere. Internally vented carburettor may enrich the mixture which inturn increases exhaust emission. Carburettor losses are significant only during hot condition when the vehicle is in operation. The fuel volatility also affects the carburettor emissions.

7.1.3 Crankcase Emission

It consists of engine blow by-gases and crank case lubricant fumes. From pullution point of view, blow by gases are most important. The blow-by is the phenomenon of leakage past the piston from the cylinder to the crankcase because of pressure difference. The blow of HC emissions are about 20% of total HC emission from the engine. This is further increased to 30% if the piston-rings are worn.

7.1.4 Lead Emission

Lead emissions come only from S.I. engines. The lead is present in the fuel as lead tetraethyl or tetramethyl, to control the self ignition tendency of fuel-air mixtures that is responsible for knock.

7.2 CONTROL OF EMISSIONS IN SI ENGINE

An emission control programme aims at reducing the concentration of CO, HC and NO. The main approaches adopted are :

(a) Engine design modification

The following steps reduce the exhaust emission by engine design modification :

(i) Use leanear air-fuel ratio.

(ii) Retard ignition timing.

(iii) Avoid flame quenching time by reducing the surface to volume ratio of the combustion chamber

(iv) Lower compression ratio.

(v) Reduce valve overlap.

(vi) Modify the induction system by using high velocity carburettors or multi-chock carburettors.

(b) Exhaust gas oxidation

The devices used to reduce HC/CO emissions are :

(i) Use of after-burner.

(ii) Use of exhaust manifold reactor.

(iii) Use of catalytic converter.

(c) Fuel Modification

Air-fuel ratios leaner than stoichiometric results in almost insignificant amount of CO and reduce HC with reduced specific fuel consumption.

(d) Blow by Control

The crankcase blow by control is the recirculation of the vapours back to the intake air cleaner.

Now we study the emission control measures in S.I. engines in more detail.

To reduce atmospheric pollution, two different approaches are followed :

1. To reduce the formation of pollutants in the emission by redesigning the engine system, fuel system, cooling system and ignition system.
2. By destroying the pollutants after these have been formed.

In petrol engines, the main pollutants which are objectionable and are to be reduced are HC, CO and NOx. The methods used are :

(a) Modifications in the Engine design

(b) Modifying the fuel used.

(c) Exhaust Gas treatment Devices

(d) Evaporative emissive control devices.

(a) Modifications in the Engine Design

Engine modifications improve the emission quality. A few parameters which improve an emission are listed below :

1. **Combustion Chamber Configuration :** Modification in the combustion chamber as reducing surface/volume ratio can reduce quenching zone and can reduce HC emission. This can also be achieved by reducing dead space around piston ring.
2. **Lower Compression Ratio :** Lower compression ratio also reduces the quenching area and thus reduces HC emission. Lower compression ratio also reduces NOx emission due to lower maximum temperature. But lowering compression ratio reduces thermal efficiency and increases fuel consumption. But using petrol of lower Octane number, it is possible to phase the lead out of petrol, i.e., use of unleaded petrol.
3. **Induction System :** The supply of designed A : F ratio mixture to multi-cylinder engine is always difficult under all operating conditions of load and power. This can be achieved by proper desing of induction system or using high velocity or multip-choke carburettors.
4. **Ignition Timing :** Retarding spark ignition allows increased time for the fuel to burn. Retarding the spark reduces NOx formation by decreasing NOx emission. It also reduces HC emission by causing higher exhaust temperature. However, retarding the ignition results in loss of power and consumption of fuel. The controls are designed to retard the spark timing during idling and providing normal spark advance during acceleration.
5. **Reduced Valve Overlap :** Increased overlap carries fresh mixture with the exhaust and increases emission level. This can be avoided by reducing the valve overlap.

(b) Modifying the Fuel Used

Most of the vehicles use S.I. engines. To reduce the pollution from the exhaust of these engines, the emission of Olefins should be obviated as far as possible. For this, we can change the fuel itself. LPG and CNG be used instead of gasoline as they produce less pollution than present petrol engines.

(c) Exhaust Gas Treatment

Exhaust Gas Oxidation : The exhaust gas coming out of exhaust mainfold are treated to reduce HC and CO emission. A few devices are discussed below.

1. **Use of After Burner :** After burner is nothing but a burner where air is supplied to the exhaust gases and the mixture is burned with the help of ignition system. The HC and CO which are formed in the engine combustion chamber because of inadequate O_2 and inadequate time to the burn are further burned by providing air in a separate box, known as after burner. The after-burner is located very near to the exhaust manifold with an intention that the temperature of exhaust should not fall. The oxidation of HC in the after burner depends upon the temperature of exhaust and mixing provided in the after burner. Air injection does nothing to the NOx emission.

   A simple arrangement of an after burner is shown in Fig. 7.3.

   The performance of this system was not satisfactory as combustion was not sustained during low HC emission.

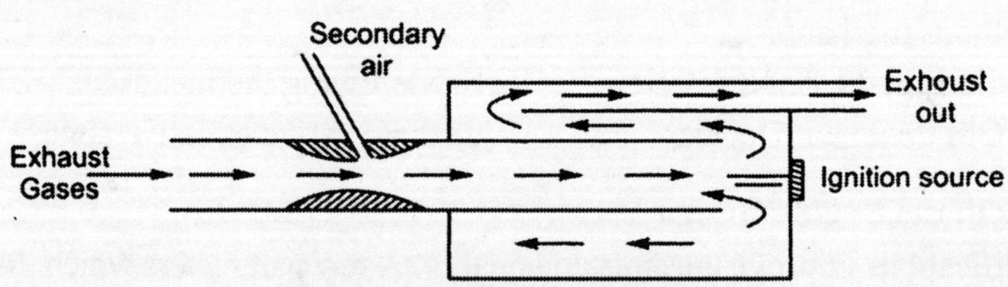

Fig. 7.3 Typical after burner.

2. **Exhaust Manifold Reactor :** This is a further development of after-burner where high temperature exhaust gases and secondary air are mixed properly and burn. Where HC carried with exhaust combines with O_2 and forms non-objectionable gases.
 There are different types of after burners where heat losses are minimised and sufficient time and mixing of exhaust and secondary air are provided.
 A special after-burner designed by Du-Point, where the entry of exhaust gases is radial and air flow is peripheral is shown in Fig. 7.4.

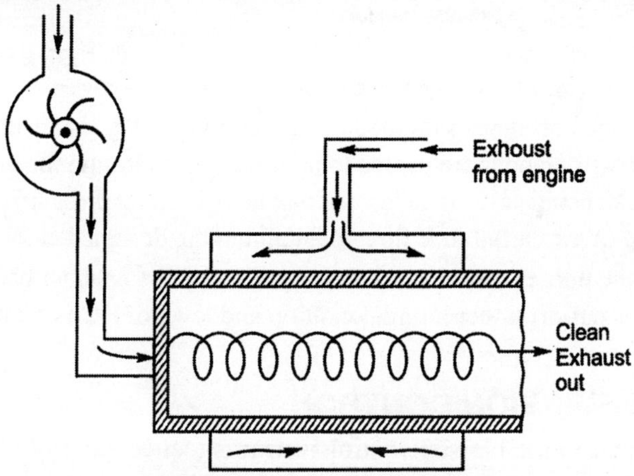

Fig. 7.4 A typical after burner.

3. **Catalytic Converters :** Catalytic oxidation of the exhausted HC and CO is accomplished by placing a common device called catalytic converter in the vehicle exhaust system. The catalytic converter is filled with catalytic material. Exhaust gas hydrocarbons and CO are oxidised while passing through the bed. The catalytic material itself does not enter into the reaction but only promotes the oxidation process at lower temperature. Usually air compressor is used to supply additional oxygen necessary for complete oxidation of the exhaust gas stream.

A catalyst is an agent that aids or speeds a process or chemical reaction without becoming a part of reaction during the process—sort of a chemical middleman. In a modern car's emissions control system, the so-called 3-way catalyst helps the three major evil elements of exhaust, —hydrocarbons (HC), carbon monoxide (CO) and oxides of nitrogen (NO_x)—react with oxygen and each other. The catalyst helps the HC and CO become non-poisonous carbon dioxide (CO_2) and water vapor, while the NO_x is converted into CO_2, nitrogen and water vapor.

(d) Evaporation Emission Control Device (EECD)

The purpose of this device is to collect all evaporative emissions (vapours) and recirculating them at a proper time.

The device is shown in Fig. 7.5. It consists of an absorbent chamber, pressure balancing valve and purge control valve. The absorbent chamber contains charcoal which can hold the hydrocarbon vapour before escaping to atmosphere. The fuel tank and carburettor float, which are main sources of HC emission in form of vapour are directly connected to absorbent chamber when the engine is turned off i.e., under hot soak-condition. This causes the petrol to boil from carburettor float and large amout of petrol vapour comes out. All these vapours are during stopping or running the engine are absorbed in the absorber chamber.

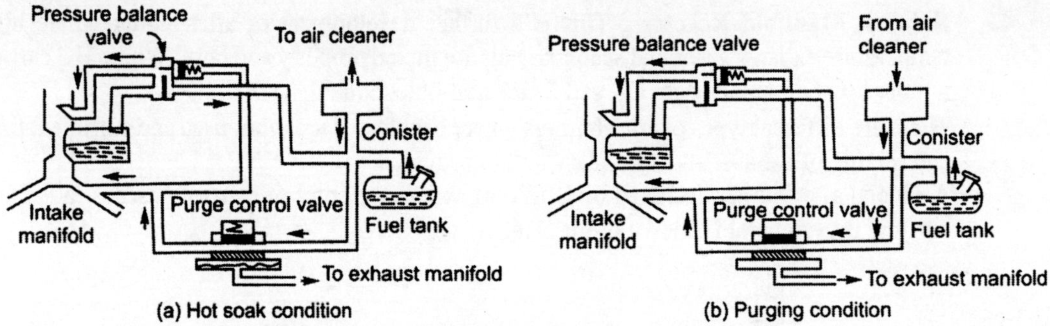

Fig. 7.5 Fuel-system evaporation loss control device.

When the absorber bed becomes saturated, the air coming out from air-cleaner is passed through absorber bed and the air-with-vapour are passed to inlet manifold through the purge valve. At this time, the seat of the pressure balancing valve is so located that there is direct pressure communication between the internal vent and top of the carburettor float maintaining the designed carburettor metering forces.

The operation of the purge control valve is controlled by the exhaust back pressure as shown in figure. The fuel supply is cut-off under idling condition and level of HC is reduced.

7.3 CRANK CASE EMISSION CONTROL

The basic principle of crank case blowby control system is recirculation of vapours back to the inlet manifold. Fig. 7.6 shows a typical closed type known as positive crank case ventilation (PCV) system. The gases escaping past the piston and enterning into the crankcase are returned back to the inlet manifold and then to the engine. During compression, the HC enters into the crankcase due to the pressure difference between the engine cylinder and crank case (small vacuum) and the HC from the crank case is taken back into the intake manifold during the expansion stroke and back side of the piston compresses the crank cases gases.

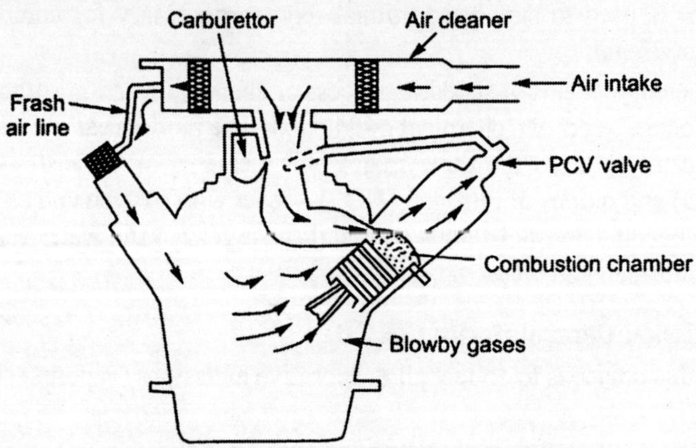

Fig. 7.6 Positive crank case ventilation system.

7.4 C.I. ENGINE EMISSIONS

The diesel engine is used more than any other type of engine for transportation, thermal power generation and many other industrial and agricultural applications. The exhaust emissions from combustion in diesel engine are no different from those of combustion processes in petrol engine, the difference being only in the level of concentration of individual pollutants. The sample of diesel exhaust may be free from smoke, odour and HC or may be heavily smoke laden highly malodorous and can have heavy concentration of unburned HC.

The pollutants from diesel engine can be classified into two types as visible and invisible emissions. Visible emission is the smoke which is objected more by the public. The invisible emissions include, CO, HC, NOx, SO_2, partially oxidised organics (as aldehydes and ketones) and odours. An unpleasant odour is also heavily objected by the public. The smoke and odour are not harmful to the public health but they are objectionable because of its unsightliness, unbearable smell and possible reduction in visibility. Other invisible emissions mentioned above have similar effect on health as they are also emitted by petrol engines.

(a) Effect of Engine Type on Diesel Emission

The type of the engine and speed of the engine are two main factors which influence the exhaust emission from a diesel engine. It has been observed that there is a significant difference in emission levels from different engines except the odour level.

The following observations can be summarised :

1. A two stroke air-scavenged engine produces high HC and intermediate NOx. The smoke level remains low at all load conditions.
2. A 4-stroke medium speed engine has lowest emissions of all constituents except high smoke intensity.
3. A 4-stroke high speed engine has high HC emissions.
4. A turbo-charged 4-stroke engine is notably low in HC but high in NOx. The smoke level is also considerably low compared with other engines.

1. Formation of smoke and Affecting Factors

Engine exhaust smoke is the result of incomplete combustion. Smoke from exhaust is a visible indicator of the combustion process within the engine. It is generated at any volume in the engine where mixture is rich. The fuel air ratio greater than 1.5 and at pressures developed in diesel engine produces soot. Once soot is formed, it can burn if it finds sufficient O_2 otherwise it comes out with exhaust. It becomes visible if it is dense. The size of the soot particles affects the appearance of smoke. The soot particles agglomerate into bigger particles which have an objectionable darkening effect on diesel exhaust.

The turbo-charged engine emits loss smoke within increasing fuel air ratio as the mixing is much better in this engine as well as sufficient O_2 is also available in the engine cylinder at all times.

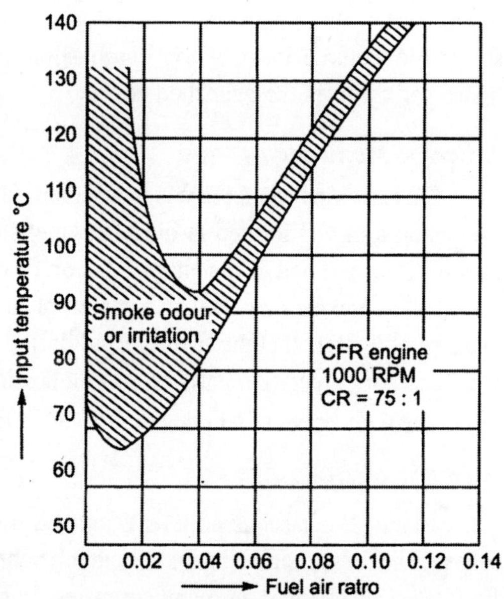

Fig. 7.7 Effect of fuel air ratio on odour in diesel exhaust.

2. Diesel Odour

It has been observed from the experiments that the products of partial oxidation are the main cause of odour in diesel exhaust. This partial oxidation may be because of very lean mixture during idling or due to wall quenching effect. The effect of fuel-air ratio and odour is shown in Fig. 7.7.

The members of the aldehyde family are considered responsible for the pungent odours of diesel exhaust. The aldehydes in the exhaust are found maximum 30 ppm but it is observed that even 1 ppm can cause irritation to the nose and eyes. There is no standard method yet developed for measuring the odour. However, a number of odour producing components like nepthaldehyde, *n*-butylbenzene, etc., are given standard rating and trained person can give odour rating for the diesel exhaust sample by comparison.

The factors which affect the odour formation in the diesel engine are discussed below :

(1) **Fuel Air Ratio :** It is already mentioned that, lean mixtures produce odours.

(2) **Mode of Engine operation :** The mode of operation of the engine affects the exhaust odour significantly. Maximum odour occurs when engine is accelerated from idling.

(3) **Engine Type :** The odour intensity does not change with the type of engine as two strokes or four-strokes engine.

It is also claimed by few research workers that the intensity of odour is reduced by additive compounds.

3. Unburned Hydro-Carbons (HC)

The concentration of hydro-carbons in diesel exhaust varies from a few ppm to several thousand ppm depending upon the load on the engine and its speed. The hydro-carbons in diesel exhaust are composed of a mixture of many individual hydrocarbons in the fuel supplied to the engine as well as partly burned hydrocarbons produced during combustion process.

During the normal operation of the engine, the relatively cold wall quench the fuel-mixture and inhibit the combustion leaving a thick layer of unburned fuel air mixture over the entire surface of the combustion chamber. The thickness of this layer depends upon combustion pressure, temperature, mixture ratio, turbulence and residual gases in the engine at the end of the exhaust stroke. A greater surface to volume ratio of the combustion chamber leads to the formation of greater fraction of hydro-carbon from the quenched zone.

4. Carbon Monoxide

CO is formed when there is insufficient O_2 to oxidize the fuel fully during the combustion of fuel. The amount of CO formed in diesel engine is considerably lower than pertrol engine because always excess air is supplied to the engine. Theoretically, diesel engine should not emit CO at all as it always operates with excess air. But CO is present in small quantities in diesel exhaust and this is, possibly, because of the fact that the fuel injected during later part of injection does not find sufficient O_2 due to local depletion in certain parts of the combustion chamber.

The percentage of CO in exhaust varies from 0.1% to 0.75% which is easily acceptable level.

5. Oxides of Nitrogen

Among the gaseous pollutants emitted by the diesel engine, NOx are the most significant. In this respect, the diesel engine is not very much behind the gasoline engine. NOx being most hazardious, the limit is set to 350 ppm in many countries. In may diesel engines, NOx varies from a few hundreds to 1000 ppm.

The mechanism of formation of NOx is same as discussed in the petrol engine. The conditions which create highest local-temperature (2000 K) and have sufficient O_2 give highest NOx concentration in diesel engine also.

The pre-combustion chamber diesel engines produce less NOx than a direct injection engines because of low peak temperature. High fuel-air ratio (rich mixture), an additional fuel tends to cool the charge and localised peak temperature falls and reduces the NOx emission.

In addition to above, injection pattern, injection period, Centane number of the fuel, viscosity and rate of burning also affect NOx formation significantly. The effect of load on NOx emision rates for 4-stroke normally aspirated engine and 4-cycle turbo-charged engines are shown in Fig. 7.8. The NOx emission for turbocharged engine is considerably high compared with normally aspirated engine at all load conditions.

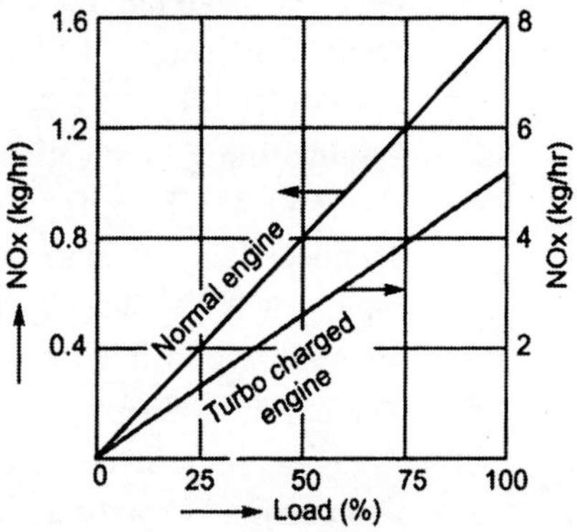

Fig. 7.8

6. Smog

The increase in the levels of SO_2, SO_3, NO_x and suspended particulate matter creates havoc with the surrounding atmosphere. Out of these, SO_2 is the most potent as it causes bronchitis spasms. O_3 causes inflammations of the inner lining of air passages. A cocktail of these lethal substances narrows and inflames the air passages, leads to smelling in the lining of the throat and finally wheezing and difficulty in breathing.

7.5 CONTROL OF EMISSION FROM DIESEL ENGINE

The main pollutants from diesel engine as mentioned earlier are HC, CO, NO_x smoke, odour and SO_2. The methods of reducing HC, CO and NO_x from petrol engine are already discussed in details and same methods are also used for reducing the pollutants from diesel engine. Therefore, the methods used to reduce the smoke and odour which are additional pollutants from diesel engine are discussed here.

1. Smoke and Control of Smoke

Formation of smoke is basically a process of conversion of molecules of hydrocarbon fuels into particles of soot. It should be noted that soot is not carbon but simply an aggomeration of very large polybenzenoid free radicals. It is also observed that soot formation during the early part of the actual combustion process is common to all diesel engines but it is consumed during later part of combustion.

Prolysis of fuel molecules themselves is thought to be responsible for soot formation. Fuel heated with insufficient O_2 will give carbonaceous deposits. It is believed that the ''heavy ends'' of diesel fuel may pyrolyze to yield the type of smoke that is observed from the diesel engine. This is believed to be the path of formation of polycyclic aromatic hydrocarbons (benzo-pyrene) found in soot.

Many theories have been put forward for the formation of smoke but the basic reactions leading to the formation of smoke are not fully known.

There is hardly any successful method to control the formation of soot except the engine has to run at lower load and maintain the engine at best possible condition.

Some methods suggested for the control of smoke are listed below.

1. **Smoke Suppressing Additives :** It has been found that some berium compounds added in fuel reduce the temperature of combustion and avoids the soot formation. It is further observed that if the soot is found, the berium compounds break them in very fine particles and reduces the smoke. But berium salts added in fuel form the deposit on engine parts and reduces the filtering capacity of the filter.

2. **Fumigation :** Fumigation is a method of injecting a small amount of fuel in the intake manifold. This helps for pre-combustion reactions during compression stroke and reduces the chemical delay because the intermediate products like peroxides and aldehydes react more rapidly with O_2 than hydrocarbons. Reducing the chemical delay curbs thermal cracking which is responsible for soot formation.

   The cracking even may not happen which is mainly responsible for soot formation when fumigation is used because it requires about 350 kJ/mole to break C-C bond and 425 kJ/mole to break C—H bond and this much energy may not be available due to easy oxidation during precombustion reaction.

3. **Catalytic Convertors :** The catalytic convertors are not effective like in petrol engine because of large soot formation which interfere in the oxidation of HC, CO and NOx. Therefore these catalysts have very small effect on engine smoke. Lot of research work is going on presently to use catalysts for the effective removing of soot as well as for removing other emissions.

2. Odour Control

It is claimed by many manufacturers that odour additive compounds can reduce the odour intensity. But it is observed that by using additives, there is hardly any effect on odour formation and carried by exhaust gases.

The control of odours by using catalyst are under development and it is found by experiment that a few oxidation catalysts reduce odour intensity.

Unfortunately, study of exhaust odour is hampered by lack of standard tests and lack of standard units to measure the intensity of odour and type of odour.

Lot of work on suppressing the odours by different methods is carried out in different countries.

3. Cleaning up Diesel Emissions with Plasma and a Converter

While the diesel engines are more economical, they produce NO_x during combustion and puts engine designers in a catch-22 situation. Presently the permitted emission of NO_x in Europe is 0.7 g per km and it will be further reduced to 0.57 g/km. This limit is already in force in U.S.A.

The major problem faced to reduce NO_x by Catalytic converter which has proved successful in conventional engines, cannot be applied to lean burn S.I. engines or diesel engines because these engines burn their fuel with high excess air (A : F = 30 : 1) and O_2 in the exhaust prevents the catalytic decomposition of NO_x.

An efficient exhaust gas purification proces for diesel engine ($SINO_x$) has been developed by Siemens and Partners. This system comprises a catalytic converter, a control system and dosing device for urea. The urea undergoes hydrolysis into CO_2 and NH_3, which acts as a reducing agent, transforming the NO_x into environmentally compatible N_2 and water. However, in order to chemically reduce NO_x effectively in catalytic converter, a minimum temperature of 200°C is essential—a condition that is usually met in trucks. In passenger cars, on other hand, the low efficiency of the process at temperatures below 200°C causes problems typical of urban driving and cold start phase.

7.6 NO_x–EMISSION CONTROL

The concentration of NO_x in the exhaust is closely related to the peak cycle temperature. There are different methods by which peak cycle temperature can be reduced and NO_x emission can be controlled.

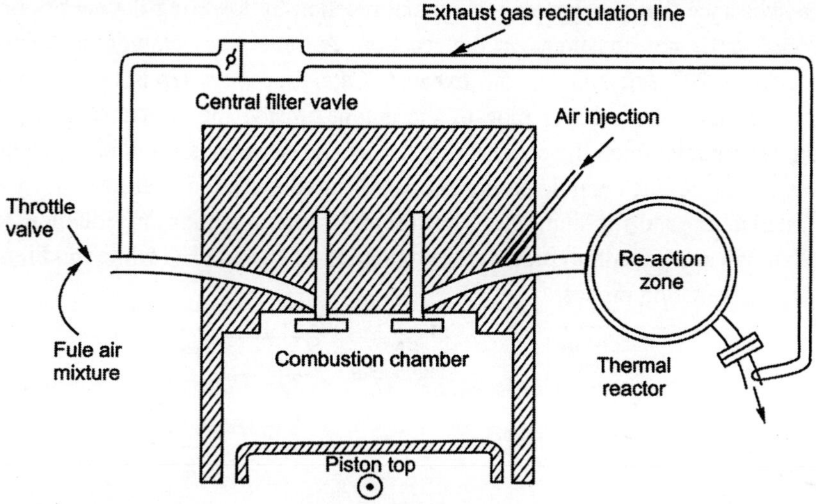

Fig. 7.9 EGR-system.

There are mainly three methods which are commony used as catalyst (which is already discussed) water injection (rarely used) and exhaust gas recirculation (EGR) method. This method is commonly used to reduce NO_x. This method is used in petrol as well as diesel engines. In S.I. engines, about 10% recirculation reduces NO_x emission by 50%. Unfortunately, the consequently poorer combustion directly increases hydrocarbon emission and calls for mixture enrichment to restore combustion regularity which gives a further indirect increase of both HC and CO.

Fig. 7.9 shows the arrangement of EGR system. A portion of the exhaust gases is recirculated to the cylinder intake charge. This reduces the quantity of O_2 available for combustion.

The exhaust gas for recirculation is taken as shown in figure through an orifice and passed through control valve for the regulation of the quantity of recirculation.

The effect of A : F ratio on NOx emission take EGR as a parameter is shown in Fig. 7.10. It can be seen from the figure that, maximum emission of NO_x occurs during

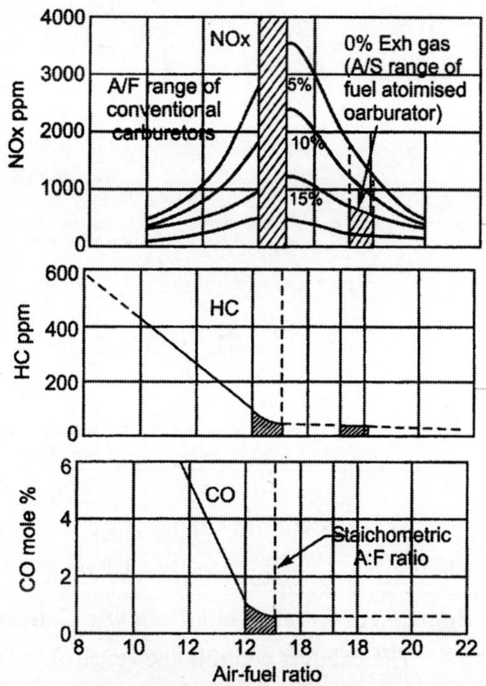

Fig. 7.10 Effect of recycling of gas on NO_x concentration.

lean mixture when gas recirculation is least effective. Whereas, for less emission of CO and HC, lean mixture is preferred. 15% recycling reduces NO_x by 80% but increases HC and CO by 50 to 80%. These two are conficting requirements of this emission control system, can be solved by adopting package system to control all emissions.

7.7 THREE-WAY CATALYTIC CONVERTER

A catalyst is a substance that accelerates a chemcial reaction by lowering the energy needed for it to proceed. The catalyst is not consumed in the reaction. A three-way catalytic converter reduces the concentration of CO, HC and NO_x in the exhaust. Catalytic converter is usually a stainless steel container mounted along the exhaust pipe of the engine. Inside the container is a porous ceramic (Al_2O_3) structure through which the exhaust gas flows. The ceramic is a single honey-comb structure with many flow passages. The ceramic passages contain small embedded particles of catalytic material like platinum, palladium and rhodium, etc. That promotes the oxidation or reduction reaction in the exhaust gas. Platinum and palladium promote the oxidation of CO and HC while rhodium promotes the reaction of NO_x by reduction process.

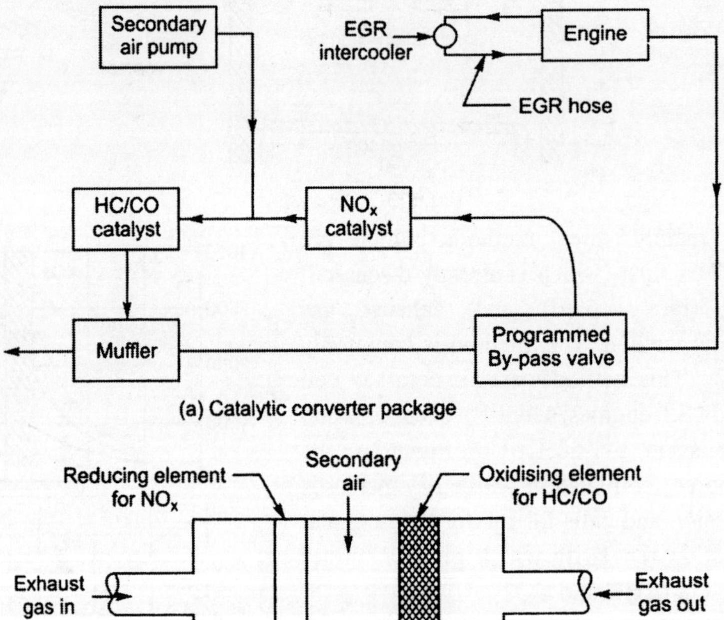

(a) Catalytic converter package

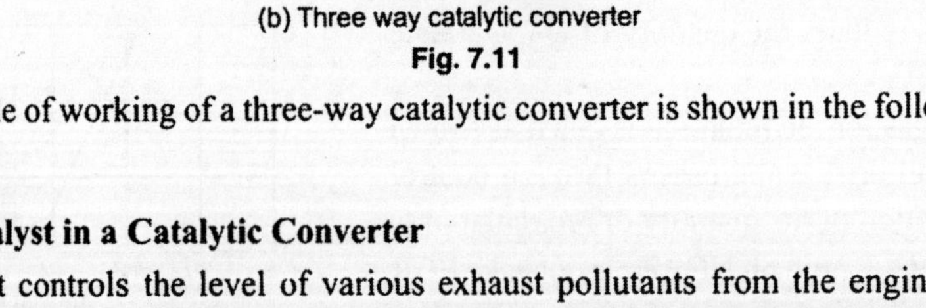

(b) Three way catalytic converter

Fig. 7.11

The principle of working of a three-way catalytic converter is shown in the following Fig. 7.11 (a) and (b).

Function of Catalyst in a Catalytic Converter

The catalyst controls the level of various exhaust pollutants from the engine by changing the chemical characteristics of the exhaust gases. Catalyst materials such as platinum or platinum and palladium, rhodium are used in the converter. CO and HC oxidized the CO_2 and H_2O by Palladium and platinum and NO_x is reduced by rhodium. An oxidation catalyst is placed down stream of the reduction catalyst.

Oxidation		**Reactions**
$2CO + O_2$	$\Rightarrow$	$2CO_2$
$CO + H_2O$	$\Rightarrow$	$CO_2 + H_2$
$C_xH_y + zO_2$	$\Rightarrow$	$xCO_2 + yH_2O$

Where $z = x + 0.25y$

Reduction		**Reactions**
$2NO + 2CO$	$\Rightarrow$	$N_2 + 2CO_2$
$2NO + 5CO + 3H_2O$	$\Rightarrow$	$2NH_3 + 5CO_2$
$2NO + CO$	$\Rightarrow$	$N_2O + CO_2$
$2NO + 2H_2$	$\Rightarrow$	$N_2 + 2H_2O$
$2NO + 5H_2$	$\Rightarrow$	$2NH_3 + 2H_2O$
$2NO + H_2$	$\Rightarrow$	$N_2O + H_2O$

7.8 ENVIRONMENTAL PROBLEMS CREATED BY EXHAUST EMISSION FROM I.C. ENGINES

The emissions exhausted into the surroundings pollute the atmosphere and cause the following problems :

1. Global warming
2. Acid rain
3. Smog
4. Odours
5. Respiratory and other health hazards.

7.9 USE OF UNLEADED PETROL

It is desirable that catalytic converter has an effective life time equal to half to the car life or at least 2 lack km. Converters lose their efficiency with age due to thermal degradation and poisoning of the active catalyst material. Just a small amount of lead on a catalyst site can reduce HC emission reduction by a factor of two or three. Use of leaded gasoline filled two times (full tank) would completely poison a converter and make it totally useless. Therefore, leaded gasoline cannot be used in engines equipped with catalytic converters.

Use of Additives

To improve the combustion performance of fuels some compounds called additives or dopes are used. The requirements of an additive are :

1. It must be knock resistant or surface ignition resistant or both.
2. It should be stable in storage and have no'adverse effect on fuel stability.
3. It should be soluble in fuel under all conditions.
4. It should be in the liquid phase at normal temperature and volatile to give rapid vaporisation in manifold.
5. It must not produce harmful deposits.
6. Its water solubility must be minimum to minimize handling losses.

The most commonly used additives are :

- Tetraethyl lead (TEL), ethylene dibromide (EDB) or ethylene dichloride (EDC) are added to TEL to avoid lead deposits.
- Tetra-methyl lead (TML)

7.10 ALTERNATIVE FUELS FOR I.C. ENGINES

The various alternative fuels for I.C. engines are :

1. Alcohols : Ethanol and methanol
2. Hydrogen
3. Biogas
4. Producer (or water) gas
5. Biomass generated gas
6. LPG and LNG
7. CNG
8. Coal gasification and coal liquification
9. Non-edible negetable oils
10. Non-edible wild oils
11. Ammonia.

1. Alcohols

Alcohols are of two types, ethanol (C_2H_5OH) and methanol (CH_3OH) which can be produced from sugarcane waste and other agricultural products. Ethanol can be produced by termination of vegetables and plant materials like sugarcane, molasses, starchy matter, cellulose material (wood, sulphite waste liquor from paper manufacture), hydrocarbon gases, etc.

The use of ethanol in I.C. engines gives better performance than gasoline engine. The power output, brake specific heat consumption, maximum thermal efficiency, and engine torque is higher as compared to petrol engines. Alcohols are not suitable fuels for C.I. engines.

Methanol behaves much like petroleum. It can be used directly or mixed with gasoline. Some of the important features of methanol as fuel are :

(i) The specific heat consumption with methanol as fuel is 50% less than petrol engine.

(ii) Exhaust CO and HC are decreased continuously with blends.

(iii) Methanol can be used as supplementary fuel in heavy vehicles.

Methanol can be produced from coal, natural gas, oil shell, farm waste, etc. Methanol emits less amount of CO_2 and other polluting gases as compared to gasoline fueled vehicles. However, methanol has high lelent heat and gives cold starting problems, is more corrosive and more toxic as compared to petrol, gives high levels of formaldehyde emission. The important objectionable emissions are : CO, HC, NO_x and aldehydes.

2. Use of Hydrogen in C.I. Engines

There are two methods by which hydrogen can be used in C.I. engines.

(i) By introducing hydrogen with air and using a spray of diesel oil to ignite the mixture that is by the dual fuel mode. The limiting conditions are when the diesel quantity is too small to produce effective ignition, that is failure of ignition and when the hydrogen air mixture is so rich that the combustion becomes unacceptability violent. It between these limits a wide range of diesel to hydrogen proportions can be tolerated.

(ii) By introducing hydrogen directly into the cylinder at the end of compression. Since the self-ignition temperture of hydrogen is very high, the gas spray is made to impinge on a hot glow plug in the combustion chamber, that is by surface ignition. It is also possible to fed a very lean hydrogen air mixture during the intake into an engine and then inject the bulk of the hydrogen towards the end of the compression stroke.

Hydrogen as I.C. Engine Fuel

Advantages :

1. Low emission. Essentially no CO or HC in the exhaust as there is no carbon in the fuel. Most exhaust would be H_2O and N_2.
2. Fuel availability. There are a number of different ways of making hydrogen, including electrolysis of water.
3. Fuel leakage to environment is not a pollutant.
4. High energy content per volume when stored as a liquid. This would give a large range vehicle range for a given fuel tank capacity.

Disadvantages :

1. Requirement of heavy, bulky fuel storage bothin vehicle and at the service station. Hydrogen can be stored as a cryogenic liquid or as a compressed gas. It stored as a liquid, it would have to be kept under pressure at a very low temperature. This would require a thermally super-insulated fuel tank. Storing in a gas phase would require a high pressure vessel with limited capacity.
2. Difficult to refuel.
3. Poor engine volumetric efficiency. Any time a gaseous fuel is used in an engine, the fuel will displace some of the inlet air and poorer volumetric efficiency will result.
4. Fuel cost would be high at present–day technology and availability.
5. High NO_x emissions because of high flame temperature.
6. Can detonate.

3. Biogas

Biogas is produced from organic waste (cow dung) which is easily available. The primary advantages of biogas is its ability to operate the engine on lean mixture, which reduces exhaust HC and CO_2 concentration in the engine exhaust. It gives less deposits and shows clean burning characteristics as compared to petrol and diesel oil.

S.I. engines can be operated on biogas after starting the engine by using petrol. Biogas can be used in S.I. engine in two forms as :

(i) To run the engine entirely on biogas.
(ii) Dual fuel engine operation on biogas and petrol.

The advantages of using biogas as fuel in C.I. engine are :

(i) The gas-air mixture provides uniform mixture in multi-cylinder engine at all times.
(ii) There is virtually no CO emission in exhaust due to lean operation of the engine.
(iii) NO_x emissions are reduced by about 60 percent.
(iv) Soot is virtually eliminated and exhaust is found to have less pungent odour than diesel oil.

4. Producer (or water) gas

Producer gas is a mixture of CO, H_2 and N_2. It is obtained from the partial oxidation of coal, coal or peat when they are burnt with an insufficient quantity of air. In using the producer gas in S.I. engine, the problem remains as to how and with which fuel to produce it economically. The producer gas can be used with S.I. engine but with a lesser output.

5. Biomass generated gas

Biomass comparises of all kinds of agricultural waste like stalks, husks of different crops, residues like bagasse, etc. Biomass gasification gives producer gas.

6. LPG as SI Engine Fuel

LPG consists of hydrocarbons of such volatility that they exist as gases under atmospheric conditions, but can be readily liquefied under pressure. LPG contains mainly members of paraffin hydrocarbons upto C_4. LPG consists of propane and to some extent butane.

Use of LPG in S.I. Engine

Advantages :

(i) LPG contains less carbon than petrol. Therefore emission is much reduced by its use.

(ii) LPG mixes with air at all temperatures.

(iii) In multi-cylinder engines a uniform mixture can be supplied to all cylinders.

(iv) There is no crankcase dilution, because the fuel is in the form of vapour.

(v) Automobile engines can use propane, if they have high compression ratio.

(vi) LPG has high antiknock characteristics.

(vii) Its heat energy is about 80% of gasoline, but its high octane value compensates the thermal efficiency of the engine.

(viii) Running on LPG translates into a cost saving of about 50%.

(ix) The engine may have a 50% longer life.

Disadvantages :

(i) Engines are normally designed to take in a fixed volume of the mixture and air. Therefore LPG will produce 10% less power for a given engine at full throttle.

(ii) The ignition temperature of LPG is somewhat higher than petrol. Therefore running on LPG could lead to a 5% reduction in valve life.

(iii) A good cooling system is quite necessary, because LPG vaporizer uses engine coolant to provide the heat to convert the liquid LPG to gas.

(iv) The vehicle weight is increased due to the use of heavy pressure cylinders for storing LPG.

(v) A special fuel feed system is required for liquid petroleum gas.

Liquified Natural Gas (LNG) comes from dry natural reservoires mainly CH_4 (methane) with very small percentage of ethane and propane. The major difficulty in using this gas is its very low boiling point (–161.5°C).

7. Compressed Natural Gas (CNG)

CNG is composed of methane, ethane and propane with traces of other gases. The main constituent is methane. Methane has very good anti-knock qualities and burns at much higher temperature as compared with petrol. It can be safely used in engines with a compression ratio upto 12 : 1. It gives higher thermal efficiency as compared to gasoline engine. The pollutants emitted by CNG are also reduced. It is nontoxic. CNG has been widely used in three wheelers and buses.

Advantages of CNG :

(i) It is cheaper than petrol.

(ii) It is engine friendly–gives lower maintenance cost and longer life.

(iii) It is safe–it is lighter than air and safely dissipates to atmosphere.

(iv) It mixes easily with air completely and gives proper combustion.

(v) It is odorless.

8. Coal gasification and coal liquification

These fuels are mainly used for power generation than I.C. engines.

9. Non-edible negetable oils

The vegetable oils including soyabean oil, cotton sead oil, sunflower oil, rapeseed oil, palm oil, coconut oil, linseed oil, it have cetane numbers and calorific values comparable with those of diesel oil and can be used as alternative fuels for diesel engine. The main disadvantage is their high viscosity which gives difficulties with fuel injection and cold flow pumping.

10. Non-edible wild oils

Of all the non-edible wild oils, mahua and neem oils have the highest production. Their use as I.C. engine fuels has not become popular yet.

11. Ammonia

Ammonia has been demonstrated as a practical fuel for engine applications. Ammonia does not contain carbon and hence absence of two major pollutants, HC and CO in the exhaust are its added advantages. The performance characteristics of engine using NH_3 as fuel are similar to the performance characteristics obtained with petroleum fuels.

The pros and cons of alternative fuels are given in Table 7.1.

Table 7.1 : Pros and Cons of Alternative Fuels

Fuel	*Advantages*	*Disadvantages*
Methanol	Familiar liquid fuel Vehicle development relatively advanced Organic emission (ozone precursors) will have lower reactivity than gasoline emissions Lower emissions of toxic pollutants, except formaldehyde Engine efficiency should be greater Abundant natural gas feedstock Less flammable than gasoline Can be made from coal or wood (as can gasoline), though at higher cost. Flex fuel ''transition'' vehicle available	Range as much as 1/2 less, or larger fuel tanks Would likely be imported from overseas Formaldehyde emissions a potential problem, esp. at higher mileage, requires improved controls More toxic than gasoline M100 has non-visible flame, explosive in enclosed tanks Costs likely somewhat higher than gasoline, esp. during transition period Cold starts a problem for M100 Greehouse problem if made from coal.
Ethanol	Familiar liquid fuel Organic emissions will have lower reactivity than gasoline emissions (but higher than methanol) Lower emissions of toxic pollutants Engine efficiency should be greater Produced from domestic sources Flex fuel ''transition'' vehicle available Lower CO with gasohol (10 percent ethanol blend) Enzyme-based production from wood being developed	Much higher cost than gasoline Food/fuel competition at high production levels Supply is limited, esp, if made from corn Range as much as 1/3 less, or larger fuel tanks Cold starts a problem for E100

Natural Gas	Though imported, likely North American source for moderate supply (1 mmbd or more gasoline displaced) Excellent emission characteristics except for potential of somewhat higher NOx emissions Gas is abundant world-wide Modest greenhouse advantage Can be made from coal	Dedicated vehicles have remaining development needs Retail fuel distribution system must be built Range quite limited, need large fuel tanks w/added costs, reduced space (LNG range not as limited, comparable to methanol) Dual fuel ''transition'' vehicle has moderate performance, space penalties Slower refueling Greenhouse problem it made from cool
Hydrogen	Excellent emission characteristics—minimal hydrocarbons Would be domestically produced Big greenhouse advantage if derived from photovoltaic energy Possible fuel cell use	Range very limited, need heavy, bulky fuel storage Vehicle and total costs high Extensive research and development effect required Needs new infrastructure
Reformulated gasoline	No infrastructure change except refineries Probable small to moderate emission reduction Engine modifications not required May be available for use by entire fleet, not just new vehicles	Emission benefits remain highly uncertain Costs uncertain, but will be significant No energy security or greenhouse advantage

7.11 CURRENT SCENARIO ON THE POLLUTION FRONT

The Central Pollution Control Board has been continuously playing a key role in abatement and control of pollution in the country by generating relevant data, providing scientific information, rendering technical inputs for formation of national policies and programmes, training and development of manpower, through activities for promoting awareness at different levels of the Government and Public at large. The Central Pollution Control Board is continuously growing organization with consistent output and continuous improvements in its infrastructure facilities, human resource development and development of expertise to meet the challenges of unending task of environment protection in the country.

7.11.1 Vehicular Pollution Control

Automobile Pollution Control initiatives included marked enforcement of a variety of control measures ranging from notification of advanced Euro-IV equivalent emission norms and commensurate fuel for new vehicles to stricter exhaust emission limits for in-use vehicles, augmentation of infrastructures for alternative fuels, mass transits and other urban planning and management options. The implementation of the road map as recommended by the Auto Fuel Policy of India has been continued for implementation. The vehicular pollution control framework in the country has now shifted its focus towards integrated control and management options and has extended its domain to cover all major metro cities. Important measures pertaining to vehicular pollution control initiated during the year are as follows :

Mass Emission Standards

Mass emission standards are the primary technical policy for controlling emissions from vehicles. The Motor Vehicle Act, 1988, and the Central Motor Vehicles Rules (CMVR), 1989, are the principal instruments for regulation of motor vehicular traffic/emissions throughout the country. The implementation of various provisions of this Act rests with the state governments. The Ministry of Shipping, Road Transport and Highways (MoSRTH) Acts as a nodal agency for the formulation and implementation of various provisions of the Motor Vehicle Act and CMVR.

- Mass Emission Standards-Bharat Stage IV (equivalent to Euro IV standards) have been notified for all categories of 4-wheelers to be implemented from 1st April, 2010 in 11 cities—Delhi, Mumbai, Kolkata, Chennai, Bangalore, Hyderabad, Ahmedabad, Pune, Surat, Kanpur and Agra.
- Mass Emission Standards-Bharat Stage III (equivalent to Euro III standards) have been notified for all categories of 4-wheelers to be implemented from 1st April, 2010 in rest of the country.
- Mass Emission Standards-Bharat Stage III have been notified for 2- and 3- wheelers to be implemented from 1st April, 2010 all over the country.

7.11.2 Fuel Quality Specifications

- Auto-Fuels commensurate to Euro III (whole country) and Euro IV (for 11 cities) specifications is proposed to be made available in the respective cities from 1st April, 2010.
- The Research Octane Number (RON) for premium petrol available in 11 mega cities has been boosted to 95 with lead content being reduced to 0.005 g/l and benzene content of maximum 1%. From 1st April, 2010, the content of sulphur in gasoline is proposed to be reduced to 0.005% (50 mg/kg) from existing 0.015% (150 mg/kg). However, all over the country, content of sulphur in gasoline is proposed to be 0.015% (150 mg/kg) with effect from 01.04.2010.
- For diesel the Cetane Number has been enhanced to 51 with Sulphur content proposed to be reduced further to 0.005% (50 mg/kg) in the 11 mega cities by 1st April, 2010. The amount of sulphur in diesel is propsed to be 0.035% (350 mg/kg) all over the country.
- Important fuel specification of Diesel and Gasoline as available in metro cities from 1st April 2010 are as given in Table 7.2

Table 7.2 : Fuel Specifications of Diesel and Gasoline

Fuel specifications	*Requirements*
Diesel	
Cetane Number (CN), min	51
Total Sulphur, max	0.005% (50 mg/kg)
Distillation, 95% vol. recovery at 0°C, max	360°C
Polycyclic Aromatic Hydrocarbon (PAH), max	11% mass
Gasoline	
Research Octane Number (RON), min	95
Reid Vapour Pressure (RVP), max	60 kpa

Benzene content, max	1% volume
Lead content (as Pb), max	0.005 g/l
Sulphur, total, max	0.005% (50 mg/kg)
Aromatics content, max	35% volume
Oxygen content, max	2.7% volume

7.11.3 In-Use Vehicles

- The tourist transport operators shall not engage or use any vehicle for the purpose of journey, the origin and destination of which falls within the National Capital Region (NCR), unless such vehicle conforms to the mass emission standards (Bharat Stage III), notified vide GSR 58(E) dated January 30, 2009.
- MSRTH has constituted a task force to introduce auditing system in PUC centres all over the country, to look into various aspects related to procedure, implementation and suggest effective institutional mechanism for the same, in which Central Pollution Control Board is one of the members.

7.11.4 Alternate Fuels Initiatives

There has been lot of developments in this front, when various organizations including the Planning Commission, Oil Companies, Auto Sectors, Central Pollution Control Board and other research agencies intiated various demonstration and feasibility studies with alternative fuels like LPG and bio-diesel (B2O) in the country. Some of the developments are depicted below :

- Bio-fuels mainly Ethanol and Biodiesel (in B20 form) are the prospective options for the country. Pilot studies on ethanol and biodiesel have been complted and many are on-going.
- Efficacy of B20 biodiesel from Jatropa feedstock has been established and experiences gained through some pilot studies. Introduction of biodiesel starting with lower blends like B5, B10, etc. is a possibility now.
- In Kolkata all three wheelers have been odered to switch over to LPG mode from September, 2005 vide notification No. 2421-WT/3M-73/2005 dated May 24, 2005.
- Besides Delhi and Mumbai, the supply of CNG as automotive fuel has been extended to the cities of Ankleshwar, Vadodra and Surat in Gujarat and Kanpur, Barelly, Agra and Lucknow in Uttar Pradesh. The total CNG vehicles in the country touching over 3.54 Lakh, as per the industry estimates.
- Work is on to introduce bio-diesel in the form of B20 as an automotive fuel. Several research studies and field trials have been initiated by Organizations like—IITs, IOC, Mercedes, Railway, etc. Already "Jatropha Carcus" has been identified and earmarked to be the prominent source of biodiesel in the country.
- Efforts for developing and popularizing electric vehicles also gained momentum during this year. Already "Reva Motors" have commercialized a small electric/battery car. Many three-wheeler manufacturers are also contemplating electric driven OEM for Indian markets.

ENVIRONMENTAL STANDARDS

Emission

Motor Vehicles : Environmental Standards

Vehicular Exhaust Emission Standards (effective for 1990-1996)

Standards for emission of smoke, vapour etc. from motor vehicles

1. Every motor vehicle shall be manufactured and maintained in such condition and shall be so driven that smoke, visible vapour, grit, sparks, ashes, cinders or only substances or oily substance do not emi there from.
2. On and from the 1st day of March, 1990, every motorvehicle shall comply with the following standards.
   (a) idling CO (carbon monoxide) emission limit for all four wheeled petrol driven vehicles shall not exceed 3 per cent by volume, and
   (b) idling CO emission limit for all two and three wheeled petrol driven vehicles shall not exceed 4.5 per cent by volume.
   (c) smoke density for all diesel driven vehicles shall to as follows :

Method of Test	*Maximum Smoke Density*		
	Light absorption coefficient K(l/m)	*Bosch units*	*Hartridge units*
Full load at a speed of 60% to 70% of maximum engine rated speed declared by the manufacture	3.1	5.2	75
Free acceleration	2.3		65

3. On and from the 1st day of April, 1991, all petrol driven vehicles shall be so manufactured that they comply, with the mass emission standards as specified at Annexure-I. The breakdown of the operting cycle used for the test shall be as specified at Annexure-II and the reference fuel of such tests shall be as specified in Annexure-III to this schedule.
4. On and from the 1st day of April, 1991 all diesel driven vehicles shall be so manufacutred that they comply, with the mass emission standards based on exhaust gas capacity specified in Annexure-IV to this schedule.
5. On and from the 1st day of April, 1992 all diesel driven vehicles shall be so manufactured that they comply with the follwing levels of emissions under the Indian driving cycle.

Mass of Carbon monoxide (CO), max, gm/kWH	Mass of Hydrocarbons (HC), max. gm/kWH	Mass of Nitrogen Oxides (NOx). max. gm/kWH
14	305	18

6. Each motor vehicle manufacutred on land after the dates specified in paragraphs (2), (3), (4) and (5), shall be certified by the manufacturers to be conforming to the standard specified in the said paragraph and the manufacturer shall further certify that the components liable to effect the emission of gaseous pollutants are so designed, constructed and assembled as to enable to vehicle, in normal use, despite the vibration to which it may be subjected, to comply with the provisions of the said paragraphs.

7. Test for smoke emission level and carbonmonoxide level for motor vehicles

   (a) Any officer not below the ran of a sub inspector of police or an Inspector of motor vehicles, who has reason to believe that a motor vehicles is by virtue of smoke emitted from it or other pollutants like carbonomonoxide emitted from it, is likely to cause environmental pollution, endangering the health or safety of any other user of environmental pollution, endagering the health or safety of any other use of the road or the public, may direct the driver of any person incharge of the vehicle to submit the vehicle for undergoing a test to measure the standard of black smoke or the standard of any of the other pollutants.

   (b) The driver or any person incharge of the vehicle shall upon demand by any officer referred to sub paragraph (a), submit the vehicle for testing for the purpose of measuring the standard of smoke or the levels of other pollutants of both.

   (c) The measurement of standard of smoke shall be done with a smoke metre of a type approved by the state government and the measurement of the other pollutants like carbon monoxide shall be done with instruments of a type approved by the state government.

Annexure-1

Mass Emission Standards for Petrol Driven Vehicles

Type Approval Tests

Two and Three Wheeler Vehicles :

Reference mass, R (kg)	*Co (g/km)*	*HC (g/km)*
R < 150	12	6
150 R<350	$12 + \left\lvert \frac{18(R-150)}{200} \right.$	$8 + \left\lvert \frac{4(R-150)}{200} \right.$
R > 350	30	12

Light Duty Vehicles :

Reference mass, rw (kg)	*Co (g/km)*	*HC (g/km)*
1	*2*	*3*
rw < 1020	14.3	2.0
1020 < 1250	16.5	2.1
1250 < 1470	18.8	2.1
1470 < 1700	20.7	2.3
1700 < 1930	22.9	2.5
1930 < 2150	24.9	2.7
rw < 2150	27.1	2.9

Conformity of Production Tests

Two and Three Wheeler Vehicles :

Reference mass, R (kg)	*Co (g/km)*	*HC (g/km)*
R < 150	12	6
150 R<350	$15 + \left\vert \frac{18(R-150)}{200} \right.$	$15 + \left\vert \frac{5(R-150)}{200} \right.$
R > 350	40	12

Light Duty Vehicles :

Reference mass rw (kg)	*Co (g/km)*	*HC (g/km)*
1	*2*	*3*
rw < 1020	17.3	2.7
1020 < 1250	19.7	2.2
1250< 1470	22.5	2.8
1470 < 1700	24.9	3.0
1700 < 1930	27.6	3.3
1930 < 2150	29.9	3.5
rw < 2150	32.6	3.7

Note : For any of the pollutants referred to above of the three results obtained may exceed the limit specified for the vehicles by not more than 10 per cent.

Mass emission standards refers to the gm. of pollutants emitted per km. run of the vehicles, as determined by a chassis dynamometer test using the Indian Driving Cycle.

SOLVED QUESTIONS FROM UNIVERSITY EXAMINATIONS

■ **Example 7.1** What is exhaust gas recirculation device? Explain its working through a neat sketch. [MDU, Dec. 2008]

Solution : Refer to Art. 7.6.

■ **Example 7.2** What are the causes of formation of NO_x in petrol engine? Explain the effect of A.F. ratio and spark advance on emission of NO_x. [MDU, Dec. 2008]

Solution : Refer to Art. 7.1.1

■ **Example 7.3** What are the major constituents which contribute to air pollution in S.I. engines? [MDU, Dec. 2009]

Solution : Refer to Art. 7.1.1.

■ **Example 7.4** State the current pollution scenario. [MDU, Dec. 2009]

Solution : Refer to Art. 7.11.

■ **Example 7.5** Name the categories of S.I. emissions. Explain various factors affecting exhaust emission. [MDU, Dec. 2009]

Solution : Refer to Art. 7.1.1.

■ **Example 7.6** Explain briefly various alternative fuels for I.C. engine. [MDU, Dec. 2009]

Solution : Refer to Art. 7.10.

■ **Example 7.7** (a) Explain the various alternatives fuels for an IC engine. [MDU, Dec. 2010]

(b) What are the major constituents which contribute to air-pollution in SI engines?

(c) State the current pollution scenario. [MDU, Dec. 2010]

Solution : (a) Refer to Art. 7.10.

(b) Refer to Art. 7.1.

(c) Refer to Art. 7.11.

■ **Example 7.8** Explain the standards for emission of pollutants from motor vehicles as per centre motor vehicles rules 1989. [MDU, Dec. 2005]

Solution : Refer to Art. 7.11.

■ **Example 7.9** Attempt any three of the following :

(a) Lubrication System used in I.C. Engines.

(b) Octane Rating of Fuel

(c) Methods of Emission Control

(d) Performance Curves of I.C. Engines [KU, Dec. 2006]

Solution : (a) Refer to Art. 5.7.1.

(b) Refer to Art. 4.10

(c) Refer to Art. 7.2

(d) Refer to Art. 6.8.

■ **Example 7.10** Attempt any three of the following :

(a) Wankle Engine

(b) Method of Emission Control [KU, Jan., 2008]

(c) Requirement of Diesel Ignition System

(d) Knocking in C.I. Engines [KU, Dec. 2007]

Solution : (a) Refer to Art. 1.

(b) Refer to Art. 7.2.

(c) Refer to Art. 3.4.1

(d) Refer to Art. 4.7.4

■ **Example 7.11** Explain methods of pollution control. [KU, Dec. 2009]

Solution : Refer to Art. 7.2.

■ **Example 7.12** Describe briefly pollutants in the exhaust of SI engine. [KU, Dec. 2010]

Solution : Refer to Art. 7.1.

SHORT ANSWER TYPE QUESTIONS

1. What are the categories of S.I. engine emissions?
2. Show the emissions from a 4-wheeler having S.I. engine on a neat sketch.
3. List the major exhaust emissions from a S.I. engine.
4. What is evaporative emission?
5. Name four methods for control of emissions in S.I. engines.
6. What is an after burner? Give its use.
7. What is a catalytic converter?
8. What is an evaporation emission control device?
9. Name the emissions from a C.I. engine.
10. Define smog and fumigation.
11. Explain SO_x and NO_x.
12. Name the catalyst materials used in a catalytic converter.
13. What are the environmental problems created by exhaust emission from I.C. engines?
14. What are additives used in I.C. engine fuels?
15. Name four alternative fuels for I.C. engines.

EXERCISES

1. What are the main pollutants emitted by S.I. engine?
2. What are the various types of exhaust emissions from S.I. engine? Discuss briefly.
3. How exhaust emissions in S.I. engines can be controlled? Explain briefly.
4. What are the various types of exhaust emissions from C.I. engine? Explain briefly.
5. Describe the methods for control of emission from diesel engine.
6. Write a short note on No_x emission control.
7. What is the function of a catalytic converter? Explain the working of a three-way catalystic converter with the help of a neat sketch.
8. Write short notes on the following :
   (a) Environmental problems created by exhaust emissions from I.C. engines.
   (b) Use of unleaded petrol.
   (c) Use of additives in I.C. engine fuels.
9. Explain the various alternative fuels for I.C. engines.

❑❑❑

Chapter 8

Reciprocating Air Compressors

Inside This Chapter

8.1 INTRODUCTION

Compressors are the mechanical devices used for increasing the pressure of a gas. Compressors used for producing high pressure air are called air compressors. Air is drawn from the atmosphere during suction, compressed to the required pressure and then delivered to the receiver. If the compression is done in a conventional cylinder with a closely fitted piston making reciprocating motion, the compressor then is called a reciprocating compressor. External work must be spplied from a prime mover to the compressor to achieve the required compression. The general arrangement of the compressor and prime- mover used

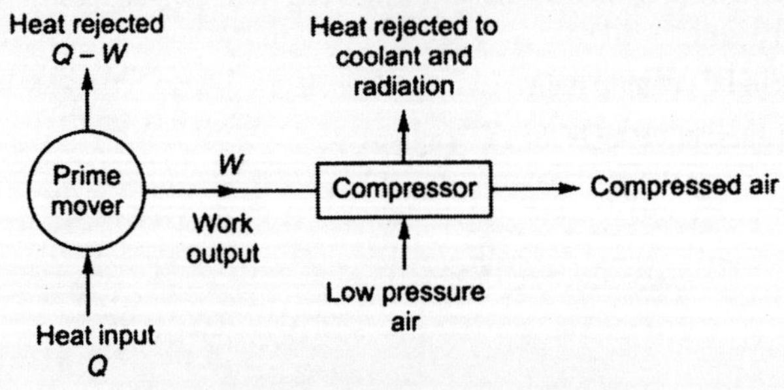

Fig. 8.1 Energy flow in an air-compressor.

to run the compressor is shown in Fig. 8.1. Part of the energy supplied to the prime mover is converted into work to run the compressor and the remaining part is lost in friction, heat radiation and to the coolant used to cool the compressor.

8.2 USES OF COMPRESSED AIR IN INDUSTRY

The several uses of compressed air are :

1. To drive a compressed air engine.
2. For producing an air blast for a workshop.
3. For sraying the fuel (atomizing) into a boiler furnace.
4. For operating pneumatic drills and tools.
5. For operating pneumatic brakes for locomotives and rolling stock.
6. Pneumatic conveying and for pumping of water by compressed air.
7. For working compressed air engines especially in mines.
8. In gas turbine power platns and air-conditioning plants.

8.3 OPERATION OF SINGLE-STAGE RECIPROCATING COMPRESSOR

The schematic arrangement of a single-stage reciprocating compressor is shown in Fig. 8.2. It consists of a cylinder with colling jacket, piston, connecting rod, crank, suction and delivery values. The piston will be making to-and-fro motions through the crankshaft arrangement and generally run by an electric motor, diesel engine, petrol engine or steam engine. During the outward motion of the piston, the pressure inside the cylinder falls below the atmospheric pressure and the suction value is opened due to the pressure difference. The air taken into the cylinder until the piston reaches the bottom dead centre position.

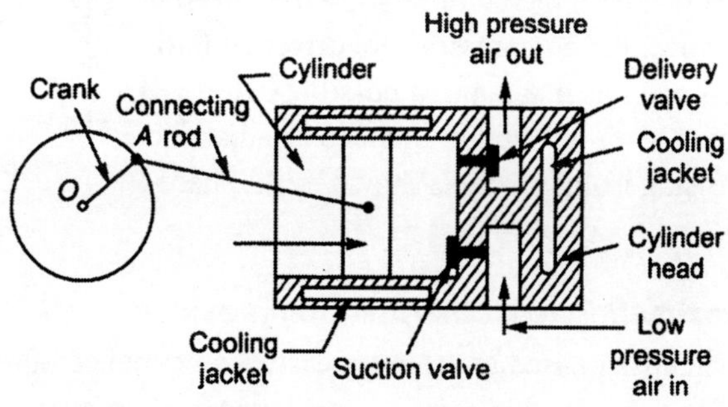

Fig. 8.2 Single-stage reciprocating air-compressor.

As the piston stacks moving inwards, the suction value is closed and the pressure starts increasing until the pressure einside the cylinder is more than the pressure of the delivery side. Then the delivery valve opens and high pressure air is delivered to the receiver till the piston reaches the top dead centre. At the end of the delivery stroke the small volume of high pressure air left in the clearance volume expands as the piston strarts moving outwards. Thus, the cyle is repeated.

8.4 TERMINOLOGY

- **Single Acting Compressor :** If the air admission from the atmosphere is on only one side of the cylinder.

- **Double Acting Compressor :** When the air from the atomosphere is drawn on both sides of the piston.
- **Single Stage Compressor :** If the total compression is done fully in one cylinder.
- **Multi-stage Compressor :** If the compression is carried out in more than one cylinder, every cylinder carrying out a part of the compression.
- **Free Air Delivered (FAD) :** It is the actual volume delivered at the stated pressure, reduced to intake pressure and temperature and expressed in cubic metre/minute (cmm).
- **Displacement of the Compressor (or swept volume) :** For a single acting compressor, swept volume,

$$V_s = \frac{\pi}{4} d^2 L, \text{ where } d = \text{diameter of piston and } L = \text{length of stroke}$$

$$\text{Volumetric efficiency, } \eta_v = \frac{\text{Actual free air delivered per stroke, } V_a}{\text{Swept volume, } V_s}$$

8.5 TYPES OF COMPRESSION

The p–v line 4-1 represents the suction stroke, the air is then compressed adiabatically, as shown by the curve 1-2. It is then forced out of the cylinder at constant pressure p_2, as shown by the line 2-3. The work done is represented by the area 1-2-3-4-1.

If the air had been compressed isothermally, as represented by the curve 1-2′, then the workdone on the air would be the area 1-2′-3-4-1, which is considerably less than that due to adiabatic compression. However it is not possible in practice to compress the air isothermally because in that case the compressor would need to run extremely slow. In practice, the compressors are driven at fairly high speeds in order to compress as much air as possible in a given time. Hence, the compression of air will approximate to an adiabatic. The work saved by compressing isothermally is shown by the shaded area 1-2-2′-1.

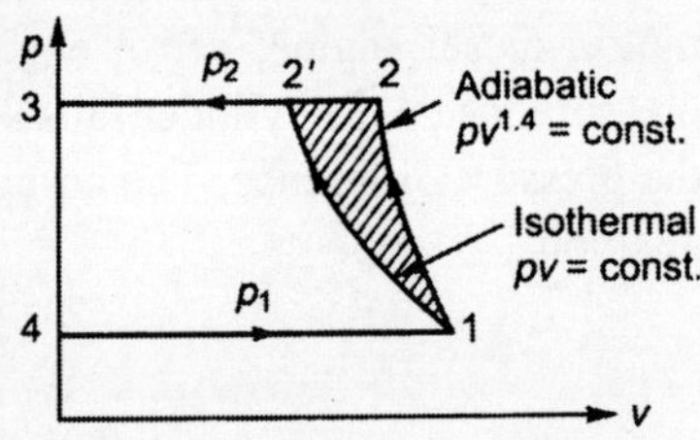

Fig. 8.3 Theoretical p-v diagram for single acting compressor.

8.5.1 Methods for Approximating to Isothermal Compression

The following practical methods are used to achieve nearly isothermal compression :

1. **Cold Water Spray :** In this method, cold water is sprayed onto the cylinder during compression, thus reducing the temperature of the air. Without the cold water spray, the compression would have been adiabatic, or $pv^{1.4}$ = constant. This effect is shown in Fig. 8.4, where now the compression would be between adiabatic and isothermal, or $pv^{1.2}$ = const. The work saved is represented by 1-2-2″-1.
2. **Water jacketing :** The water is circulated around the cylinder through the water jacket to cool the air during compression. This method is commonly used for all types of reciprocating compressors.

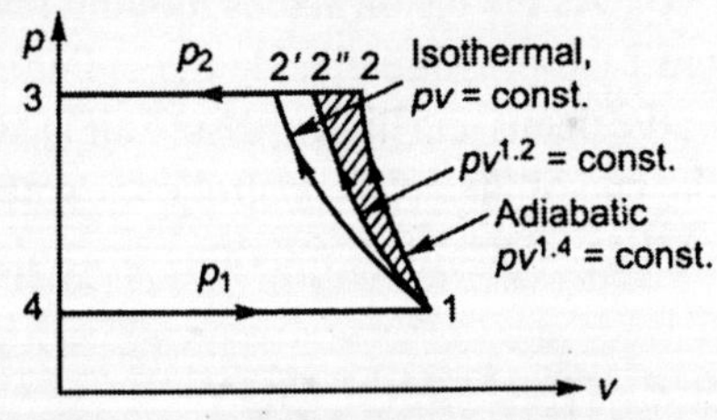

Fig. 8.4 Effect of type of compression on work done.

3. **Inter-cooling by using multi-stage compression :** In a multi-stage compressor, the air is compressed in several stages. In principle, it is equivalent to number of compressors in series where the air passes from one cylinder to the next and the pressure increases in each cylinder. The p-v diagram for a four-stage compressor is shown in Fig. 8.5. The dotted line 1-4′ is the isothermal. In the first stage the air is compressed adiabatically to '2', cooled at constant pressure to 2′ in the inter-cooler, possibly to the initial temperature. For comlete inter cooling, the point 2′ is on the isothermal line 1-4′. The air is then drawn into the second cylinder for the second stage of its compression and the process is repeated for the subsequent cylinders. Line 1-a represents the adiabatic compression in a single cylinder. The compressor work saved by inter-cooling is represented by the area 2-a-4-3′-3-2′-2.

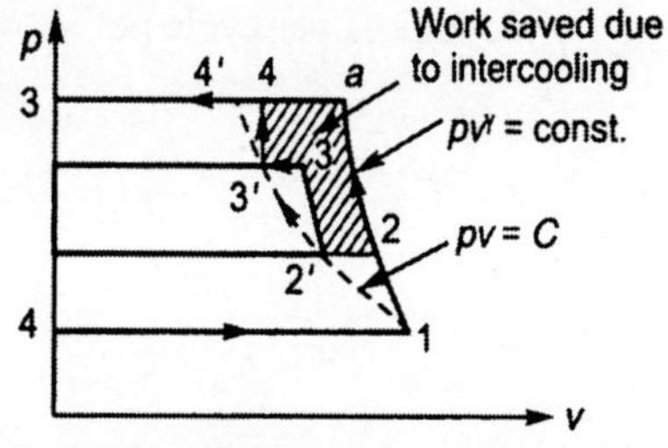

Fig. 8.5 Multi-stage compression.

4. **External Fins :** Small capacity compressors are provided with fins to increase the heat transfer from the surface of the cylinder.

8.6 SINGLE-STAGE COMPRESSION

8.6.1 Required Work

(a) Without Clearance

Consider the theoretical p-v diagram for a single stage air-compressor as shown in Fig. 8.6. The work done on air per cycle is represented by the area '1-2-3-4-1'.

$$W = \text{area}\,(a-b-2-3-a) + \text{area}\,(b-c-1-2-b) - \text{area}(a-c-1-4-a)$$

$$= p_2v_2 + \frac{p_2v_2 - p_1v_1}{\gamma - 1} - p_1v_1$$

$$= (p_2v_2 - p_1v_1)\left[1 + \frac{1}{\gamma - 1}\right]$$

$$= \left(\frac{\gamma}{\gamma - 1}\right)(p_2v_2 - p_1v_1)$$

$$= \left(\frac{\gamma}{\gamma - 1}\right) mR(T_2 - T_1)$$

$$= \left(\frac{\gamma}{\gamma - 1}\right) p_1v_1\left[\frac{p_2v_2}{p_1v_1} - 1\right]$$

$$= \left(\frac{\gamma}{\gamma - 1}\right) p_1v_1\left[\left(\frac{p_2}{p_1}\right)^{\frac{\gamma-1}{\gamma}} - 1\right]$$

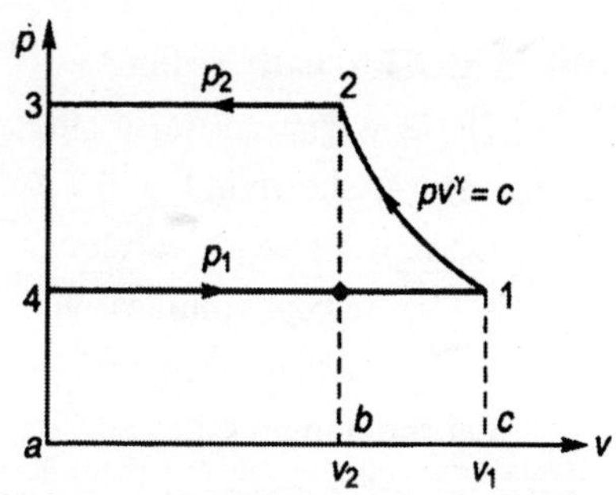

Fig. 8.6 p-v diagram without clearance.

$$= \left(\frac{\gamma}{\gamma - 1}\right) mR\, T_1 \left[\left(\frac{P_2}{p_1}\right)^{\frac{\gamma-1}{\gamma}} - 1\right] \qquad \text{...(8.1)}$$

where m = mass of air delivered per cycle.

Work done per cycle per kg of air delivered,

$$w = \left(\frac{\gamma}{\gamma - 1}\right) RT_1 \left[\left(\frac{p_2}{p_1}\right)^{\frac{\gamma-1}{\gamma}} - 1\right] \qquad \text{...(8.2a)}$$

$$= \left(\frac{\gamma}{\gamma - 1}\right) RT_1 \left(\frac{T_2}{T_1} - 1\right) \qquad \text{...(8.2b)}$$

If the compression had followed the law, $pv^n = c$, then

$$w = \left(\frac{n}{n - 1}\right) RT_1 \left[\left(\frac{p_2}{p_1}\right)^{\frac{n-1}{n}} - 1\right] \qquad \text{...(8.3a)}$$

$$= \left(\frac{n}{n - 1}\right) RT_1 \left[\frac{T_2}{T_1} - 1\right] \qquad \text{...(8.3b)}$$

Power required in driving the compressor $= \dfrac{WN}{60 \times 10^3}$ kW ...(8.4)

where N = number of complete cycles per minute

= *rpm*, if single acting

= number of strokes per minute, if double acting.

(b) With Clearance Volume

The *p-v* diagram for a single stage and single acting air compressor is shown in Fig. 8.7 with clearance volume.

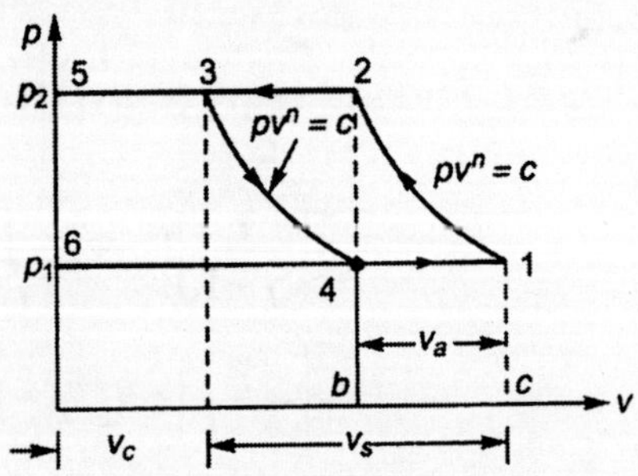

Fig. 8.7 *p-v* diagram with clearance.

Let v_c = clearance volume

v_s = swept volume = $v_1 - v_c$

v_a = actual volume = $v_1 - v_1$

Let the compression and expansion processes follow the same low pv^n = const.

Work done per cycle,

W = area (1-2-3-4-1)

= area(1-2-5-6-1) – area(3-4-6-5-3)

$$= \left(\frac{n}{n-1}\right) p_1 v_1 \left[\left(\frac{p_2}{p_1}\right)^{\frac{n-1}{n}} - \right] - \left(\frac{n}{n-1} p_4 v_4 \left[\left(\frac{p_3}{p_4}\right)^{\frac{n-1}{n}} - 1\right.\right.$$

Now $p_3 = p_2$ and $p_4 = p_1$

$$\therefore \quad W = \left(\frac{n}{n-1}\right) p_1 (v_1 - v_4) \left[\left(\frac{p_2}{p_1}\right)^{\frac{n-1}{n}} - 1\right]$$

$$= \left(\frac{n}{n-1}\right) p_1 v_a \left[\left(\frac{p_2}{p_1}\right)^{\frac{n-1}{n}} - 1\right] \qquad \text{...(8.5)}$$

$$= \left(\frac{n}{n-1}\right) m_1 R T_1 \left[\left(\frac{p_2}{p_1}\right)^{\frac{n-1}{n}} - 1\right] \qquad \text{...(8.6)}$$

$$= \left(\frac{n}{n-1}\right) R T_1 \left[\frac{T_2}{T_1} - 1\right] \text{ for 1 kg of actual air} \qquad \text{...(8.7)}$$

where m_1 = actual mass of air delivered per cycle.

8.6.2 Volumetric Efficiency

The ratio of the actual free air delivered to the swept volume of the compressor is called the volumetric efficiency.

$$\eta_v = \frac{v_a}{v_s} = \frac{v_1 - v_4}{v_1 - v_c}$$

Let
$$c = \frac{v_c}{v_s} = \text{clearance ratio}$$

Then
$$\eta_v = \frac{(v_c + v_s) - v_4}{v_s} = \frac{v_s - (v_4 - v_c)}{v_s} = \frac{v_s - v_c\left(\dfrac{v_4}{v_c} - 1\right)}{v_s}$$

Now
$$p_3 v_3^n = p_4 v_4^n$$

or
$$p_2 v_c^n = p_1 v_4^n$$

$$\frac{v_4}{v_c} = \left(\frac{p_2}{p_1}\right)^{1/n}$$

$$\eta_v = 1 - c\left[\left(\frac{p_2}{p_1}\right)^{1/n} - 1\right] \qquad \text{...(8.8a)}$$

$$= 1 + c - c\left(\frac{p_2}{p_1}\right)^{1/n} \quad \text{...(8.8 b)}$$

$$= 1 + c - c\left(\frac{v_1}{v_2}\right) \quad \text{...(8.8 c)}$$

When referred to ambient conditions,

$$\eta_v = \frac{\text{Volume of air sucked referred to ambient conditions}}{\text{Swept volume}}$$

$$= \frac{p_1 T_0}{p_0 T_1}\left(\frac{v_1 - v_4}{v_s}\right) = \frac{p_1 T_0}{p_0 T_1}\left[1 + c - c\left(\frac{p_2}{p_1}\right)^{1/n}\right] \quad \text{...(8.9)}$$

Fctors affecting volumetric efficiency

The volumetric efficiency of a compressor is lowered by any of the folowing conditions :

1. Very high speed.
2. Leakage through piston seals.
3. Too large a clearnace volume.
4. Obstruction at inlet valves.
5. Overheating of air by contact with cylinder walls.
6. Inertia effect of air.

Volumetric efficiency decreases with increase in clearance ratio. Higher the delivery pressure, lower the volumetric efficiency. Further if $\frac{p_2}{p_1}$ increases, volumetric efficiency decreases.

8.6.3 Isothermal Efficiency

The p-v and T-S diagrams for isothermal and polytropic compression of air in the compressor are shown in Fig. 8.8.

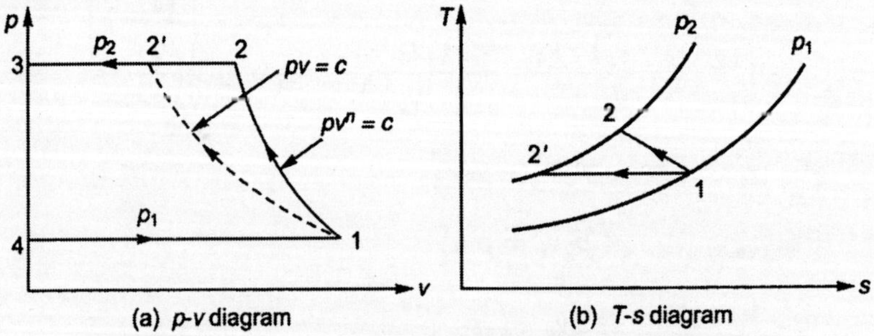

Fig. 8.8 Isothermal and polytrophic compression in a compressor.

Polytropic workdone, $$W_p = \left(\frac{n}{n-1}\right) RT_1\left[\left(\frac{p_2}{p_1}\right)^{\frac{n-1}{n}} - 1\right]$$

Isothermal workdone $$W_i = p_1 v_1 \ln \frac{v_1}{v_{2'}} + p_2 v_{2'} - p_1 v_1$$

But $$p_1 v_1 = p_2 v_{2'}$$

$\therefore$ $$W_i = p_1 v_1 ln \frac{v_1}{v_{2'}} = mRT_1 \ln\left(\frac{p_2}{p_1}\right)$$

$$= RT_1 \ln\left(\frac{p_2}{p_1}\right) \text{per kg of air delivered}$$

Work saved, $\Delta W = W_p - W_i$

$$= \left(\frac{n}{n-1}\right) RT_1 \left[\left(\frac{p_2}{p_1}\right)^{\frac{n-1}{n}} - 1\right] - RT_1 \ln\left(\frac{p_2}{p_1}\right)$$

$$= RT_1 \left[\left(\frac{n}{n-1}\right)\left\{\left(\frac{p_2}{p_1}\right)^{\frac{n-1}{n}} - 1\right\} - \ln\left(\frac{p_2}{p_1}\right)\right] \quad ...(8.10)$$

Isothermal efficiency, $$\eta_i = \frac{W_i}{W_p} = \frac{\ln\left(\frac{p_2}{p_1}\right)}{\left(\frac{n}{n-1}\right)\left[\left(\frac{p_2}{p_1}\right)^{\frac{n-1}{n}} - 1\right]} \quad ...(8.11)$$

Isothermal power, $$P_i = \frac{p_1 v_1 \ln\left(\frac{p_2}{p_1}\right) \times N}{60 \times 10^3} \text{ kW/kg of air} \quad ...(8.12)$$

Compressor efficiency, $$\eta_c = \frac{\text{Isothermal power}}{\text{Indicated power}} \quad ...(8.13)$$

8.6.4 Adiabatic Efficiency

It is defined as the ratio of the work done on the compressor with reversible adiabatic compression to the work done with irreversible polytropic compression.

Adiabatic efficiency, $$\eta_a = \frac{\left(\frac{\gamma}{\gamma-1}\right)\left[\left(\frac{p_2}{p_1}\right)^{\frac{\gamma-1}{\gamma}} - 1\right]}{\left(\frac{n}{n-1}\right)\left[\left(\frac{p_2}{p_1}\right)^{\frac{n-1}{n}} - 1\right]} \quad ...(8.14)$$

8.6.5 Calculation of Main Dimensions

The actual volume of air drawn in by the compressor per stroke,

$$v = (v_1 - v_3)\eta_v \qquad ...(8.15)$$

Capicity of a single acting compressor,

$$Q = vN = (v_1 - v_3)\eta_v N \qquad ...(8.16)$$

Swept volume, $$v_s = v_1 - v_3 = \frac{Q}{N\eta_v} \qquad ...(8.17)$$

Also $$v_s = \frac{\pi}{4} d^2 L$$

where d = piston diameter

L = length of stroke

$\frac{L}{d} \simeq 0.6 - 0.9$

Thus, L and d can be determined.

8.7 MULTI-STAGE COMPRESSION

Single stage compression suffers from many disadvantages like (*i*) handling of very high pressure range is one cylinder resulting in leakage past the piston, (*ii*) ineffective colling of the gas, (*iii*) necessitating robust construction of the cylinder to withstand the high delivery pressure.

The volumetric efficiency of a single stage compressor with the fixed clearance decreases with an increase in pressure ratio and thus reduce capacity.

Thus the necessity of multi-stage compression arises. The advantages of multi-stage compression with inter colling between stages are listed below :

1. The air can be cooled perfectly at pressures intermediate between the suction and delivery pressures resulting in less power required as compared to a single stage compressor for the same pressure limits and quantity of free air delivered.
2. The mechanical balance of the machine is better due to phasing of the operations.
3. The pressure range and hence the temprature range in each stage can be kept within desirable limits. This results in :
   (a) less loss of air due to leakage past the piston,
   (b) more perfect lubrication due to lower temperatures,
   (c) better volumetric efficiency, and
   (d) the lighter construction of the machine which reduces the cost.

8.7.1 Two-Stage Compressor

The schematic arrangement of two-stage compressor with inter cooler and *p-v* diagram are shown in Fig. 8.9. The air is first taken into the low pressure (L.P.) cylinder at pressure p_1. After compression to intermediate pressure p_2, the air at condition 2 is passed through the intercooler and leaves it at point 3 where its temperature is reduced from T_2 to T_3. The air may be cooled to point 3′ such that $T_{3'} = T_1$. Finally the air is compressed in high pressure (H.P.) cylinder from condition 3 to 4 and is discharged to the receiver. Area 2-3-4-6-2 represents the work saved due to inter cooling.

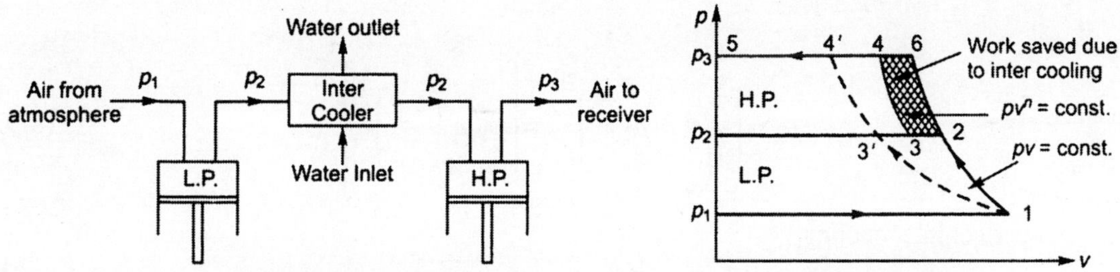

(a) Schematic arrangement (b) *p-v* diagram

Fig. 8.9 Two stage compression with intercooling.

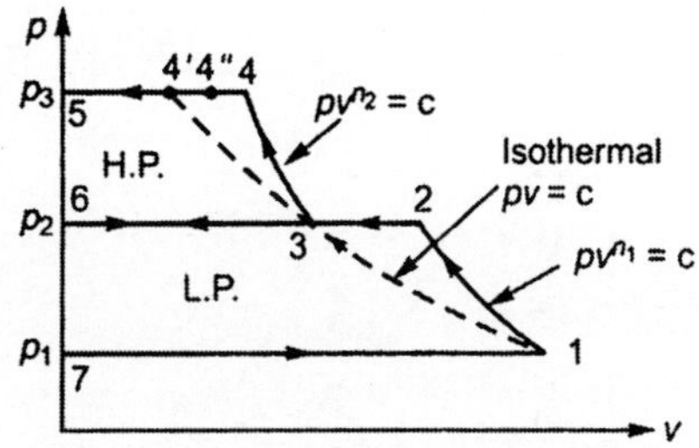

Fig. 8.10 Two stage compression with perfect intercooling.

Fig. 8.10 shows the *p-v* diagram for two-stage compression with perfect inter colling. Process 1-2 represents the polytropic compression in the L.P. cylinder with $pv^{n_1} =$ const. Process 2-3 represents inter cooling from T_2 to $T_3 = T_1$. Process 3-4 represents, the polytropic comrpession in the H.P. cylinder with $pv^{n_2} =$ const. Curve 1-3-4′ represents the isothermal compression.

Total work done one the air,

$$W_t = \left(\frac{n_1}{n_1 - 1}\right) RT_1 \left[\left(\frac{p_2}{p_1}\right)^{\frac{n_1 - 1}{n_1}} - 1\right] + \left(\frac{n_2}{n_2 - 1}\right) RT_3 \left[\left(\frac{p_3}{p_2}\right)^{\frac{n_2 - 1}{n_2}} - 1\right] \qquad \text{...(8.18a)}$$

Let

$$\frac{n_1 - 1}{n_1} = a \text{ and } \frac{n_2 - 1}{n_2} = b$$

$$W_t = \frac{RT_1}{a}\left[\left(\frac{p_2}{p_1}\right)^n - 1\right] + \frac{RT_3}{b}\left[\left(\frac{p_3}{p_2}\right)^b - 1\right] \qquad \text{...(8.18 b)}$$

For p_1 and p_3 to be constant, intermediate pressure p_2 must be determined for minimum work. Thus,

$$\frac{dW_t}{dp_2} = 0$$

$$\frac{T_1}{a} \cdot a\left(\frac{p_2}{p_1}\right)^{n-1} \cdot \frac{1}{p_1} + \frac{T_3}{b} \cdot b\left(\frac{p_3}{p_2}\right)^{n-1} \cdot \left(\frac{-p_3}{p_2^2}\right) = 0$$

$$T_1 \frac{p_2^{a-1}}{p_1^a} - T_3 \frac{p_3^b}{p_2^{b+1}} = 0$$

$$T_1 p_2^{a+b} = T_3 p_1^a p_3^b$$

$$p_2 = \left[\frac{T_3}{T_1}(p_1^a p_3^b)\right]^{\frac{1}{a+b}} \qquad \text{...(8.19)}$$

For $n_1 = n_2 = n, a = b.$ Then

$$p_2 = \left[\frac{T_3}{T_1}(p_1 p_3)^a\right]^{\frac{1}{2a}} = \left[p_1 p_3 \left(\frac{T_3}{T_1}\right)^{\frac{1}{a}}\right]^{\frac{1}{2}} \quad \text{...(8.20)}$$

For perfect inter cooling,

$$T_3 = T_{1'} \text{ and } p_2 = \sqrt{p_1 p_3} \quad \text{...(8.21)}$$

$$(W_t)_{\min} = \frac{RT_1}{a}\left[\left\{\frac{\sqrt{p_1 p_3}}{p_1}\right\}^a + \left\{\frac{p_3}{\sqrt{p_1 p_3}}\right\}^a - 2\right]$$

$$= \frac{2RT_1}{a}\left[\left(\frac{p_3}{p_1}\right)^{\frac{a}{2}} - 1\right]$$

$$= \left(\frac{2n}{n-1}\right) RT_1 \left[\left(\frac{p_3}{p_1}\right)^{\frac{n-1}{2n}} - 1\right] \quad \text{...(8.22 a)}$$

$$= \left(\frac{2n}{n-1}\right) p_1 v_1 \left[\left(\frac{p_3}{p_1}\right)^{\frac{n-1}{2n}} - 1\right] \quad \text{...(8.22 b)}$$

In general, with i number of stages, we have

$$\frac{p_2}{p_1} = \frac{p_3}{p_2} = \frac{p_4}{p_3} = \ldots = \frac{p_{i+1}}{p_i}$$

Minimum shaft work, $$W_t = \left(\frac{in}{n-1}\right) RT_1 \left[\left(\frac{p_{i+1}}{p_i}\right)^{\frac{n-1}{in}} - 1\right] \quad \text{...(8.23)}$$

The minimum incdicated power (I.P.), with imperfect inter cooling is :

$$\text{I.P.} = \left(\frac{n}{n-1}\right) mR \left[2\sqrt{T_1 T_3}\left(\frac{p_3}{p_1}\right)^{\frac{n-1}{2n}} - (T_1 + T_3)\right] \times \frac{1}{10^3} \quad \text{...(8.24)}$$

where m = mass of air delivered in kg per second.

8.7.2 Heat Rejected to the Intercooler

Let m = mass of air in the cylinder

Then $$p_1 V_1 = mRT_1$$

or
$$m = \frac{p_1 V_1}{RT_1} \quad ...(1)$$

From the compression 1-2, we have

$$p_1 V_1^n = p_2 V_2^n$$

or
$$V_2 = V_1 \left(\frac{p_1}{p_2} \right)^{\frac{1}{n}} \quad ...(2)$$

Then
$$p_2 V_2 = mRT_2$$

or
$$T_2 = \frac{p_2 V_2}{mR} \quad ...(3)$$

From the constant pressure line 2-3, we have

$$\frac{p_2 V_2}{T_2} = \frac{p_3 V_3}{T_3}$$

$$T_3 = T_2 \left(\frac{p_3 V_3}{p_2 V_2} \right) \quad ...(4)$$

Heat rejected to inter cooler $= mc_p (T_2 - T_3)$ ...(8.25)

8.7.3 Cylinder Dimensions

For steady state flow the mass of air passing through each cylinder per stroke must be same.

Let V_a = actual volume of air per stroke taken during suction.

p = density of air.

Then
$$V_{a_1} \rho_1 = V_{a_2} \rho_2 = V_{a_3} \rho_3 = ... = \text{const.}$$

or
$$\frac{V_{a_1} p_1}{RT_1} = \frac{V_{a_2} p_2}{RT_2} = \frac{V_{a_3} p_3}{RT_3} = ...$$

For perfect inter cooling, $T_1 = T_2 = T_3 = ...$

$\therefore$
$$V_{a_1} p_1 = V_{a_3} p_2 = V_{a_3} p_3 = ...$$

$$V_{s_1} \eta_{v_1} p_1 = V_{s_2} \eta_{v_2} p_2 = V_{s_3} \eta_{v_3} p_3 = ...$$

where η_v = volumetric efficiency of a cylinder

V_s = stroke volume of the cylinder.

$$\frac{\pi}{4} d_1^2 L_1 \eta_{v_1} p_1 = \frac{\pi}{4} d_2^2 L_2 \eta_{v_2} p_2 = \frac{\pi}{4} d_3^2 L_3 \eta_{v_3} p_3 = ...$$

or
$$d_1^2 L_1 \eta_{v_1} p_1 = d_2^2 L_2 \eta_{v_2} p_2 = d_3^2 L_3 \eta_{v_3} p_3 = ... \quad ...(8.26a)$$

If
$$\eta_{v_1} = \eta_{v_2} = \eta_{v_3} = ... \text{ and } L_1 = L_2 = L_3 = ... \quad ...(8.26b)$$

Then
$$d_1^2 p_1 = d_2^2 p_2 = d_3^2 p_3 = ...$$

8.7.4 Inter-cooler and After-cooler

Inter-Cooler : An inter-cooler is a simple heat exchnager in which heat is removed from the air after it has been compressed and its temperature has risen as a result of compression. The inter-cooler commonly used is of the counter flow type as shown in Fig. 8.11 because it gives high effectiveness.

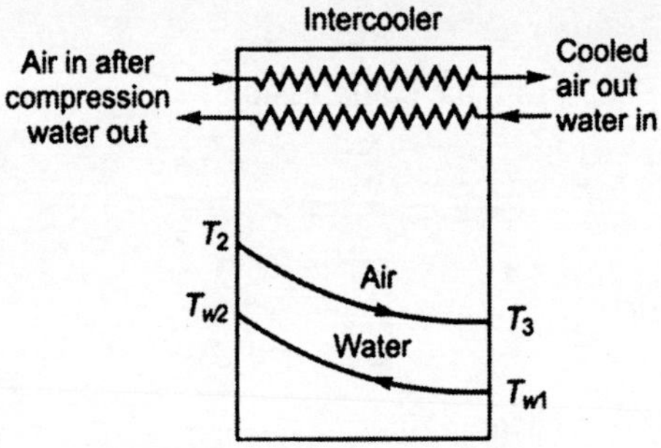

Fig. 8.11 Counter flow intercooler.

Effectiveness of inter cooler,

$$\varepsilon = \frac{\text{Actual drop in temperature of air}}{\text{Maximum possible drop in temperature of air}}$$

$$\frac{T_2 - T_3}{T_{w_2} - T_{w_1}} = \frac{T_2 - T_3}{T_2 - T_1} \quad \ldots(8.27)$$

A simple section showing the principles of construction of an inter cooler is shown in Fig. 8.12. The coolant which may be water or any other fluid, passes through the tubes secured between two end plates and the air circulates over the tubes trhrough a system of baffles. Two passes are used for water flow and air is made to flow partly parallel and partly cross with the help of baffles. This type of inter-cooler gives better effectiveness than the ordinary counter-flow type. The purpose of the inter-cooler is to reduce the work done on the air.

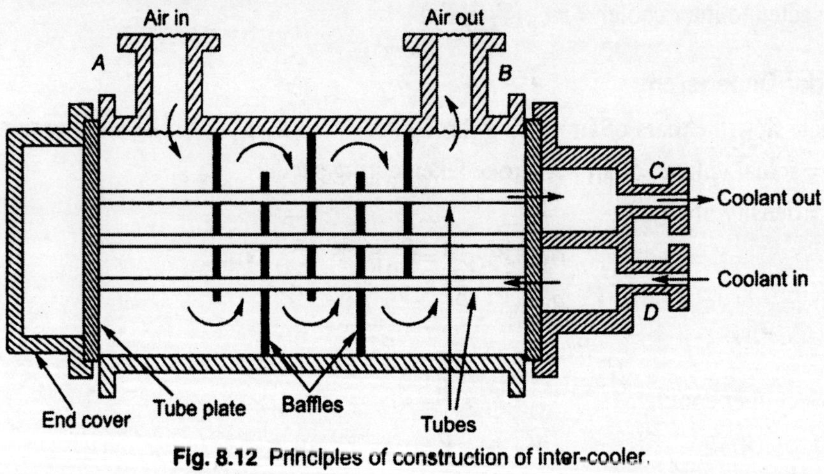

Fig. 8.12 Principles of construction of inter-cooler.

After-Cooler : An after-cooler is used to cool the air coming out from the compressor before it enters the receiver. The air coming out from the compressor at pressure p_3 is sufficiently hot at temperature T_4 (see Fig. 8.10). If this air is cooled in the after-cooler then the pressure will remain p_3 but the temperature will fall from T_4 to $T_{4''}$. Therefore, the volume of air leaving the after coller will be given by,

$$\frac{p_{4''}v_{4''}}{T_{4''}} = \frac{p_4 v_4}{T_4}. \text{ But } p_{4''} = p_4.$$

Thus

$$v_{4''} = v_4 \left(\frac{T_{4''}}{T_4}\right)$$

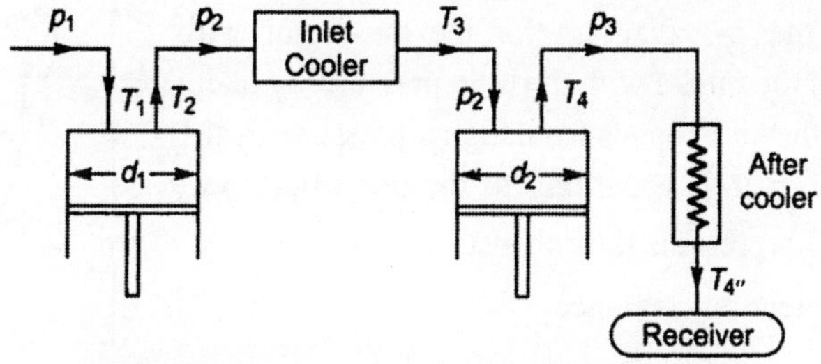

Fig. 8.13 Position of inter-cooler and after-cooler.

As $$T_{4''} < T_{4'} v_{4''} < v_4$$

Therefore, the purpose of the after-cooler is to reduce the size of the receiver. The position of the inter-cooler and after-cooler is shown in Fig. 8.13.

8.8 INDICATED POWER OF A COMPRESSOR

Indicated power, $$\text{I.P.} = \frac{p_m LAN}{60 \times 10^3} \text{ kW} \quad \text{...(8.28)}$$

where A = area of the cylinder, m^2

L = length of stroke, m

N = rpm of compressor crank

p_m = mean effective pressure of air, P_a (N/m^2)

Theoretically, *m.e.p.* for a single-acting, single-stage compressor is,

$$p_m = \left(\frac{n}{n-1}\right) p_1 \eta_v \left[\left(\frac{p_2}{p_1}\right)^{\frac{n-1}{n}} - 1\right] \quad \text{...(8.29 a)}$$

where η_v = volumetric efficiency of compressor

Using indicator card, the *m.e.p.* is :

$$p_m = \frac{\text{Area of indicator card, cm}^2}{\text{Measured length of the indicator card, cm} \times \text{Vertical scale, N/m}^2\text{/cm}} \text{ N/m}^2 \quad \text{...(8.29 b)}$$

$$\text{Mechanical efficiency, } \eta_{mech} = \frac{\text{Indicated power of compressor}}{\text{Brake power of compressor}} \quad \text{...(8.30)}$$

8.9 AIR MOTORS

The working of an air motor is similar to that of air compressor. High pressure air is admitted to the motor cylinder through a mechanically operated inlet value and drives the piston in the forward direction. After a part of the stroke of the piston has been performed, the air supply is cut-off and the stroke is completed after decreasing pressure as the air expands in the cylinder. After the expansion stroke is completed the air is allowed to escape into the atmosphere through a mechanically operated discharge valve. The return stroke is performed by compressed air acting on the other side of the piston in double acting cylinder.

Fig. 8.14 shows the *p-v* diagram for the air motor with clearance. Process 4-1 is the intake with the high pressure p_2 at cut off point 1. From 1 to 2 the air expands from higher pressure p_2 at 1 to atmospheric pressure p_1 at 2, according to the polytropic law, $pv^n = c$. The process 2-3 represents the exhaust.

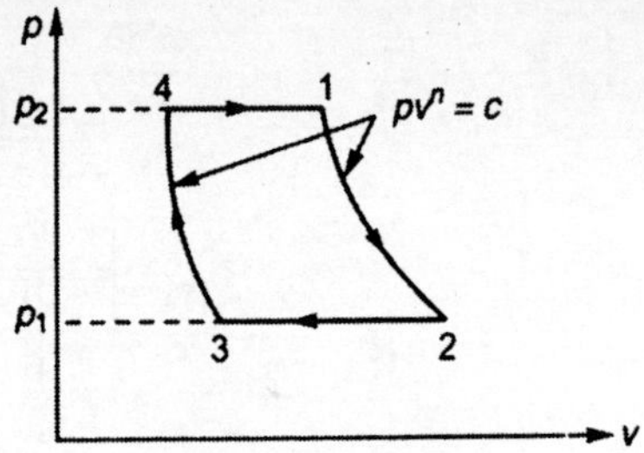

Fig. 8.14 *p-v* diagram for an air motor.

Work obtained neglecting clearance,

$$W = \left(\frac{n}{n-1}\right) p_1 V_1 \left[1 - \left(\frac{p_2}{p_1}\right)^{\frac{n-1}{n}}\right] \quad \text{...(8.31 a)}$$

$$= \left(\frac{n}{n-1}\right) mRT_1 \left[1 - \left(\frac{p_2}{p_1}\right)^{\frac{n-1}{n}}\right] \quad \text{...(8.31 b)}$$

Work done considering clearance,

$$'V = \left(\frac{n}{n-1}\right) p_1 (V_1 - V_4) \left[1 - \left(\frac{p_2}{p_1}\right)^{\frac{n-1}{n}}\right] \quad \text{...(8.32 a)}$$

$$= \left(\frac{n}{n-1}\right) m_1 RT_1 \left[1 - \left(\frac{p_2}{p_1}\right)^{\frac{n-1}{n}}\right] \quad \text{...(8.32 b)}$$

where m_1 = actural mass of air dlivered per cycle.

8.10 INDICATOR DIAGRAM

The theoretical of *p-v* diagram for the single stage reciprocating compressor is represented by 1-2-3-4-1 in Fig. 8.15. The actual indicator diagram is 1-2′-3-4′-1. The difference between the actual and theoretical indicator diagrams is due to the intake and discharge losses. The intake losses include the friction loss in pipe, friction less in inlet valve and valve inertia loss. Theoretically, thc inlet valve should open at 4, but actually it opens little afterwards at 4′ due to the inertia of the valve and the pressure inside the cylinder falls below atmospheric pressure. The oscillating part of the curve indicates valve flutter due to vibration of spring loaded valve. On the delivery side, the discharge value should open at 2, but actually it opens little afterwards at 2′ due to the inertia of the spring loaded discharge valve. The effect of these losses is to increase the work required by the compressor.

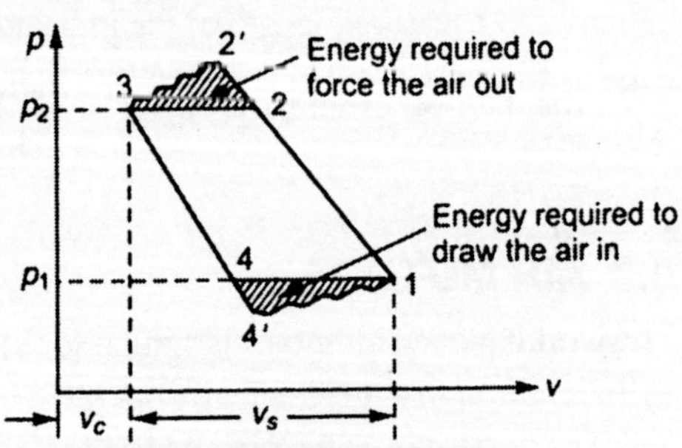

Fig. 8.15 Actual *p-v* diagram for single stage compressor.

The actual indicator diagram for two-stage compressor with inter cooler is shown in Fig. 8.16.

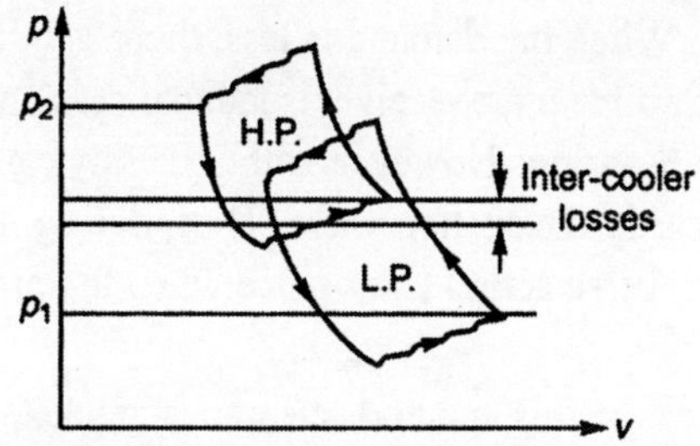

Fig. 8.16 Actual *p-v* diagram for two-stage compressor.

8.11 HEAT REJECTED PER KG OF AIR

If the air is cooled to the initial temperature, there is no change in internal energy per kg mass of air and all the work done is resjected to the cooling medium partly during the compression proces and the remaining after compression at constant pressure.

Now $$q = du + w$$

But $$du = 0$$

For a single-stage compressor, the heat rejected is given by work done $-\int v \cdot dp$.

$$\text{Heat rejected} = w = \left(\frac{n}{n-1}\right) p_1 v_1 \left[\left(\frac{p_2}{p_1}\right)^{\frac{n-1}{n}} - 1\right]$$

Also $$p_1 v_1 = RT_1 \text{ and } \frac{T_2}{T_1} = \left(\frac{p_2}{p_1}\right)^{\frac{n-1}{n}}$$

$\therefore$ $$w = \left(\frac{n}{n-1}\right) RT_1 \left(\frac{T_2}{T_1} - 1\right)$$

$$= \left(\frac{n}{n-1}\right) R(T_2 - T_1) \quad \text{...(8.33 a)}$$

$$= \left[c_p + \frac{c_v(\gamma - n)}{n-1}\right](T_2 - T_1) \quad \text{...(8.33 b)}$$

= Heat rejected to the coolant in the inter cooler

+ Heat rejected to the coolant during compression process

8.12 CONTROL OF COMPRESSOR

In order to balance the demand and supply of air, it is desirable to incorporate devices for the compressor control. The common methods of control are :

1. Throttle control
2. Clearance control
3. Blowing air to waste

1. **Throttle Control :** When the demand is less, there is build up of pressure in the receiver and the high pressure air from the receiver is led to piston and cylinder. The movement of the piston is resisted by a spring. However with excessive pressure the piston depresses the spring, thus closing partly the suction vlave. Thus, during the suction stroke the air intake is partly throttled. The reverse action takes palce when the pressure in the receiver falls due to increase in demand.
2. **Clearance Control :** In this method, clearance pockets are provided for increasing or decreasing clearance. Thus, the volumetric efficiency is reduced in proper proportion to control output.
3. **Blowing air to Waste :** In case of excessive pressure built-up in the receiver due to decrease in demand, a by-pass valve from the high pressure cylinder delivers air directly to atomosphere. When the pressure in the receiver falls, the relay piston closes the valve.

■ **Example 8.1** A single stage air compressor is required to compress 72 m^3 of air per minute from 15°C and 1.0 bar to 8 bar pressure. Find the temperature at the end of the compression, work done, power and heat rejected during each of the following processes : (a) isothermal, (b) adiabatic, and (c) polytropic compression following the law $pV^{1.25}$ = constant.

Solution : (a) Isothermal Compression :

$$p_1V_1 = p_2V_2$$

$$V_2 = \frac{1 \times 72}{8} = 9 \text{ m}^3/\text{min}$$

$$T_2 = T_1 = 273 + 15 = 288 \text{ K}$$

Word done, $$W_i = p_1V_1 \ln \frac{V_1}{V_2} = 1 \times 10^5 \times 72 \ln\left(\frac{72}{9}\right) = 150 \times 10^5 \text{ J/min}$$

Power, $$P_i = \frac{150 \times 10^5}{60 \times 10^3} = 250 \text{ kW}$$

Heat rejected, $$Q = W = 250 \text{ kJ/s} \qquad [\because dU = 0]$$

(b) Adiabatic Compression :

$$p_1V_1^\gamma = p_2V_2^\gamma$$

$$\frac{T_2}{T_1} = \left(\frac{p_2}{p_1}\right)^{\frac{\gamma-1}{\gamma}} = (8)^{\frac{0.4}{1.4}} = 1.81$$

$$T_2 = 2.88 \times 1.81 = 521 \text{ K}$$

Work done, $$W_a = \left(\frac{\gamma}{\gamma-1}\right) p_1V_1 \left[\left(\frac{p_2}{p_1}\right)^{\frac{\gamma-1}{\gamma}} - 1\right]$$

$$= \left(\frac{1.4}{0.4}\right) \times 1 \times 10^5 \times 72 \left[(8)^{\frac{1}{3.5}} - 1\right] = 204.5 \times 10^5 \text{ kJ/min}$$

Power, $P_a = \dfrac{2045 \times 10^5}{60 \times 10^3} = 340.8 \text{ kW}$

Heat rejected = 0

(c) Polytropic Compression :

$$p_1 V_1^{1.25} = p_2 V_2^{1.25}$$

$$\frac{T_2}{T_1} = \left(\frac{p_2}{p_1}\right)^{\frac{n-1}{n}} = (8)^{\frac{0.25}{1.25}} = 15157$$

$$T_2 = 288 \times 15157 = 43653 \text{ K}$$

Work done, $W_p = \left(\dfrac{n}{n-1}\right) p_1 V_1 \left[\left(\dfrac{p_2}{p_1}\right)^{\frac{n-1}{n}} - 1\right]$

$$= \left(\frac{125}{025}\right) \times 1 \times 10^5 \times 72[(8)^{0.2} - 1] = 185.66 \times 10^5 \text{ kJ/min}$$

Power, $P_p = \dfrac{185.66 \times 10^5}{60 \times 10^3} = 309.43 \text{ kW}$

Heat rejected $= \left(\dfrac{\gamma - n}{\gamma - 1}\right) \times W_p = \left(\dfrac{1.4 - 125}{1.4 - 1}\right) \times 185.66 \times 10^5$

$= 69622$ kJ/min.

■ **Example 8.2** Determine the size of cylinder for a double acting compressor of 40 kW in which the air is drawn in at 1 bar and 10°C, and compressed according to the law $pV^{1.2}$ = const. to 6 bar at 100 rpm. Average piston speed is 150 m/min. Neglect clearance volume.

Solution : Work doen/stroke $= \left(\dfrac{n}{n-1}\right) p_1 V_1 \left[\left(\dfrac{p_2}{p_1}\right)^{\frac{n-1}{n}} - 1\right] = p_m V_s$

As clearance volume $V_c = 0$, therefore $V_1 = V_s$ = swept volume

$$\therefore \quad p_m = \left(\frac{n}{n-1}\right) p_1 \left[\left(\frac{p_2}{p_1}\right)^{\frac{n-1}{n}} - 1\right]$$

$$= \left(\frac{12}{02}\right) \times 1[(6)^{\frac{0.2}{1.2}} - 1] = 2.088 \text{ bar}$$

Power, $P = \dfrac{2 p_m \text{ LAN}}{60 \times 10^3}$

$$40 = \frac{2 \times 2.088 \times 10^5 \; A \times 100}{60 \times 10^3}$$

$$A = 766 \times 10^{-4} m^2 = \frac{\pi}{4} d^2$$

$$d = 0.3123 \text{ m } or \ 312.3 \text{ mm}$$

Now
$$2LN = 150$$

$$L = \frac{150}{2 \times 100} = 0.75 \text{ m or } 750 \text{ mm}$$

■ **Example 8.3** A double acting compressor with a piston displacement of 0.05 m^3 per stroke operates at 500 rpm. The clearance is 5 percent and it receives air at 100 kPa, discharges it at 600 kPa. The compression is polytropic, $pV^{1.35} = c$. Determine the power required and the air discharged in m^3/s.

Solution :

$$W = \left(\frac{2n}{n-1}\right) p_1 V_a \left[\left(\frac{p_2}{p_1}\right)^{\frac{n-1}{n}} - 1\right]$$

$$V_a = V_c + V_5 - V_4$$

$$V_c = 0.05 \times 0.05 = 0.0025 \text{ m}^3$$

$$V_4 = V_c \left(\frac{p_3}{p_4}\right)^{\frac{1}{n}} = 0.0025 \left(\frac{600}{100}\right)^{\frac{1}{1.35}} = 0.0094 \text{ m}^3$$

$$V_a = 0.0025 + 0.05 - 0.0094 = 0.0431 \text{ m}^3$$

$$W = \left(\frac{2 \times 1.35}{0.35}\right) \times 100 \times 0.0431 \left[\left(\frac{600}{100}\right)^{\frac{0.35}{1.35}} - 1\right]$$

$$= 19.659 \text{ kJ/cycle}$$

$$\text{Total work} = 19.659 \times \frac{500}{60} = 163.823 \text{ kJ/s or kW}$$

$$\text{Volume discharged} = 2(V_2 - V_3) \times \frac{N}{60}$$

$$V_2 = V_1 \left(\frac{p_1}{p_2}\right)^{\frac{1}{n}} = 1.05 \times 0.05 \left(\frac{100}{600}\right)^{\frac{1}{1.35}} = 0.0139 \text{ m}^3$$

$$V_3 = V_c$$

$$\text{Volume discharged} = 2(0.0139 - 0.0025) \times \frac{500}{60} = 0.19 \text{ m}^3/\text{s}$$

■ **Examle 8.4** A two stage air compressor with comlete intercooling delivers air to the main at a pressure of 30 atm, the suction condition being 1 atm and 27°C. If both cylinders have the same stroke, find the ratio of the cylinder diameters for the efficiency of compression to be a maximum. Assume the index of compression to be 1.3.

Solution : Volume of L.P. cylinder $= V_1$

Volume of H.P. cylinder $= V_3$

$$L_1 = L_3$$

$$\therefore \quad \frac{V_1}{V_3} = \frac{d_1^2}{d_3^2} \text{ or } \frac{d_1}{d_3} = \sqrt{\frac{V_1}{V_3}}$$

Now

$$pV^{1.3} = c$$

$$p_1 V_1^{1.3} = p_2 V_2^{1.3}$$

or

$$\frac{V_1}{V_2} = \left(\frac{p_2}{p_1}\right)^{\frac{1}{1.3}}$$

For maximum efficiency,

$$p_2 = \sqrt{p_1 p_3} = \sqrt{1 \times 30} = 5.477 \text{ atm}$$

$$\frac{V_1}{V_2} = \left(\frac{5.477}{1}\right)^{\frac{1}{1.3}} = 3.699$$

$$\frac{T_2}{T_1} = \left(\frac{p_2}{p_1}\right)^{\frac{n-1}{n}}, T_2 = 300\left[\frac{5.477}{1}\right]^{\frac{0.3}{1.3}} = 444 \text{ K}$$

From constant pressure process 2-3 : (See Fig. 8.10)

$$\frac{V_3}{V_2} = \frac{T_3}{T_2} = \frac{T_1}{T_2} \text{ for perfect cooling}$$

$$= \frac{300}{444}$$

or

$$\frac{V_2}{V_3} = 1.48$$

$$\frac{V_1}{V_3} = \frac{V_1}{V_2} \times \frac{V_2}{V_3} = 3.699 \times 1.48 = 5.475$$

$$\frac{d_1}{d_3} = \sqrt{3.475} = 2.34$$

■ **Example 8.5** A two stage air compressor with perfect inter cooling takes in air at 1 bar and 27°C. The law of compression in both the stages is $pV^{1.3}$ = const. The compressed air is delivered at 9 bar from the H.P. cylinder to an air receiver. Calculate per kg of air (a) the minimum work done, (b) the heat rejected to the intercooler, and (c) the minimum work done in a three stage compressor working under the same conditions. Take $c_p = 1.005$ kJ/kg.K.

Solution :

$$T_1 = 273 + 27 = 300 \text{ K}$$

For 1 kg of air,

$$p_1 v_1 = RT_1, \; v_1 = \frac{287 \times 300}{1 \times 10^5} = 0.861 \text{ m}^3/\text{kg}$$

(a) The minimum work required in two stage compressor with perfect intercooling is,

$$W = \left(\frac{2n}{n-1}\right) p_1 v_1 \left[\left(\frac{p_3}{p_1}\right)^{\frac{n-1}{2n}} - 1\right]$$

$$= \left(\frac{2 \times 1.3}{0.3}\right) \times 2 \times 10^5 \times 0.861 \left[\left(\frac{9}{1}\right)^{\frac{0.3}{2.6}} - 1\right]$$

$$= 215.3 \text{ kJ/kg of air}$$

(b)
$$p_2 = \sqrt{p_1 p_3} = \sqrt{1 \times 9} = 3 \text{ bar}$$

$$\frac{T_2}{T_1} = \left(\frac{p_2}{p_1}\right)^{\frac{n-1}{n}} = \left(\frac{3}{1}\right)^{\frac{0.3}{1.3}} = 1.289$$

$$T_2 = 300 \times 1.289 = 386.6 \text{ K}$$

Heat rejected to intercooler,

$$q = c_p (T_2 - T_3) = c_p (T_2 - T_1) \qquad [\because T_3 = T_1]$$

$$= 1.005\,(386.6 - 300) = 87 \text{ kJ/kg of air}$$

(c) The least work done in case of three stage air compressor working between the same pressure limits is given by,

$$W = \left(\frac{3n}{n-1}\right) p_1 V_1 \left[\left(\frac{p_4}{p_1}\right)^{\frac{n-1}{3n}} - 1\right]$$

$$p_4 = 9 \text{ bar}, \; p_1 = 1 \text{ bar}$$

$$W = \left(\frac{3 \times 1.3}{0.3}\right) \times 1 \times 10^5 \times 0.861 \left[\left(\frac{9}{1}\right)^{\frac{0.3}{3.9}} - 1\right]$$

$$= 206 \text{ kJ/kg of air}$$

■ **Example 8.6** The bore and stroke of a single cylinder single acting air compressor are 11 cm and 14 cm respectively. Air is taken from the atmosphere at 1 bar and elivered through a valve, which maintains a delivery pressure of 7.5 bar in a reservoir of 1.5 m^3 capacity. A motor giving 2 kW drives the compressor at 400 rpm and indicated mean effective pressure is 2 bar. The reservoir is initially at atmospheric pressure of 1 bar and 17°C and after 19 minutes running reaches 7.5 bar and 75°C. Calculate the delivery in m^3 of free air per minute at 17°C, the volumetric efficiency at atmospheric conditions, the isothermal and mechanical efficiency.

Solution : Swept volume, $V_s = \frac{\pi}{4} d^2 L = \frac{\pi}{4} \times 11^2 \times 14 \times 10^{-6} = 0.001331 \text{ m}^3$

Initial mass of air in the reservoir $= \dfrac{p_1 V_1}{R T_1}$

Final mass of air after 19 minutes $= \dfrac{p_2V_2}{RT_2}$

Mass of air delivered in 19 minutes,

$$m = \frac{p_2V_2}{RT_2} - \frac{p_1V_1}{RT_1} = \frac{V}{R}\left(\frac{p_2}{T_2} - \frac{p_1}{T_1}\right) \qquad (\because V_1 = V_2 = V)$$

$$\text{Volume delivered at initial condition} = \frac{mRT_1}{p_1} = \frac{RT_1}{p_1} \times \frac{V}{R}\left(\frac{p_2}{T_2} - \frac{p_1}{T_1}\right)$$

$$= \frac{VT_1}{p_1}\left(\frac{p_2}{T_2} - \frac{p_1}{T_1}\right)$$

$$= \frac{1.5 \times 290}{1}\left(\frac{7.5}{348} - \frac{1}{290}\right) = 7.875 \text{ m}^3$$

$$\text{Volume delivered per minute} = \frac{7.875}{19} = 0.4145 \text{ m}^3/\text{min}$$

$$\text{Volume delivered per stroke} = \frac{0.4145}{400} = 0.001036 \text{ m}^3$$

$$\text{Volumetric efficiency} = \frac{0.0001036}{0.001331} \times 100 = 77.84\%$$

$$\text{Indicated power} = \frac{p_m (LA) N}{60 \times 10^3}$$

$$= \frac{2 \times 0.001331 \times 400 \times 10^5}{60 \times 10^3} = 1.774 \text{ kW}$$

Mechanical efficiency, $\eta_{mech} = \dfrac{1.774}{2} \times 100 = 88.7\%$

$$\text{Isothermal power} = p_1V_1 \ln \frac{p_2}{p_1}$$

$$= \frac{1 \times 10^5 \times 0.4248}{60 \times 10^3} \times \ln\left(\frac{8}{1}\right)$$

$$= 1.472 \text{ kW}$$

Isothermal effciency, $\eta_{iso} = \dfrac{1.472}{1.774} \times 100 = 83\%$

■ **Example 8.7** A single stage, single acting air compressor delivers 15 m^3/min of free air from 1 bar to 8 bar at 300 rpm. The clearance volume is 6% of the stroke volume and compression and expansion follow the law $pv^{1.3}$ = const. Calculate the diameter and stroke of the compressor. Take $L = 1.5\,D$. The temperature and pressure of air at suction are the same as atmospheric air.

Solution : Clearance ratio, $c = 0.06$

Volumetric efficiency, $\eta_v = 1 + c - c\left(\dfrac{p_2}{p_1}\right)^{\frac{1}{n}}$

$$= 1 + 0.06 - 0.06\left(\frac{8}{1}\right)^{\frac{1}{1.3}}$$

$$= 0.763 \text{ or } 76.3\%$$

F.A.D. per minute,
$$V_s \eta_v \times 300 = 15$$

$$V_s = \frac{15}{0.763 \times 300} = 0.06553 \text{ m}^3$$

Also
$$\frac{\pi}{4} D^2 L = V_s$$

$$\frac{\pi}{4} D^2 \times 15D = 0.06553$$

$$D = 0.3817 \text{ m } 94\ 381.7 \text{ mm}$$

$$L = 15 \times 03817 = 05726 \text{ m or } 572.6 \text{ mm}$$

■ **Example 8.8** A four cylinder double acting compressor is rquired to compress 25 m^3/min of air at 1 bar and 25°C to a pressure of 15 bar. Determine the size of motor required and the cylinder dimensions if the following additional data is given :

Clearance Volume = 5%

L/D = 1.2

RPM = 300

$$\eta_{mech} = 80\%$$

Value of index,
$$n = 135$$

Assume no pressure change in suction valves and that the air gets heated by 10°C during suction stroke.

Solution : Net work done,
$$W = \left(\frac{n}{n-1}\right) p_1 (v_1 - v_4)\left[\left(\frac{p_2}{p_1}\right)^{\frac{n-1}{n}} - 1\right]$$

where $v_1 - v_4$ = suction volume $= \frac{25}{60}$ m^3 / s

$$W = \left(\frac{135}{035}\right) \times 1 \times 10^5 \times \frac{25}{60}\left[\left(\frac{15}{1}\right)^{\frac{0.35}{1.35}} - 1\right]$$

$$= 163600 \text{ N.m/s}$$

Theoretical power,
$$P_{th} = \frac{W}{10^3} = 1636 \text{ kW}$$

Motor power,
$$P_m = \frac{P_{th}}{\eta_{mech}} = \frac{1636}{08} = 2045 \text{ kW}$$

$$\text{Volumetric efficiency} = 1 + c - c\left(\frac{p_2}{p_1}\right)^{\frac{1}{n}} \times \frac{p_1 T_0}{p_0 T_1}$$

$$= 1 + 0.05 - 0.05\left(\frac{15}{1}\right)^{\frac{1}{1.35}} \times \left(\frac{1 \times 298}{1 \times 308}\right) = 0.69$$

$$\text{Swept volume of one cylinder} = \frac{25}{4} \times \frac{1}{2 \times 300} \times \frac{1}{0.69} = 0.0151 \text{ m}^3$$

$$\frac{\pi}{4} D^2 L = 0.0151$$

$$\frac{\pi}{4} D^2 \times 1.2D = 0.0151$$

$$D = 0.252 \text{ m or } 252 \text{ mm}$$

$$L = 1.2 \times 0.252 = 0.302 \text{ m or } 302 \text{ mm}$$

■ **Example 8.9** A two-cylinder, single acting air compressor is to deliver 15 kg of air per minute at 6.5 bar from suction conditions 1 bar and 13°C. Clearance may be taken as 4 percent of stroke volume and the index for both compression and re-expansion as 1.3. The compressor is directly coupled to a four cylinder, four stroke petrol engine which runs at 1800 rpm with b.m.e.p. of 6 bar. Assuming a stroke-bore ratio of 1.1 for both engine and compressor and a mechanical efficiency of 85% for the compressor, calculate the required cylinder dimensions.

Solution : Amount of air delivered per cylinder $= \frac{15}{2} = 7.5$ kg/min

Now

$$p_1(V_1 - V_4) = mRT_1$$

$$V_1 - V_4 = \frac{7.5 \times 287 \times 286}{1 \times 10^5} = 6.156 \text{ m}^3/\text{min}$$

$$= \frac{6.156}{1800} = 3.42 \times 10^{-3} \text{ m}^3/\text{stroke}$$

From the expansion curve, we have

$$\frac{V_4}{V_3} = \left(\frac{p_3}{p_4}\right)^{\frac{1}{n}} = \left(\frac{6.5}{1}\right)^{\frac{1}{1.3}} = 4.22$$

$$V_3 = 0.04 V_s$$

$$V_4 = 0.1688 V_s$$

Now

$$V_1 = V_3 + V_s = 0.04 V_s + V_s = 1.04 V_s$$

$$V_1 - V_4 = (1.04 - 0.1688) V_s = 0.8712 V_s = 3.42 \times 10^{-3}$$

$$V_s = 3.9256 \times 10^{-3} \text{ m}^3$$

$$\frac{\pi}{4} D^2 L = V_s$$

$$\frac{\pi}{4} D^2 \times 1.1 D = 3.9256 \times 10^{-3}$$

$$D = 0.1656 \text{ m or } 165.6 \text{ mm}$$

$$L = 1.1\,D = 0.1822 \text{ m or } 182.2 \text{ mm}$$

Now I.P. of the compressor $= \dfrac{\text{Work done per minute}}{60 \times 10^3}$

Work done per minute $= \left(\dfrac{n}{n-1}\right) mRT_1 \left[\left(\dfrac{p_2}{p_1}\right)^{\frac{n-1}{n}} - 1\right]$

$$= \frac{1.3}{0.3} \times 15 \times 287 \times 286 \left[\left(\frac{6.5}{1}\right)^{\frac{0.3}{1.3}} - 1\right]$$

$$= 2882491 \text{ N.m/min}$$

I.P. of compressor $= \dfrac{2882491}{60 \times 10^3} = 48.04$ kW

B.P. of engine $= \dfrac{\text{I.P.}}{\eta_{mech}} = \dfrac{48.04}{0.85} = 56.52$ kW

Now $\quad$ B.P. $= \dfrac{(b.m.e.p.) \times \text{LAN} \times i}{60 \times 10^3}$

$$56.52 = \frac{6 \times 10^5 \times (\text{LA}) \times 1800 \times 4}{60 \times 10^3 \times 2}$$

$$\text{LA} = 15.70 \times 10^{-4} \text{ m}^3$$

$$\frac{\pi}{4} D^2 \times 1.1D = 15.70 \times 10^{-4}$$

$$D = 0.122 \text{ m or } 122 \text{ mm}$$

$$L = 0.134 \text{ m or } 134 \text{ mm}$$

■ **Example 8.10** A multistage compressor has to be designed to supply air at 135 bar, while atmospheric condition is 1.03 bar and 15°C. The value of compression index may be assumed as 1.35. Due to practical reasons the intercoolers are not able to cool the air below 45°C, while the maximum temperature allowable in the system is 120°C. Calculate the number of stages that are necessary in the compression and the rate of cooling water circulated per kg of air. Take $c_p = 1$ kJ/kg. K.

Solution : For the first stage of compression,

$$T_1 = 273 + 15 = 288 \text{ K}, T_2 = 273 + 120 = 393 \text{ K}$$

$$p_1 = 1.03 \text{ bar}, n = 1.35$$

$$\frac{p_2}{p_1} = \left(\frac{T_2}{T_1}\right)^{\frac{n}{n-1}}, \; p_2 = 1.03 \left(\frac{393}{288}\right)^{\frac{1.35}{0.35}} = 3.0416 \text{ bar}$$

The intercoolers cools the air from $T_2 = 393$ K to $T_3 = 273 + 45 = 318$ K. For the second stage inlet temperature is $T_3 = 318$ K and outlet temperature is $T_4 = 393$ K (see Fig. 7.9 b).

For each stage, $$\frac{T_{outlet}}{T_{inlet}} = \left(\frac{P_{outlet}}{P_{inlet}}\right)^{\frac{n-1}{n}}$$

For the second stage compression, outlet $= p_4 = p_3\left(\frac{T_4}{T_3}\right)^{\frac{n}{n-1}}$ $[\because\ p_3 = p_2]$

$$= p_2\left(\frac{T_4}{T_3}\right)^{\frac{n}{n-1}} = 3.416\left(\frac{393}{318}\right)^{\frac{0.135}{0.35}} = 7.73 \text{ bar}$$

Pressure ratio $= \frac{p_4}{p_3} = \frac{7.73}{3.416} = 2.26$

The pressure ratio is same for all the subsequent stage till the pressure reaches 135 bar. Hence, the number of stage i can be calculated from,

$$(2.26)^i = \frac{135}{3.416}$$
$$i \ln 2.26 = \ln 39.52$$
$$i = 4.51$$

Therefore, the minimum number of stages required including the first stage

$$= 4.51 + 1 = 5.51 \simeq 6$$

Heat received by stage coolers per kg of air compressed

$$= c_p(T_2 - T_3)(i-1) = 1(393 - 318) \times 5 = 375 \text{ kJ}$$

Cooling water circulated per kg of air $= \frac{375}{4.18} = 89.7$ kg

■ **Example 8.11** A single acting two stage comressor with complete intercooling delivers 10 kg/min of air at 16 bar. The sunction occurs at 1 bar and 27°C. The comression and expansion processes are reversible with polytropic index $n = 1.3$. The compressor runs at 450 rpm. Calculate :

(a) The power required to drive the compressor,
(b) the isothermal efficiency,
(c) the free air delivered,
(d) the heat transferred in intercooler, and
(e) the swept and clearance volumes for each cylinder if the clearance ratio for L.P. and H.P. cylinders are 0.04 and 0.06 respectively.

Solution : The p-V diagram is shown in Fig. 8.17

$$p_1 = \text{bar},\ p_3 = 16 \text{ bar},\ T_1 = 273 + 27 = 300 \text{ K},\ N = 450 \text{ rpm},\ n = 1.3,\ c_1 = 0.04,\ c_2 = 0.06$$

For perfect intercooling, $p_2 = \sqrt{p_1 p_3} = \sqrt{1 \times 16} = 4$ bar

(a) Power required :

Work done in two stages with perfect intercooling,

$$W = \left(\frac{2n}{n-1}\right) mRT_1\left[\left(\frac{p_3}{p_1}\right)^{\frac{n-1}{2n}} - 1\right]$$

$$= \frac{2 \times 1.3}{0.3} \times 10 \times 0.287 \times 300 \left[\left(\frac{16}{1} \right)^{\frac{0.3}{2.6}} - 1 \right]$$

$$= 2813.24 \text{ kJ/min}$$

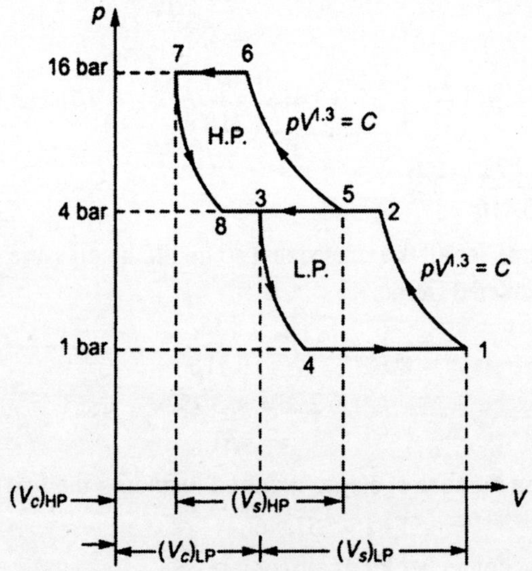

Fig. 8.17

Power developed, $P_p = \frac{W}{60} = \frac{2813.24}{60}$

$$= 46.887 \text{ kW}$$

(b) Isothermal efficiency :

Isothermal power, $P_i = mRT_1 \ln \frac{p_3}{p_1} = \frac{10}{60} \times 0.287 \times 300 \times \ln 16 = 390.787 \text{ kW}$

Isothermal efficiency $= \frac{P_i}{P_p} = \frac{39.787}{46.887} \times 100 = 84.85\%$

(c) Free air delivered (F.A.D.) :

Free air delivered $= \frac{mRT_1}{p_1} = \frac{10 \times 0.287 \times 300}{1 \times 10^2} = 8.61 \text{ m}^3/\text{min}$

(d) Heat transferred in intercooler :

$$T_2 = T_1 \left(\frac{p_2}{p_1} \right)^{\frac{n-1}{n}} = 300 \left(\frac{4}{1} \right)^{\frac{0.3}{1.3}} = 413.1 \text{ K}$$

Heat transferred in intercooler with perfect intercooling

$$= mc_p (T_2 - T_3) = \frac{10}{60} \times 1.005(413.1 - 300) = 18.44 \text{ kW}$$

(e) L.P. Stage :

$$\eta_v = 1 + c_1 - c_1\left(\frac{p_2}{p_1}\right)^{\frac{1}{n}} = 1 + 0.04 - 0.04\left(\frac{4}{1}\right)^{\frac{1}{1.3}} = 0.9238 \text{ or } 92.38\%$$

Swept volume, $$V_s = V_1 - V_3 = \frac{\text{F.A.D.}}{\text{Speed of compressor } \times \eta_v}$$

$$= \frac{8.61}{450 \times 0.9238} = 0.0207 \text{ m}^3$$

Clearance volume $= c_1(V_1 - V_3) = c_1 V_s = 0.04 \times 0.0207$

$= 8285 \times 10^{-4} \text{ m}^3$

(f) H.P. Stage :

$$\eta_v = 1 + c_2 - c_2\left(\frac{p_3}{p_2}\right)^{\frac{1}{n}}$$

$$= 1 + 0.06 - 0.06(4)^{\frac{1}{1.3}} = 0.8857 \text{ or } 88.57\%$$

Swept volume, $$V_s = V_5 - V_7 = \frac{\text{F.A.D.}}{\text{Stage pressure ratio } \times N \times \eta_v}$$

$$= \frac{8.61}{4 \times 450 \times 0.8857} = 5.4 \times 10^{-3} \text{ m}^3$$

Clearance volume $= c_2 V_s = 0.06 \times 5.4 \times 10^{-3} = 324 \times 10^{-4} \text{ m}^3$

■ **Example 8.12** A reciprocating compressor is to be designed to deliver 4.5 kg/min of air from 100 kPa and 27°C to compress through an overall pressure ratio of 9. The law of comprssion is $pV^{1.3} = \text{const}$. Calculate the saving in power consumption and gain in isothermal efficiency, when a two-stage compressor with complete inter-cooling is used in place of a single-stage compressor. Assume equal pressure ratio in both the stages of the two-stage compressor. (Gas constant = 0.287 kJ/kg. K).

Solution : $m = 4.5$ kg/min, $p_1 = 100$ kPa, $T_1 = 273 + 27 = 300$ K, $r_p = 9$, $n = 13$

Work done in single-stage compression $$= \left(\frac{n}{n-1)}\right) mRT_1 \left[r_p^{\frac{n-1}{n}} - 1\right]$$

$$= \frac{13}{03} \times \frac{45}{60} \times 0.287 \times 300 \left[9^{\frac{0.3}{1.3}} - 1\right]$$

$= 5.54$ kW

$$\text{Work done in two-stage compression} = \left(\frac{2n}{n-1}\right) mRT_1 \left(r_p^{\frac{n-1}{2n}} - 1\right)$$

$$= \frac{2 \times 13}{03} \times \frac{45}{60} \times 0287 \times 300 \left[9^{\frac{0.3}{2.6}} - 1\right]$$

$$= 4.84 \text{ kW}$$

Saving in power = 5.57 – 4.84 = 0.70 kW

■ **Example 8.13** A double-acting single cylinder reciprocating air compressor has a piston displacement of 0.015 m^3 per revolution, oerates at 500 rpm and has 5% clearance. The air is received at 1 bar and delivered at 6 bar. The compression and expansion are polytropic with $n = 13$. Detemine (i) the volumetric efficiency, (ii) the power required, and (iii) the heat transferred and its direction, during compression and inlet temperature of air at 20°C.

Solution : $v_s = 0015 \text{ m}^3/\text{rev}$, $N = 500$ rpm, $c = v_c / v_s = 005$

$p_1 = 1$ bar, $p_2 = 6$ bar, $n = 13$, $T_1 = 273 + 20 = 293$ K

(i) Volumetric efficiency, $\eta_v = 1 + c - c\left(\frac{p_2}{p_1}\right)^{\frac{1}{n}}$

$$= 1 + 005 - 005\left(\frac{6}{1}\right)^{\frac{1}{1.3}} = 08516 \text{ or } 85.16\%$$

(ii) Power required, $P = \left[\frac{n}{n-1}\right] p_1 v_a \left[\left(\frac{p_2}{p_1}\right)^{\frac{n-1}{n}} - 1\right]$

$\eta_v = \frac{v_a}{v_s}$, $\qquad v_a = 0015 \times 08516 \times 500 \times 2$

$$= 12.774 \text{ m}^3/\text{min}$$

$$p = \left(\frac{13}{03}\right) \times 1 \times 10^2 \times 12.774 \times \left[\left(\frac{6}{1}\right)^{\frac{0.3}{1.3}} - 1\right]$$

$$= 283454 \text{ kJ/min or } 47.242 \text{ kW}$$

(iii) Swept volume, $V_s = v_s \times 2\text{ N} = 0015 \times 2 \times 500 = 15 \text{ m}^3/\text{min}$

Actual air sucked, $V_{as} = 15 \times 08516 = 12.774 \text{ m}^3/\text{min}$

Mass of air sucked,

$$\dot{m} = \frac{p_1 V_{as}}{RT_1} = 1 \times 10^5 \times 12.774/(287 \times 293) = 1519 \text{ kg/min}$$

$$T_2 = T_1 \left(\frac{p_2}{p_1}\right)^{\frac{(n-1)}{n}} = 293\left(\frac{6}{1}\right)^{\frac{0.3}{1.3}} = 443 \text{ K}$$

Heat transferred during compression,

$$Q_c = \dot{m}C_v\left[\frac{\gamma - n}{n-1}\right][T_2 - T_1]$$

$$= 15.19 \times 0.718 \times \left[\frac{(1.4 - 13)}{(13 - 1)}\right][443 - 293]$$

$$= 545321 \text{ kJ/min}$$

■ **Example 8.14** A single acting, two stage air compressor running at 300 rpm delivers air at 20 bar, while the intake conditions are 0.98 bar 305 K. The intermediate pressure is 5 bar and the clearance volume for low pressure compressor is 4 percent of its stroke volume. The compressor delivers 3 m^3/min at 1 bar and 15°C. Estimate

(i) Power required to drive the compresor in kW.

(ii) Low pressure cylinder dimensions, if $L = D$.

(iii) Isothermal efficiency when the intercooling is perfect and the index $n = 13$ for compression and expansion.

Solution : $N = 300$ rpm, $p_3 = 20$ bar, $p_2 = 5$ bar, $p_1 = 0.98$ bar, $T_1 = 305$ K,

$$r_c = 0.04,\ n = 13$$

$$T_2 = T_1\left(\frac{p_2}{p_1}\right)^{\frac{(n-1)}{n}} = 305\left(\frac{5}{0.98}\right)^{\frac{0.3}{1.3}} = 4442 \text{ K}$$

(i) Work input in the L.P. cylinder, (Fig. 8.18)

$$W_{LP} = \left[\frac{n}{n-1}\right]R(T_2 - T_1)$$

$$= \left(\frac{13}{03}\right) \times 0287(4442 - 305) = 173 \text{ kJ/kg}$$

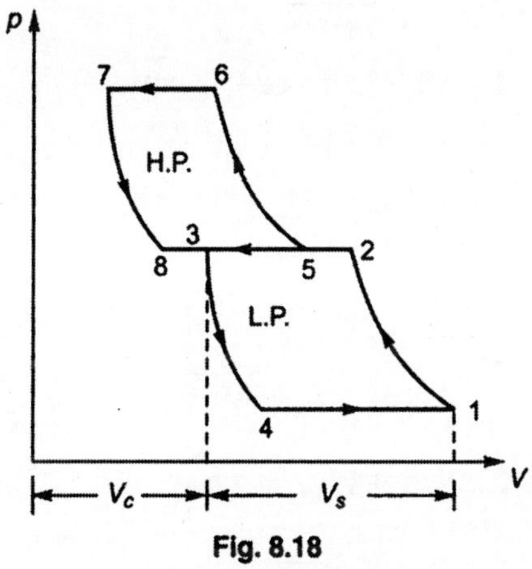

Fig. 8.18

For perfect intercooling

$$T_5 = T_1 = 305 \text{ K}$$

$$T_6 = T_5\left(\frac{p_3}{p_2}\right)^{\frac{(n-1)}{n}} = 305\left(\frac{20}{5}\right)^{\frac{0.3}{1.3}} = 419.94 \text{ K}$$

Work input in HP cylinder,

$$W_{HP} = \left(\frac{1.3}{0.3}\right) \times 0.287(419.94 - 305) = 142.95 \text{ kJ/kg}$$

Total work $= 173 + 142.95 = 315.95$ kJ/kg

$$\dot{m} = \frac{pV}{RT} = \frac{1\times 10^5 \times 3}{287\times 288\times 60} = 0.0605 \text{ kg/s}$$

Power required $= 315.95 \times 0.0605 = 19.115$ kW

(ii) Actual volume sucked,

$$V_1 = \frac{p_0 V_0 T_1}{T_0 p_1} = \frac{1\times 3\times 305}{288\times 0.98} = 3.24 \text{ m}^3/\text{min}$$

Volumetric efficiency, $\eta_v = 1 + c - c\left(\frac{p_2}{p_2}\right)^{\frac{1}{n}}$

$$= 1 + 0.04 - 0.04\left(\frac{5}{0.98}\right)^{\frac{1}{1.3}} = 0.899$$

Swept volume, $V_s = \frac{V_i}{\eta_v} = \frac{3.24}{0.899} = 3.604 \text{ m}^3/\text{min}$

$$V_s = \frac{\pi D_1^2 \times D_1 N}{4}$$

$$3.604 = \frac{\pi D_1^3 \times 300}{4}$$

$$D_1 = 0.2482 \text{ m or } 24.82 \text{ cm}$$

$$L = D_1 = 24.82 \text{ cm}$$

(iii) Isothermal work done $= RT_{atm} \ln\left(\frac{p_3}{P_{atm}}\right)$

$$= 0.287 \times 288 \times \ln\left(\frac{20}{1}\right)$$

$$= 247.615 \text{ kJ/kg}$$

Isothermal efficiency $= \frac{247.615}{315.95} = 0.7837$ or 37.37%

■ **Example 8.15** Find the optimum intermediate pressure of a two stage reciprocating compressor, if intercooling is done up to a temperature $T\omega$ which is greater than the inlet temperature.

Solution : Work done in the LP cylinder, (Fig. 8.19)

$$W_1 = \left[\frac{n}{n-1}\right].RT_1\left[\left(\frac{p_2}{p_1}\right)^{\frac{(n-1)}{n}} - 1\right]$$

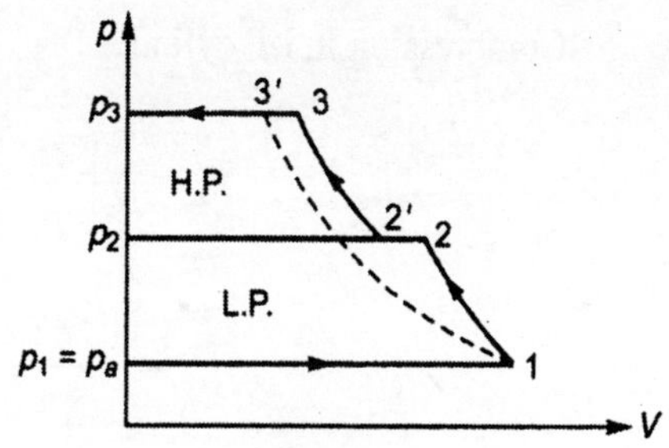

Fig. 8.19

Work done in the HP cylinder,

$$W_2 = \left[\frac{n}{n-1}\right].RT_2'\left[\left(\frac{p_3}{p_2}\right)^{\frac{(n-1)}{n}} - 1\right]$$

Total work done, $W = W_1 + W_2$

$$= \left[\frac{n}{n-1}\right]\cdot R \times \left[T_1\left(\frac{p_2}{p_1}\right)^{\frac{(n-1)}{n}} + T_2'\left(\frac{p_3}{p_2}\right)^{\frac{(n-1)}{n}} - T_1 - T_2'\right]$$

For maximum work, $$\frac{dW}{dp_2} = 0$$

$$\left[\frac{n-1}{n}\right]\frac{T_1 p_2^{-1/n}}{p_1^{(n-1)/n}} - T_2' \cdot p_3^{(n-1)/n}\left[\frac{n-1}{n}\right]p_2^{(2n-1)/n} = 0$$

or $$p_2 = [p_1 p_3]^{\frac{1}{2}}\left[\frac{T_2'}{T_1}\right]^{\frac{n}{(2n-2)}}$$

■ **Example 8.16** A two stage air compressor takes in air at 1.013 bar and 15°C and delivers at 43.4 bar. The intercooler pressure is 7.56 bar. The intercooling is perfect and the index of compression is 1.3. Calculate the work done in compressing 1 kg of air. If both cylinders have the same stroke and the piston diameters are 9 cm and 3 cm and the volumetric efficiency of the compressor is 90%, will the intercooler pressure be steady or will rise or fall as the compressor continues working?

Solution : $p_1 = 1.013$ bar, $T_1 = 273 + 15 = 288$ K,

$p_3 = 43.4$ bar, $p_2 = 7.56$ bar, $n = 1.3$,

$m = 1$ kg, $d_1 = 9$ cm, $d_2 = 3$ cm, $\eta_v = 0.9$

Input work per kg for perfect intercooling $T_3 = T_1$),

$$W = \left[\frac{n}{n-1}\right]RT_1\left[\left(\frac{p_2}{p_1}\right)^{\frac{(n-1)}{n}} + \left(\frac{p_3}{p_2}\right)^{\frac{(n-1)}{n}} - 2\right]$$

$$= \left(\frac{1.3}{0.3}\right) \times 0.287 \times 288\left[\left(\frac{7.56}{1.013}\right)^{\frac{0.3}{1.3}} + \left(\frac{43.4}{7.56}\right)^{\frac{0.3}{1.3}} - 2\right] = 389.3 \text{ kJ/kg}$$

Compression in LP cylinder :

$$v_2 = v_1\left(\frac{p_1}{p_2}\right)^{\frac{1}{n}} = \left(\frac{RT_1}{p_1}\right)\left(\frac{p_1}{p_2}\right)^{\frac{1}{n}}$$

$$= \left[\frac{0.287\times 288}{1.013\times 10^2}\right]\left(\frac{1.013}{7.56}\right)^{\frac{1}{1.3}} = 0.17386\ \text{m}^3/\text{kg}$$

$$T_2 = T_1\left(\frac{p_2}{p_1}\right)^{\frac{(n-1)}{n}} = 288\left(\frac{7.56}{1.0130}\right)^{\frac{0.3}{1.3}} = 457.97\ \text{K}$$

For constant pressure in the intercooler, the volume of air leaving the intercooler and entering the HP cylinder,

$$v_2' = \frac{v_2 T_3}{T_2} = 0.17386\times\frac{288}{457.97} = 0.10933\ \text{m}^3/\text{kg}$$

$$\left(\frac{v_1}{v_2}\right)_{\text{th}} = \frac{\left[\dfrac{0.287\times 288}{1.013\times 10^2}\right]}{0.17386} = 4.693$$

Ratio of effective cylinder volumes,

$$\frac{v_{1e}}{v_{2e}} = \left(\frac{v_{LP}}{v_{HP}}\right)\times\eta_v = \left(\frac{d_1}{d_2}\right)^2 \eta_v = \left(\frac{9}{3}\right)^2\times 0.9 = 0.81$$

As $\left(\dfrac{v_{1e}}{v_{2e}}\right) > \left(\dfrac{v_1}{v_2}\right)_{\text{th}}$, hence more air is supplied to HP cylinder than it can hold and consequently the pressure in the intercooler will rise.

■ **Example 8.17** The clearance volume of a single cylinder single acting air compressor is 10% of the swept volume. The compressor has a bore of 120 mm and a stroke of 152 mm and runs at 400 rpm. Suction conditions are 1 bar and 24°C and the delivery pressure is 10 bar. Compression and re-expansion curves follow the low $pV^{1.3} = \text{const}$. Calculate the diagram volumetric efficiency, the rate of air delivery, and the power input required to drive the compressor assuming a mechanical efficiency of 74%. Take $R = 0.287$ kJ/kg. K.

Solution : Clearance ratio, $c = \dfrac{V_c}{V_s} = 0.1, D = 0.12\ \text{m}, L = 0.152\ \text{m}, N = 400\ \text{rpm}$

$$p_1 = 1\ \text{bar},\ p_2 = 10\ \text{bar},\ T_1 = 273 + 24 = 297\ \text{K},\ n = 1.3,\ \eta_{\text{mech}} = 0.74$$

The p-V diagram for the compressor with clearance is shown in Fig. 8.20.

Volumetric efficiency, $\eta_v = 1 + c - c\left(\dfrac{p_2}{p_1}\right)^{\frac{1}{n}} = 1 + 0.1 - 0.1\,(10)^{\frac{1}{1.3}}$

$$= 0.5122 \text{ or } 51.22\%$$

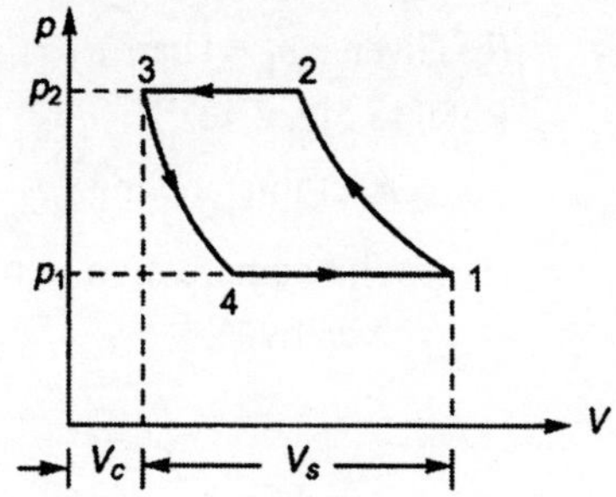

Fig. 8.20

Swept volume,

$$V_s = \frac{\pi}{4} D^2 L$$

$$= \frac{\pi}{4} (0.12)^2 \times 0.152 = 1.719 \times 10^{-3}\ \text{m}^3$$

$$V_1 = V_c + V_s = 1.1\, V_s = 1.1 \times 1.719 \times 10^{-3} = 1.891 \times 10^{-3}\ \text{m}^3$$

Volume flow rate at inlet,

$$\dot{V}_i = \frac{V_1 N}{60} = 1.891 \times 10^{-3} \times \frac{400}{60}$$

$$= 12.607 \times 10^{-3}\ \text{m}^3/\text{s}$$

Specific volume at inlet,

$$v_s = \frac{RT_1}{p_1} = \frac{287 \times 297}{1 \times 10^5} = 0.85239\ \text{m}^3/\text{kg}$$

Mass flow rate of air at inlet,

$$\dot{m} = \frac{\dot{V}_1}{v_s} = \frac{12.607 \times 10^{-3}}{0.85239} = 0.01479\ \text{kg/s}$$

Theoretical Work done on the compressor,

$$W_{th} = \left(\frac{n}{n-1}\right) p_1 V_a \left[\left(\frac{p_2}{p_1}\right)^{\frac{n-1}{n}} - 1\right]$$

Now

$$V_a = V_1 - V_4$$

$$= \eta_v (V_1 - V_c) = \eta_v V_s$$

$$W_{th} = \frac{1.3}{0.3} \times 1 \times 10^5 \times 1.719 \times 10^{-3} \times 0.5122 \left[10^{\frac{0.3}{1.3}} - 1\right]$$

$$= 267.55\ \text{J}$$

Actural work done,

$$W_a = \frac{W_h}{\eta_{mech}} = \frac{267.55}{0.74} = 361.55\ \text{J}$$

SOLVED QUESTIONS FROM UNIVERSITY EXAMINATIONS

■ **Example 8.18**

(a) Derive an expression for the work done for a two stage reciprocating air compressor.

(b) Find the percentage saving in work done by compressing air in two stages from 1 bar to 7 bar instead of one stage. Assume compression index 1.35 in both the cases and optimum pressure and complete enter cooling in two-stage compression. [KU., Dec., 2005]

Solution : (a) Refer to Art. 8.7.1.

(b) Given : $p_1 = 1$ bar, $p_3 = 7$ bar, $n_1 = n_2 = 1.35$

Refer to Fig. 8.10.

For perfect intercooling, $p_2 = \sqrt{p_1 p_3} = \sqrt{1 \times 7} = 2.646$ bar and $T_3 = T_1$

(i) With compression in one stage (see Fig. 8.6) without clearance, work done per kg of air is given by,

$$W_1 = \left(\frac{n}{n-1}\right) RT_1 \left[\left(\frac{p_3}{p_1}\right)^{\frac{n-1}{n}} - 1\right] \qquad \text{...(1)}$$

(ii) Work done with perfect intercooling,

$$W_2 = \left(\frac{2n}{n-1}\right) RT_1 \left[\left(\frac{p_3}{p_1}\right)^{\frac{n-1}{2n}} - 1\right] \qquad \text{...(2)}$$

$$\text{Percentage saving in work} = \frac{W_1 - W_2}{W_1} = 1 - \frac{W_2}{W_1}$$

$$= 1 - 2\left[\frac{\left(\frac{p_3}{p_1}\right)^{\frac{n-1}{2n}} - 1}{\left(\frac{p_3}{p_1}\right)^{\frac{n-1}{n}} - 1}\right]$$

$$= 1 - 2\left[\frac{\left(\frac{7}{1}\right)^{\frac{0.35}{2.70}} - 1}{\left(\frac{7}{1}\right)^{\frac{0.35}{1.35}} - 1}\right]$$

$$= 1 - 2\left(\frac{0.2869}{0.6561}\right)$$

$$= 0.1255 \text{ or } 12.55\%$$

■ Examploe 8.19

(a) Derive an expression for the volumetric efficiency of a reciprocating compressor.

(b) Estimate the work required for a two stage reciprocating single acting air-compressor to compress 2.8 m^3 of air per min at 1.05 bar and 10°C to a final pressure of 35 bar. The intermediate receiver cools the air to 30°C and 5.6 bar pressure. For air take $\gamma = 1.4$.

[KU, Dec., 2006]

Solution : (a) Refer to Art. 8.6.2.

(b) Given : $V_1 = 2.8\ m^3/\text{min}$, $p_1 = 1.05$ bar,

$T_1 = 273 + 10 = 283$ K, $p_3 = 35$ bar, $T_3 = 30 + 273 = 303$ K, $p_2 = 5.6$ bar, $\gamma = 1.4$

Refer to Fig. 8.10.

$$p_1\dot{V}_1 = \dot{m}RT_1$$

$$\dot{m} = \frac{p_1\dot{V}_1}{RT_1} = \frac{1.05\times 10^5 \times 2.8}{0.287\times 10^3 \times 283}$$

$$= 3.62 \text{ kg/min or } 0.06 \text{ kg/s}$$

Work done on air per kg of air for a two-stage compressor is given by,

$$W_t = \left(\frac{\gamma}{\gamma-1}\right)RT_1\left[\left(\frac{p_2}{p_1}\right)^{\frac{\gamma-1}{\gamma}} - 1\right] + \left(\frac{\gamma}{\gamma-1}\right)RT_3\left[\left(\frac{p_3}{p_2}\right)^{\frac{\gamma-1}{\gamma}} - 1\right]$$

$$= \left(\frac{1.4}{0.4}\right)\times 287\times 283\left[\left(\frac{5.6}{1.05}\right)^{\frac{0.4}{1.4}} - 1\right] + \left(\frac{1.4}{0.4}\right)\times 287\times 303\left[\left(\frac{35}{5.6}\right)^{\frac{0.4}{1.4}} - 1\right]$$

$$= 174343.66 + 209427.93 = 383771.59 \text{ J/kg}$$

Total work done on air, $\dot{W}_t = \dot{m}\times W_t$

$$= 0.06\times 383771.59 = 230263 \text{ J/s or } 23.026 \text{ kW}$$

■ **Example 8.20**

(a) Prove that the condition for maximum work required for a two stage reciprocating compressor is

$$p_2 = \sqrt{P_1 P_3}$$

(b) A 2-stage single acting reciprocating compressor, takes in air at the rate of 0.2 m^3/sec. The intake pressure and temperature of air are 0.1 MPa and 16°C. The air is compressed to a final pressure of 0.7 MPa. The intermediate pressure is ideal and intercooling is perfect. The compression index in both the stages is 1.25 and compressor runs at 600 rpm. Neglect the clearance, determine :

(i) Intermediate pressure

(ii) Total volume of each cylinder

(iii) The power required to derive the compressor

(iv) Rate of heat rejection in intercooler.

Take $c_p = 1.005, R = 287$ J/kg K [KU, Dec., 2007]

Solution : (a) Refer to Art. 8.7.1

(b) Given : $\dot{V}_1 = 0.2 \text{ m}^3/\text{s},\ p_1 = 0.1 \text{ mPa},\ T_1 = 273 + 16 = 289 \text{ K},$

$$p_3 = 0.7 \text{ mPa},\ n = n_1 = n_2 = 1.25,\ N = 600 \text{ rpm},$$

$$c_p = 1.005 \text{ kJ/kg. K},\ R = 287 \text{ J/kg. K}$$

Refer to Fig. 8.10. For perfect intercooling, total work done per kg of air,

$$w_t = \left(\frac{2n}{n-1}\right)RT_1\left[\left(\frac{p_3}{p_1}\right)^{\frac{n-1}{2n}} - 1\right]$$

$$= \left[\frac{2\times 1.25}{1.25-1}\right]\times 2.87\times 2.89\left[\left(\frac{0.7}{0.1}\right)^{\frac{1.25-1}{2\times 1.25}}-1\right]$$

$$= 178173.2 \text{ J/kg}$$

$$p_1\dot{V}_1 = \dot{m}RT_1$$

$$\dot{m} = \frac{0.1\times 10^6\times 0.2}{287\times 289} = 0.241 \text{ kg/s}$$

(i) Intermediate pressure,

$$p_2 = \sqrt{p_1 p_3} = \sqrt{0.1\times 0.7} = 0.26457 \text{ MPa}$$

(ii) Compression pressure 1-2 :

$$p_1\dot{V}_1^n = p_2\dot{V}_2^n$$

$$\dot{V}_2 = \dot{V}_1\left(\frac{p_1}{p_2}\right)^{\frac{1}{n}} = 0.2\left(\frac{0.1}{0.26457}\right)^{\frac{1}{1.25}} = 0.0918 \text{ m}^3/\text{s}$$

Constant pressure process 2-3 :

$$\frac{p_2\dot{V}_2}{T_2} = \frac{p_3\dot{V}_3}{T_3}$$

For perfect intercooling, $T_3 = T_1$ and $p_2 = p_3$

$$\therefore \qquad \dot{V}_3 = \dot{V}_2\left(\frac{T_1}{T_2}\right)$$

$$\frac{T_2}{T_1} = \left(\frac{p_2}{p_1}\right)^{\frac{n-1}{n}} = \left(\frac{0.26457}{0.1}\right)^{\frac{0.25}{1.25}} = 1.2148$$

$$T_2 = 1.2148\times 289 = 351 \text{ K}$$

$$\therefore \qquad \dot{V}_3 = 0.0918\left(\frac{289}{351}\right) = 0.0756 \text{ m}^3/\text{s}$$

(iii) Power required to drive the compressor,

$$P = \frac{\dot{m}w_t N}{60\times 10^3} = \frac{0.241\times 178173.2\times 600}{10^3\times 60} = 429.4 \text{ kW}$$

(iv) Rate of heat rejection in intercooler $= \dot{m}c_p\ (T_2 - T_3)$

$$= 0.241\times 1.005\ (351-289) = 15 \text{ kJ/s}$$

■ **Example 8.21**

(a) Prove that the volumetric efficiency of a reciprocating compressor is given by

$$\eta_v = \frac{p_1 T_a}{p_a T_1}\left[1+c-c\left(\frac{p_2}{p_1}\right)^{\frac{1}{n}}\right], \text{ where } c = \text{clearance ratio.}$$

(b) A single stage double acting air compressor delivers 3 m^3 of free air/min at 1.013 bar and 20°C to 8 bar with following data :
RPM = 300, $\eta_{mech} = 0.9$, pressure loss in passing through intake valve = 0.04 bar, temperature rise of air during suction stroke = 12°C, clearance volume = 5% of stroke volume. Index of compression and expression = 1.35, length of stroke = 1.2 times of the cylinder diameter. Calculate : (a) Volumetric efficiency, (b) cylinder diamension, (c) indicated power, and (d) isothermal efficiency of the compressor. Take for air, $R = 0.287$ kJ/kg.K. [KU, Jan., 2008]

Solution : (a) Refer to Art. 8.6.2.

(b) Given : $V_a = 3\ m^3/min$, $p_a = 1.013$ bar,

$$T_a = 273 + 20 = 293\ K,\ p_1 = 1.013 - 0.04 = 0.973\ bar,$$
$$T_1 = T_a + 12 = 293 + 12 = 305\ K,\ V_c = 0.05\,V_s,$$
$$n = 1.35,\ L = 1.2D,\ R = 0.287\ kJ/kg.K,$$
$$N = 300\ rpm$$

The p-V diagram is shown in Fig. 8.21.

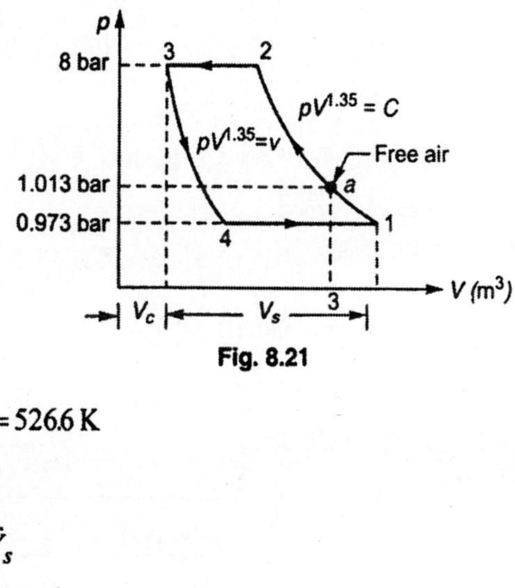

Fig. 8.21

Mass of free air delivered,

$$\dot{m} = \frac{p_1 \dot{V}_a}{RT_a} = \frac{1.013 \times 10^5 \times 3}{0.287 \times 10^3 \times 293} = 3.614\ kg/min$$

Compression process 1-2 :

$$\frac{T_2}{T_1} = \left(\frac{p_2}{p_1}\right)^{\frac{n-1}{n}}$$

$$T_2 = 305\left(\frac{8}{0.973}\right)^{\frac{0.35}{1.35}} = 526.6\ K$$

Expansion process 3-4 :

$$\dot{V}_3 = \dot{V}_c = 0.05\,\dot{V}_s$$

$$\frac{\dot{V}_4}{\dot{V}_3} = \left(\frac{p_3}{p_4}\right)^{\frac{1}{n}},\ \dot{V}_4 = 0.05\,\dot{V}_s\left(\frac{8}{0.973}\right)^{\frac{1}{1.35}} = 0.238\,\dot{V}_s$$

$$\dot{V}_1 - \dot{V}_4 = (\dot{V}_c + \dot{V}_s) - \dot{V}_4 = (0.05\,\dot{V}_s + \dot{V}_s) - 0.238\,\dot{V}_s = 0.812\,\dot{V}_s$$

Corresponding value of F.A.D. per cylce,

$$\dot{V} = \frac{p_1 T_a}{p_a T_1}(\dot{V}_1 - \dot{V}_4) = \frac{0.973 \times 293}{1.013 \times 305} \times 0.812\,\dot{V}_s = 0.749\,\dot{V}_s$$

(a) Volumetric efficiency,

$$\eta_{cool} = \frac{\dot{V}}{\dot{V}_s} = \frac{0.749\,\dot{V}_s}{\dot{V}_s} = 0.749 \text{ or } 74.9\%$$

(b) Volume of air inhaled per cycle

$$= \frac{V_a}{2 \times N} = \frac{3}{2 \times 300} = 0.005 \text{ m}^3\text{/cycle}$$

Stroke volume, $V_s = \frac{\pi}{4} D^2 L = \frac{\pi}{4} D^2 \times 1.2D = 0.94248 D^3 \text{ m}^3\text{/cycle}$

Also stroke volume $= \frac{0.005}{\eta_{cool}} = \frac{0.005}{0.749} = 6.6756 \times 10^{-3} \text{ m}^3\text{/cycle}$

$$\therefore \qquad 0.94248 D^3 = 6.6756 \times 10^{-3}$$

$$D = 0.192 \text{ m or } 192 \text{ mm}$$

$$L = 1.2 \times 192 = 230.4 \text{ mm}$$

(c) Indicated power, $IP = \left(\frac{n}{n-1}\right) \dot{m} R\,(T_2 - T_1) \times \frac{1}{60}$

$$= \frac{1.35}{0.35} \times \frac{3.614}{60} \times 0.287(526.6 - 305)$$

$$= 14.776 \text{ kW}$$

(d) Isothermal indicated power,

$$(IP)_{isoth} = \dot{m} R T_1 \ln\left(\frac{p_2}{p_1}\right)$$

$$= \frac{3.614}{60} \times 0.287 \times 305 \ln\left(\frac{8}{0.973}\right) = 11.108 \text{ kW}$$

Isothermal efficiency,

$$\eta_{isoth} = \frac{11.108}{14.776} = 0.7517 \text{ or } 75.17\%$$

SHORT ANSWER TYPE QUESTIONS

1. List at least six uses of compressed air in industry.
2. Explain the working of a single stage reciprocating air compressor.
3. What do you mean by free air delivered (FAD)?
4. Define volumetric efficiency.
5. What are the factors on which volumetric effciency depends ?
6. What are the methods generally adoped for approximating to isothermal compression?
7. Define clearance ratio. How the volumetric efficiency is dependent on clearance ratio?
8. Define isothermal efficiency and compressor efficiency.
9. Differentiate between isothermal power and indicated power.
10. Define adiabatic efficiency.
11. What are the advantages of multi-stage compression?

12. What is the condition for minimum compressor work with perfect-inter cooling?
13. Distinguish between the function of inter-cooler and after-cooler.
14. What is an air motor?
15. How heat is rejected in air compressor?

EXERCISES

1. A single-stage single-acting compressor delivers 15 m^3 of free air per minute from 1 bar to 8 bar at 300 rpm. Assuming n = 1.3 for both compression and expansion and clearance $\frac{1}{16}$th of swept volume, find the diameter and stroke of the compressor. Take L = 1.5 D. **[Ans.** 383 mm, 587.5 mm]
2. A single-stage double-acting air compressor delivers air at 7 bar. The pressure and temperature at the end of suction stroke are 1 bar and 27°C. It delivers 2 m^3 of free air per minute when the compressor is running at 300 rpm. The clearance volume is 5% of the stroke volume. The ambient pressure and temperature are 1.03 bar and 20°C. Index of compression and expansion is 1.30 and 1.35 respectively. Calculate (a) the volumetric efficiency, (b) indicated power and brake power if mechanical effciency is 80%, and (c) diameter and stroke of the cylinder if both are equal. **[Ans.** 83.9%, 8.5 kW, 10.64 kW, 174.6 mm]
3. A two-stage single-acting air compressor delivers air at 20 bar. The pressure and temperature of air before compression in L.P. cylinder are 1 bar and 27°C. The discharge pressure of L.P. cylinder is 4.7 bar. The pressure of air leaving the intercooler is 4.5 bar and the air is cooled to 27°C. The diameter and stroke of L.P. cylinder are 0.4 m and 0.5 m respectively. The clearance volume is 4% of stroke in both cylinders. The speed of compressor is 200 rpm. The index of compression and re-expansion in both cylinders is 1.3. Determine (a) the indicated power to run the compressor, and (b) the heat rejected to inter-cooler per minute. **[Ans.** 68.9 kW, 1716 kJ/min]
4. A single-stage, double-acting air compressor delivers 15 m^3 of air per minte measured at 1.013 bar, 27°C and delivers at 7 bar. At the end of the suction stroke the pressure and temperature are 0.98 bar and 40°C. The clearance volume is 4% of the swept volume and the stroke is 1.3 times the bore. The compressor runs at 300 rpm. Calculate (a) the volumetric effciency, (b) cylinder dimensions, (c) indicated power and (d) isothermal efficiency. n = 1.3 for both compression and expansion. R = 0.287 kJ/kg.K **[Ans.** 79.6%; 313.3 mm, 407.3 mm; 65.83 kW; 78.92%]
5. A single-stage, double acting air compressor takes air at 0.98 bar and 32°C and delivers at 6.32 bar. The clearance is 5% of the stroke volume. The compression and expansion follow the law $pv^{1.32} = c$. The air handled by the compressor is 17 m^3 min when measured at 1 bar and 15°C. Determine the temperature of air delivered, stroke volume and indicated power of compressor if it turns at 500 rpm. Neglect the area of piston and take R = 0.287 kJ/kg.K.
6. The free air delivered of a single cylinder single stage reciprocating air compressor is 2.5 m^3/min. The ambient air is at 0°C and 1.013 bar and delivery pressure is 7 bar. The clearance ratio is 5% and law of compression and expansion is $pv^{1.25} = c$. If L = 1.2 D and the compressor runs at 150 rpm, determine the size of the cylinder. **[Ans.** 275.5 mm, 330 mm]
7. A single stage double acting air compressor delivers air at 7 bar. The amount of free air delivered is 2 m^3 at 300 rpm. The pressure and temperature at the end of suction stroke are 1 bar and 27°C. The ambient conditions are 1.03 bar and 20°C. The clearance is 5% stroke. Dertermine (a) the brake power required to run

the compressor if the mechanical efficiency is 80%, and (b) the diameter and stroke of the cylinder if both are equal.

Assume compression and re-expansion follow the law $pv^{1.3}$ = const. **[Ans.** 10.8 kW, 179.5 mm]

8. A two-stage double acting air compressor delivers air at the rate of 1.35 kg/s. The suction pressure is 1 bar and inter-stage pressure is 7 bar and delivery pressure 42 bar. Air enters the L.P. cylinder at 17°C and cooled in the inter cooler to 32°C. The clearances in L.P. and H.P. cylinders are 6% and 8% of respective strokes. The law of compression and re-expansion is $pv^{1.21} = c$ in both cylinders. Speed of compressor is 500 rpm. Calculate (a) the amount of cooling water required per minute in inter-cooler, if rise in temperature of water is limited to 20°C, (b) power required, and (c) size of cylinder if length of stroke and bore are equal.
[Ans. 2259 kg/min, 1068 kW, 483 mm]

9. Determine the size of a cylinder for a double acting air compressor of 37 kW, in which air is drawn in at 1 bar, 15°C and compressed according to the law $pV^{1.2}$ = const. to 6 bar. The compressor runs at 100 rpm with average piston speed of 152.5 m/min. Neglect clearance. **[Ans.** 298.5 mm, 762.5 mm]

10. A single stage double acting air compressor handles 17 m^3/min of air, measured at 1 bar and 15°C. The pressure and temperature at the end of suction are 0.98 bar and 32°C. The air is delivered at 6.325 bar. Assuming a clearance factor of 5% and the compression and expansion processes to follow the law $pV^{1.32}$ = const., determine the stroke volume of the compressor running at 500 rpm. Also calculate the indicated power of the compressor. Take $R = 0.287$ kJ/kg.K. **[Ans.** 0.2175 m^3, 70.73 kW]

11. A single acting compressed air motor works on compressed air at 10.5 bar and 37°C supplied at the rate of 1 kg/min. Cut-off takes place at 25% of the stroke and the expansion follows the adiabatic frictionless law down to 1.01325 bar. Determine the mean effective pressure, the indicated power and the cylinder volume if the machine runs at 250 rpm. Neglect clearance. **[Ans.** 3.37 bar, 1.94 kW, 0.001383 m^3]

12. The cylinder of an air motor has a bore of 63.5 mm and a stroke of 114 mm. The supply pressure and temperature are 6.3 bar and 24°C and exhaust pressure is 1.013 bar. The clearance volume is 5% of the swept volume and the cut-off ratio is 0.5. The air is compressed by the returning piston after it has travelled through 0.95 of its stroke. The law of compression and expansion is $pV^{1.3}$ = const. Calculate the temperature at the end of expansion and the indicated power of the motor which runs at 300 rpm. Also calculate the air supplied per minute. Take $R = 0.287$ kJ/kg.K. **[Ans.** 244.4 K, 0.746 kW, 0.418 kg/min]

13. A water cooled air compressor requires a work input of 200 kN.m/kg of air delivered. The enthalpy of air leaving the compressor is 75 kJ/kg greater than that entering. Heat lost to the cooling water is 105 kJ/kg. From the first law analysis estimate the heat lost by compressor to atmosphere.

14. A two-stage single acting air compressor equipped with an inter-cooler draws air at 1.013 bar and 30°C and delivers it at 20 bar. The mass of air delivered by the compressor is 10 kg/min. The ratio of clearance volume to swept volume in the low pressure and high pressure cylinders is 0.03 and 0.04 respectively. If the compressor has a mechanical efficiency of 82% and it operates under maximum efficiency conditions, determine :
(a) power supplied at the shaft of the compressor;
(b) isothermal efficiency;
(c) ratio of cylinder diameters for identical stroke length; and
(d) heat rejected in the inter-cooler.
Assume $R = 0.287$ kJ/kg.K, $c_p = 1.005$ kJ/kg.K and adiabatic index = 1.4.

15. A single acting two-stage reciprocating air compressor is to compress air from 1 bar and 30°C to 12 bar. The base of the low pressure cylinder is 30 cm. The stroke length of both the low and the high pressure cylinders is the same and is equal to 40 cm. The compressor runs at 180 rpm. The clearance volume in both the cylinders is

3% of the stroke volume. Index of adiabatic compression and expansion is 1.3 in both the cylinders. Determine the shaft power required to drive the compressor when (i) the air is cooled to its original temperature before entering the H.P. cylinder, (ii) when the air is cooled to 45°C in the inter-cooler. Assume mechanical efficiency to be 85% in both cases. Also assume $R = 0.287$ kJ/kg.K.

16. A multi-stage compressor has to be designed to supply air at 135 bar, while atmospheric condition is 1.03 bar and 15°C. The value of compression index may be assumed as 1.35. Due to practical reasons the intercoolers are not able to cool the air below 45°C, while the maximum temperature allowable in the system is 120°C. Calculate the number of stages that are necessary in the compression and the rate of cooling water circulated per kg of air. Take $c_p = 1$ kJ/kg.K.

17. An air compressor has eight stages of equal pressure ratio 1.35. The flow rate through the compressor and its overall efficiency are 50 kg/s and 82% respectively. If the air enters the compressor at a pressure of 1.0 bar and temperature of 313 k, determine :
    (i) State of air at the exit of the compressor
    (ii) Polytropic or small stage efficiency
    (iii) Efficiency of each stage
    (iv) Power required to drive the compressor assuming the overall efficiency as 90%.

❑❑❑

Chapter 9

Rotary Compressors

Inside This Chapter

9.1 CLASSIFICATION OF COMPRESSORS

Air compressors may be classified as shown in Fig. 9.1.

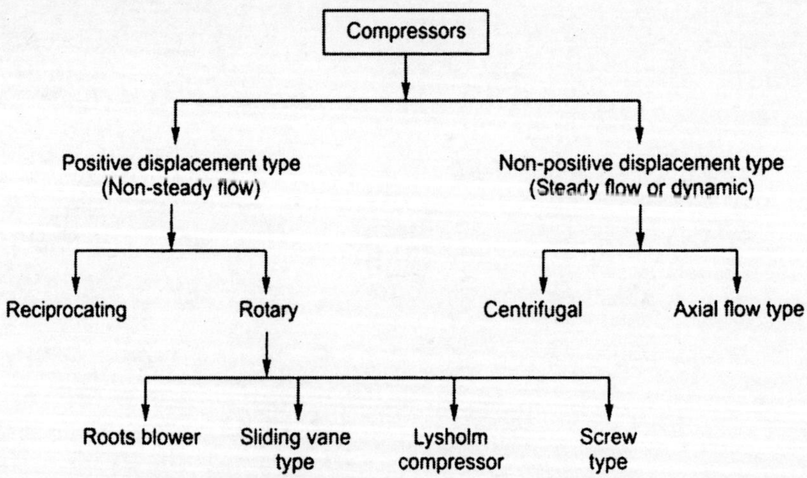

Fig. 9.1 Classification of compressors.

In a rotary compressor, the compression of air is achieved due to the rotating blades fitted in a rotor.

9.2 POSITIVE DISPLACEMENT ROTARY COMPRESSORS

In the positive displacement type rotary compressors, the air is compressed by entrapping it between the reduced space of two sets of engaging surfaces. The pressure rise is either by back flow of air, as in the case of roots blower, or by both squeezing action and back flow of air, as in the case of vane type.

1. Roots Blower

The schematic arrangement of a roots blower is shown in Fig. 9.2(a). It consists of two rotors which are driven externally. One rotor is connected to the driver and the second rotor is gear driven from the first in the opposite direction. The lobes of the rotors are of epicycloid, hypercycloid or involute profiles, to ensure a seal between high pressure region and the low pressure region, at all angular positions. A small clearance provided between the rotors are the cylinder surface to reduce wear reduces the efficiency of the compressor due to leakage.

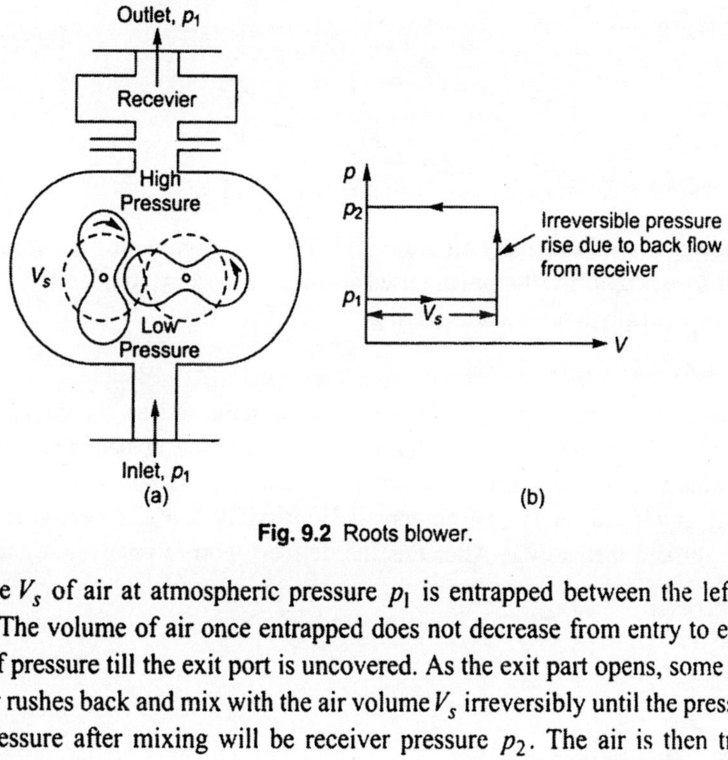

Fig. 9.2 Roots blower.

The volume V_s of air at atmospheric pressure p_1 is entrapped between the left hand rotor and cylinder casing. The volume of air once entrapped does not decrease from entry to exit and therefore there is no rise of pressure till the exit port is uncovered. As the exit part opens, some high pressure air from the receiver rushes back and mix with the air volume V_s irreversibly until the pressure is equalised. The resulting pressure after mixing will be receiver pressure p_2. The air is then transferred to the receiver. This happens four times in one revolution in case of two lobed rotor and six times in case of three lobed rotor. The p–V diagram for the roots blower is shown in Fig. 9.2(b).

Work done per revolution, $$W = 2i\,(p_2 - p_1)\,V_s \quad \text{...(9.1)}$$

where i = number of lobes

Let V_1 = number handled per minute at p_1 and T_1.

Then workdone per minute is,

$$W_{\text{roots}} = (p_2 - p_1)\,V_1 \quad \text{...(9.2)}$$

If the compression is isentropic then ideal work required per minute is,

$$W_{\text{ideal}} = \left(\frac{\gamma}{\gamma-1}\right) p_1 V_1 \left[\left(\frac{p_2}{p_1}\right)^{(\gamma-1)/\gamma} - 1\right] \quad \text{...(9.3)}$$

Roots efficiency, $$\eta_{\text{roots}} = \frac{W_{\text{ideal}}}{W_{\text{roots}}} = \frac{\left(\frac{\gamma}{\gamma-1}\right) p_1 V_1 \left[\left(\frac{p_2}{p_1}\right)^{(\gamma-1)/\gamma} - 1\right]}{(p_2 - p_1) V_1}$$

$$= \frac{\left(\frac{\gamma}{\gamma-1}\right)[r_p^{(\gamma-1)/\gamma} - 1]}{r_p - 1} \quad \text{...(9.4)}$$

where $r_p = \frac{p_2}{p_1}$, is the pressure ratio.

Also $$\frac{\gamma}{\gamma-1} = \frac{c_p}{R}$$

$\therefore$ $$\eta_{\text{roots}} = \frac{c_p}{R}\left[\frac{r_p^{(\gamma-1)/\gamma} - 1}{r_p - 1}\right] \quad \text{...(9.5)}$$

The roots blower is used to supply air from 0.15–1500 m^3/mm with pressure ratio upto 3.6 per blower. Rotational speeds upto 12500 rpm are used and can be directly coupled to a steam turbine or a gas turbine shafts without any intermediate gearing.

2. Vane Type Blower

The schematic diagram of a vane type blower is shown in Fig. 9.3(a). It consists of a rotor located eccentrically in a cylindrical outer casing. The rotor carries a set of spring loaded vanes in the slots of the rotor. The air of volume V_1 at atmospheric pressure p_1 is entrapped between two vanes. As the rotation proceeds CCW the entrapped air is first compressed reversibly to V_a. Afterwards the compression continues to V_c (say) and then to V_d. After this the delivery port is uncovered and the irreversible

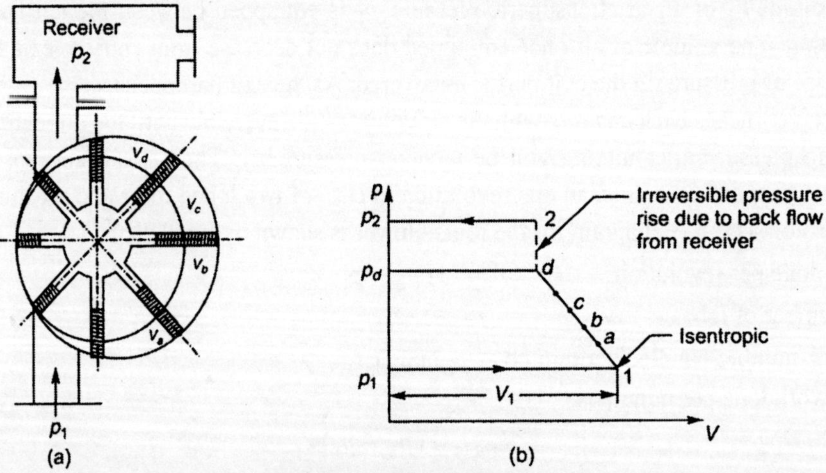

Fig. 9.3 Vane type blower.

compression takes place from pressure p_d to p_2. The p-V diagram for the compression process is shown in Fig. 9.3(b). The workdone per revolution with i vanes is given by :

$$W = i\left(\frac{\gamma}{\gamma-1}\right)p_1V_1\left[\left(\frac{p_d}{p_1}\right)^{(\gamma-1)/\gamma}-1\right]+i\,(p_2-p_d)\,V_d \qquad \text{...(9.6)}$$

The vane blowers require less power than the root blowers for the same capacity and pressure rise. They are used to deliver upto 150 m^3/min of air at pressure ratio upto 8.5. The speed is limited to 3000 rpm.

3. Lysholm Compressor

The Lysholm compressor is shown in Fig. 9.4. Its principle of working is similar to that of the roots blower. The air is admitted at one end of the compressor and trapped between helical rotors and the casing. The screw action of the rotors displaces the air axially. It produces constant compression internally. Its main disadvantages is mechanical complexity.

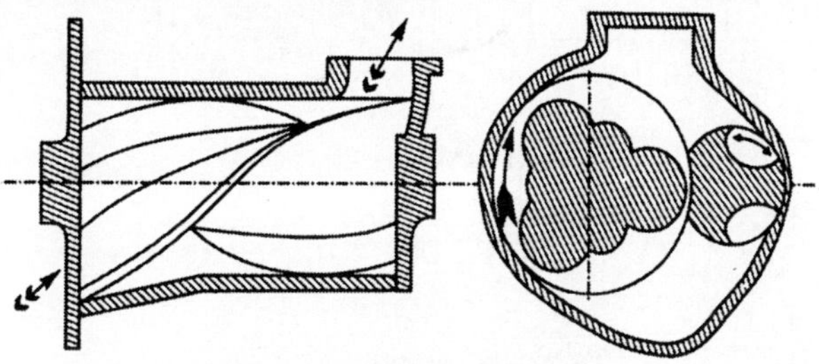

Fig. 9.4 Lysholm compressor.

4. Screw Compressor

The screw may be single helical or double helical. The advantage of double helical is that they are balanced axially. The air is carried forward to the discharge along the rotor in pockets formed between teeth and the casing. Such a compressor is shown in Fig. 9.5. The principle of working is similar to that of the roots blower.

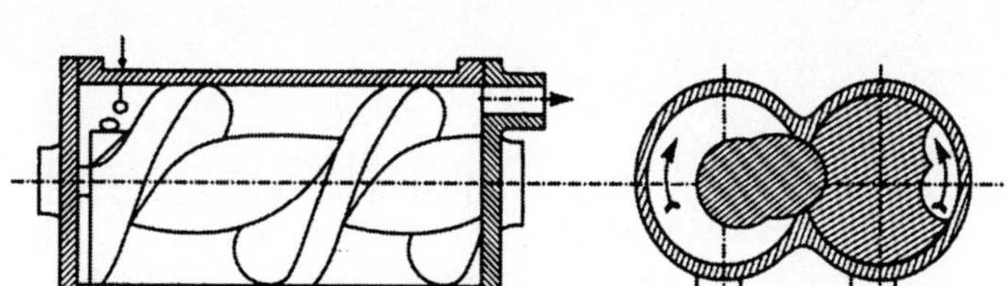

Fig. 9.5 Screw compressor.

■ **Example 9.1** Free air of 25 m^3/min is compressed from 1 bar to 2.5 bar. Calculate the indicated power required if the compression is carried out in (a) roots blower, and (b) vanes blower. Assume that there is 20% of reduction in volume before the back flow occurs. Also calculate the isentropic efficiency in each case.

Solution : The p-V diagrams for roots blower and vanes blower are shown in Fig. 9.6(a) and (b) respectively.

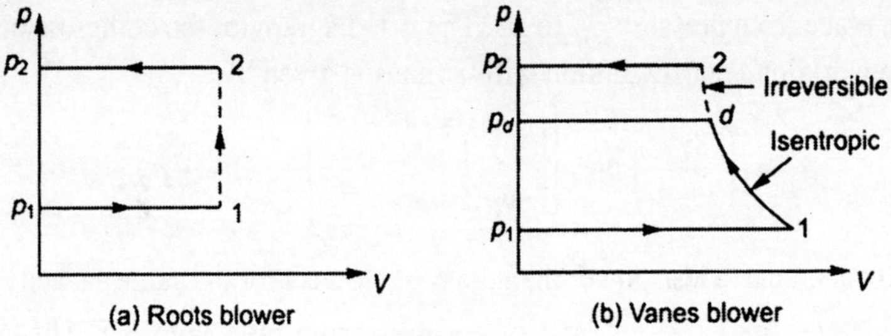

Fig. 9.6

(a) Roots blower :

$$\text{I.P.} = \frac{(p_2 - p_1)V_1}{60 \times 10^3} = \frac{(2.5 - 1) \times 10^5 \times 25}{60 \times 10^3} = 62.5 \text{ kW}$$

(b) Vanes blower :

$$\text{I.P.} = \frac{1}{10^3 \times 60}\left[\left(\frac{\gamma}{\gamma - 1}\right) p_1 V_1 \left\{\left(\frac{p_d}{p_1}\right)^{(\gamma - 1)/\gamma} - 1\right\} + (p_2 - p_1)V_d\right]$$

$$\frac{p_d}{p_1} = \left(\frac{V_1}{V_d}\right)^{\gamma} - \left(\frac{V_1}{0.8V_1}\right)^{\gamma}$$

$$= \left(\frac{1}{0.8}\right)^{1.4} = 1.367$$

$$p_d = 1 \times 1.367 = 1.367 \text{ bar}$$

$$\text{I.P.} = \frac{1}{10^3 \times 60}\left[\frac{1.4}{0.4} \times 1 \times 10^5 \times 25 \{(1.367)^{0.4/1.4} - 1\} + (2.5 - 1.367) \times 10^5 \times 0.80 \times 25\right]$$

$$= \frac{1}{10^3 \times 60}[8.175 \times 10^5 + 22.66 \times 10^5] = 51.39 \text{ kW}$$

I.P. if whole compression is carried out isentropically,

$$\text{I.P.} = \frac{1}{10^3 \times 60}\left[\left(\frac{\gamma}{\gamma - 1}\right) p_1 V_1 \left\{\left(\frac{p_2}{p_1}\right)^{(\gamma - 1)/\gamma} - 1\right\}\right]$$

$$= \frac{1}{10^3 \times 60}\left[\frac{1.4}{0.4} \times 1 \times 10^5 \times 25 \left\{\left(\frac{2.5}{1}\right)^{0.4/1.4} - 1\right\}\right]$$

$$= 43.64 \text{ kW}$$

$$\text{Isentropic efficiency of roots blower} = \frac{43.64}{62.5} \times 1 = 69.83\%$$

$$\text{Isentropic efficiency of vanes blower} = \frac{43.64}{51.39} \times 100 = 84.92\%$$

■ **Example 9.2** A rotary vanes blower works between the pressure limits of 1 bar and 1.8 bar and gives 5 m^3/min of free air delivered when running at 240 rpm. Determine the power required to drive the blower when (a) ports are so placed that there is no internal compression, and (b) when the ports are so placed that there is 50% pressure rise due to internal adiabatic compression before back flow occurs. Assuming mechanical efficiency to be 98%, calculate the blower efficiency.

Solution : The *p-V* diagrams without and with internal compression are shown in Fig. 9.7(a) and (b) respectively.

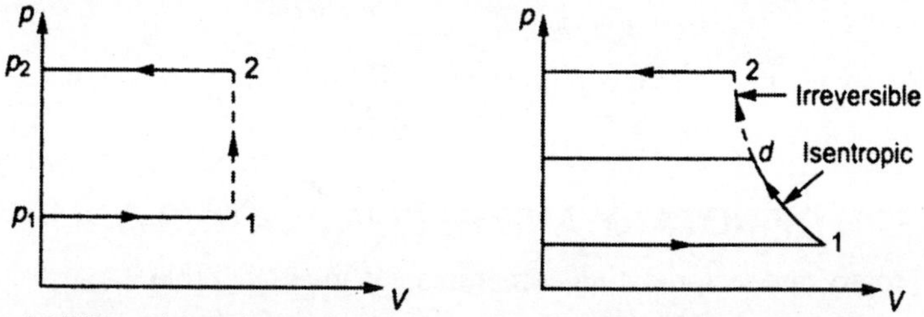

(a) Without internal compression (b) With internal compression

Fig. 9.7

(a) Without internal compression,

$$\text{I.P.} = (p_2 - p_1)V_1 = (1.8-1)\times 10^5 \times \frac{5}{60\times 10^3} = 6.67\ \text{kW}$$

(b) With internal compression,

$$p_d = p_1 + \frac{p_2 - p_1}{2} = 1 + \frac{1.8-1}{2} = 1.4\ \text{bar}$$

$$V_d = V_1\left(\frac{p_1}{p_d}\right)^{1/\gamma} = 5\left(\frac{1}{1.4}\right)^{1/1.4} = 3.932\ \text{m}^3/\text{min}$$

$$\text{I.P.} = \frac{1}{10^3\times 60}\left[\left(\frac{\gamma}{\gamma-1}\right)p_1V_1\left\{\left(\frac{p_d}{p_1}\right)^{(\gamma-1)/\gamma} - 1\right\} + (p_2 - p_d)V_d\right]$$

$$= \frac{1}{10^3\times 60}\left[\left(\frac{1.4}{0.4}\right)\times 1\times 10^5\times 5\left\{\left(\frac{1.4}{1}\right)^{0.4/1.4} - 1\right\} + (1.8-1.4)\times 105\times 3.932\right]$$

$$= \frac{1}{10^3\times 60}[1.766\times 10^5 + 1.5728\times 10^5]$$

$$= 5.565\ \text{kW}$$

For power required to be minimum, the whole compression should be a reversible adiabatic process.

$$\text{Ideal I.P.} = \left(\frac{\gamma}{\gamma-1}\right)p_1V_1\left[\left(\frac{p_2}{p_1}\right)^{(\gamma-1)/\gamma} - 1\right]$$

$$= \frac{1}{10^3 \times 60} \times \frac{1.4}{0.4} \times 1 \times 10^5 \times 5 \left[\left(\frac{1.8}{1} \right)^{0.4/1.4} - 1 \right]$$

$$= 5.333 \text{ kW}$$

$$\text{Actual I.P.} = \text{Ideal I.P.} \times \eta_{mech}$$

$$= 5.333 \times 0.98 = 5.226 \text{ kW}$$

$$\text{Isentropic efficiency without internal compression} = \frac{5.226}{6.67} \times 100 = 78.35\%$$

$$\text{Isentropic efficiency with internal compression} = \frac{5.226}{5.333} = 98\%$$

9.3 COMPARISON OF ROTARY AND RECIPROCATING COMPRESSORS

The comparison of rotary and reciprocating compressors is given in Table 9.1.

Table 9.1 : Comparison of Rotary and Reciprocating Compressors

Item	*Reciprocating*	*Rotary*
1. Air discharge	Low at very high pressure, 300 m^3/min upto 1 k bar	Large at low pressure, 3000 m^3/min upto 10 bar
2. Speed	Low, require some reduction gear	High, upto 50,000 rpm, can be directly coupled turbines, etc.
3. Receiver required	Yes, as supply is intermittent	No, as supply is continuous
4. Size	Large for the same discharge	Small
5. Balancing problem	Yes	No
6. Quality of air supplied	Less clean	More clean as it does not come into contact with lubricating oil
7. Lubrication system	More complicated	Simple
8. Compression process	May approach isothermal process	Isentropic with $n \simeq 1.4$

■ **Example 9.3** A rotary air compressor draws air at atmospheric pressure and 27°C. It compresses the air to a pressure of 5 atmospheres, and at the end of compression the temperature is 300°C. What is the polytropic efficiency of the compressor? Assume $\gamma = 1.4$ for air.

Solution :

$$\frac{p_2}{p_1} = \left(\frac{T_2}{T_1} \right)^{n/(n-1)}$$

$$\left(\frac{n}{n-1} \right) \ln \left(\frac{T_2}{T_1} \right) = \ln \left(\frac{p_2}{p_1} \right)$$

$$\frac{n}{n-1} = \frac{\ln 5}{\ln \left(\frac{573}{300} \right)} = 2.487$$

$$n = 1.67$$

$$\text{Polytropic efficiency} = \left(\frac{\gamma - 1}{\gamma}\right)\left(\frac{n}{n-1}\right) \times 100$$

$$= \frac{0.4}{1.4} \times \frac{1.67}{0.67} \times 100 = 71.21\%$$

■ **Example 9.4** A rotary compressor compresses air from static pressure of 1 bar and 20°C to a static pressure of 2 bar. While compressing air at the flow rate of 1 kg/s it absorbs 110 kW. Calculate (a) the static and stagnation temperatures in the delivery pipe, and (b) the static isentropic efficiency. Velocity of air at inlet and heat lost to the surroundings may be neglected and velocity in the delivery pipe may be taken as 100 m/s.

Solution : Because the inlet velocity is negligible, therefore the static and stagnation values of temperature and pressure are same at the inlet.

$$\therefore \quad p_1 = p_{01} \text{ and } T_1 = T_{01}$$

$$\text{Work done per second} = \dot{m} c_p (T_{02} - T_{01}) = P$$

$$1 \times 1.005\,(T_{02} - T_{01}) = 110$$

$$T_{02} - T_{01} = 109.45°\text{C}$$

Now

$$T_{01} = T_1 = 273 + 20 = 293 \text{ K}$$

$$T_{02} = 293 + 109.45 = 402.45 \text{ K} = T_2 + \frac{v_2^2}{2c_p}$$

$$T_2 = 402.45 - \frac{100^2}{2 \times 1.005 \times 10^3} = 397.48 \text{ K}$$

Isentropic static temperature in delivery pipe,

$$T_{isen} = T_1 \left(\frac{p_2}{p_1}\right)^{(\gamma - 1)/\gamma} = 293\,(2)^{1/3.5} = 357.3 \text{ K}$$

$$\text{Isentropic efficiency} = \frac{T_{isen} - T_1}{T_2 - T_1} = \frac{357.3 - 293}{397.48 - 293} = \frac{64.3}{104.48} = 0.6154 \text{ or } 61.54\%$$

SOLVED QUESTIONS FROM UNIVERSITY EXAMINATIONS

■ **Example 9.5** Attempt any three of the following :

(a) Brayton cycle

(b) Root and vane blower

(c) Heat balance sheet of I.C. engine

(d) Wet and dry sump lubrication system [MDU, May, 2008]

Solution :

(a) Refer to Art. 2.10

(b) Refer to Art. 9.2

(c) Refer to Art. 6.3

(d) Refer to Art. 5.7.1

■ **Example 9.6** Explain various types of rotary compressors with diagrams. [MDU, Dec., 2009]

Solution : Refer to Art. 9.2.

SHORT ANSWER TYPE QUESTIONS

1. What are the various types of compressors?
2. What is a positive displacement compressor?
3. Explain the working of a roots blower.
4. Derive an expression for the workdone and efficiency of a roots blower.
5. Explain the working of a vane type blower.
6. Explain the working of a screw compressor.
7. Compare rotary and reciprocating compressors.

❑❑❑

Chapter 10

Centrifugal Compressors

Inside This Chapter

10.1 INTRODUCTION

A centrifugal compressor is of the dynamic type in which the air flows continuously and steadily through the various parts and the rise in pressure is primarily due to the centrifugal action. They are used to supply large quantities of air but at a lower pressure ratio.

10.2 CONSTRUCTIONAL FEATURES

Essentially a centrifugal compressor consists of four elements, inlet buckets, impeller, diffuser, and casing as shown in Fig. 10.1. The inlet buckets are attached to the shaft and rotate with it, which guide air on the impeller. The impeller consists of a disc on which radial blades are attached. The diffuser surrounds the impeller and provides diverging passages for air flow thus increasing the air pressure. The air coming out from the diffuser in collected in the casing and taken out from the outlet of the compressor.

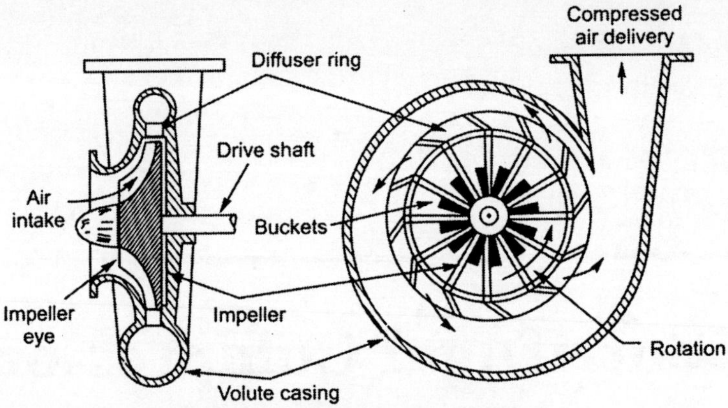

Fig. 10.1 Centrifugal compressor.

10.3 WORKING PRINCIPLE

The air enters the eye of the compressor at atmospheric pressure and low velocity. The inlet buckets guide the air to the impeller where it moves radially outward and is guided by the impeller blades. The impeller increases the momentum of the air flowing thorugh it, causing a rise in pressure and temperature of air. The air leaving the impeller enters the diffuser where its velocity is reduced by providing more cross-sectional area for the flow of air. Part of the kinetic energy of air is converted into pressure energy and further increases the pressure of air.

10.4 VARIATION OF VELOCITY AND PRESSURE

The variation of velocity and pressure air through the impeller and diffuser is shown in Fig. 10.2. Nearly, half of the total pressure rise in the impeller and the remaining half occurs in the diffuser. A pressure ratio

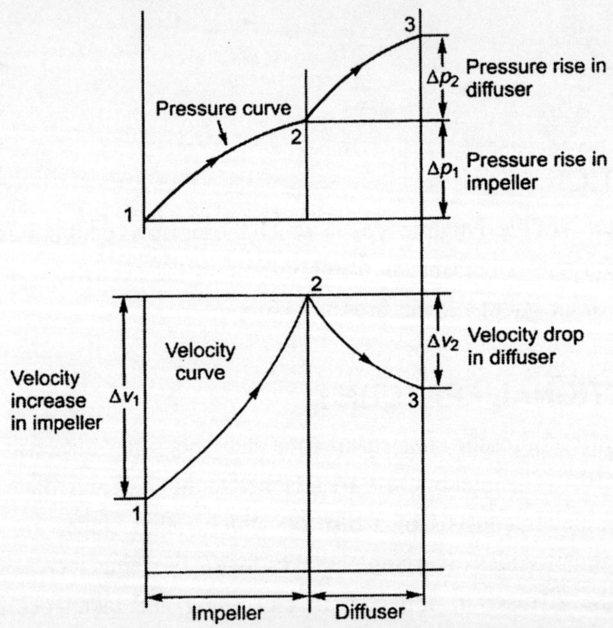

Fig. 10.2 Variation of velocity and pressure of air passing through impeller and diffuser.

of 4 can be achieved in single-stage compressors. For higher pressure ratios, multiple-stage compressors are used. A pressure ratio of 12 :1 is possible with multi-stage compressors.

10.5 TYPES OF IMPELLER

The impellers which are generally used are of two types, single eye type and double eyed type as shown in Fig. 10.3(a) and (b). In the single eye type, the air enters into the compressor from one side only and in the double eyed type, the air enters from both sides. The double eyed type impeller sucks in more air and has the advantage of self balancing over the single eye impeller.

In multi-stage compressors, the output of the first stage is passed on to the second stage and so on, as shown in Fig. 10.3(c).

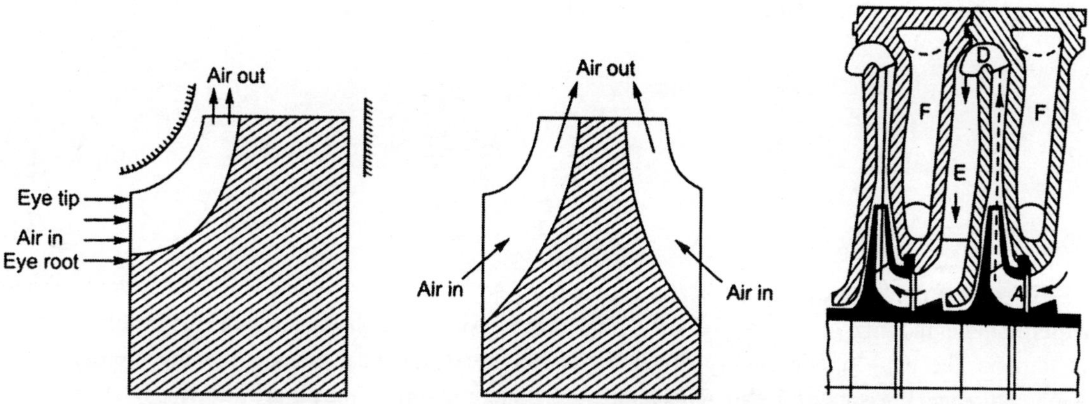

(a) Single eye impeller (b) Double eye impeller (c) Multi-stage compressor

Fig. 10.3 Types of impellers.

10.6 CENTRIFUGAL V'S RECIPROCATING COMPRESSORS

The comparison of centrifugal and reciprocating compressors is given in Table 10.1.

Table 10.1 : Comparison of Centrifugal and Reciprocating Compressors

Aspect	*Reciprocating Compressor*	*Centrifugal Compressor*
1. Vibration problems	More due to unbalanced reciprocating parts	Less as there are no reciprocating parts
2. Mechanical efficiency	Low due to many sliding parts	High as sliding parts are less
3. Installation cost	Higher	Lower
4. Pressure ratio per stage	5 to 8	3 to 4.5
5. Delivery pressure	High upto 5000 atm	Medium upto 400 atm.
6. Volume delivered	Small upto 300 m^3/min	High upto 3000 m^3/min
7. Maintenance cost	Higher	Lower
8. Isentropic Efficiency	About 80%	About 70%
9. Suitability	High pressure low volumes of air	Low pressure high volumes of air
10. Uses	Driving pneumatic tools and air-operated controlling equipments	Superchanging of I.C. engines, refrigerants and industrial gases.

10.7 STATIC AND TOTAL HEAD VALUES

The velocities of air encountered in centrifugal compressors are very high as compared to that in reciprocating compressors. Therefore, total head quantities should be taken into account in the analysis of centrifugal compressors. The total head quantities account for the kinetic energy of the air passing through the compressor.

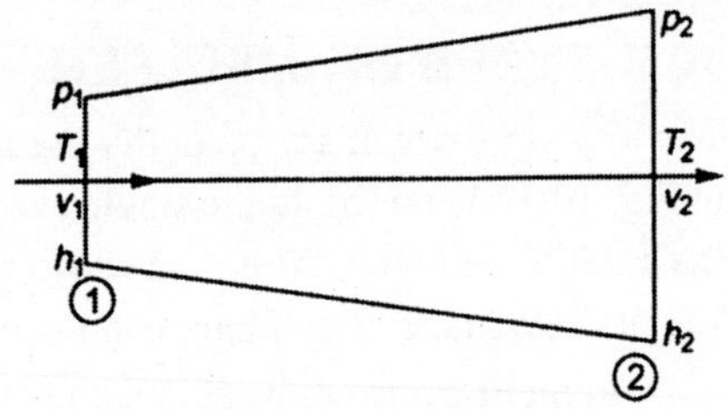

Fig. 10.4 Passage of air through varying area of cross-section.

Consider a horizontal passage of varying area of cross-section, as shown in Fig. 10.4, through which air is flowing from left to right. Assuming no external heat transfer and work transfer to the system, the steady flow energy equation for one kg mass of air flow can be written as :

$$h_1 + \frac{v_1^2}{2} = h_2 + \frac{v_2^2}{2}$$

or
$$c_p T_1 + \frac{v_1^2}{2} = c_p T_2 + \frac{v_2^2}{2}$$

or
$$c_p T + \frac{v^2}{2} = \text{const.}$$

Temperature 'T' is called the "static temperature", which is the temperature of the air measured by a thermometer moving with the air velocity. If the moving air is brought to rest under reversible adiabatic conditions, the total kinetic energy of the air is converted into heat energy, thereby increasing the temperature and pressure of the air. This temperature and pressure of the air are known as "stagnation" or "total head" temperature and pressure. The total head quantities are denoted by a suffix notation 'o'.

$$c_p T + \frac{v^2}{2} = c_p T_0 \quad \text{...(10.1)}$$

where T_0 = stagnation or total head temperature

$$T_0 = T + \frac{v^2}{2c_p} \quad \text{...(10.2)}$$

Similarly
$$h_0 = h + \frac{v^2}{2} \quad \text{...(10.3)}$$

where h_0 = stagnation or total head enthalpy

To find the total head pressure, apply the following equation

$$\frac{p_0}{p} = \left(\frac{T_0}{T}\right)^{\frac{\gamma}{\gamma - 1}} \quad \text{...(10.4)}$$

where p, T = static pressure and temperature

p_0, T_0 = stagnation pressure and temperature.

10.8 ADIABATIC AND ISENTROPIC PROCESSES

The process is said to be adiabatic if it does not exchange heat with the surroundings. The ideal reversible adiabatic process is called isentropic process in which the entropy of the system does not change.

However, in a rotary compressor, during adiabatic compression there are losses due to friction in air and blade passages, eddies formation and shocks at entry and exit. These factors cause internal heat generation and consequently the maximum temperature reached would be higher than that for adiabatic comrpession. This results in a progressive increase in entropy. Such a process although adiabatic is not reversible adiabatic or isentropic.

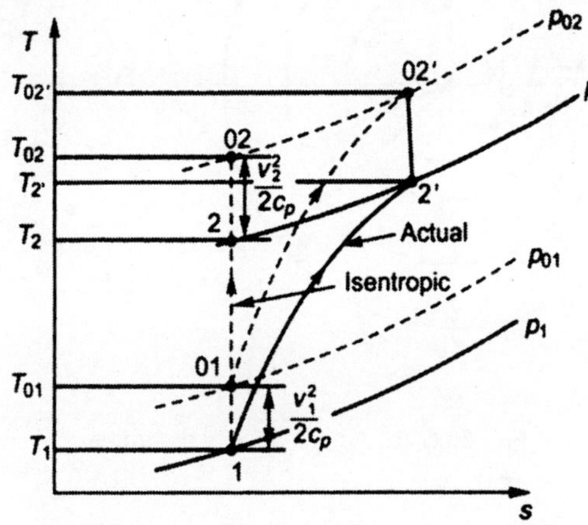

Fig. 10.5 Isentropic and actual compression.

The isentropic and adiabatic processes for the static and total head values are shown in Fig. 10.5 on the T-s diagram. Process 1-2 is the isentropic process for the static temperature and 01-02 for the total head temperature. Processes 1-2′ and 01-02′ are the adiabatic process for the static and total head temperatures respectively.

10.8.1 Isentropic Efficiency

Isentropic efficiency may be defined as the ratio of isentropic temperature rise to actual temperature rise.

Isentropic effieicncy,
$$\eta_{\text{isen}} = \frac{(\Delta T)_{\text{isen}}}{(\Delta T)_{\text{actual}}} = \frac{T_{02} - T_{01}}{T_{02'} - T_{01}} \quad \text{...(10.5 a)}$$

$$= \frac{r_p^{(\gamma-1)/\gamma} - 1}{r_p^{(n-1)/n} - 1} \quad \text{...(10.5 b)}$$

where
$$r_p = \frac{p_2}{p_1} \text{ and } n > \gamma.$$

During compression process work has to be done on the impeller. The energy balance then gives,

$$c_p T_{01} = c_p T_{02} - W \quad \text{or} \quad W = c_p\,(T_{02} - T_{01}) \quad \text{...(10.6)}$$

Then
$$\eta_{\text{isen}} = \frac{\text{Isentropic work}}{\text{Actual work}} = \frac{c_p\,(T_{02} - T_{01})}{c_p\,(T_{02'} - T_{01})} = \frac{T_{02} - T_{01}}{T_{02'} - T_{01}} \quad \text{...(10.7)}$$

10.9 VELOCITY DIAGRAMS

The velocity diagrams at inlet and outlet of impeller for the centrifugal compressor are shown in Fig. 10.6(a).

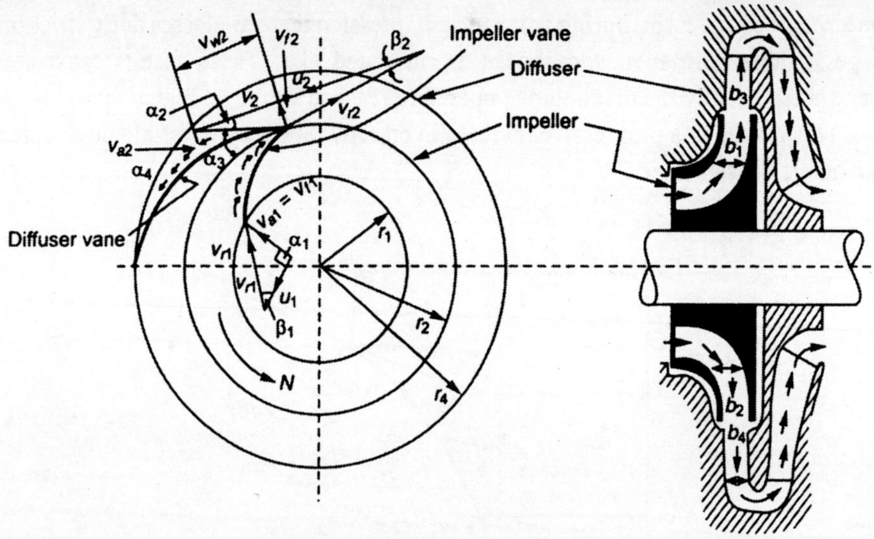

Fig. 10.6 (a) Actual velocity diagrams.

Let

u_1 = mean blade velocity at entrance

u_2 = mean blade velocity at exist

v_{a1} = absolute velocity at inlet to rotor

v_{a2} = absolute velocity at outlet to rotor

v_{r1} = relative velocity at inlet to rotor

v_{r2} = relative velocity at outlet to rotor

v_{w1} = velocity of whirl at inlet

v_{w2} = velocity of whirl at outlet

v_{f1} = velocity of flow at inlet

v_{f2} = velocity of flow at inlet

D_1, D_2 = mean diameter of rotor at inlet and outlet respectively

α_1 = exit angle from stator or guide vanes at entrance

β_1 = inlet angle to rotor in impeller blade angle at inlet

α_2 = inlet angle to the diffuser or the stator

β_2 = outlet angle from rotor or impeller blade angle at outlet.

If it is assumed that the entry of the air to the rotor is axial, then whirl component $v_{w1} = 0$ and $v_{a1} = v_{f1}$.

10.9.1 Theory of Operation

(a) Ideal Velocity Diagrams

By Newton's second law of motion, the rate of change of angular momentum of air is equal to the torque applied to the body causing that change. Considering 1 kg/s of mass flow rate of air,

Torque, $$T = v_{w2}\, r_2 - r_{w1} r_1 \qquad \ldots(10.8)$$

where $$r_1 = \frac{D_1}{2} \text{ and } r_2 = \frac{D_2}{2}$$

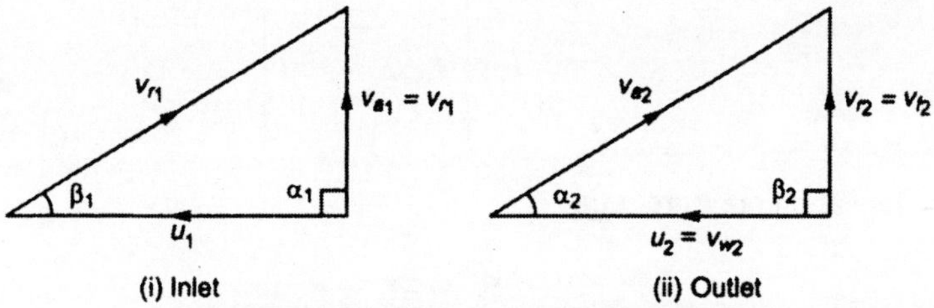

Fig. 10.6 (b) Ideal velocity diagrams.

For ideal case, let us assume that the impeller is radial vaned, and there is no frictional loss, no heat transfer and the air leaves the impeller with a tangential velocity $v_{w2} = u_2$. Also for axial entry, $v_{w1} = 0$.

Torque, $$T = u_{\omega_2} = u_2 r_2 \qquad \text{...(10.9)}$$

Work done by impeller for 1 kg/s air flow rate,

$$w = u_2 r_2 \omega \qquad \text{...(10.10)}$$

where ω = angular speed of rotor, rad/s

Now, $$r_2 \omega = u_2$$

$\therefore$ $$w = u^2 \qquad \text{...(10.11)}$$

Since air cannot leave the impeller at a velocity greater than the impeller tip velocity, therefore Eq. (10.11) gives the maximum work capacity of the impeller.

Also by steady flow energy equation, we have

$$w = \left(h_2 + \frac{v_{a2}^2}{2}\right) - \left(h_1 + \frac{v_{a1}^2}{2}\right) = c_p\left[T_2 + \frac{v_{a2}^2}{2c_p}\right] - c_p\left[T_1 + \frac{v_{a1}^2}{2c_p}\right]$$

$$= c_p\,(T_{02} - T_{01}) \qquad \text{...(10.12a)}$$

$$= h_{02} - h_{01} \qquad \text{...(10.12b)}$$

In most practical problems,

$$v_{a1} = v_{a2}$$

$\therefore$ $$w = c_p\,(T_2 - T_1) \qquad \text{...(10.13a)}$$

$$= h_2 - h_1$$

Now, $$w = c_p T_{01}\left(\frac{T_{02}}{T_{01}} - 1\right) \qquad \text{...(10.13b)}$$

$$T_{01} = T_1\left[\frac{p_{01}}{p_1}\right]^{\frac{\gamma-1}{\gamma}}$$

$$T_{02} = T_2\left[\frac{p_{02}}{p_2}\right]^{\frac{\gamma-1}{\gamma}}$$

$$w = c_p T_{01}\left[\left[\frac{p_{02}}{p_{01}}\right]^{\frac{\gamma-1}{\gamma}} - 1\right] \qquad \text{...(10.14)}$$

$$= c_p T_{01}\left[(r_{op})^{\frac{\gamma-1}{\gamma}} - 1\right] \qquad \text{...(10.15)}$$

where $r_{op} = \dfrac{p_{o2}}{p_{o1}}$ = stagnation pressure ratio

If $v_{a1} = v_{a2}$, then

$$w = c_p T_1\left[r_p^{\frac{\gamma-1}{\gamma}} - 1\right] \qquad \text{...(10.16)}$$

where, $r_p = \dfrac{p_2}{p_1}$ = static pressure ratio

Comparing Eq. (10.11) and (10.14), we have

$$u_2^2 = c_p T_{01}\left[\left(\frac{p_{02}}{p_{01}}\right)^{\frac{\gamma-1}{\gamma}} - 1\right] \qquad \text{...(10.17)}$$

$$\frac{p_{02}}{p_{01}} = \left[\frac{u_2^2}{c_p T_{01}} + 1\right]^{\frac{\gamma}{\gamma-1}} \qquad \text{...(10.18)}$$

If $v_{a1} = v_{a2}$, then

$$\frac{p_2}{p_1} = \left[\frac{u_2^2}{c_p T_1} + 1\right]^{\frac{\gamma}{\gamma-1}} \qquad \text{...(10.19)}$$

Total temperature increase across an impeller,

$$\Delta T_0 = \frac{u_2^2}{c_p} = \text{const.} \qquad \text{...(10.20)}$$

Indicated power,

$$\text{I.P.} = \frac{\dot{m}\, u_2^2}{10^3}\ \text{kW} \qquad \text{...(10.21)}$$

where $\dot{m}$ = mass flow rate of air, kg/s

Thus, for an ideal isentropic centrifugal compressor, the power input depends on the following :

1. The mass flow rate of air.
2. The total pressure ratio of the compressor which in turn depends upon the square of the impeller velocity.
3. The total inlet temperature.
4. The total inlet temperature rise between inlet and outlet.
5. There is a maximum work capacity of an impeller depending upon tip velocity. The maximum pressure produced depends upon the following :
   1. Inlet temperature.
   2. Square of the impeller velocity.
   3. It is independent of impeller diameter.

(b) Actual Velocity Diagrams

The actual velocity diagrams are shown in Fig. 10.7.

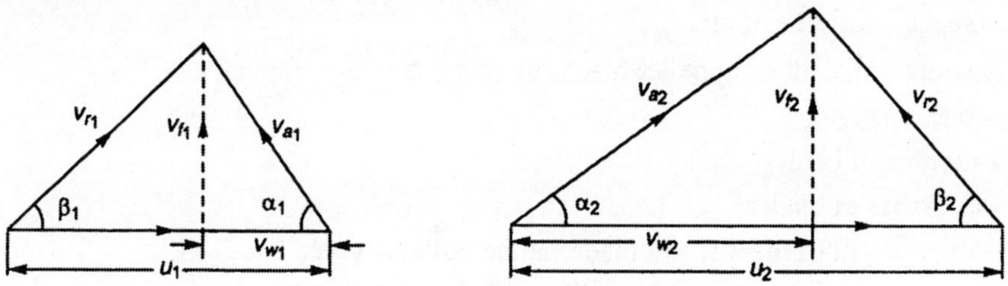

Fig. 10.7 Actual velocity diagrams for centrifugal compressor.

Theoretical torque, $$T = v_{w2}r_2 - v_{w1}r_1$$

Work done on 1 kg/s of air.

$$w = (v_{w2}r_2 - v_{w1}r_1)\,\omega$$

where, ω = angular speed of rotor, rad/s

$$\omega = v_{\omega_2} u_2 = v_{\omega_1} u_1 \qquad \text{...(10.22)}$$

Now,
$$v_{r1}^2 = u_1^2 + v_{a1}^2 - 2u_1 v_{a1} \cos\alpha_1 = u_1^2 + v_{a1}^2 - 2u_1 v_{w1}$$

$$v_{r2}^2 = u_2^2 + v_{a2}^2 - 2u_2 v_{a2} \cos\alpha_2 = u_2^2 + v_{a2}^2 - 2v_2 v_{w2}$$

$$u_1 v_{w1} = \frac{1}{2}(u_1^2 + v_{a1}^2 - v_{r1}^2)$$

$$u_2 v_{w2} = \frac{1}{2}(u_2^2 + v_{a2}^2 - v_{r2}^2)$$

$$w = \frac{v_{a2}^2 - a_{a1}^2}{2} + \frac{v_{r1}^2 - v_{r2}^2}{2} + \frac{u_2^2 - u_1^2}{2} \qquad \text{...(10.23)}$$

$= \Delta KE + \Delta p$ due to diffusion action + Δp due to centrifugal action.

Thus, the fraction of kinetic energym imparted to air and converted into pressure energy in impeller is given by,

$$\frac{v_{r1}^2 - v_{r2}^2}{2} + \frac{u_2^2 - u_1^2}{2} = \int_1^2 \frac{dp}{\rho} \qquad \text{...(10.24)}$$

where ρ = density of air.

If v_{a1} = diffuser outlet velocity, then

$$\frac{v_{a2}^2 - v_{a1}^2}{2} = \int_1^2 \frac{dp}{\rho} + (\Delta h)_{\text{diff. loss}} \qquad \text{...(10.25)}$$

10.9.2 Width of Blades of Impeller and Diffuser

Air mass flow rate per second,

$$\dot{m}_a = \frac{\text{Circumferential area of flow} \times v_{f1}}{\text{Specific volume of air}}$$

$$= \frac{(\pi D_1 - nt)\, b_1 \times v_{f1}}{v_{s1}} = \frac{\pi D_1 k_b b_1 v_{f1}}{v_{s1}} \qquad \text{...(10.26a)}$$

$$= \frac{(\pi D_2 - nt)\, b_2 \times v_{f2}}{v_{s2}} = \frac{\pi D_2 k_b b_2 v_{f2}}{v_{s2}} \quad \text{...(10.26b)}$$

where,. v_s = specific volume of air.

b_1, b_2 = width or height of impeller blades at inlet and outlet respectively.

k_b = blade factor

n = number of blades

t = thickness of blades.

The width or height of the diffuser blades at the outlet is given by,

$$m_a = \frac{(\pi D_d - nt_d)\, b_d \times v_{fd}}{v_{sd}} \quad \text{...(10.27)}$$

where suffix 'd' represents the quantities at the diffuser outlet. The width or height of the diffuser blades at the inlet is the same as that of the impeller blades at outlet.

10.10 SLIP FACTOR AND PRESSURE COEFFICIENT

In the ideal velocity diagrams, we had assumed that $v_{w2} = u_2$, but this condition is not satisfied in actual practice due to secondary flow effets. Actually, $v_{w2} < u_2$. The difference $(u_2 - v_{w2})$ is called the slip. Slip factor is defined as the ratio of actual whirl component to the ideal whirl component and is denoted by ϕ_s.

Slip factor,

$$\phi_s = \frac{v_{w2}}{u_2} \quad \text{...(10.28)}$$

The total head pressure ratio is given by,

$$\frac{p_{02}}{p_{01}} = \left(\frac{T_{02}}{T_{01}}\right)^{\frac{\gamma-1}{\gamma}} = \left[1 + \frac{T_{02} - T_{01}}{T_{01}}\right]^{\frac{\gamma-1}{\gamma}}$$

Substituting the value of $(T_{02} - T_{01})$ from Eq. (10.5), we have

$$\frac{p_{02}}{p_{01}} = \left[1 + \frac{\eta_{isen}\,(T_{02'} - T_{01})}{T_{01}}\right]^{\frac{\gamma-1}{\gamma}} \quad \text{...(10.29)}$$

The actual work done per kg of air is given by :

$$c_p\,(T_{02'} - T_{01}) = u_2 v_{w2}$$

The actual work done per kg of air by the compressor is always greater than $u_2 v_{w2}$ due to fluid friction and windage losses. Therefore, the actual work is obtained by multiplying $u_2 v_{w2}$ by the factor ϕ_w, called the 'work factor' or power input factor.

$$c_p\,(T_{02'} - T_{01}) = \phi_w u_2 v_{w2}$$

$$T_{02'} - T_{01} = \frac{\phi_w u_2 v_{w2}}{c_p} \quad \text{...(10.30)}$$

Eq. (10.29) becomes

$$\frac{p_{02}}{p_{01}} = \left[1 + \frac{\eta_{isen}\,\phi_w\,u_2 v_{w2}}{c_p T_{01}}\right]^{\frac{\gamma-1}{\gamma}}$$

Now, $v_{\omega 2} = \phi_s u_2$

$$\frac{p_{02}}{p_{01}} = \left[1 + \frac{\eta_{isen}\, \phi_w\, \phi_s v_2^2}{c_p T_{01}}\right]^{\frac{\gamma-1}{\gamma}} \qquad ...(10.31)$$

The pressure coefficient is defined as the ratio of isentropic work to Euler work and is denoted by ϕ_p.

$$\text{Pressure coefficient,} \quad \phi_p = \frac{\text{Isentropic work}}{\text{Euler work}} = \frac{c_p\,(T_{02} - T_{01})}{u_2 v_{w2}} \qquad ...(10.32)$$

Using Eq. (10.5) and assuming radial vanes of impeller, *i.e.*, $v_{w2} = u_2$, we get

$$\phi_p = \frac{c_p\, \eta_{isen}\, (T_{02'} - T_{01})}{u_2^2}$$

Using Eqs. (10.28) and (10.30) we get

$$\phi_p = \phi_w \phi_s \eta_{isen} \qquad ...(10.33)$$

10.11 LOSSES

There are two types of losses in a centrifugal compressor : internal and external. The internal losses sustained in the compressor are manifested by an increase of enthalpy of air. The internal losses which occur in compressor are :

1. Friction between air and walls of flow passage
2. Disc Friction
3. Leakage between impeller and casing
4. Turbulence
5. Shock.

The compressor sustains an external loss in the form of bearing friction and windage loss. The friction losses are proportional to v_a^2 and hence proportional to $\dot{m}_a^2$. The incidence loss in terms of drag coefficient C_D is proportional to $C_D v_a^2$.

10.12 EFFECT OF IMPELLER BLADE SHAPE ON PERFORMANCE

The following types of blade shapes are used for impellers of centrifugal compressors :

1. Backward-curved blades ($\beta_2 < 90°$)
2. Radial blades ($\beta_2 = 90°$)
3. Forward curved blades ($\beta_2 > 90°$).

The characteristics of these blades are shown in Fig. 10.8. Centrifugal effects on the curved blades create bending moment and produce increased stresses which limit the maximum speed at which the impeller can run.

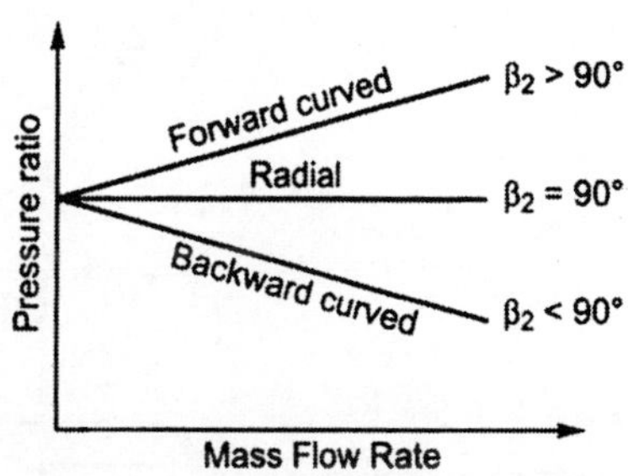

Fig. 10.8 Characteristics of blades.

Slightly backward curved impeller blades (*i.e.*, $\beta_2 < 90°$) gives optimum efficiency because the degree of reaction increases with decreasing blade tip angles. With increasing degree of reaction increasing part of kinetic energy is transferred into pressure energy within the impeller. By selecting very small tip blade angles β_2, the length of impeller channel increases and consequently the friction loss increases considerably, thereby decreasing the beneficial effect and reducing the efficiency. The optimum blade angle is about 45°.

Radial blades are used for aircraft centrifugal compressors because they can be manufactured easily, lowest centrifugal stresses, free from bending stresses and equal energy conversion in impeller and diffuser giving higher pressure ratio and efficiency.

10.13 THE DIFFUSER

The diffuser converts the kinetic energy imparted to air by the impeller into pressure rise. In the vaned diffuser the vanes are used to remove the whirl of the fluid at a higher rate than is possible by a simple increase in radius, there by reducing the length of flow path and diameter. A typical vaned type diffuser is shown in Fig. 10.9. A pressure gradient exist in the diffuser in the direction opposite to that of flow, therefore the flow stream lines will tend to break away from the diverging passage walls, reversing their direction and resulting into turbulence. About 20° is accepted as the maximum included angle of divergence for satisfactory diffusion.

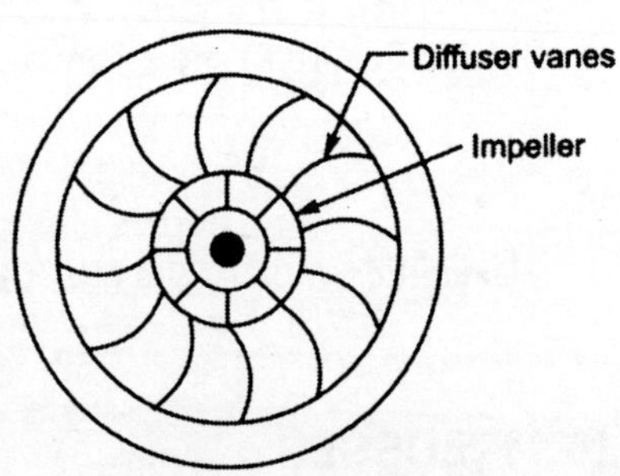

Fig. 10.9 Vaned diffuser.

The clearance between the impeller and the vanes leading edges constitutes a vaneless diffuser. This helps to smooth out velocity variation between the impeller tip and diffuser vanes. This also reduces the circumferential pressure gradient at impeller tip.

The flow follows an approximately logarithmic spiral path to the vanes after which it is constrained by the diffuser channels.

Diffuser efficiency,
$$\eta_d = \frac{p_4 - p_3}{\frac{1}{2}\rho\,(v_{a3}^2 - v_{a4}^2)} \qquad \ldots(10.34)$$

where suffices 3 and 4 denote upstream and down stream conditions of diffuser and ρ is the mass density of air.

10.14 PRE-WHIRL

In order to avoid acceleration to sonic or supersonic velocities the Mach number of flow entering the impeller eye should be kept below the value of 0.9. When the absolute velocity of approach is high

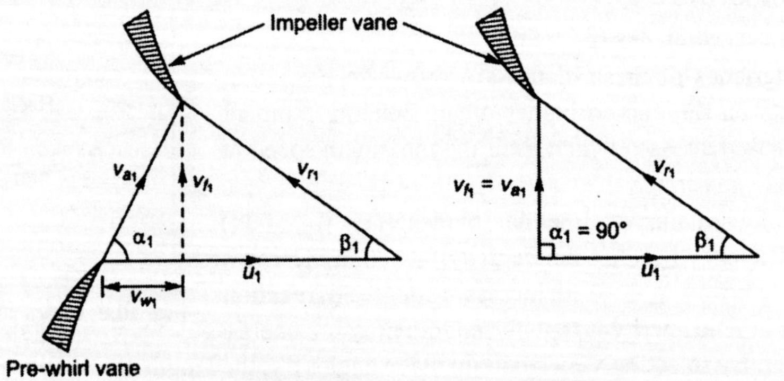

Fig. 10.10 Effect of pre-whirl on inlet velocity triangle.

enough or the static absolute temperature of the entering air is low enough to make the Mach number more than 0.9, the method of providing a pre-whirl and thereby reducing the relative velocity entering the impeller eye is quite helpful.

Pre-whirl is provided by intake guide vane as shown in Fig. 10.10, which shows the velocity triangles for blades without pre-whirl and with pre-whirl. It may be observed that the magnitude of the relative velocity with pre-whirl vane gets reduced at the entrance, but the flow enters with a large relative angle. Thus, the flow inlet Match number is lowered by pre-whirl vane. But since $v_{w1} \neq 0$ with pre-whirl vane, the work done on the air must be greater for the same compressor inlet and outlet conditions.

10.15 PERFORMANCE CHARACTERISTICS

Let us consider the following variables for the performance of the compressor.

p_3 = pressure at the outlet of diffuser
p_1 = inlet pressure
T_1 = inlet temperature
D = diameter
G = mass rate of flow
N = r.p.m. of rotor

Then, $$p_3 = f(p_1, T_1, D, G, N)$$

The following dimensions are chosen :

$$M = \text{mass (kg)}, L = \text{length } (m), t = \text{time } (s)$$

Therefore, $$p = M^1 L^{-1} t^{-2} \left(\frac{kg \times m}{s^2 \times m^2}\right)$$

$$T = L^2 t^{-2} \left[\frac{\text{Heat}}{\text{Mass}} = \frac{\text{Nm}}{\text{kg}} = \frac{\text{kg} \times m \times m}{\text{kg} \times s^2}\right]$$

$$D = L; G = M^1 t^{-1}; N = t^{-1} \text{ (rpm = rad/s)}$$

Number of variables, $n = 6$

Basic dimensions, $m = 3$

According to Buckingham's *Pi* theorem, the number of dimensionless groups

$$= n - m = 6 - 3 = 3$$

The three dimensionless groups are formed by combining p_1, T_1 and D with each of the remaining variables, so that

$$\pi_1 = p_1^{a_1} T_1^{b_1} D^{c_1} p_3$$

$$\pi_2 = p_1^{a_2} T_1^{b_2} D^{c_2} G$$

$$\pi_3 = p_1^{a_3} T_1^{b_3} D^{c_3} N$$

Substituting the basic dimensions for the variables in π_1, we have

$$\pi_1 = (M^1 L^{-1} t^{-2})^{a_1} (L^2 t^{-2})^{b_1} L^{c_1} (M^2 L^{-1} t^{-2})^1$$

For M, $0 = a_1 + 1$

For L, $0 = -a_1 + 2b_1 + c_1 - 1$

For t, $0 = -a_1 - 2b_1 - 2$

From which, $a_1 = -1, b_1 = 0, c_1 = 0$

$$\therefore \qquad \pi_1 = p_1^{-1} T_1^0 D^0 p_3^1 = \frac{p_3}{p_1}$$

Similarly,

$$\pi_2 = p_1^{-1} T_1^{0.5} D^{-2} G^{-1} = \frac{G\sqrt{T_1}}{p_1 D^2}$$

$$\pi_3 = p_1^0 T_1^{-0.5} D^1 N^1 = \frac{DN}{\sqrt{T_1}}$$

Thus,

$$\pi_1 = f_1(\pi_2, \pi_3)$$

$$\frac{p_3}{p_1} = f_2\left(\frac{G\sqrt{T_1}}{p_1 D^2}, \frac{DN}{\sqrt{T_1}}\right) \qquad \text{...(10.35)}$$

For a given compressor under consideration, D = const.

$$\frac{p_3}{p_1} = f_3\left(\frac{G\sqrt{T_1}}{p_1}, \frac{N}{\sqrt{T_1}}\right) \qquad \text{...(10.36)}$$

Likewise the ratio $\frac{T_3}{T_1}$ can be formed to be function of the dimensionless groups of Eqs. (10.35) and (10.36). Similarly, if stagnation temperture and pressure are considered, the result will again be the same except that stagnation temperatures and pressures are substituted in place of static pressures and temperatures.

The performance of a given compressor or geometrically similar compressor can be plotted in terms of three dimensionless parameters provided dynamic similarity is maintained Fig. 10.11 shows a plot of $\frac{G\sqrt{T_1}}{p_1}$ against $\frac{p_3}{p_1}$ and $\frac{N}{\sqrt{T_1}}$. The operating line is the locus of the points of maximum efficiency at various values of $\frac{N}{\sqrt{T_1}}$. The surge line represents the stable operation, the region to the right of the surge line being for stable conditions.

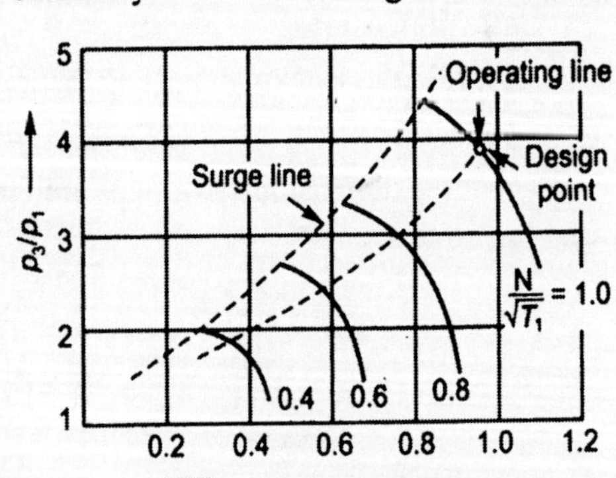

Fig. 10.11 $\frac{G\sqrt{T_1}}{p_1}$ v's $\frac{p_3}{p_1}$

The performance of the compressor has been plotted in terms of volume flow rate and pressure ratio in Fig. 10.12.

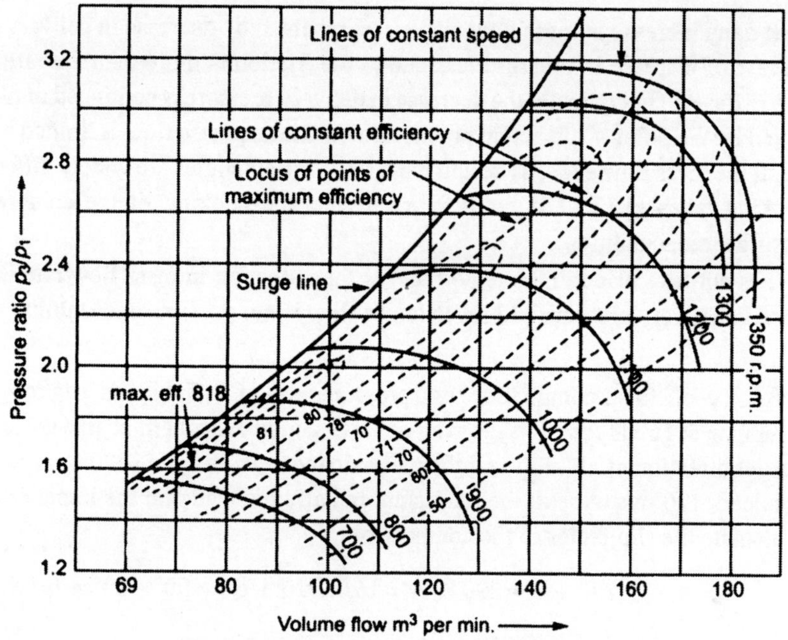

Fig. 10.12 Pressure ratio v's volume flow rate.

10.16 SURGING AND CHOKING

The delivery pressure v's mass flow rate in a dynamic compressor is shown in Fig. 10.13. Suppose a discharge valve is put in the line for this compressor. The mass rate of flow will be zero when the discharge valve is closed and the static pressure developed is that delivered by the impeller to the air contained in the compressor. Such a situation is shown by point '*a*' in Fig. 10.13. If the valve is now opened, the flow of air takes place and the diffuser becomes effetive in increasing the static pressure. This is shown by points *b* and *c*. The maximum delivery presusre is obtained at point *c*. As the valve is opened more, the mass flow rate increases beyond point *c* and the efficiency of compressor decreases with decrease in delivery pressure. When the designed mass rate is greatly exceeded the incidence between the vane and air angle becomes so large that flow separation and shock occur, accompanied by rapid decrease in efficiency.

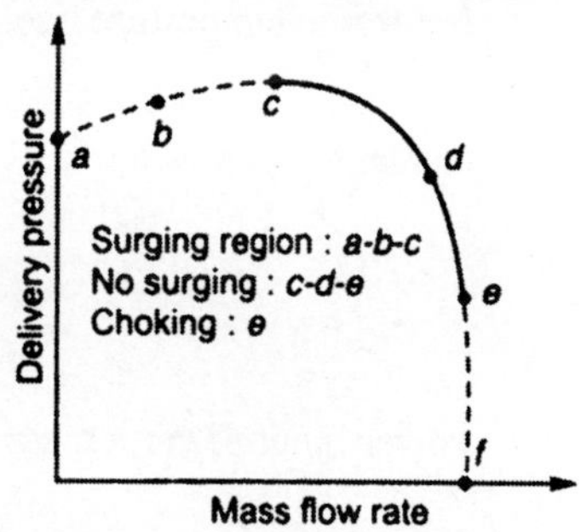

Fig. 10.13 Surging and choking phenomenas.

If the compressor operates at some point *b* to the left of point *c*, then a decrease in mass rate of flow is accompanied by decrease in pressure developed by the compressor. If the static pressure of the air at compressor outlet does not decrease as rapidly as the developed pressure, there is a natural tendency of the air to flow back into the compressor in the direction of pressure gradient. With the drop in pressure at the compressor outlet, the pressure gradient is reversed and also the direction of flow is reversed. Thus, this leads to unstable condition of cyclic reversal taking place at extremely high frequencies. This pulsating air flow phenomena is called 'surging'. Such a situation is avoided by keeping the operating point to the right of point *c*, because in this region a decrease in mass flow rate is accompanied by increase in delivery pressure, leading to stability.

After ponit *c* any increase in mass flow rate is accompanied by decrease in delivery pressure. This is because, the rate of increase in pressure loss due to the friction is more than the rate of increase in pressure gain by diffuser. Theoretically the decrease in delivery pressure is continued until the mass flow rate is represented by the point '*f*'. In actual practice the maximum mass flow is limited by the point '*e*'. This is because, if the mass flow exceeds design mass flow, the air angles are widely different from vane angles and choking takes place. The point '*e*' represents the choking of the compressor, *i.e.*, the maximum mass flow rate condition.

Surging does not take place in the region '*ce*' as the reduction in mass flow rate is accompanied with the increase in delivery pressure and reversion of flow is not possible and stability of operation is maintained.

■ **Example 10.1** A centrifugal compressor has a pressure ratio of 4 : 1 with an isentropic efficiency of 82% when running at 16000 rpm. It takes in air at 17°C. Guide vanes at inlet give the air a pre whirl of 20° to the axial direction at all radii and the mean diameter of the eye is 200 mm, the absolute air velocity at inlet is 120 m/s. At exit the blades are radially inclined and the impeller tip diameter is 550 mm. Calculate the slip factor of the compressor.

Solution : $\frac{p_2}{p_1} = 4, T_1 = 273 + 17 = 290\text{ K}, N = 16{,}000\text{ rpm}, \alpha_1 = 90° - 20° = 70°$

$$d_1 = 200\text{ mm}, v_{a_1} = 120\text{ m/s}, \eta_{isen} = 0.82, d_2 = 550\text{ mm}, \beta_2 = 90°$$

Temperature after isentropic compression,

$$T_2 = T_1\left(\frac{p_2}{p_1}\right)^{\frac{\gamma-1}{\gamma}} = 290\,(4)^{\frac{0.1}{0.4}} = 430.94\text{ K}$$

Isentropic temprature rise,

$$(\Delta T) = 430.94 - 390 = 140.94\text{ K}$$

Actual temperature rise,

$$(\Delta T)_a^{isen} = \frac{(\Delta T)_{isen}}{\eta_{isen}} = \frac{140.94}{0.82}$$

$$= 171.88\text{ K}$$

Power input per unit mass flow rate

$$= c_p \times (\Delta T)_a = 1.005 \times 171.88 = 172.74\text{ kJ/kg}$$

An inlet to impeller (Fig. 10.14 (a)),

$$v_{a1} = 120\text{ m/s}$$

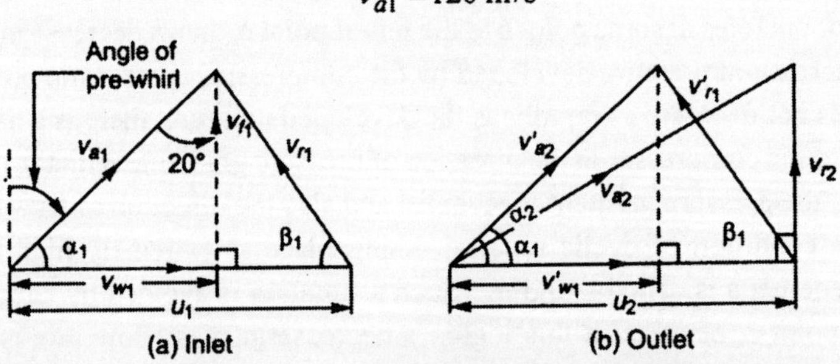

(a) Inlet (b) Outlet

Fig. 10.14

Angle of pre-whirl = 20°

$$u_1 = \frac{\pi d_1 N}{60 \times 10^3} = \frac{\pi \times 200 \times 16000}{60 \times 10^3} = 167.58 \text{ m/s}$$

$$u_{w1} = v_{a1} \sin 20° = 120 \sin 20° = 41.04 \text{ m/s}$$

At exit of impeller (Fig. 10.14 (b))

$$u_2 = \frac{\pi d_2 N}{60 \times 10^3} = \frac{\pi \times 550 \times 16000}{60 \times 10^3} = 460.77 \text{ m/s}$$

For radial discharge, $v_{w2} = u_2 = 460.77$ m/s

Power input per unit mass flow rate

$$= u_2 v'_{w_2} - u_1 v_{w1}$$

$$= 460.77\, v'_{w_2} - 167.55 \times 41.04 = 172.74 \times 10^3$$

$$v'_{w_2} = 389.82 \text{ m/s}$$

$$\text{Slip factor} = \frac{v'_{w_2}}{v_{\omega 2}} = \frac{389.82}{460.77} = 0.846$$

■ **Example 10.2** The free air delivered by a centrifugal compressor is 25 kg/min. The suction condition is 1 bar and 25°C. The velocity of air at inlet is 50 m/s. The isentropic efficiency of the compressor is 70%. If the total head pressure ratio of the compressor is 4, determine (*a*) the total head temperature of air at the exit of the compressor, and (*b*) brake power required to run the compressor assuming mechanical efficiency of 96%. Pressure and temperature of air at inlet are static. For air $\gamma = 1.4$ and $R = 0.287$ kJ/kg. K.

Solution :

$$c_p = R + c_p = R + \frac{c_p}{\gamma}$$

$$c_p = \frac{\gamma R}{\gamma - 1} = \frac{1.4 \times 0.287}{0.4}$$

$$= 1.005 \text{ kJ/kg. K}$$

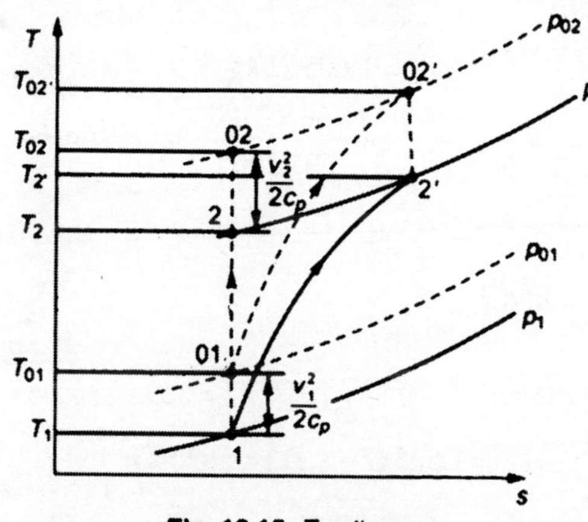

Fig. 10.15 *T-s* diagram.

$$T_{01} = T_1 + \frac{v_1^2}{2c_p}$$

$$= 298 + \frac{50^2}{2 \times 1.005 \times 10^3}$$

$$= 299.24 \text{ K}$$

$$\eta_{isen} = \frac{T_{02} - T_{01}}{T_{02'} - T_{01}}$$

$$T_{02} = T_{01}\left(\frac{p_{02}}{p_{01}}\right)^{\frac{\gamma-1}{\gamma}} = 299.24\,(4)^{\frac{1}{3.5}} = 444.67 \text{ K}$$

$$0.7 = \frac{444.67 - 299.24}{T_{02'} - 299.124}$$

$$T_{02'} = 507 \text{ K}$$

$$\text{I.P. required to run the compressor} = \frac{\dot{m}_a c_p\,(T_{02'} - T_{01})}{\eta_{mech}}$$

$$= \frac{25 \times 1.005\,(507 - 299.24)}{0.96 \times 60} = 90.62 \text{ kW}$$

■ **Example 10.3** Air at a temperature of 20°C flows into the centrifugal compressor running at 21,000 rpm. The folowing data is given :

Slip factor	= 0.85
Work input factor	= 1.0
Isentropic efficiency	= 72%
Outer diameter of blade tip	= 60 cm

Assuming the absolute velocities of air entering and leaving the compressor are same, determine :

(a) the temperature rise of air passing through compressor, and

(b) the static pressure ratio

Take $c_p = 1.005$ kJ/kg. K.

Solution :

$$u_2 = \frac{\pi D_2 N}{60 \times 100} = \frac{\pi \times 60 \times 21000}{60 \times 100} = 659.7 \text{ m/s}$$

Work done per kg of air $= u_2 v_{w2} - u_1 v_{w1}$

$$= u_2^2 \phi_s \phi_w \qquad \left[\because v_{w1} = 0 \text{ and } \phi_s = \frac{v_{w2}}{u_2}\right]$$

$$= (659.7)^2 \times 0.85 \times 10^{-3} \times 1 = 369.923 \text{ kJ/s}$$

Also work done $= c_p\,(T_2' - T_1) = 1.005\,(T_{2'} - T_1) = 369.923$ [as $v_{a1} = v_{a2}$]

$$T_{2'} - T_1 = 368.1 \text{ K}$$

$$\eta_{isen} = \frac{T_2 - T_1}{T_{2'} - T_1}$$

$$T_2 = 293 + 0.72 \times 368.1 = 558 \text{ K}$$

Static pressure ratio,

$$\frac{p_2}{p_1} = \left(\frac{T_2}{T_1}\right)^{\frac{\gamma}{\gamma-1}} = \left(\frac{558}{293}\right)^{3.5} = 9.53$$

■ **Example 10.4** Air at a temperature of 300 K flows in a centrifugal compressor running at 25,000 rpm. Slip factor = 0.82, isentropic total head efficiency = 0.75, outer diameter of blade tip = 550 mm. Assuming the absolute velocities of air at inlet and exit of the compressor to be same, calculate (*a*) the temperature rise of air passing through the compressor, and (*b*) static pressure ratio. $c_p = 1.005$ kJ/kg.K.

Solution :

$$u_2 = \frac{\pi D_2 N}{60} = \frac{\pi \times 0.55 \times 25{,}000}{60} = 719.95 \text{ m/s}$$

Work done per kg of air,

$$W = u_2 v_{w2} - u_1 v_{w1}$$

$$v_{w1} = 0 \text{ and } v_{w2} = u_2 \phi_s$$

$$W = u_2^2 \phi_s = (719.95)^2 \times 0.82 = 425.027 \text{ kJ/s}$$

Also work done $= c_p \ (T_2' - T_1)$

$$1.005\,(T_2' - T_1) = 425.027$$

$$T_2' - T_1 = 422.9$$

$$\eta_{isen} = \frac{T_2 - T_1}{T_2' - T_1} = \frac{T_2 - 300}{422.9} = 0.15$$

$$T_2 = 717.2 \text{ K}$$

$$\frac{p_2}{p_1} = \left(\frac{T_2}{T_1}\right)^{\frac{\gamma}{\gamma-1}} = \left(\frac{617.2}{300}\right)^{3.5} = 12.49$$

■ **Example 10.5** Explain the concept of polytropic efficiency, η_∞, as applied to a multistage rotary compressor. Derive the relationship between isentropic efficiency η_i and polytropic efficiency η_∞.

$$n_i = \frac{r^{\frac{(\gamma-1)}{\gamma}} - 1}{r^{\left\{\frac{(\gamma-1)}{(\gamma\eta_\infty)}\right\}} - 1}$$

where r is the pressure ratio of the compressor.

Solution : Polytropic efficiency is the isentropic efficiency of one small stage of a multistage compressor. This small stage efficiency is constant for all stages of a compressor with infinite number of stages.

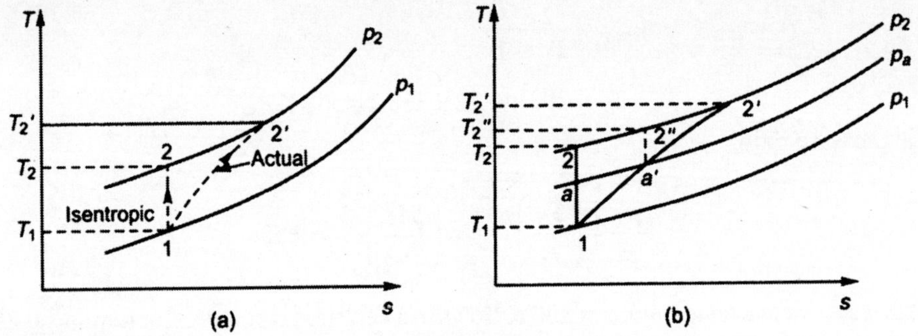

Fig. 10.16

The actual and isentropic compression of a single stage compressor is shown in Fig. 10.16 (a) on T-s diagram.

Isentropic efficiency, n_i = Isentropic work/Actual work

$$= \frac{(h_{02} - h_{01})}{(h'_{02} - h_{01})} = \frac{(T_2 - T_1)}{(T'_2 - T_1)} = \frac{\left(\frac{T_2}{T_1} - 1\right)}{\left(\frac{T'_2}{T_1} - 1\right)} = \frac{r^{\frac{(\gamma-1)}{\gamma}} - 1}{r^{\frac{(n-1)}{n}} - 1}$$

where $r = \frac{p_{02}}{p_{01}}$.

For a multistage rotary compressor, as shown in Fig. 10.16 (b), actual compression path is $1-2'$ and isentropic path is $1-2$.

Polytropic efficiency,
$$\eta_\infty = \left[\left(\frac{\gamma-1}{\gamma}\right)\right]\left[\frac{n}{(n-1)}\right]$$

or
$$\frac{(n-1)}{n} = \left[\left(\frac{\gamma-1}{\gamma\eta_\infty}\right)\right]$$

Thus,
$$\eta_i = \frac{r^{\frac{(\gamma-1)}{\gamma}} - 1}{r^{\left(\frac{(\gamma-1)}{(\gamma\eta_\infty)}\right)} - 1}$$

■ **Example 10.6** A centrifugal compressor handles 180 kg/min of air. The suction pressure and temperature are 1 bar and 300 K. The suction velocity is 90 m/s. The delivery conditions are 2 bar, 400 K and 240 m/s. Calculate (a) the isentropic efficiency, (b) the power required to drive the compressor, and (c) the over all efficiency of the unit. Take $c_p = 1.005$ kJ/kg. K

Assume that entire kinetic energy gained in the impeller is converted into pressure in the diffuser.

Solution : $\dot{m} = \frac{180}{60} = 3$ kg/s, $p_1 = 1$ bar, $T_1 = 300$ K, $v_{a1} = 90$ m/s, $p_2 = 2$ bar

$$T_2 = 350 \text{ K}, \; v_{a2} = 240 \text{ m/s}$$

(a) Isentropic efficiency, $\eta_{isen} = \frac{\text{Isentropic work}}{\text{Actual work}}$

$$\frac{T_2}{T_1} = \left(\frac{p_2}{p_1}\right)^{\frac{\gamma-1}{\gamma}} = \left(\frac{2}{1}\right)^{\frac{1}{3.5}} = 1.219$$

$$T_2 = 300 \times 1.219 = 365.7 \text{ K}$$

Isentropic work done, $$w_{isen} = c_p\,(T_2 - T_1) + \frac{v_{a2}^2 - v_{a1}^2}{2 \times 10^3}$$

$$= 1.005\,(365.7 - 300) + \frac{240^2 - 90^2}{2 \times 10^3}$$

$$= 66.028 + 24.75 = 90.778 \text{ kJ/kg}$$

Actual work done in the impeller,

$$w_a = 1.005\,(400 - 300) + \frac{240^2 - 90^2}{2 \times 10^3}$$

$$= 100.5 + 24.75 = 125.25 \text{ kJ/kg}$$

$$\eta_{isen} = \frac{90.778}{125.25} = 0.7248 \text{ or } 72.48\%$$

(b) Power required to drive the compressor,

$$\text{I.P.} = \dot{m} \times w_a = 3 \times 125.25 = 375.75 \text{ kW}$$

(c) Let T_3 = temperature at the exit of diffuser

$$c_p\,(T_3 - T_2) = \frac{v_{a2}^2 - v_{a1}^2}{2 \times 10^3}$$

$$1.005\,(T_3 - 400) = 24.75$$

$$T_3 = 424.6 \text{ K}$$

$$\frac{p_3}{p_2} = \left(\frac{T_3}{T_2}\right)^{\frac{\gamma}{\gamma-1}} = \left(\frac{424.6}{400}\right)^{3.5} = 1.2326$$

$$p_3 = 2 \times 1.2326 = 2.465 \text{ bar}$$

After isentropic compression, the delivery temperature from diffuser is,

$$\frac{T_3'}{T_1} = \left(\frac{p_3}{p_1}\right)^{\frac{\gamma-1}{\gamma}} = \left(\frac{2.465}{1}\right)^{\frac{1}{3.5}} = 1.294$$

$$T_3' = 300 \times 1.294 = 388.2 \text{ K}$$

Overall efficiency,

$$\eta_{overall} = \frac{T_3' - T_1}{T_3 - T_1} = \frac{388.2 - 300}{424.6 - 300}$$

$$= \frac{88.2}{124.6} = 0.7078 \text{ or } 70.78\%$$

■ **Example 10.7** A centrifugal compressor handles 540 kg/min of air at 1 bar and 20°C ambient conditions. The air is compressed from 1 bar static pressure to 4.5 bar total pressure. The air enters the impeller eye with a velocity of 150 m/s with no pre-whirl. The ratio of whirl speed to tip speed is 0.9. Calculate (a) the rise in total temperature during compression if the change in kinetic energy is negligible, (b) the tip diameter of the impeller, (c) the power required, and (d) the eye diameter if the hub diameter is 130 mm.

Assume that the compressor runs to 20,000 rpm with isentropic efficiency of 82%.

Solution : $\dot{m} = \dfrac{540}{60} = 9$ kg/s, $p_{01} = p_1 = 1$ bar, $T_1 = 273 + 20 = 293$ K, $N = 20{,}000$ rpm, $\eta_{isen} = 0.82$,

$p_{02} = 4.5$ bar, $v_{a1} = 150$ m/s, $\phi_s = \dfrac{v_{w2}}{u_2} = 0.9$, $d_h = 130$ mm $= 0.13$ m

The *T-s* diagram is shown in Fig. 10.17.

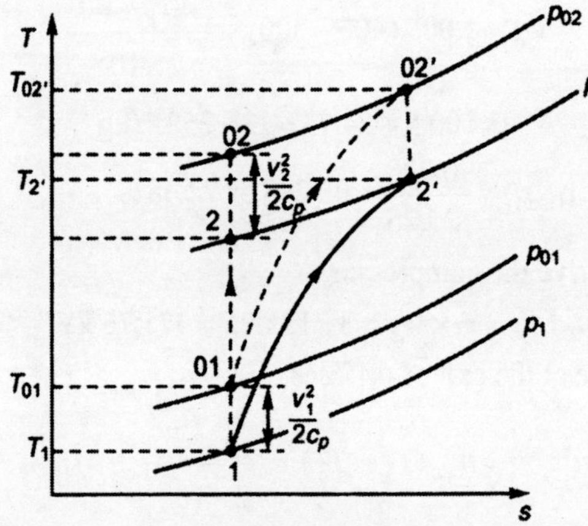

Fig. 10.17

(a)
$$T_{01} = T_1 + \frac{v_{a1}^2}{2c_p} = 293 + \frac{150^2}{2 \times 10^3 \times 1.005} = 304.2 \text{ K}$$

$$p_{01} = p_1 \left(\frac{T_{01}}{T_1}\right)^{\frac{\gamma}{\gamma-1}} = 1\left(\frac{304.2}{293}\right)^{3.5} = 1.14 \text{ bar}$$

$$\frac{T_{02}}{T_{01}} = \left(\frac{p_{02}}{p_{01}}\right)^{\frac{\gamma-1}{\gamma}} = \left(\frac{4.5}{1.14}\right)^{\frac{1}{3.5}} = 1.48$$

$$T_{02} = 304.2 \times 1.48 = 450.3 \text{ K}$$

Isentropic rise in total temperature,

$$(\Delta T)_{isen} = 450.3 - 304.2 = 146.13 \text{ C}$$

Actual rise in total temperature,

$$(\Delta T)_{actual} = \frac{(\Delta T)_{isen}}{\eta_{isen}} = \frac{146.13}{0.82} = 178.2°\text{C}$$

(b) Actual work consumed by the compressor,

$$w_{actual} = c_p\,(\Delta T)_{actual} = 1.005 \times 178.2 = 179.1 \text{ kJ/kg}$$

Work consumed given by Euler's equation without pre-whirl,

$$w_{actual} = \frac{u_2 v_{w2}}{10^3} \text{ kJ/kg}$$

$$v_{w2} = \phi_s u_2 = 0.9 \times u_2$$

$$179.1 = \frac{0.9 u_2^2}{10^3}$$

$$u_2 = 446.08 \text{ m/s}$$

$$= \frac{\pi D_2 N}{60}$$

Tip diameter of impeller,

$$D_2 = \frac{446.08 \times 60}{\pi \times 20{,}000}$$

$$= 0.426 \text{ m or } 426 \text{ mm}$$

(c) Power required $= \dot{m} \times w_{actual} = 9 \times 179.1 = 1611.9$ kW

(d) $\dot{m} = \frac{\pi}{4}(D_1^2 - d_b^2) \times v_{a1} \times \rho_1$

$$\rho_1 = \frac{p_1}{RT_1} = \frac{1 \times 10^5}{287 \times 293} = 1.189 \text{ kg/m}^3$$

$$9 = \frac{\pi}{4}[D_1^2 - (0.13)^2] \times 150 \times 1.189$$

$$D_1 = 0.2848 \text{ m or } 284.8 \text{ mm}$$

■ **Example 10.8** A centrifugal compressor delivers 720 m^3/min of free air while running at 12,000 rpm. The ambient air conditions are 1 bar and 27°C. The air is compressed to 4.2 bar with an isentropic efficiency of 0.85. Blades are radial at outlet of impeller and flow velocity is 60 m/s, which may be assumed constant throughout. The outer radius of impeller is twice the inner and the slip factor is 0.9. The blade area coefficient is 0.9 at inlet. Determine (a) final temperature of air, (b) theoretical power, (c) inlet and outlet diameter of impeller, (d) breadth of impeller at inlet, (e) impeller blade angle at inlet, and (f) diffuser blade angle at inlet.

Solution : $p_1 = 1$ bar, $T_1 = 273 + 27 = 300$ K, $p_2 = 4.2$ bar, $\eta_{isen} = 0.85$, $v_{f2} = 60$ m/s, $r_2 = 2r_1$, $\phi_s = 0.9$, $k_b = 0.9$.

(a) $\frac{T_2}{T_1} = \left(\frac{p_2}{p_1}\right)^{\frac{\gamma-1}{\gamma}}$, $T_2 = 300\left(\frac{4.2}{1}\right)^{\frac{1}{3.5}} = 452$ K

$$\eta_{isen} = \frac{T_2 - T_1}{T_{2'} - T_1}, \; T_{2'} = 300 + \frac{452 - 300}{0.85} = 478.8 \text{ K}$$

(b) Mass flow rate, $\dot{m} = \frac{p_1 V}{RT_1} = \frac{1\times10^5 \times 720}{60\times287\times300} = 13.94 \text{ kg/s}$

Theoretical power, $P_{th} = \dot{m}\, c_p\, (T_{2'} - T_1)$

$$= 13.94 \times 1.005\,(478.8 - 300) = 2504.4 \text{ kW}$$

(c) $$\text{Work done} = \frac{\phi_s u_2^2}{10^3} = c_p\,(T_{2'} - T_1)$$

$$0.9 u_2^2 = 10^3 \times 1.005\,(478.8 - 300)$$

$$u_2 = 446.8 \text{ m/s} = \frac{\pi D_2 N}{60}$$

$$D_2 = \frac{60 \times 446.8}{\pi \times 12000} = 0.711 \text{ m or } 711 \text{ mm}$$

$$D_1 = \frac{D_2}{2} = 355.6 \text{ mm}$$

(d) Volume flow rate $= \pi D_1 b_1 k_b v_{f_1}$

$$b_1 = \frac{720}{60\times\pi\times0.711\times0.9\times60} = 0.099 \text{ m or } 99 \text{ mm}$$

(c) $$\tan\beta_1 = \frac{v_{f_1}}{u_1} = \frac{60}{\left(\frac{446.8}{2}\right)} = 0.26858$$

$$\beta = 15^\circ.03$$

(f) $$\tan\alpha_2 = \frac{v_{f_2}}{\phi_s u_2} = \frac{60}{0.9\times446.8} = 0.1492$$

$$\alpha_2 = 8.48^\circ$$

■ **Example 10.9** A centrifugal blower compresses 5.4 m^3/s of air from 1 bar and 25°C to 1.5 bar. The index of compression is 1.5. The flow velocity of 70 m/s is constant throughout. The inlet and outlet impeller diameters are 0.30 m and 0.60 m respectively. The blower rotates at 9000 rpm. Determine (a) the blade angles at inlet and outlet of the impeller, (b) the absolute angle at the tip of the impeller, and (c) the breadth of blade at inlet and outlet.

Assume that no diffuser is employed and the whole pressure increase occurs in the impeller and the blades have negligible thickness.

Solution : $\dot{V}_1 = 5.4 \text{ m}^3/\text{s}$, $p_1 = 1$ bar, $T_1 = 273 + 25 = 298$ K, $n = 1.5$, $p_2 = 1.5$ bar, $v_{f_1} = v_{f_2} = 70$ m/s, $D_1 = 0.3$ m, $D_2 = 0.6$ m, $N = 9000$ rpm

$$T_2 = T_1\left(\frac{p_2}{p_1}\right)^{\frac{n-1}{n}} = 298\left(\frac{1.5}{1}\right)^{\frac{0.5}{1.5}} = 341.1 \text{ K}$$

$$u_1 = \frac{\pi D_1 N}{60} = \frac{\pi\times0.3\times9000}{60} = 141.37 \text{ m/s}$$

$$u_2 = 2u_1 = 282.74 \text{ m/s}$$

$$\text{Work done} = c_p\,(T_2 - T_1) = \frac{u_2 v_{w2}}{10^3}$$

$$v_{w2} = \frac{1.005\,(341.1 - 298) \times 10^3}{282.74} = 153.2 \text{ m/s}$$

(a)

$$\tan\beta_1 = \frac{v_{f1}}{u_1} = \frac{70}{141.37} = 0.4951$$

$$\beta_1 = 26.34°$$

$$\tan\beta_2 = \frac{v_{f2}}{u_2 - v_{w2}} = \frac{70}{282.74 - 153.2} = 0.5404$$

$$\beta_2 = 28.38°$$

(b)

$$\tan\alpha_2 = \frac{v_{f2}}{v_{w2}} = \frac{70}{153.2} = 0.4569$$

$$\alpha_2 = 24.56°$$

(c) $\dot{V}_1 = \pi D_1 b_1 v_{f1}$

$$b_1 = \frac{5.4}{\pi \times 0.3 \times 70} = 0.0818 \text{ m or } 81.8 \text{ mm}$$

$$\dot{V}_2 = \frac{p_1 \dot{V}_1}{T_1} \times \frac{T_2}{p_2} = \frac{1 \times 5.4 \times 341.1}{298 \times 1.5} = 4.12 \text{ m}^3/\text{s}$$

$$\dot{V}_2 = \pi D_2 b_2 v_{f2}$$

$$b_2 = \frac{4.12}{\pi \times 0.6 \times 70} = 0.0312 \text{ m or } 31.2 \text{ mm}$$

■ **Example 10.10** A centrifugal compressor delivers 18 kg/s of air with a total head pressure ratio of 4 : 1. The speed of the compressor is 16000 rpm. Total head temperature at inlet is 22°C, slip factor 0.9, power input factor 1.05 and isentropic efficiency 0.82. Calculate (a) over all diameter of the impeller, and (b) power input.

Solution : $\dot{m} = 18$ kg/s, $r_{op} = 4$, $N = 16000$ rpm,

$$T_{01} = 273 + 22 = 295 \text{ K},$$

$$\phi_s = 0.9,\ \phi_w = 1.05,\ \eta_{isen} = 0.82$$

(a)

$$\frac{T_{02}}{T_{01}} = \left(\frac{p_{02}}{p_{01}}\right)^{\frac{\gamma - 1}{\gamma}} = (4)^{\frac{1}{3.5}} = 1.486$$

$$T_{02} = 295 \times 1.486 = 438.4 \text{ K}$$

$$\eta_{isen} = \frac{T_{02} - T_{01}}{T_{02'} - T_{01}}$$

$$0.82 = \frac{438.4 - 295}{T_{02'} - 295}$$

$$T_{02'} = 469.8 \text{ K}$$

$$\text{Work done} = \frac{\phi_w \phi_s u_2^2}{10^3} = c_p\,(T_{02'} - T_{01})$$

$$1.05 \times 0.9 \times u_2^2 = 10^3 \times 1.005\,(469.8 - 295)$$

$$u_2 = 431.2 \text{ m/s}$$

$$D_2 = \frac{60 \times 431.2}{\pi \times 16000} = 0.5147 \text{ m or } 514.7 \text{ mm}$$

(b) Power input, $P = \dot{m}\, c_p\,(T_{02'} - T_{01})$

$$= 18 \times 1.005\,(469.8 - 295) = 3162.1 \text{ kW}$$

■ **Example 10.11** A single sided impeller centrifugal compressor fitted with a diffuser, but with no water cooling, requires 2000 kW when running at 16500 rpm. The outlet diameter of impeller is 0.5 m and the width of the casing of the vortex chamber between impeller and the diffuser is 50 mm. During a test on this compressor, static temperature and pressure were measured at a radius of 0.28 m and were found to be 120°C and 3 bar respectively. Assuming whirl speed at the tip of the impeller to be 0.95 of the peripheral speed and neglecting friction in vortex chamber, calculate (a) the flow rate, (b) resultant speed at the section given, and (c) the total temperature at the section given. Take $c_p = 1.005$ kJ/Kg. K, $\gamma = 1.4$ and intake pressure = 1 bar.

Solution : $p = 2000$ kW, $N = 16500$ rpm, $D_2 = 0.5$ m, $b = 50$ mm

$$\phi_s = 0.95,\ p_1 = 1 \text{ bar}$$

(a) Work done = $v_{w2} u_2$

$$u_2 = \frac{\pi D_2 N}{60} = \frac{\pi \times 0.5 \times 16500}{60} = 432 \text{ m/s}$$

$$v_{w2} = \phi_s u_2 = 0.95 \times 432 = 410.4 \text{ m/s}$$

$$\text{Power per kg of air} = \frac{v_{w2} u_2}{10^3} = \frac{410.4 \times 432}{10^3} = 177.28 \text{ kW}$$

$$\text{Air mass flow rate, } \dot{m} = \frac{P}{177.28} = \frac{2000}{177.28} = 11.28 \text{ kg/s}$$

(b) Using suffix 3 for the given section

Area of flow, $A_{f3} = 2\pi r_3 b_3 = 2\pi \times 0.28 \times 50 \times 10^{-3} = 0.088 \text{ m}^2$

Volume rate of flow at the given section,

$$\dot{V}_3 = \frac{\dot{m} R T_3}{p_3} = \frac{11.28 \times 287 \times 393}{3 \times 10^5}$$

$$= 4.24 \text{ m}^3/\text{s}$$

$$v_{r3} = v_{f3} = \frac{V_3}{A_{f3}} = \frac{4.24}{0.088} = 48.19 \text{ m/s}$$

For free vortex in vaneless space, $v_{w1} \times r_1 = \text{const.}$

i.e., $v_{w1} r_1 = v_{w2} r_2 = v_{w3} r_3$

$$v_{w3} = \frac{v_{w2} r_2}{r_3} = \frac{410.4 \times 0.25}{0.28} = 366.4 \text{ m/s}$$

$$v_{a3} = \sqrt{v_{f3}^2 + v_{w3}^2} \qquad [\because v_{f3} = v_{r3}]$$

$$= \sqrt{(48.19)^2 + (366.4)^2} = 369.6 \text{ m/s}$$

(c) $$T_{03} = T_3 + \frac{v_{a3}^2}{2c_p} = 393 + \frac{(369.6)^2}{2 \times 1.005 \times 10^3} = 461 \text{ K or } 188°\text{C}$$

SOLVED QUESTIONS FROM UNIVERSITY EXAMINATIONS

■ **Example 10.12** A rotary compressor working between 1 bar and 2.2 bar has internal and external diameter of impeller as 300 mm and 600 mm respectively. The vane angle at inlet and outlet are 30° and 45° respectively. If the air enters the impeller at 15 m/s, find speed of impeller in rpm, work done on compressor per kg of air. [MDU, Dec., 2007]

Solution : Given : $p_1 = 1$ bar, $p_2 = 2.2$ bar, $D_1 = 300$ mm, $D_2 = 600$ mm,

$$\beta_1 = 30°, \beta_2 = 45°, v_{r1} = 15 \text{ m/s}$$

Let

$$T_1 = 20° + 273 = 293 \text{ K}, \gamma = 1.4$$

$$\frac{T_2}{T_1} = \left(\frac{p_2}{p_1}\right)^{\frac{\gamma-1}{\gamma}}$$

$$T_2 = 293\left(\frac{2.2}{1}\right)^{\frac{0.4}{1.4}} = 367 \text{ K}$$

Refer to Fig. 10.18.

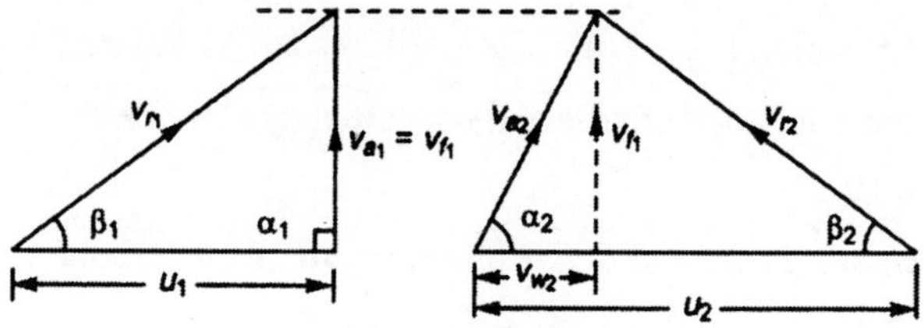

Fig. 10.18 Velocity triangles.

$$\frac{u_1}{v_{r1}} = \cos \beta_1$$

$$u_1 = 15 \cos 30° = 12.99 \text{ m/s}$$

$$u_1 = \frac{\pi D_1 N_1}{60}$$

$$12.99 = \frac{\pi \times 0.3 \times N_1}{60}$$

$$N_1 = 827 \text{ rpm}$$

$$v_{\omega 1} = 0, v_{f_1} = v_{r1} \sin \beta = 15 \sin 30° = 7.5 \text{ m/s}$$

$$v_{f_2} = v_{f_1} = 7.5 \text{ m/s}$$

$$v_2 = \frac{\pi D_2 N_2}{60} = \frac{\pi \times 0.6 \times 827}{60} = 25.98 \text{ m/s}$$

$$v_{\omega 2} = u_2 - v_{f_2} \tan \beta_2 = 25.48 - 7.5 \tan 45° = 18.48 \text{ m/s}$$

$$\text{Work done on the compressor} = c_p (T_2 - T_1) - \frac{u_2 u_{\omega 2}}{10^3}$$

$$= 1.005 (367 - 293) - \frac{25.98 \times 18.48}{10^3}$$

$$= 74.37 - 0.48$$

$$= 73.89 \text{ kJ/kg}$$

■ **Example 10.13** Derive an expression for the work done for a centrifugal air compressor. [MDU. May, 2008]

Solution : Refer to Art. 10.9.1.

■ **Example 10.14** A centrifugal air compressor having a pressure ratio of 5, compresses air at a rate of 10 kg/s. If the initial pressure and temperature is 1 bar and 20°C, find the final temperature of air and power required to drive the compressor. Take $\gamma = 1.4$ and $c_p = 1$ kJ/kg.K. [MDU, May, 2008]

Solution : Given, $r_p = 5$, $\dot{m} = 10$ kg/s, $p_1 = 1$ bar, $T_1 = 273 + 20 = 293$ K, $\gamma = 1.4$, $c_p = 1$ kJ/kg.K

$$\frac{T_2}{T_1} = (r_p)^{\frac{\gamma - 1}{\gamma}}$$

$$T_2 = 293 (5)^{\frac{0.4}{1.4}} = 464 \text{ K or } 191°\text{C}$$

Power required to drive the compressor,

$$P = \dot{m} c_p (T_2 - T_1) = 10 \times 1 \times (464 - 293) = 1710 \text{ kW}$$

■ **Example 10.15** What are the blade shapes and how are they classified? [MDU, May, 2009]

Solution : Refer to Art. 10.12.

■ **Example 10.16** Define slip factor and derive an expression for the same. [MDU, May, 2009, Dec., 2009]

Solution : Refer to Art. 10.10.

■ **Example 10.17** With a neat sketch explain the inlet and exit velocity triangles for various types of blades. [MDU. May, 2009]

Solution : Refer to Art. 10.9.1 and Art. 10.12.

■ **Example 10.18** (a) With a neat sketch, explain the inlet and exit velocity triangles for various types of blades.

(b) Explain the following :

(i) Degree of Reaction

(ii) Surging
(iii) Choking
(iv) Polytropic Efficiency. [MDU. Dec., 2010]

Solution : (a) Refer to Example 10.17.

(b) (i) Refer to Art. 10.6.
(ii) Refer to Art. 10.11.
(iii) Refer to Art. 10.11.
(iv) Refer to Art. 10.8.

■ **Example 10.19** What is the difference between rotary and centrifugal compressors? [MDU, May, 2008]

Solution :

Aspect	*Rotary Compressors*	*Centrifugal Compressors*
Pressure Ratio	2 to 3	About 3 to 6
Suitability for multi-staging	No	Yes
Operational speed	Low	High
Mechanical efficiency	Low	High
Starting torque required	High	Low
Delivery pressure	Low	Medium and with multistaging upto 400 atm.
Suitability	For handling large volumes of air at low pressures.	For handling large volumes of air at low and medium pressures.

SHORT ANSWER TYPE QUESTIONS

1. What are the constructional featus of a centrifugal compressor?
2. Explain the principle of working of a centrifugal compressor.
3. How the velocity and pressure of air vary through the impeller and the diffuser?
4. What are the various types of impellers?
5. Define total head values.
6. Define isentropic efficiency.
7. Define stagnation pressure ratio.
8. Define slip factor.
9. What is work factor and pressure coefficient?
10. What is the relationship between ϕ_p, ϕ_w, ϕ_s and ϕ_{isen}?
11. List the losses which generally occur in a centrifugal compressor.
12. What are the effects of impeller blade shapes on performance?
13. What is the role of diffuser?
14. Explain pre-whirl and its role.
15. Explain the phenomena of surging and choking.

EXERCISES

1. A single eye, single stage centrifugal compressor delivers 20 kg/s of air with a pressure ratio of 4 when running at 15000 rpm. The pressure and temperature of air at the inlet are 1 bar and 17°C. Assuming the following :
   Slip factor = 0.9, work input factor = 1.04, isentropic efficiency = 75%.
   Calculate (a) the I.P. required to drive the compressor, and (b) the blade angles at the impeller eye if the root and tip diameters are 20 cm and 40 cm respectively.
2. A centrifugal compressor delivers 600 m^3/min of free air when running at 900 rpm. The following data is given :

Inlet pressure and temperature of air	= 1 bar and 20°C
Compression ratio	= 3.5
Isentropic efficiency	= 83%
Flow velocity throughout the impeller	= 60 m/s
Tip diameter	= 2 × eye diameter
Blade area coefficient	= 0.94

   The blades are radial at outlet of impeller.
   Calculate (a) the I.P. required to run the compressor, (b) impeller diameters at inlet and outlet, (c) breadth of impeller at inlet, and (d) impeller blade angle at inlet.
3. The first stage of a centrifugal compressor is single-inlet with an eye 60 cm, hub 20 cm and an over all diameter of 100 cm. When supplied with 20°C air at 1 bar and running at 6000 rpm, it compresses 30 kg/s of air through a compression ratio of 2.0.
   Calculate (a) the isentropic efficiency of rotor based on impeller tip velocity with slip factor as 0.9, and (b) the rotor power and shaft power. Take $\eta_{mech} = 0.97$.
4. A single-sided centrifugal compressor is to deliver 14 kg/s of air when operating at a pressure ratio of 4 : 1 and a speed of 12000 rpm. The total head inlet conditions may be taken as 288 K and 1 bar. Assuming a slip factor of 0.9, a power input factor of 1.04 and an isentropic efficiency (based on the total head) of 80%, estimate the over-all diameter of the impeller. If the Mach number is not to exceed unity at the impeller tip and 50% of the losses are assumed to occur in the impeller, find the minimum possible depth of the diffuser. **[Ans. 5.4 cm]**
5. The following data refer to a single-sided centrifugal compressor :
   Air mass flow rate = 9 kg/s, Eye tip diameter = 0.3 m, Slip factor = 0.9, Isentropic efficiency = 80%. Eye root diameter = 0.15 m, over-all diameter of impeller = 0.5 m, Power input factor = 1.04, Rotational speed = 18000 rpm, Inlet stagnation pressure and temperature = 1.1 bar, 295 K.
   Calculate (a) pressure ratio of the compressor, (b) inlet angle of the impeller vane of the root and tip radii of the eye, and (c) the axial depth of the impeller channel at the periphery of the impeller.
6. An air-craft fitted with a single-sided centrifugal compressor flies with a speed of 850 km/h at an altitude where the pressure is 0.23 bar and the temperature 217 K. The inlet duct of the impeller eye contains fixed vanes which give the air pre-whirl of 25° at all radii. These inner and outer diameters of the eye are 180 mm and 330 mm respectively, the diameter of the impeller tip is 540 mm and the rotational speed 16000 rpm. Estimate the stagnation pressure at the compressor outlet when the mass flow is 216 kg/min.
   Neglect losses in inlet duct and fixed vanes, and assume that the isentropic efficiency of the compressor is 0.80. Take the slip factor as 0.9 and the power input factor as 1.04. **[Ans. 1.81 bar]**

7. A centrifugal compressor compresses 30 kg/s of air at a rotational speed of 1500 rpm. The air enters the compressor axially, and the conditions at the exit section are radius = 0.3 m, relative velocity of air at the tip = 100 m/s at an angle of 80°. Find the torque and power required to drive the compressor and work done.

8. Air at a temperature of 300 K flows in a centrifugal compressor running at 18,000 rpm. Slip factor = 0.8; isentropic total head efficiency = 0.75; Outer diameter of blade tip = 0.5 m. Determine (a) the temperature rise of air, and (b) static pressure ratio.

   Assume that the velocities of air at inlet and exit of the compressor are same.

9. A single inlet type centrifugal compressor delivers 8 kg/s of air. The ambient air conditions are 1 bar and 20°C. The compressor runs at 22,000 rpm with isentropic efficiency of 82%. The air is compressed in the compressor from 1 bar static pressure to 4.2 bar total head pressure. The air enters the impeller eye with a velocity of 150 m/s with no pre-whirl. Assuming that the ratio of whirl speed to tip speed is 0.9, calculate (a) the rise in total head temperature during compression if the change in kinetic energy is negligible, (b) the top diameter of the impeller, (c) power required, and (d) the eye diameter if the hub diameter is 10 cm.

10. A centrifugal compressor is to have a pressure ratio of 3.5 : 1. The inlet eye of the impeller is 30 cm in diameter. The axial velocity at inlet is 130 m/s and the mass flow is 10 kg/s. The velocity in the delivery duct is 115 m/s. The tip speed of the impeller is 450 m/s and runs at 16000 rpm with total head isentropic efficiency of 78% and pressure coefficient of 0.72. The ambient conditions are 1 bar and 15°C. Calculate : (a) the static pressure ratio, (b) the static pressure and temperature at inlet and outlet of compressor, (c) work of compressor per kg of air, and (d) the theoretical power required.

11. The following details refer to a centrifugal air compressor of a total head pressure ratio of 3 and isentropic efficiency of 70%. Find the power required to run the compressor.

    Suction conditions :

Pressure	= 1 bar,
Temperature	= 20°C
Velocity of air at inlet	= 60 m/s

    Pressure and temperature are static values.

Free air delivered	= 20 kg/min
Mechanical efficiency	= 95%
Ratio of specific heats for air	= 1.4
Specific gas constant for air	= 0.287 kJ/kg.K

12. A centrifugal compressor has a pressure ratio of 4 : 1 with an isentropic efficiency of 82% when running at 16000 rpm. It takes in air at 17°C. Guide vanes at inlet give the air a pre-whirl of 20° to the axial direction at all radii and the mean diameter of the eye is 200 mm; the absolute air velocity at inlet is 120 m/s. At exit the blades are radially inclined and the impeller tip diameter is 550 mm. Calculate the slip factor of the compressor.

□□□

Chapter 11

Axial Flow Compressors

Inside This Chapter

11.1 INTRODUCTON

In an axial flow compressor, the air flows, throughout the compressor, parallel to its axis. Axial flow compressor has many merits over centrifugal compressor and is particularly suitable for super charging of internal combustion engines, turbojets and gas turbines.

11.2 CONSTRUCTIONAL FEATURES

The cross-section of a typical axial flow compressor is shown in Fig. 11.1 (a) and a photograph of the moving and fixed rings of blading is shown in Fig. 11.1 (b). It consists of adjacent rows of moving and fixed blades. The moving blades are mounted on the rotating drum and fixed blades are fixed to the casing or stator. One stage of the compressor comprises of a row of moving blades followed by a row of fixed blades.

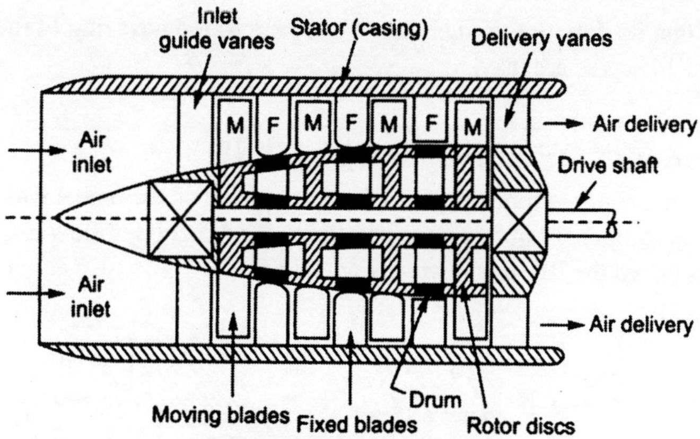

(a) Typical axial flow compressor

(b) Photographic view of moving and fixed rings of blading

Fig. 11.1

The blades are of aero-foil section designed on the basis of aerodynamic theory. Such a design prevents losses due to shock and turbulence and are free from stalling problems. The blades are said to be stalled when the air stream fails to follow the blade profile.

The blades of a reaction gas turbine have profile formed by a number of circular arcs. This is so because the acceleration process carried out in the converging blade passages of a reaction turbine is much more efficient and stable process as compared with the diffusing or decelerating process carried out in the diverging passage between the blades of an axial flow compressor.

To keep the flow velocity constant throughout the compressor length the annular area is reduced from inlet to outlet of the compressor. In the diverging passage of the moving blades, there is rise in temperature due to diffusion and the absolute velocity is also increased due to work input.

The fixed blades serve to convert a part of the kinetic energy of air into pressure energy by diffusion process and also guide and redirect the air stream to enter the next stage without shock.

11.3 WORKING PRINCIPLE

The axial compressor is generally driven by an internal combustion engine or a turbine. The work input to the rotor shaft is transferred by the rotor moving blades to the air, thus accelerating it. The space between the moving blades form diffuser passages so that the velocity of air relative to the blades is decreased as the air passes through them and consequently the pressure rises. The air is then further diffused in the fixed blades which also form diffuser passages. In the fixed blades the air is turned through an angle so that its direction is such that it can pass to the next ring of moving blades without shock. Generally 5 to 14 stages are used.

11.4 SIMPLE THEORY OF AEROFOIL BLADING

When an object is placed within a fluid stream, a force is exerted on the object which may be inclined to the flow direction. The component of this force parallel to the direction of flow is called the drag and the normal component is called the lift, as shown in Fig. 11.2.

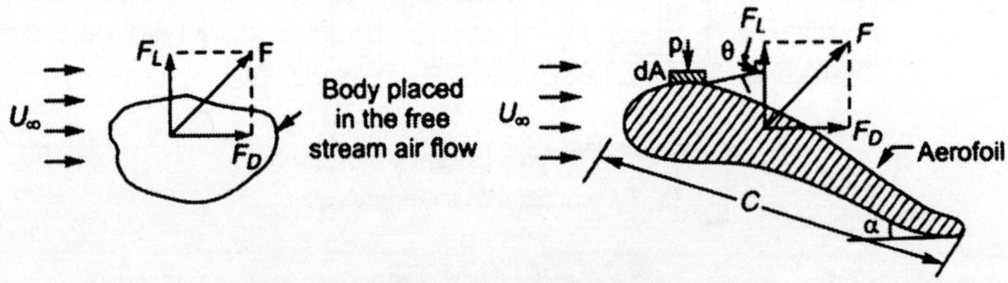

Fig. 11.2 Definition sketch of lift and drag.

Consider an aerofoil section shown in Fig. 11.3, which has an angle of attack α, the chord length C and the length of the foil perpendicular to the plane of the paper, termed as the span, l. The resultant force F executed by the fluid on the aerofoil has two mutually perpendicular components : the lift F_L and the drag F_D. Consider a surface element of elmentary area dA of aerofoil on which a pressure p act. Let the tangent to dA make an angle θ with the flow direction. The differential lift and drag on this area element are :

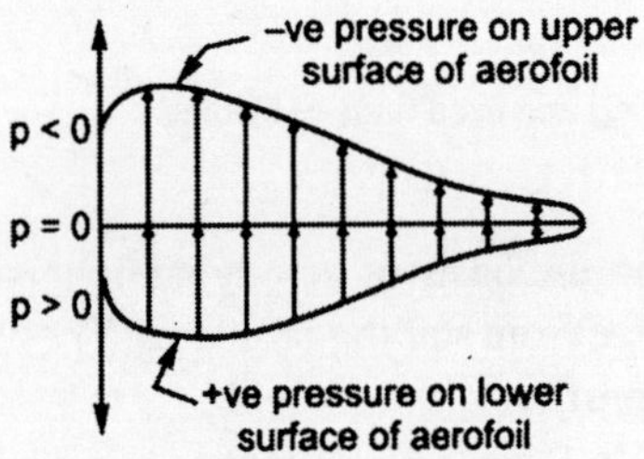

Fig. 11.3 Pressure distribution on an aerofoil section.

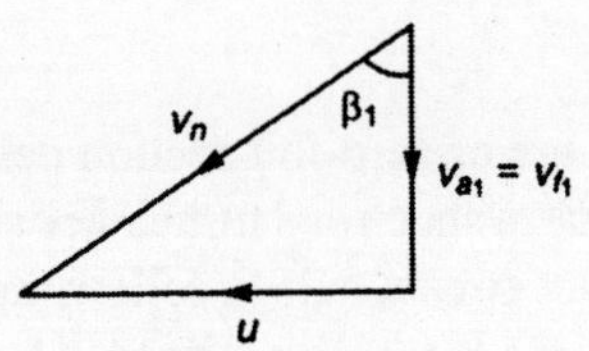

Fig. 11.4

$$dF_L = pdA \cos\theta$$

$$dF_D = pdA \sin\theta$$

Lift,
$$F_L = \int_A pdA \cos\theta$$

Drag,
$$F_D = \int_A pdA \sin\theta$$

where $\int_A$ represents the integration over the entire body surface area A.

We define the following lift and drag coefficients :

Lift coefficient,
$$C_L = \frac{F_L}{\frac{1}{2}\rho U_\infty^2 A} \quad \text{...(11.1)}$$

Drag coefficient,
$$C_D = \frac{F_D}{\frac{1}{2}\rho U_\infty^2 A} \quad \text{...(11.2)}$$

where ρ = density of air stream

U_{∞} = free stream velocity of air

A = projected area of the body perpendicular to the oncoming flow, *i.e.*, the frontal area of the body.

Let

$\dot{V}$ = free air delivered by the compressor, m^3/min

i = number of moving blades per blade ring

p_1, T_1 = ambient air conditions

D_m = blade ring mean diameter, m

k_b = blade factor

h = blade height, m

N = rotor speed, rpm

C_L, C_D = lift and drag coefficients respectively

A = projected area of blades, m^2

Then density of air,
$$\rho = \frac{p_1}{RT_1}$$

$$\dot{V} = v_a A$$

$$A = k_b \pi D_m h$$

$$v_a = \frac{\dot{V}}{60\,A} = \frac{\dot{V}}{60 \times k_b \times \pi D_m h} \text{ m/s}$$

$$u = \frac{\pi D_m N}{60} \text{ m/s}$$

$$\tan \beta_1 = \frac{u}{v_a} \qquad \text{(see Fig. 11.4)}$$

$$v_{r1} = \sqrt{u^2 + v_a^2}$$

Lift force,
$$F_L = \frac{1}{2} C_L \rho v_r^2 A \qquad ...(11.3)$$

Drag force,
$$F_D = \frac{1}{2} C_D \rho v_r^2 A \qquad ...(11.4)$$

Power input/stage,
$$P = \frac{1}{2}(F_L \cos \beta_1 + F_D \sin \beta_1)\, u \times 10^{-3} \text{ kW} \qquad ...(11.5)$$

Mass flow rate,
$$\dot{m} = \frac{\dot{V}\rho}{60} \text{ kg/s} \qquad ...(11.6)$$

Also
$$P = \frac{\dot{m} c_p T_1}{\eta_c} \left[r_p^{\frac{\gamma - 1}{\gamma}} - 1 \right] \qquad ...(11.7)$$

where $r_p = \frac{p_2}{p_1}$ and η_c = efficiency of the compressor.

11.5 VELOCITY DIAGRAMS

Figure 11.5 shows the velocity triangles for one stage of an axial flow compressor. All angles are measured from the axial direction. The blade velocity u is taken to be same at blade entry and exit as the air enters and leaves the blades at almost equal radii. Due to the diffusion process in the moving blades, $v_{r2} < v_{r1}$ but $v_{a2} > v_{a1}$. The air leaves the fixed blades with velocity v_{a3} at an angle α_3 and is redirected to the next page. It is assumed that $v_{a1} = v_{a3}$.

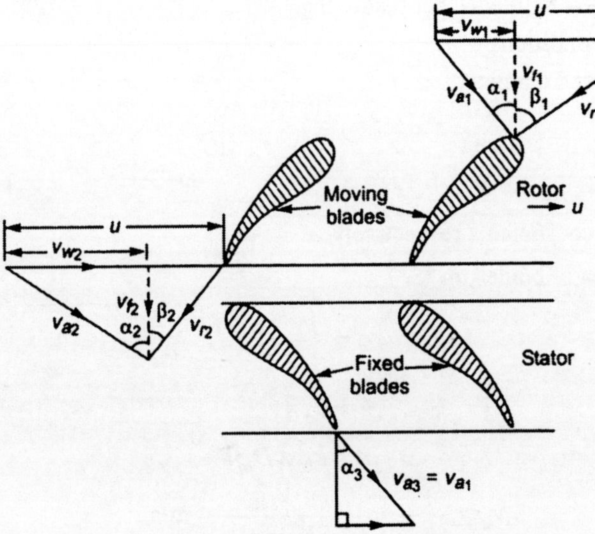

Fig. 11.5 Velocity diagrams for axial flow compressor.

From the inlet velocity triangle, we have

$$\tan\alpha_1 = \frac{v_{w1}}{v_{f1}}, \quad \tan\beta_1 = \frac{u - v_{w1}}{v_{f1}}$$

$$\therefore \quad \frac{u}{v_{f1}} = \tan\alpha_1 + \tan\beta_1 \qquad \text{...(11.8)}$$

From the outlet velocity triangle, we have

$$\tan\alpha_2 = \frac{v_{w2}}{v_{f2}}, \quad \tan\beta_2 = \frac{u - v_{w2}}{v_{f2}}$$

$$\therefore \quad \frac{u}{v_{f2}} = \tan\alpha_2 + \tan\beta_2 \qquad \text{...(11.9)}$$

Assume 1 kg flow of air through the compressor stage, we have

Tangential force per kg of air $= v_{w2} - v_{w1}$

Work required by the stage per kg of air,

$$w = u(v_{w2} - v_{w1}) = c_p(T'_{02} - T_{01}) \qquad \text{...(11.10)}$$

Now $\quad v_{f1} = v_{f2} = v_f$

$$\therefore \quad w = uv_f(\tan\alpha_2 - \tan\alpha_1) \qquad \text{...(11.11)}$$

$$= uv_f(\tan\beta_1 - \tan\beta_2) \quad \text{...(11.12)}$$

$$= u[u - v_f(\tan\alpha_1 + \tan\beta_2)] \quad \text{...(11.13)}$$

Also isentropic efficiency of the stage,

$$\eta_{isen} = \frac{T_{02} - T_{01}}{T'_{02} - T_{01}}$$

$$T_{02} = T_{01} + \eta_{isen}(T'_{02} - T_{01})$$

$$\frac{T_{02}}{T_{01}} = 1 + \eta_{isen}\frac{(T'_{02} - T_{01})}{T_{01}}$$

and

$$\frac{p_{02}}{p_{01}} = \left(\frac{T_{02}}{T_{01}}\right)^{\frac{\gamma}{\gamma-1}} = \left[1 + \eta_{isen}\frac{(T'_{02} - T_{01})}{T_{01}}\right]^{\frac{\gamma}{\gamma-1}}$$

$$= \left[1 + \eta_{isen}\frac{u(v_{w2} - v_{w1})}{c_p T_{01}}\right]^{\frac{\gamma}{\gamma-1}} \quad \text{...(11.14)}$$

11.6 DEGREE OF REACTION, R_d

Degree of reaction, R_d, is defined as the ratio of pressure rise in the rotor blades to the pressure rise in the compressor stage.

Pressure rise in the rotor blades $= \frac{1}{2}(v_{r1}^2 - v_{r2}^2)$

But

$$v_{r1}^2 = v_f^2 + (v_f \tan\beta_1)^2$$

and

$$v_{r2}^2 = v_f^2 + (v_f \tan\beta_2)^2$$

$$\frac{1}{2}(v_{r1}^2 - v_{r2}^2) = \frac{v_f^2}{2}(\tan^2\beta_1 - \tan^2\beta_2)$$

Pressure rise in the compressor stage

$$= uv_f(\tan\beta_1 - \tan\beta_2)$$

$$R_d = \frac{\frac{v_f^2}{2}(\tan^2\beta_1 - \tan^2\beta_2)}{uv_f(\tan\beta_1 - \tan\beta_2)}$$

$$= \frac{v_f}{2u}(\tan\beta_1 + \tan\beta_2) \quad \text{...(11.15)}$$

For 50% degree of reaction, $R_d = \frac{1}{2}$

$$\therefore \quad \frac{u}{v_f} = \tan\beta_1 + \tan\beta_2 \quad \text{...(11.16)}$$

From Eqs. (11.8) and (11.9), we have

$$\frac{u}{v_f} = \tan\alpha_1 + \tan\beta_1 = \tan\alpha_2 + \tan\beta_2$$

$$\therefore \quad \alpha_1 = \beta_2 \text{ and } \alpha_2 = \beta_1 \qquad \text{...(11.17)}$$

Therefore with 50% degree of reaction, the blades are symmetrical. This reduces the friction losses considerably.

11.7 PRESSURE RISE IN ISENTROPIC FLOW THROUGH A CASCADE

Consider the incompressible isentropic and steady flow through a cascade from condition 1 to condition 2. Using Bernoulli's equation, we have

$$\frac{p_1}{\rho} + \frac{v_1^2}{2} = \frac{p_2}{\rho} + \frac{v_2^2}{2}$$

$$(\Delta p)_{\text{isen}} = (p_2)_{\text{isen}} - \frac{\rho}{2}[(v_{f1}^2 + v_{w1}^2) - (v_{f2}^2 + v_{w2}^2)]$$

Now $\quad v_{f1} = v_{f2} = v_f$

$$\therefore \quad (\Delta p)_{\text{isen}} = \frac{\rho}{2}(v_{w1}^2 - v_{w2}^2) = \frac{\rho}{2} v_f^2 (\tan^2\alpha_1 - \tan^2\alpha_2) \qquad \text{...(11.18)}$$

11.8 POLYTROPIC EFFICIENCY

It is the isentropic efficiency of one stage of a multistage compressor. This small stage efficiency is constant for all stages of a compressor with infinite number of stages.

Consider the compression process of a multistage compressor on T-s diagram shown in Fig. 11.6 from total pressure p_{01} to p_{02} in four stages of equal pressure ratio with intermediate pressures p_{0a}, p_{0b} and p_{0c}.

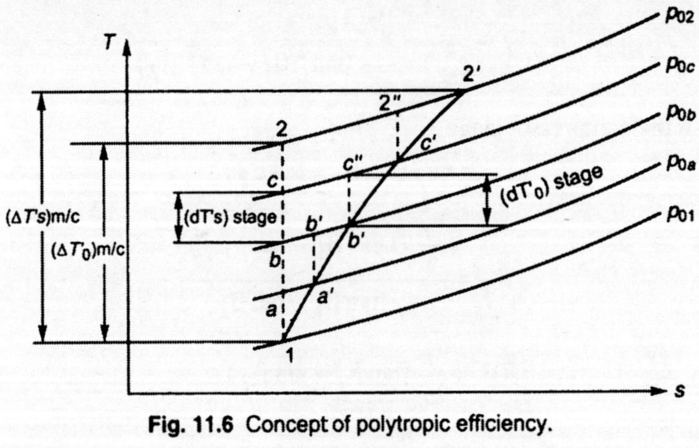

Fig. 11.6 Concept of polytropic efficiency.

Overall isentropic stagnation efficiency of machine is,

$$\eta_{\text{isen}\,(m/c)} = \frac{(\Delta T_0)_{m/c}}{(\Delta T_0')_{m/c}}$$

Stagnation isentropic efficiency for the stage,

$$\eta_{isen\,(st)} = \frac{(dT_0)_{st}}{(dT_0')_{st}}$$

Total actual temperature rise,

$$\Delta T_0' = \frac{(\Delta T_0)_{m/c}}{\eta_{isen\,(m/c)}}$$

Also

$$\Delta T_0' = \frac{\Sigma (dT_0)_{st}}{\eta_{isen\,(st)}}$$

Thus

$$\frac{(\Delta T_0)_{m/c}}{\eta_{isen\; m/c}} = \frac{\Sigma (dT_0)_{st}}{\eta_{isen\,(st)}}$$

or

$$\frac{\eta_{isen\,(m/c)}}{\eta_{isen\,(st)}} = \frac{(\Delta T_0)_{m/c}}{\Sigma (dT_0)_{st}}$$

Now

$$(\Delta T_0)_{m/c} = (1-a) + (a-b) + (b-c) + (c-2)$$

$$\Sigma (dT_0)_{st} = (1-a) + (a'-b'') + (b'-c'') + (c'-2'')$$

On the T-s plot, the constant pressure lines diverge towards the right, therefore,

$$(a'-b'') > (a-b)$$

$$(b'-c'') > (b-c)$$

and so on.

So we can say that

$$\Sigma (dT_0)_{st} > (\Delta T_0)_{m/c}$$

$$\therefore \qquad \eta_{isen\,(st)} > \eta_{isen\,(m/c)}$$

The small stage efficiency $\eta_{isen\,(st)}$ which is constant for all stages is called polytropic efficiency and is denoted by η_p.

Let the law of compression for the irreversible adiabatic path $1-2'$ be,

$$p_0 v_0^n = c_1 \qquad ...(1)$$

and for the isentropic path 1-2 is,

$$p_0 v_0^\gamma = c_2$$

Eq. (1) can be written as :

$$p_0 = c_1 \rho_0^n$$

Differentiating, we get

$$dp_0 = c_1 n \rho_0^{n-1}\, d\rho_0 = n \frac{p_0}{\rho_0}\, d\rho_0 \qquad ...(2)$$

Now

$$p_0 = \rho_0 R T_0'$$

or

$$\rho_0 = \frac{p_0}{R T_0'}$$

Differentiating, we get

$$d\rho_0 = \frac{1}{R}\left[\frac{T_0' \cdot dp_0 - p_0 \cdot dT_0'}{T_0'^2}\right] \qquad \text{...(3)}$$

Eq. (2) becomes

$$dp_0 = n\frac{p_0}{\rho_0} \cdot \frac{1}{R}\left[\frac{T_0' \cdot dp_0 - p_0 \cdot dT_0'}{T_0'^2}\right]$$

$$= nRT_0' \cdot \frac{1}{R}\left[\frac{T_0' \cdot dp_0 - p_0 \cdot dT_0'}{T_0'^2}\right]$$

$$= n\left[\frac{T_0' \cdot dp_0 - p_0 \cdot dT_0'}{T_0'}\right] = n \cdot dp_0 - np_0\frac{dT_0'}{T_0'}$$

or

$$dp_0(n-1) = np_0\frac{dT_0'}{T_0'}$$

$$dT_0' = dp_0\left(\frac{n-1}{n}\right)\frac{T_0'}{p_0} \qquad \text{...(4)}$$

Similarly for the ideal compression process $p_0 v_0^{\gamma} = c_2$, we have

$$dT_0 = dp_0\left(\frac{\gamma-1}{\gamma}\right)\frac{T_0'}{p_0}$$

Thus

$$\eta_p = \frac{dT_0}{dT_0'} = \left(\frac{\gamma-1}{\gamma}\right)\left(\frac{n}{n-1}\right) \qquad \text{...(11.19)}$$

From Eq. (4), we have

$$\frac{n}{n-1} = \frac{dp_0}{T_0} \cdot \frac{T_0'}{dT_0'}$$

$$\eta_p = \frac{\gamma-1}{\gamma} \times \frac{dp_0}{p_0} \times \frac{T_0'}{dT_0'}$$

$$\eta_p \frac{aT_0'}{T_0'} = \frac{\gamma-1}{\gamma} \cdot \frac{dp_0}{p_0}$$

Integrating between the end states 1 and 2′, we get

$$\eta_p \ln\left(\frac{T_{02}'}{T_{01}'}\right) = \left(\frac{\gamma-1}{\gamma}\right)\ln\left(\frac{p_{02}'}{p_{01}'}\right) = \left(\frac{\gamma-1}{\gamma}\right)\ln\left(\frac{p_{02}}{p_{01}}\right)$$

$$\eta_p = \frac{\left(\frac{\gamma-1}{\gamma}\right)\ln\left(\frac{p_{02}}{p_{01}}\right)}{\ln\left(\frac{T_{02}'}{T_{01}'}\right)} = \frac{\ln\left(\frac{p_{02}}{p_{01}}\right)^{\frac{\gamma-1}{\gamma}}}{\ln\left(\frac{T_{02}'}{T_{01}}\right)} \qquad \text{...(11.20)}$$

11.9 FLOW COEFFICIENT, HEAD OR WORK COEFFICIENT, DEFLECTION COEFFICIENT AND PRESSURE COEFFICIENT

Flow Coefficient : It is defined as the ratio of flow (or axial) velocity at inlet to the blade velocity.

Flow coefficient, $$\phi_f = \frac{v_{f_1}}{u}$$

Now $$u = v_{f_1}(\tan\alpha_1 + \tan\beta_1)$$

$\therefore$ $$\phi_f = \frac{1}{\tan\alpha_1 + \tan\beta_1} \quad \text{...(11.21a)}$$

Also $$v_{f_1} = v_{f_2} = v_f$$

$$\phi_f = \frac{v_{f_2}}{u} = \frac{1}{\tan\alpha_2 + \tan\beta_2} \quad \text{...(11.21b)}$$

Head or Work Coefficient : It is defined as the ratio of actual work done to the kinetic energy corresponding to the mean peripheral velocity.

Head or work coefficient,

$$\phi_h = \frac{c_p \cdot \Delta T}{\frac{u^2}{2}} = \frac{2(v_{w2} - v_{w1})\,u}{u^2}$$

$$= \frac{2(v_{w2} - v_{w1})}{u} = \frac{2(\tan\alpha_2 - \tan\alpha_1)}{\tan\alpha_2 + \tan\beta_2} \quad \text{...(11.22)}$$

Deflection Coefficient : ϕ_{def} : It is defined as,

$$\phi_{def} = \frac{u(v_{w2} - v_{w1})}{u^2} = \frac{v_{w2} - v_{w1}}{u} = \frac{\phi_h}{2} \quad \text{...(11.23)}$$

Pressure coefficient : It is defined as the ratio of isentropic work done to kinetic energy corresponding to the peripheral velocity.

Pressure coefficient, $$\phi_p = \frac{c_p \Delta T_{isen}}{\frac{u^2}{2}} = \eta_{isen} \cdot \phi_h \quad \text{...(11.24)}$$

11.10 PRESSURE RISE IN A STAGE AND NUMBER OF STAGES

The static pressure ratio from Eq. (11.14) is,

$$\frac{p_2}{p_1} = \left[1 + \eta_{isen}\left\{\frac{T_2' - T_1}{T_1}\right\}\right]^{\frac{\gamma}{\gamma-1}}$$

Temperature rise in rotating blades,

$$\Delta T_R = T_2' - T_1 = \frac{v_{r1}^2 - v_{r2}^2}{2c_p}$$

$$\therefore \qquad \frac{p_2}{p_1} = \left[1 + \eta_{isen} \frac{\Delta T_R}{T_1}\right]^{\frac{\gamma}{\gamma - 1}}$$

Pressure rise in rotating blades,

$$\Delta p_R = p_2 - p_1 = p_1 \left[\left\{1 + \eta_{isen} \frac{\Delta T_R}{T_1}\right\}^{\frac{\gamma}{\gamma - 1}} - 1\right] \qquad ...(11.25)$$

$$\Delta T_s = T_3 - T_2 = \frac{v_{a2}^2 - v_{a3}^2}{2c_p}$$

$$\therefore \qquad \frac{p_3}{p_2} = \left[1 + \eta_s \frac{\Delta T_s}{T_2}\right]^{\frac{\gamma}{\gamma - 1}}$$

Pressure rise in stationary blades is,

$$\Delta p_s = p_3 - p_2 = p_2 \left[\left\{1 + \eta_s \frac{\Delta T_s}{T_2}\right\}^{\frac{\gamma}{\gamma - 1}} - 1\right] \qquad ...(11.26)$$

Pressure increases in stage is,

$$\Delta p_{st} = \Delta p_R + \Delta p_s \qquad ...(11.27)$$

and

$$\Delta T_{st} = \Delta T_R + \Delta T_s \qquad ...(11.28)$$

The stagnation pressure ratio is,

$$\frac{p_{02}}{p_{01}} = \left[1 + \eta_{0\,isen} \frac{\Delta T'_0}{T_{01}}\right]^{\frac{\gamma}{\gamma - 1}} \qquad ...(11.29)$$

If the work done per stage is assumed to be the same, then number of stage z is given by,

Number of stages,

$$z = \frac{\Delta T_0}{(dT_0)_{stage}} \qquad ...(11.30)$$

If pressure ratio per stage is the same, then

$$(r_p)_{st} = \frac{p_{02}}{p_{01}} = \frac{p_{03}}{p_{02}} = \ldots = \frac{p_{0(z+1)}}{p_{0z}}$$

The overall pressure ratio is,

$$r_p = [(r_p)_{st}]^z$$

or

$$z = \frac{\ln r_p}{\ln (r_p)_{st}} \qquad ...(11.31)$$

11.11 SURGING, CHOKING AND STALLING

1. Surging

The delivery pressure v's mass flow rate in an axial flow compressor is shown in Fig. 11.7. The mass flow is zero when the discharge value is closed, but the air entrapped in the blade area is compressed and pressure of air increases. The condition is represented by point 'a' in Fig. 11.7. As the valve is opened, the flow of air starts and the pressure rises upto point 'b'.

Any increase in air mass flow after point 'b' is accompanied by decrease in delivery pressure. This is because, the rate of increase in pressure loss due to friction is more than the rate of increase in pressure rise due to diffuser. Theoretically the decrease in delivery pressure is continued upto point 'e'. In actual practice, the maximum mass flow is limited by the point 'e' because beyond point 'e' the mass flow exceeds design mass flow, the air angles are widely different from vane angles and choking takes place.

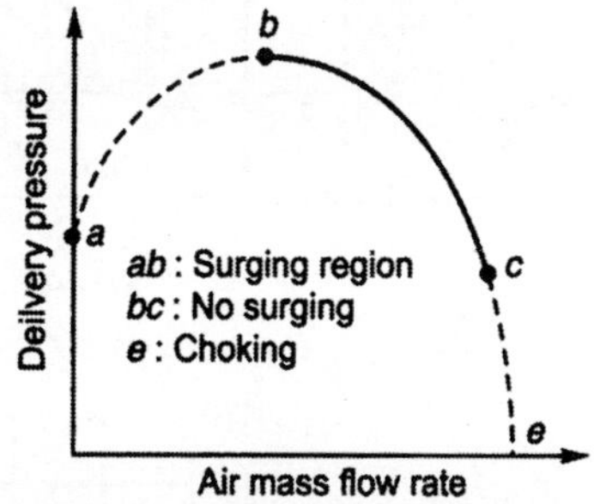

Fig. 11.7 Depicting surging and choking.

If the compressor is working between points 'a' and 'b' then as the mass flow is decreased the delivery pressure decreases. However the pressure downstream side of the compressor does not fall quickly resulting in reversal of flow from downstream side towards the resulting pressure gradient. When this occurs, the pressure to the downstream side of the compressor falls and the compressor will start delivering air and the cycle would be repeated continuously. Thus the flow would be pulsating between points 'a' and 'b'. This pulsating air flow phenomena is known as 'surging'. The surging causes overheating and stress reversal in the blades and damages the compressor. Within the region bc the flow is stable. A fall in mass flow rate will result in rise in delivery pressure to restore the fall.

2. Choking

The maximum mass flow rate possible in a compressor is known as choking. The point 'e' in Fig. 11.7 represents the choking of the compressor.

3. Stalling

It is a phenomena which occurs on the suction side of the compressor in which the flow breaks away from the aerofoil blading. It may be due to lesser flow rate than the designed value or due to the non-uniformity in the profile of the blades. Thus stalling preceeds surging. Stalling is a local phenomenon whereas surging is a complete system phenomena.

11.12 PERFORMANCE CHARACTERISTICS

The relationship between pressure rise and mass flow rate at various speeds is shown in Fig. 11.8. At a certain speed, efficiency increases as the mass flow rate increases and reaches a maximum value after which it decreases, as shown in Fig. 11.9. The power consumed increases as mass flow increases. Figure 11.10 shows the performance and constant efficiency curves. The performance curves are plotted with dimensionless parameters, $\frac{G\sqrt{T_1}}{p_1}$, $\frac{N}{\sqrt{T_1}}$ and $\frac{p_3}{p_1}$. At constant value of $\frac{N}{\sqrt{T_1}}$ there is a considerable

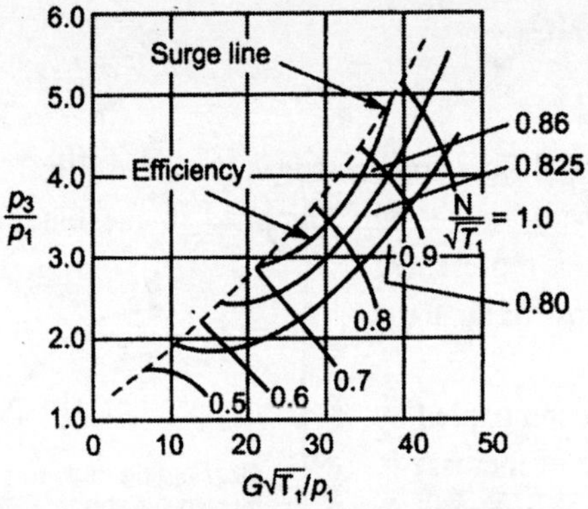

Fig. 11.8 Pressure ratio v's mass flow rate.

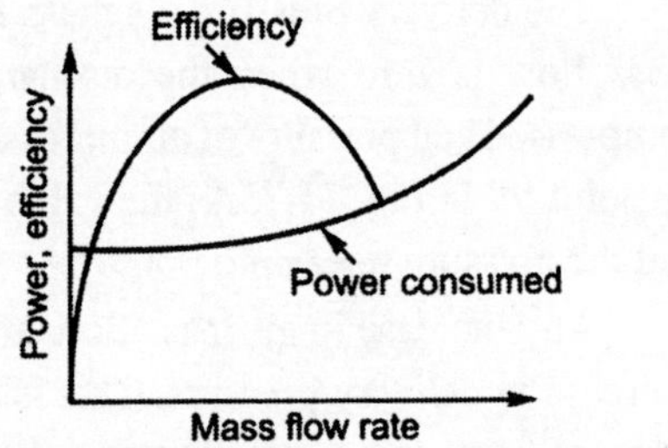

Fig. 11.9 Power and efficiency v's mass flow rate.

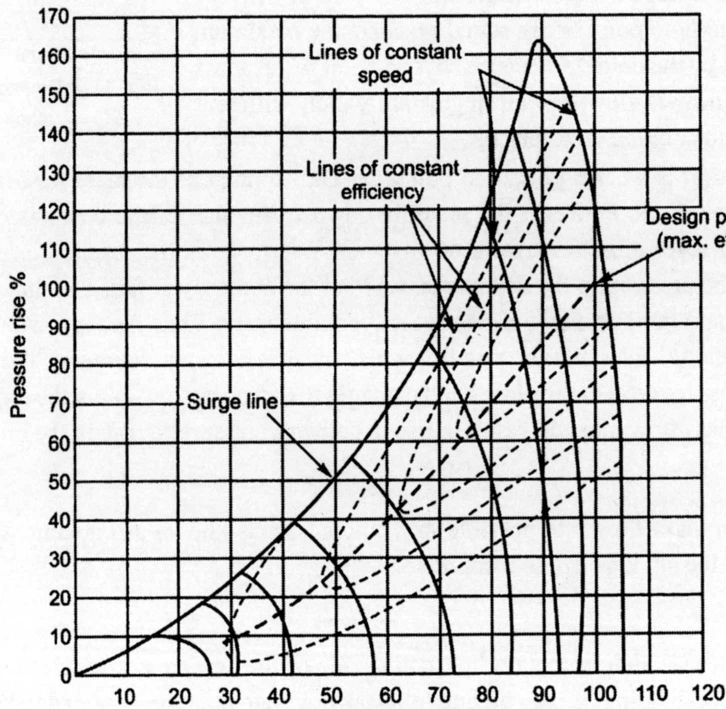

Fig. 11.10 Pressure rise v's volume flow.

narrower range of mass flow rate $\frac{G\sqrt{T_1}}{p_1}$ than in centrifugal compressor. The surge point is reached in axial compressors much before the maximum pressure ratio for given value of $\frac{N}{\sqrt{T_1}}$. Since the design usually calls for the operating line to be near the maximum point on the $\frac{N}{\sqrt{T_1}}$ curves, it follows that the operating line for axial flow compressors must be very near the surge line, thus narrowing the range of stable operation.

11.13 COMPARISON OF AXIAL FLOW AND CENTRIFUGAL COMPRESSORS

The merits and demerits of axial flow compressors are given in Table 11.1.

Table 11.1 : Merits and Demerits of axial flow compressor over Centrifugal compressor

S. No.	*Parameter*	*Axial flow*	*Centrifugal*
1.	Pressure ratio	1.2 : 1, require more number of stages for higher pressure ratio.	4 : 1 per stage
2.	Isentropic efficiency	85–88%	70%
3.	Power required per kg air flow rate	Lower than centrifugal	Higher than axial
4.	Flexibility of operation	Less	More
5.	Frontal area	More	Less
6.	Effect of deposit formation on the impeller	Effects	No effect
7.	Starting torque	High	Low
8.	Efficiency v's speed curve	More steep	More flat
9.	Suitability for supercharging	No	Suitable
10.	Applications	Large gas turbines	Refrigerators and industrial gases

11.14 APPLICATIONS OF AXIAL FLOW COMPRESSORS

Axial flow compressors have wide applications in the following :

1. Jet engines
2. Gas turbines for power plants
3. Steel mills

11.15 LOSSES IN AXIAL FLOW COMPRESSORS

The following losses occur in axial flow compressors :

1. Profile losses on the surface of blades.
2. Skin friction loss on the annulus walls.
3. Secondary flow losses.

■ **Example 11.1** An axial compressor is fitted with half-reaction blading, the blade inlet and outlet angles being 50° and 15° when measured from the axial direction. The mean diameter of a certain blade pair is 85 cm and the rotor runs at 5500 rpm. Calculate the necessary isentropic efficiency of the stage if the pressure ratio of compression is to be 1.4 and the inlet air temperature is 25°C. $c_p = 1.005$ kJ/kg. K, $\gamma = 1.4$. [CSE, 1987]

Solution :

$$\beta_1 = 50°, \beta_2 = 15°, v_{f1} = v_{f2} = v_f, \alpha_1 = 15°, \alpha_2 = 50°$$

Mean blade speed,

$$u = \frac{\pi DN}{60} = \frac{\pi \times 0.85 \times 5500}{60} = 244.78 \text{ m/s}$$

From the velocity triangles, (Fig. 11.11)

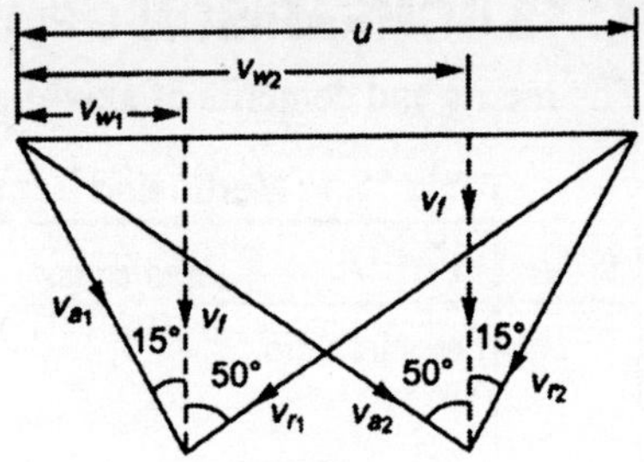

Fig. 11.11

$$\tan\beta_1 = \frac{u - v_{wl}}{v_f}$$

$$\tan\beta_2 = \frac{u - v_{w2}}{v_f} = \frac{v_{wl}}{v_f} \text{ for } R_d = 0.5$$

$$\frac{\tan\beta_1}{\tan\beta_2} = \frac{u - v_{wl}}{v_{wl}} = \frac{\tan 50°}{\tan 15°}$$

$$= \frac{1.192}{0268} = 4.448$$

$$v_{wl} = 0.1826\,u$$

$$v_{w2} = 0.8164\,u$$

$$\Delta v_w = v_{w2} - v_{wl} = 0.6328\,u = 0.6328 \times 244.78 = 155 \text{ m/s}$$

Work done/kg, $\quad w_a = u \times \Delta v_w = 244.78 \times 155 = 37940.9$ N.m/kg

Isentropic work, $\quad w_{isen} = c_p(T_2 - T_1)$

$$T_2 = T_1\left(\frac{p_2}{p_1}\right)^{\frac{\gamma-1}{\gamma}} = 298\,(1.4)^{\frac{1}{3.5}} = 328 \text{ K}$$

$$w_{isen} = 1.005\,(328 - 298) = 30.15 \text{ kJ/kg}$$

Isnetropic efficiency,

$$\eta_{isen} = \frac{w_{isen}}{w_a} = \frac{30.15 \times 10^3}{37940.9} \times 100 = 79.46\%$$

■ **Example 11.2** An axial flow compressor comprises a number of similar stages with equal workdone per stage, and the velocity of flow is uniform throughout the compressor. The following data is given :

Overall stagnation pressure ratio	= 4
Stagnation inlet temperature	= 330 K
Relative air angle at rotor inlet	= 130°
Relative air angle at rotor outlet	= 100°
Blade velocity	= 180 m/s
Degree of reaction	= 0.5
Oveall stagnation adiabatic efficiency	= 0.86

Calculate (a) the stagnation outlet temperature, and (b) number of stages.

Solution : The velocity diagrams are shown in Fig. 11.12.

$$T_{02} = T_{01}\left(\frac{p_{02}}{p_{01}}\right)^{\frac{\gamma-1}{\gamma}}$$

$$= 330\left(\frac{4}{1}\right)^{\frac{1}{3.5}} = 49038 \text{ K}$$

Stagnation adiabatic efficiency,

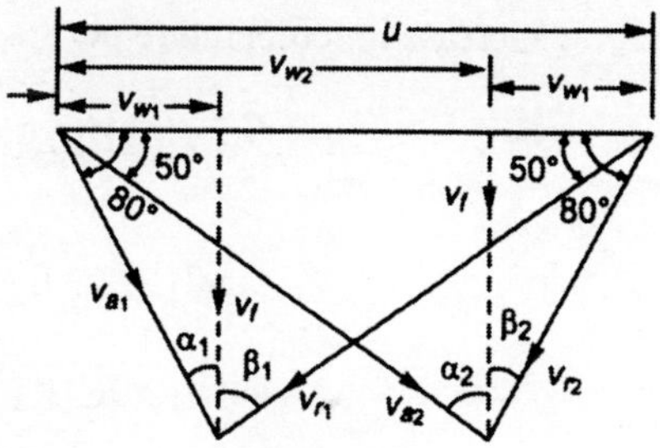

Fig. 11.12

$$\eta_a = \frac{T_{02} - T_{01}}{T'_{02} - T_{01}}$$

$$0.86 = \frac{490.38 - 330}{T'_{02} - 330}$$

$$T'_{02} = 516.5 \text{ K}$$

Also

$$\tan 80° = \frac{v_f}{u - v_{w2}} = 5.671$$

$$\tan 50° = \frac{v_f}{u - v_{w1}} = 1.192$$

For degree of reaction = 0.5, the velocity diagrams are similar.

$\therefore$

$$u - v_{w2} = v_{w1}$$

$$\frac{v_f}{v_{w1}} = 5.671 \text{ and } \frac{v_f}{u - v_{w1}} = 1.192$$

$$\frac{u - v_{w1}}{v_{w1}} = \frac{5.671}{1.192} = 4.757$$

$$v_{w1} = 0.1737\,u$$

$$v_{w2} = u - v_{w1} = 0.8263\,u$$

$$\Delta v_w = v_{w2} - v_{w1} = 0.6526\,u$$

Work done per stage $= u \cdot \Delta v_w = 180 \times 0.6526 \times 180 = 21144$ N.m/kg

Total work $= c_p(T_{02} - T_{01}) = 1.005(490.38 - 330) = 161.182$ kJ/kg

$$\text{Number of stages} = \frac{161.182 \times 10^3}{21144} = 7.62 = 8$$

■ **Example 11.3** Air at a temperature of 300 K enters a ten stage axial flow compressor at the rate of 3.5 kg/sec. The pressure ratio is 6.0 and the isentropic efficiency is 90%. The process is adiabatic and the compressor has symmetrical stages. The axial velocity of 120 m/sec is uniform across the stage and the mean blade speed of each stage is 200 m/sec. Assume that the temperature change is same in each stage.

Determine the direction of the air at entry to and exit from the rotor and the stator blades. Also find the power given to the air. Take $c_p = 1.005$ kJ/kg.K and $\gamma = 1.4$. [IES, 2001]

Solution : The velocity diagrams are shown in Fig. 11.13.

$z = 10$, $T_1 = 300$ K, $\dot{m} = 3.5$ kg/s, $r_p = 6$, $\eta_i = 0.9$,

$v_f = 120$ m/s, $u = 200$ m/s

$$(\Delta T)_{\text{stage}} = (\Delta T)_{\text{overall}}/z$$

$$= \left[\frac{T_1}{z\eta_i}\right][r_p^{(\gamma-1)/\gamma} - 1]$$

$$= \left[\frac{300}{10 \times 0.9}\right][6^{0.4/1.4} - 1] = 22.284 \text{ K}$$

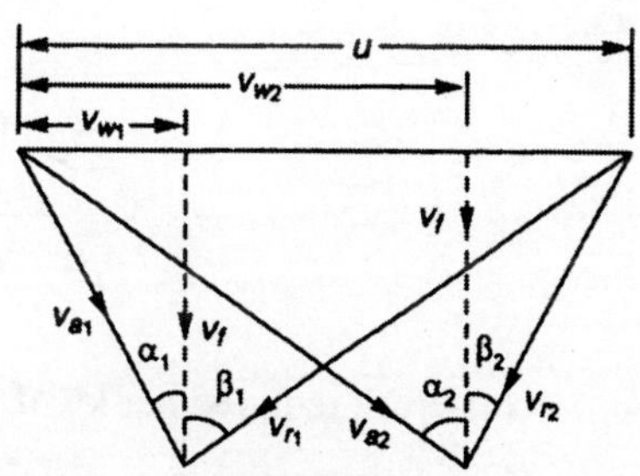

Fig. 11.13

Increase in temperature per stage,

$$(\Delta T)_{stage} = \left(\frac{uv_f}{c_p}\right)(\tan\beta_1 - \tan\beta_2)$$

$$\tan\beta_1 - \tan\beta_2 = 22.284 \times 1.005 \times 10^3 / (200 \times 120) = 0.93313$$

Also $$\tan\beta_1 + \tan\beta_2 = \frac{u}{v_f} = \frac{200}{120} = 1.6667$$

$$\tan\beta_1 = 1.2999, \beta_1 = 52.43°, \beta_2 = 20.14°$$

Power supplied $= z\dot{m}c_p(\Delta T)_{stage} = 10 \times 3.5 \times 1.005 \times 22.284 = 783.84$ kW

■ **Example 11.4** In an axial flow air compressor : $u = 250$ m/s, $v_f = 200$ m/s, $\alpha_1 = 50°, \alpha_2 = 15°$ and $\rho = 1\ \text{kg/m}^3$. Determine (a) the pressure rise, and (b) the work done per kg of air.

Solution :

(a) Pressure rise through a moving blade ring,

$$\Delta p = \frac{1}{2}\rho v_f^2 (\tan^2\alpha_1 - \tan^2\alpha_2)$$

$$= \frac{1}{2} \times 1 \times 200^2 (\tan^2 50° - \tan^2 15°) \times 10^{-5} = 0.2697 \text{ bar}$$

(b) Work done per kg of air,

$$w = u(v_{w1} - v_{w2}) = uv_f(\tan\alpha_1 - \tan\alpha_2)$$

$$= 250 \times 200 (\tan 50° - \tan 15°) \times 10^{-3} = 46.19 \text{ kW}$$

■ **Example 11.5** An axial flow air compressor having 8 stages and with 50% degree of reaction compresses air in the pressure ratio of 4.5 : 1. The air enters the compressor at 27°C and flows through it with a constant speed of 100 m/s. The moving blades of compressor rotate with a mean speed of 200 m/s. The isentropic efficiency of the compressor is 85%. Calculate (a) the work done on the compressor, and (b) blade angles.

Take $\gamma = 1.4$ and $c_p = 1.005$ kJ/kg.K for air.

Solution : $z = 8, R_d = 0.5, r_p = 4.5, T_1 = 273 + 27 = 300$ K, $v_f = 100$ m/s, $u = 200$ m/s, $\eta_{isen} = 0.85$.

$$\eta_{isen} = \frac{T_2 - T_1}{T_2' - T_1}$$

$$\frac{T_2}{T_1} = (r_p)^{\frac{\gamma-1}{\gamma}}, T_2 = 300(4.5)^{\frac{1}{3.5}} = 461 \text{ K}$$

$$T_2' = T_1 + \frac{T_2 - T_1}{\eta_{isen}} = 300 + \frac{461 - 300}{0.85} = 489.4 \text{ K}$$

(a) Work required per kg of air,

$$w = c_p(T_2' - T_1) = 1.005(489.4 - 300) = 190.36 \text{ kJ/kg}$$

Total work done $= zu(v_{w2} - v_{w1}) = zu\, v_f(\tan\alpha_2 - \tan\alpha_1)$

$$19036 = 8 \times 200 \times 100\,(\tan\alpha_2 - \tan\alpha_1) \times 10^{-3}$$

$$\tan\alpha_2 - \tan\alpha_1 = 1.1897 \qquad ...(1)$$

For $R_d = 0.5, \alpha_1 = \beta_2$ and $\alpha_2 = \beta_1$

$\therefore$

$$1.1897 = \tan\beta_1 - \tan\beta_2 \qquad ...(2)$$

Now

$$u = v_f(\tan\alpha_1 + \tan\beta_1)$$

$$\tan\alpha_1 + \tan\beta_1 = \frac{u}{v_f} = \frac{200}{100} = 2 \qquad ...(3)$$

From Eq. (2), we have

$$\tan\beta_1 - \tan\alpha_1 = 1.1897 \qquad ...(4)$$

Adding Eqs. (3) and (4), we get

$$2\tan\beta_1 = 3.1897$$

$$\tan\beta_1 = 1.5949$$

$$\beta_1 = 57.9° = \alpha_2$$

$$\tan\alpha_1 = \tan\beta_1 - 1.1897 = 0.4052$$

$$\alpha_1 = 22° = \beta_2$$

■ **Example 11.6** The overall isentropic efficiency of an axial flow air compressor is 83%. It draws air at 25°C and 1 bar and compresses to 4 bar. Assuming 50% degree of reaction, blade velocity 180 m/s, $\alpha_1 = 15°, \beta_1 = 45°$ and work input factor as 0.80, calculate (a) the flow velocity and (b) number of stages.

Solution : $\eta_{isen} = 0.83, T_1 = 273 + 25 = 298$ K, $p_1 = 1$ bar, $p_2 = 4$ bar,

$$R_d = 0.5, \alpha_1 = 15°, \beta_1 = 45°, u = 180 \text{ m/s}, \phi_w = 0.80$$

$$T_2 = T_1\left(\frac{p_2}{p_1}\right)^{\frac{\gamma-1}{\gamma}} = 298\left(\frac{4}{1}\right)^{\frac{1}{3.5}} = 442.8 \text{ K}$$

$$\eta_{isen} = \frac{T_2 - T_1}{T_2' - T_1}$$

$$T_2' = T_1 + \frac{T_2 - T_1}{\eta_{isen}} = 298 + \frac{442.8 - 298}{0.83} = 472.5 \text{ K}$$

Work required per kg of air

$$= c_p(T_2' - T_1) = 1.005(472.5 - 298) = 175.37 \text{ kJ/kg}$$

From inlet velocity triangle, we have

$$\frac{u}{v_f} = \tan\alpha_1 + \tan\beta_1 = \tan 15° + \tan 45° = 1.268$$

$$v_f = \frac{180}{1.268} = 141.96 \text{ m/s}$$

Work done per kg per stage $= \phi_w u(v_{w2} - v_{wl})$

For $\qquad R_d = 0.5, \alpha_2 = \beta_1 = 45°$ and $\alpha_1 = \beta_2 = 15°$

$$v_{w2} = v_f \tan\alpha_2 = 141.96 \tan 45° = 141.96 \text{ m/s}$$

$$v_{wl} = v_f \tan\alpha_1 = 141.96 \tan 15° = 38.04 \text{ m/s}$$

Work done/stage $= 0.80 \times 180(141.96 - 38.04) \times 10^{-3} = 14.965$ kJ/kg

Number of stages, $\qquad z = \dfrac{175.37}{14.965} = 11.72 = 12$

■ **Example 11.7** In a 10 stage axial flow air compressor, the overall stagnation pressure ratio is 4 : 1 with an overall isentropic efficiency of 90 percent. The inlet stagnation temperature and pressure are 293 K and 1 bar. The work is divided equally between the stages. The mean blade speed is 180 m/s and 50% reaction blading are used. The flow velocity throughout the compressor is 90 m/s. Calculate (a) the blade angles, and (b) the power required.

Solution : $z = 10$, $r_{op} = 4$, $\eta_{isen} = 0.9$, $T_{01} = 293$ K, $p_{01} = 1$ bar, $u = 180$ m/s, $R_d = 0.5$, $v_f = 90$ m/s

(a) For $R_d = 0.5, \alpha_1 = \beta_2$ and $\alpha_2 = \beta_1$

With isentropic compression the temperature of air leaving the compressor is,

$$T_{010} = T_{01}\left(\frac{p_{010}}{p_{01}}\right)^{\frac{\gamma-1}{\gamma}} = 293\left(\frac{4}{1}\right)^{\frac{1}{3.5}} = 435.4 \text{ K}$$

$$T'_{010} = T_{01} + \frac{T_{010} - T_{01}}{\eta_{isen}} = 293 + \frac{435.4 - 293}{0.9} = 451.2 \text{ K}$$

Work required by the compressor $= c_p(T'_{010} - T_{01})$

$$= 1.005(451.2 - 293) = 159 \text{ kJ/kg}$$

$$= zu(v_{w2} - v_{wl})$$

$$= zuv_f(\tan\alpha_2 - \tan\alpha_1)$$

$$\tan\alpha_2 - \tan\alpha_1 = \frac{159 \times 10^3}{10 \times 180 \times 90} = 0.9815 \qquad \text{...(1)}$$

From inlet velocity triangle, we have,

$$\frac{u}{v_f} = \tan\alpha_1 + \tan\beta_1 = \frac{180}{90} = 2 \qquad \text{...(2)}$$

Eq. (1) is

$$\tan\beta_1 - \tan\alpha_1 = 0.9815 \qquad \text{...(3)}$$

Adding Eqs. (2) and (3), we get

$$2\tan\beta_1 = 2.9815$$

$$\tan\beta_1 = 1.4908$$

$$\beta_1 = 56.15° = \alpha_2$$

$$\tan\alpha_1 = 2 - 1.4908 = 0.5092$$

$$\alpha_1 = 26.98° = \beta_2$$

(b) Power required $= \dot{m}c_p(T'_{010} - T_{01}) = 1 \times 1.005(451.2 - 293) = 159$ kW

■ **Example 11.8** The overall stagnation pressure ratio and isentropic efficiency for an axial flow air compressor are 5 and 82% respectively. The inlet stagnation pressure and temperature are 1 bar and 303 K. The mean blade speed is 200 m/s and degree of reaction is 0.5. The relative air angles are 12° and 30° at the rotor inlet and outlet respectively. The work factor is 0.9. Calculate (a) stagnation polytropic efficiency, (b) number of stages, (c) inlet temperature and pressure, and (d) blade height in the first stage if the hub-tip ratio is 0.42 and mass flow rate is 20 kg/s.

Solution : $r_{op} = 5, \eta_{isen} = 0.82, p_{01} = 1$ bar, $T_{01} = 303$ K, $u = 200$ m/s, $R_d = 0.5, \beta_1 = 30° = \alpha_2$, $\beta_2 = 12° = \alpha_1, \phi_w = 0.9, \dot{m} = 20$ kg/s, hub-tip ratio $= 0.42$.

(a)

$$T_{0z} = T_{01}\left(\frac{p_{0z}}{p_{01}}\right)^{\frac{\gamma-1}{\gamma}}$$

$$= 33(4)^{\frac{0.4}{1.4}} = 479.9 \text{ K}$$

$$T'_{0z} = T_{01} + \frac{T_{0z} - T_{01}}{\eta_{isen}}$$

$$= 303 + \frac{479.9 - 303}{0.82} = 518.7 \text{ K}$$

Polytropic efficiency,

$$\eta_p = \frac{\ln\left(\frac{p_{0z}}{p_{01}}\right)^{\frac{\gamma-1}{\gamma}}}{\ln\left(\frac{T'_{0z}}{T_{01}}\right)} = \frac{\ln(5)^{\frac{0.4}{1.4}}}{\ln\left(\frac{518.7}{303}\right)}$$

$$= 0.8553 \text{ or } 85.53\%$$

(b) From the inlet velocity triangle, we have

$$\frac{u}{v_f} = \tan\alpha_1 + \tan\beta_1 = \tan 12° + \tan 30° = 0.7899$$

$$v_f = \frac{200}{0.7899} = 253.2 \text{ m/s}$$

$$v_{f2} = v_{f2}\tan\alpha_2 = 253.2 \tan 30° = 146.2 \text{ m/s}$$

$$v_{w1} = v_f \tan\alpha_1 = 253.2 \tan 12° = 53.8 \text{ m/s.}$$

Work consumed per stage $= \phi_W\, u(v_{w2} - v_{w1})$

$$= 0.9 \times 200(146.2 - 53.8) \times 10^{-3} = 16.632 \text{ kJ/kg}$$

Total work consumed by the compressor $= c_p(T'_{0z} - T_{01})$

$$= 1.005(518.7 - 303) = 216.78 \text{ kJ/kg}$$

Number of stages,

$$z = \frac{216.78}{16.632} = 13$$

(c) $v_{a1} = \dfrac{v_f}{\cos \alpha_1} = \dfrac{2532}{\cos 12°} = 258.86 \text{ m/s}$

Static temperature,

$$T_1 = T_{01} - \frac{v_{a1}^2}{2c_p}$$

$$= 303 - \frac{(258.86)^2}{2 \times 1.005 \times 10^3} = 269.7 \text{ K}$$

$$\frac{p_1}{p_{01}} = \left(\frac{T_1}{T_{01}}\right)^{\frac{\gamma}{\gamma - 1}},$$

$$p_1 = 1\left(\frac{269.7}{303}\right)^{3.5} = 0.665 \text{ bar}$$

(d)
$$\rho_1 = \frac{p_1}{RT_1} = \frac{0.665 \times 10^5}{287 \times 269.7} = 0.859 \text{ kg/m}^3$$

$$m = \rho_1 A_1 v_f$$

$$20 = 0.859 \times \pi r_1^2 [1 - (0.42)^2] \times 2532$$

$$r_1 = 0.1885 \text{ m or } 188.5 \text{ mm}$$

$$\frac{r_h}{r_1} = 0.42, \; r_h = 0.42 \times 188.5 = 0.0792 \text{ m or } 79.2 \text{ mm}$$

■ **Example 11.9** A multistage axial flow air compressor delivers 25 kg/s of air. The inlet stagnation condition is 1 bar and 20°C. The power consumed by the compressor is 4500 kW. Calculate (a) the delivery pressure, (b) number of stages, and (c) overall isentropic efficiency of the compressor.

Assume that temperature rise in the first stage is 15°C, the polytropic efficiency of compression is 0.90 and the stage stagnation pressure ratio is constant.

Solution : $m = 25$ kg/s, $p_{01} = 1$ bar, $T_{01} = 273 + 20 = 293$ K, $P = 4500$ kW, $T_{02} = 293 + 15 = 308$ K, $n_p = 0.90$

(a)
$$\eta_p = \frac{\ln\left(\dfrac{p_{02}}{p_{01}}\right)^{\frac{\gamma - 1}{\gamma}}}{\ln\left(\dfrac{T_{02}}{T_{01}}\right)}$$

$$0.9 = \frac{\ln\left(\dfrac{p_{02}}{p_{01}}\right)^{\frac{1}{3.5}}}{\ln\left(\dfrac{308}{293}\right)}$$

$$\ln\left(\frac{p_{02}}{p_{01}}\right) = 3.5 \times 0.9 \times \ln\left(\frac{308}{293}\right) = 0.15727$$

$$\frac{p_{02}}{p_{01}} = 1.17$$

$$P = \dot{m}c_p(T'_{0z} - T_{01})$$

$$4500 = 25 \times 1.005\,(T'_{0z} - 293)$$

$$T'_{0z} = 472.1 \text{ K}$$

$$\eta_p = \left(\frac{\gamma - 1}{\gamma}\right)\left(\frac{n}{n-1}\right)$$

$$0.9 = \left(\frac{1.4 - 1}{1.4}\right)\left(\frac{n}{n-1}\right)$$

$$\frac{n}{n-1} = 3.15$$

$$\frac{p_{0z}}{p_{01}} = \left(\frac{T'_{0z}}{T_{01}}\right)^{\frac{n}{n-1}} = \left(\frac{472.1}{293}\right)^{3.15} = 4.4934$$

$$p_{0z} = 1 \times 4.4934 = 4.4934 \text{ bar}$$

(b) Now

$$\frac{p_{02}}{p_{01}} = \frac{p_{03}}{p_{02}} = \ldots = \frac{p_{0z}}{p_{0(z-1)}}$$

$$\left(\frac{p_{02}}{p_{01}}\right)^z = \frac{p_{0z}}{p_{01}}$$

$$z \ln\left(\frac{p_{02}}{p_{01}}\right) = \ln\left(\frac{p_{0z}}{p_{01}}\right)$$

$$z = \frac{\ln\left(\frac{4.4934}{1}\right)}{\ln 1.17} = 9.57 = 10$$

(c)

$$T_{0z} = T_{01}\left(\frac{p_{0z}}{p_{01}}\right)^{\frac{\gamma-1}{\gamma}} = 293(4.4934)^{\frac{1}{3.5}} = 450.1 \text{ K}$$

$$(\eta_{isen})_{overall} = \frac{T_{0z} - T_{01}}{T'_{0z} - T_{01}}$$

$$= \frac{450.1 - 293}{472.1 - 293} = 0.8772 \text{ or } 87.72\%$$

■ **Example 11.10** An axial compressor takes in 1000 m^3/min of free air at 15°C and 0.9 bar. The blades are of aerofoil type having projected area and blade length as 19.25 cm^2 and 6.75 cm respectively. The blade ring mean diameter is 60 cm and speed is 6000 rpm. On each blade ring there are 50 blades and the blades occupy 10% of the axial area of flow. Values of C_L and C_D are 0.6 and 0.05 respectively at zero angle of incidence. Assuming isentropic compression, calculate the pressure rise per blade ring and the power input per stage. Assume axial inlet. [CSE, 1996]

Solution : $\dot{V} = 1000\ \text{m}^3/\text{min}$, $T_1 = 273 + 15 = 288$ K, $p_1 = 0.9$ bar, $A = 1925\ \text{cm}^2$, $h = 6.75$ cm, $k_b = 0.9$, $C_L = 0.6$, $C_D = 0.05$, $D_m = 60$ cm, $i = 50$

Density of air,

$$\rho = \frac{p_1}{RT_1} = \frac{0.9 \times 10^5}{287 \times 288} = 1.09\ \text{kg/m}^3$$

Area across flow,

$$A_f = k_b \pi D_m h$$
$$= 0.9 \times \pi \times 0.6 \times 6.75 \times 10^{-2}$$
$$= 0.1145\ \text{m}^2$$

$$v_{a1} = \frac{\dot{V}}{60 A_f} = \frac{1000}{60 \times 0.1145} = 145.55\ \text{m/s}$$

Blade velocity,

$$u = \frac{\pi D_m N}{60}$$
$$= \frac{\pi \times 0.6 \times 6000}{60}$$
$$= 188.5\ \text{m/s}$$

For axial inlet, $v_{a1} = v_f = 145.55$ m/s

The velocity triangle at inlet is shown in Fig. 11.14.

$$\tan \beta_1 = \frac{u}{v_f} = \frac{188.5}{145.55} = 1.295$$

$$\beta_1 = 52.325°$$

$$v_{r1} = \sqrt{u^2 + v_{a1}^2}$$

$$= \sqrt{(188.5)^2 + (145.55)^2} = 238.15\ \text{m/s}$$

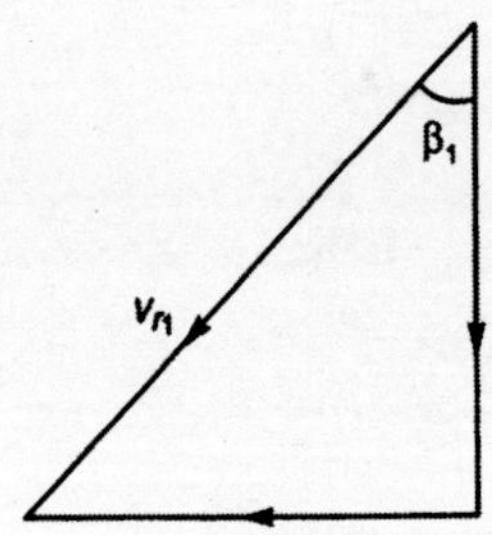

Fig. 11.14

Lift force, $F_L = \frac{1}{2} C_L \rho v_{r1}^2 A$

$$= \frac{1}{2} \times 0.6 \times 1.09 \times (238.15)^2 \times 1925 \times 10^{-4}$$
$$= 35.7\ \text{N}$$

Drag force, $F_D = \frac{1}{2} C_D \rho v_{r1}^2 A$

$$= \frac{1}{2} \times 0.05 \times 1.09 \times (238.15)^2 \times 1925 \times 10^{-4}$$
$$= 2.975\ \text{N}$$

Power input per stage,

$$P = (F_L \cos\beta_1 + F_D \sin\beta_1)\, ui$$
$$= (35.7 \cos 52.325° + 2.975 \sin 52.325°) \times 188.5 \times 50 \times 10^{-3}$$
$$= 227.839 \text{ kW}$$

Mass flow rate,

$$\dot{m} = \dot{V}\rho = \frac{1000 \times 1.09}{60} = 18.167 \text{ kg/s}$$

Now

$$P = \dot{m} c_p T_1 \left(r_p^{\frac{\gamma-1}{\gamma}} - 1 \right)$$

$$227.839 = 18.167 \times 1.005 \times 288 \left(r_p^{\frac{1}{3.5}} - 1 \right)$$

$$r_p = 1.16$$
$$p_2 = 0.9 \times 1.16 = 1.044 \text{ bar}$$

Pressure rise $= p_2 - p_1 = 1.044 - 0.9 = 0.144$ bar

■ **Example 11.11** An axial flow compressor gives a pressure rise of 4 : 1 and the total head isentropic efficiency is 86%. Stagnation inlet temperature is 17°C. The inlet and outlet air angles from the rotor blades of the above compressor are 45° and 10° respectively. The rotor and stator blades are symmetrical. The mean blade and axial velocity remain constant. Assuming blade speed of 220 m/s and work input factor 0.86, find the polytropic efficiency, number of stages required and Mach number. Take $R = 287$ J/kg.K.

Solution : $r_p = 4$, $\eta_{o\,\text{isen}} = 0.86$, $T_{01} = 273 + 17 = 290$ K, $\alpha_1 = 45°$, $\alpha_2 = 10°$, $u = 220$ m/s, $\phi_w = 0.86$

Polytropic efficiency,

$$\eta_p = \left(\frac{\gamma-1}{\gamma}\right)\left(\frac{n}{n-1}\right)$$

Head isentropic efficiency,

$$\eta_{o\,\text{isen}} = \frac{r_p^{\frac{\gamma-1}{\gamma}} - 1}{r_p^{\frac{n-1}{n}} - 1}$$

$$= \frac{r_p^{\frac{\gamma-1}{\gamma}} - 1}{r_p^{\left(\frac{\gamma-1}{\gamma}\right)\frac{1}{\eta_p}} - 1}$$

$$0.86 = \frac{4^{\frac{1}{3.5}} - 1}{4^{\frac{1}{3.5 \times \eta_p}} - 1}$$

$$= \frac{0.486}{4^{\frac{1}{3.5\eta_p}} - 1}$$

$$4^{\frac{1}{3.5\eta_p}} = 1.565$$

$$\frac{1}{3.5\eta_p} \ln 4 = \ln 1.565$$

$$\eta_p = 0.8843 \text{ or } 88.43\%$$

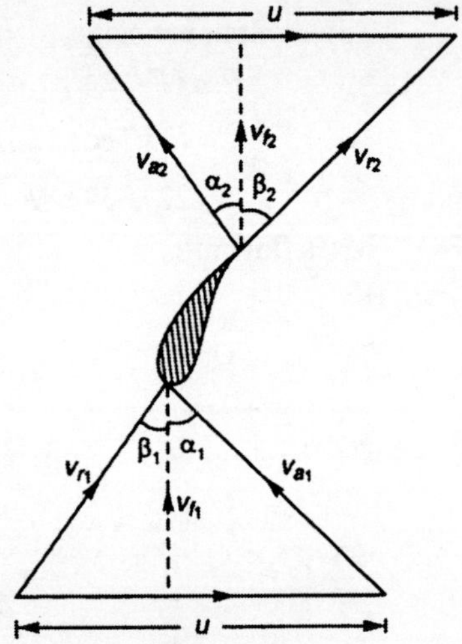

Fig. 11.15

The velocity diagrams are shown in Fig. 11.15.

$$\alpha_1 = \beta_2 = 10°, \alpha_2 = \beta_1 = 45°$$

$$u = v_{f1}(\tan \alpha_1 + \tan \beta_1)$$

$$v_{f1} = \frac{220}{\tan 10° + \tan 45°} = 187 \text{ m/s}$$

Power required per kg of air $= c_p(T_2 - T_1) = \phi_w u v_f (\tan \alpha_2 - \tan \alpha_1)$

$$T_2 - T_1 = \frac{0.86 \times 220 \times 187 \, (\tan 45° - \tan 10°)}{1.005 \times 10^3} = 29°\text{C}$$

Also $$\frac{T_{02}}{T_{01}} = r_p^{\frac{n-1}{n}} = 4^{\frac{\gamma - 1}{\gamma \eta_p}} = 4^{\frac{0.4}{1.4 \times 0.8843}} = 1.565$$

$$T_{02} = 290 \times 1.565 = 453.86 \text{ K}$$

Total temperature rise $= T_{02} - T_{01}$

$$= 453.86 - 290 = 163.86°\text{C}$$

Number of stages, $$z = \frac{163.86}{29} = 5.65 = 6$$

$$v_{r1} = \frac{v_f}{\cos \alpha_1} = \frac{187}{\cos 45°} = 264.5 \text{ m/s}$$

$$v_{a1} = \frac{v_f}{\cos \beta_1} = \frac{187}{\cos 10°} = 190 \text{ m/s}$$

$$T_1 = T_{01} - \frac{v_{a1}^2}{2c_p} = 290 - \frac{190^2}{2 \times 1.005 \times 10^3} = 272 \text{ K}$$

$$M = \frac{v_{r1}}{\sqrt{\gamma R T_1}} = \frac{264.5}{\sqrt{1.4 \times 287 \times 272}} = 0.8$$

■ **Example 11.12** Air at 1.01325 bar and 288 K enters an axial flow compressor stage with an axial velocity of 150 m/s. There are no inlet guide vanes. The rotor stage has a tip diameter of 60 cm and a hub diameter of 50 cm and rotates at 100 rps. The air enters the rotor and leaves the stator in the axial direction with no change in velocity or radius. The air is turned through 30° as it passes through rotor.

Assuming constant specific heats and that air enters and leaves are blade at the blade angles, construct the velocity diagrams at mean diameter for this stage, and calculate (a) mass flow rate, (b) power required, and (c) degree of reaction.

Solution : $p_1 = 1.01325$ bar, $T_1 = 288$ K, $v_f = 150$ m/s, $D_2 = 60$ cm, $D_1 = 50$ cm, $N = 100$ rps, $\beta_2 \sim \beta_1 = 30°$

The velocity triangles are shown in Fig. 11.16.

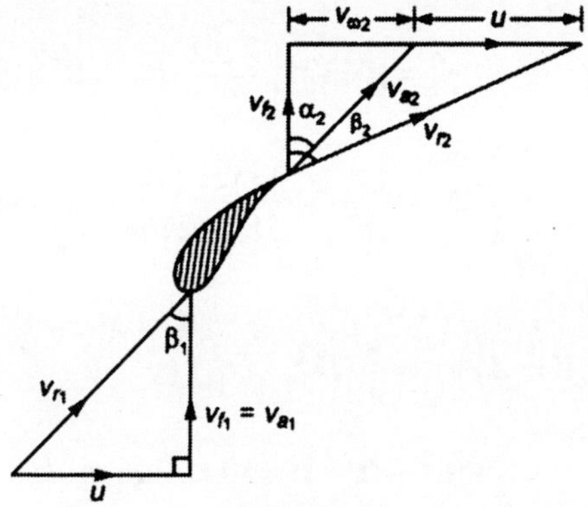

Fig. 11.16

$$D_m = \frac{D_1 + D_2}{2} = \frac{0.5 + 0.6}{2}$$

$$= 0.55 \text{ m}$$

$$u = \pi D_m N$$

$$= \pi \times 0.55 \times 100$$

$$= 172.8 \text{ m/s}$$

$$\tan \beta_1 = \frac{u}{v_{f1}} = \frac{172.8}{150}$$

$$= 1.152$$

$$\beta_1 = 49.04°$$

$$\beta_2 \sim \beta_1 = 49.04°$$

$$\beta_2 = 49.04 - 30$$

$$= 19.04° \text{ or } 79.04°$$

$$v_{f2} \tan \beta_2 = u + v_{w2}$$

$$150 \tan 19.04° = 51.77 < 172.8 \text{ m/s}$$

$\therefore v_{w2}$ is –ve

$\therefore$

$$\tan \alpha_2 = \frac{v_{w2}}{v_f} = \frac{u - v_f \tan \beta_2}{v_f} = \frac{121.03}{150} = 0.80$$

$$\alpha_2 = 38.9°$$

Now

$$T_{01} = T_1 + \frac{v_{a1}^2}{2c_p}$$

$$288 = T_1 + \frac{150^2}{2 \times 1.005 \times 10^3} = T_1 + 11.2$$

$$T_1 = 276.8 \text{ K}$$

$$\frac{p_{01}}{p_1} = \left(\frac{T_{01}}{T_1}\right)^{\frac{\gamma}{\gamma - 1}}$$

$$p_1 = 1.01325\left(\frac{276.8}{288}\right)^{3.5} = 0.8819 \text{ bar}$$

$$\rho_1 = \frac{p_1}{RT_1} = \frac{0.8891 \times 10^5}{287 \times 276.8} = 1.11 \text{ kg / m}^3$$

Volume flow rate,

$$V = A_f v_f = \frac{\pi}{4}(D_2^2 - D_1^2) v_f$$

$$= \frac{\pi}{4}(0.6^2 - 0.5^2) \times 150 = 12.96 \text{ m}^3/\text{s}$$

(a) Mass flow rate,

$$m = V_\rho = 12.96 \times 1.11 = 14.384 \text{ kg/s}$$

(b) Power required,

$$P = muv_f = 14.384 \times 172.8 \times 150 \times 10^{-3} = 372.85 \text{ kW}$$

(c) Degree of reaction,

$$R_d = \frac{v_f}{2u}(\tan \beta_1 + \tan \beta_2)$$

$$= \frac{150}{2 \times 172.8}(\tan 49.04 + \tan 19.04) = 0.65$$

SOLVED QUESTIONS FROM UNIVERSITY EXAMINATIONS

■ **Example 11.13** Attempt any three of the following :

(i) Sterling cycle.

(ii) Theories of detonation.

(iii) Performance curves of C.I. Engine.

(iv) Axial flow compressor. [MDU, Dec., 2007]

Solution :

(i) Refer to art. 2.4.

(ii) Refer to art. 4.8.

(iii) Refer to art. 6.8.

(iv) Refer to art. 11.1 and art. 11.2.

■ **Example 11.14** An axial flow compressor with pressure compression ratio 4 draws air at 20°C and delivers at 197°C. The mean blade speed and flow velocity are constant throughout the compressor. Assuming 50% reaction and blading velocity as 180 m/s, find the flow velocity and number of stages. Take work factor $= 0.82, \alpha = 20°, \beta = 42°, c_p = 1.005$ kJ/kg.K. [MDU, May, 2008]

Solution : Given : $r_p = 4, T_1 = 273 + 20 = 293$ K, $T_2 = 273 + 197 = 470$ K, $u = u_1 = u_2$ and $v_{f1} = v_{f2} = v_f, R_d = 0.5, u = 180$ m/s, $\phi_w = 0.82, \alpha_1 = 20°, \beta_1 = 42°, c_p = 1.005$ kJ/kg.K

For 50% degre of reaction,

$$\alpha_2 = \beta_1 = 42° \text{ and } \beta_2 = \alpha_1 = 20°$$

Work (or head) factor,

$$\phi_w = \frac{2(v_{w2} - v_{w1})}{u}$$

$$0.82 = \frac{2(v_{w2} - v_{w1})}{180}$$

$$v_{w2} - v_{w1} = \frac{0.82 \times 180}{2} = 73.8 \text{ m/s}$$

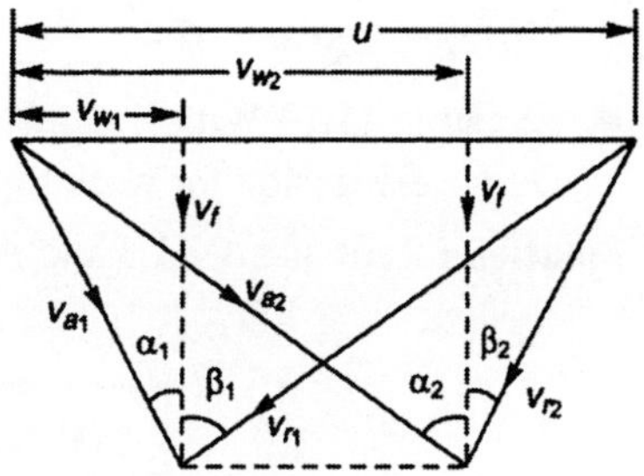

Fig. 11.17

Refer to Fig. 11.17.

$$\frac{v_{w1}}{v_f} = \tan\alpha_1 \text{ and } \frac{v_{w2}}{v_f} = \tan\alpha_2$$

$$\Delta v_w = v_{w2} - v_{w1} = v_f(\tan\alpha_2 - \tan\alpha_1)$$

$$= v_f(\tan 42° - \tan 20°) = 0.5364 v_f$$

$\therefore$

$$0.5364 v_f = 73.8$$

Flow velocity,

$$v_f = 137.6 \text{ m/s}$$

Work done per stage $= u \times \Delta v_w$

$$= 180 \times 73.8 = 13284 \text{ N.m/kg}$$

Total work $= c_p(T_2 - T_1)$

$$= 1.005 \times 10^3 (470 - 293) = 177885 \text{ N.m/kg}$$

Number of stages, $z = \frac{177885}{13284}$

$$= 13.39 \simeq 14$$

■ **Example 11.15** What are the basic requirements of compressors for aircraft application? Do axial flow compressors meet them? Explain. [MDU, Dec. 2008]

Solution : Requirements of compressors for aircraft application :

1. High efficiency
2. Smaller frontal area
3. Suitability for multistaging
4. High starting torque
5. Lesser drag

The above requirements are met by axial flow compressors.

■ **Example 11.16** Attempt any **THREE** of the following :

(a) Rotary Air Compressor
(b) Open and Closed Gas Turbine
(c) Types of Lubrication System
(d) Knocking in C.I. Engines [KU, Jan., 2006]

Solution :

(a) Refer to art. 9.2.
(b) Refer to art. 12.6.
(c) Refer to art. 5.7.1.
(d) Refer to art. 4.7.4.

■ **Example 11.17** With a suitable sketch explain the working principle of an axial flow compressor. Derive expression for work input to the compressor. [MDU, Dec. 2008]

Solution : Refer to art. 11.3 and Art. 11.4.

SHORT ANSWER TYPE QUESTIONS

1. How does axial flow compressors differ from centrifugal compressors?
2. Explain the working principle of axial flow compressors.
3. Define lift and drag coefficients.
4. Define degree of reaction and write its formula.
5. For 50% degree of reaction, what is the condition on blade angles?
6. Write the expression for pressure rise in isentropic flow through a cascade.
7. Define polytropic efficiency and write its formula.
8. Define flow coefficient and work coefficient.
9. Define deflection coefficient and pressure coefficient.
10. Differentiate between choking and stalling.
11. List the various losses in an axial flow compressor.
12. List the main applications of axial flow compressors.
13. List the merits and demerits of axial flow compressors.
14. Draw the efficiency and power consumption curves for the axial flow compressor.

EXERCISES

1. An axial flow air compressor of 50% degree of reaction has blades with inlet and outlet blade angles of 40° and 15° respectively. The pressure ratio is 5 : 1 with an overall isentropic efficiency of 85% when the air inlet temperature is 30°C. The blade speed and axial velocity are constant throughout the compressor. Assuming a blade speed of 210 m/s, find the number of stages required when the work coefficient is 0.9 for all stages.
2. In an axial flow air compressor, the overall stagnation pressure ratio is 4.5 with overall stagnation isentropic efficiency 85%. The inlet stagnation pressure and temperature are 1 bar and 310 K. The mean blade speed is 180 m/s and degree of reaction is 0.5 at the mean radius. The relative air angles are 10° and 30° at rotor inlet

and outlet respectively. The work coefficient is 0.9. Calculate (a) stagnation polytropic efficiency, (b) number of stages, (c) inlet temperature and pressure, and (d) blade height in the first stage if the hub-tip ratio is 0.45 and mass flow rate is 20 kg/s.

3. A multi-stage axial flow compressor delivers 19 kg/s of air. The inlet stagnation condition is 1 bar and 25°C. The power consumed by the compressor is 4400 kW. Calculate (a) the delivery pressure, (b) number of stages, and (c) overall isentropic efficiency of the compressor.
   Assume that temperature rise in the first stage is 21°C. The polytropic efficiency of compressor is 0.9 and stage stagnation pressure ratio is constant.

4. An axial flow compressor compresses the air upto overall stagnation pressure of 10 bar with overall stagnation isentropic efficiency of 88%. The inlet stagnation pressure and temperature are 1 bar and 300 K. The mean blade speed is 200 m/s. The degree of reaction is 0.5 at the mean radius with air angles of 30° and 10° at rotor inlet and outlet with axial direction respectively. The work done factor is 0.88. The hub-tip ratio is 0.4. The mass flow rate is 50 kg/s. Work out the following :
   (a) Draw the inlet and outlet velocity triangles and show the compression process on *T-s* diagram.
   (b) The stagnation polytropic efficiency.
   (c) The number of stages.
   (d) The blade height in the first stage of the compressor.

5. An axial flow compressor provides a total head pressure ratio of 4 : 1 with an overall total head isentropic efficiency of 85%, when the inlet total head temperature is 290 K. The compressor is designed for 50% reaction with inlet and outlet air angles from the rotor blades of 45° and 10° respectively. The mean blade speed and axial velocity are constant throughout the compressor. Assuming a valve of 201.16 m/s for the blade speed and a work done factor of 0.86, find the number of stages required. What is the inlet Mach number relative to the rotor at the mean blade height of the first stage?

6. An axial flow compressor of 50% reaction blading has isentropic efficiency of 82%. It draws air at 17°C and compresses in the pressure ratio of 4 : 1. The mean blade speed and flow velocity are constant throughout the compressor. The inlet and outlet angles of blades are 15° and 45° respectively (angles measured from axial direction). Blade speed = 180 m/s and work input factor = 0.84. Calculate (a) flow velocity, and (b) number of stages.

7. Air at a temperature of 300 K enters a ten-stage axial flow compressor at the rate of 3.5 kg/s. The pressure ratio is 6.0 and the isentropic efficiency is 90%. The process is adiabatic and the compressor has symmetrical stages. The axial velocity of 120 m/s is uniform across the stages and the mean blade speed is 200 m/s. Assume that the temperature change is same in each stage.
   Determine the direction of the air at entry and exit from the rotor and stator blades. Also find the power given to the air. For air, take $c_p = 1.005$ kJ/kg.K and $\gamma = 1.4$.

8. An axial flow compressor consists of a number of similar stages with equal work done per stage. The velocity of flow is uniform throughout the compressor. The data given is as follows :

Overall stagnation pressure ratio	= 3.5 : 1
Stagnation inlet temperature	= 60°C
Relative air angle at rotor inlet	= 130°
Relative air angle at rotor outlet	= 100°
Blade velocity	= 185 m/s
Degree of reaction	= 50%
Overall stagnation adiabatic efficiency	= 87%

   Calculate (a) stagnation outlet temperature, and (b) number of stages.

❑❑❑

Chapter 12

Gas Turbines

Inside This Chapter

12.1 INTRODUCTION

The gas turbine is the most satisfactory way of producing very large power in a self-contained and compact unit. The gas turbine obtains its power by utilizing the energy of a jet or burnt gases and air, the velocity of the jet being absorbed as it flows over several rings of moving blades which are fixed to a common shaft. The gas turbine requires an air compressor which is driven off its own shafting. This absorbs a considerable proportion of the power produced and thus lowers the overall efficiency, which is of the order of 20 to 30%.

12.2 FIELDS OF APPLICATION OF GAS TURBINE

The following are the major fields of application of gas turbines :

1. **Aviation :** A gas turbine is used in aviation because it is self-contained, light weight not requiring cooling water and generally fit into the overall shape of the aircraft structure. It is also suitable for turbo-jet and turbo-propeller engines.
2. **Power generation :** It is used for electric power generation because of its simplicity, lack of cooling water, quick installation and starting. Also suitable for supercharging, road transport and railways.
3. **Oil and gas industry :** It is used in oil and gas industry because of cheaper supply of fuel and low installation cost.
4. **Marine propulsion :** It is used in marine propulsion because it is of light weight and not requiring cooling water.

12.3 LIMITATIONS OF GAS TURBINES

The limitations of gas turbines are :

1. They are not self starting.
2. Low efficiency at part loads.
3. Non-reversible.
4. Higher rotor speeds, and
5. Low overall plant efficiency.

12.4 COMPARISON OF GAS TURBINES WITH I.C. ENGINES

(a) Advantages :

1. Perfect balancing of the rotating parts is possible.
2. Mechanical efficiency is very high (95%) compared with 85% of I.C. engine because of its large number of sliding parts.
3. Torque on the turbine shaft is continuous and uniform, therefore flywheel is not required.
4. Work developed is more because expansion of gases upto atmospheric pressure is possible.
5. The weight of gas turbine per kW of power developed is less and are therefore superior for use in aircraft.
6. The compression, combustion and expansion take place in different units and therefore these units can be designed, tested and developed individually.
7. The gas turbine can be driven at much higher speeds.
8. The maximum pressure used in gas turbines is very low, therefore its components can be made lighter.
9. The lubrication and ignition systems are much more simple.
10. Cheaper fuels can be used in gas turbine.
11. The installation and maintenance costs are less.
12. The exhaust is free from smoke and is less polluting.

(b) Disadvantages :

1. The thermal efficiency of a simple gas turbine is low.
2. Very poor thermal efficiency at part loads.

3. The temperature of gases supplied to the turbine is limited to 1100 K as there are no metals which can be used for blades to sustain this temperature at high centrifugal stresses.
4. The manufacture of blades is difficult and costly as Ni-Cr alloy is used.
5. The control of fuel is difficult with wide operating speeds and output.
6. A special cooling system is required for the blades.
7. The starting is more difficult.
8. It produces five times exhaust gases than I.C. engine for same output.
9. Speed reduction device is always required due to higher operating speeds.
10. The life of blades and combustion chamber is low due to high temperature.

12.5 ADVANTAGES OF GAS TURBINES OVER STEAM TURBINES

1. Boiler along with accessories are not required.
2. The capital and running cost is less.
3. The space required is less for the plant.
4. Starting of the plant is more easy and quick.
5. The weight of the plant per kW generated is less.
6. The control of plant is much easier.
7. The plant can be installed anywhere as cooling water is not required.
8. Gas turbine plant is used as a best type of peak load plant.

12.6 CLASSIFICATION OF GAS TURBINES

The gas turbines may be classified as :

1. Constant pressure combustion (or continuous combustion) gas turbine.
   (a) open cycle constant pressure gas turbine.
   (b) closed cycle constant pressure gas turbine.
2. Constant volume combustion (or explosion) gas turbine.

1. Constant Pressure Combustion Gas Turbine

In this type, the fuel is burned at constant pressure and the cycle used is the Joule cycle. The turbine is of the reaction type using oil fuel and is fitted with an axial flow rotary air compressor which is coupled to the rotor shaft. The heat is supplied to the working fluid at constant pressure.

The open and closed cycle gas turbines are shown in Fig. 12.1 (a) and Fig. 12.1 (b) respectively.

A simple open cycle constant pressure gas turbine consists of a compressor, combustion chamber and turbine. The air is taken from the atmosphere by the compressor and is compressed to the combustion pressure of 1 to 4 atm. It is then forced into the combustion chamber. Part of this air is used as combustion air for the oil which enters the burner, the remainder is forced through the annular space between the wall of the combustion chamber and the burner jacket. The air receives heat from the burner jacket and also mixes with the products of combustion chamber and the burner jacket. This raises the temperature and volume of the air. The use of a very large quantity of air in excess of the combustion air prevents the temperature of the mixture from racching values which are too high for the metal of the rotor blades. It also prevents the burner from becoming too hot.

The high pressure and high temperature gases now enters the turbine and flows through the blade rings, where it continuously expands, its pressure energy converted into kinetic energy, which in turn is

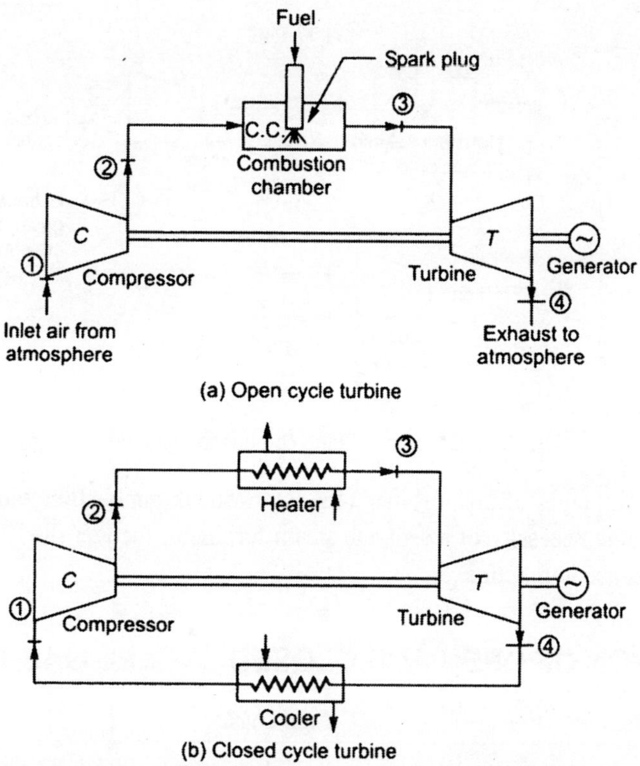

Fig. 12.1 Constant pressure combustion gas turbines.

absorbed by the rotor. On leaving turbine, the spent gases pass away to exhaust. The speed of the turbine is generated by controlling the fuel oil supply and by the safety valve which by-passes some of the mixture if the pressure becomes too high, the net power developed by the turbine is utilized in driving the generator but a considerable amount of the total power produced by the rotor is absorbed in dirving the compressor. The turbine is started by the electric starter motor.

In the closed cycle constant pressure gas turbine, the compressed fluid coming out from the compressor is heated in heater by an external source at constant pressure and then the high pressure and temperature fluid develops the work while passing through the turbine. The fluid coming out from the turbine is cooled to its original temperature in the cooler using an external cooling source before passing into the compressor. Thus, the working fluid is continuously circulated in the closed cycle. Such a turbine has now become obsolete.

2. Constant Volume Combustion Gas Turbine

The constant volume combustion gas turbine is shown diagrammatically in Fig. 12.2. Air is taken from the atmosphere and is compressed by an axial flow compressor driven by a separate steam turbine. The compressed air coming out of the compressor at about 3 bar is passed to the constant volume combustion chamber. The fuel is injected into the combustion chamber in the current of air by a separate full pump. The air-fuel mixture formed in the combustion chamber is then ignited by means of a spark plug. The combustion takes place at constant volume in the combustion chamber and the pressure inside the chamber increases to nearly 12 bar. The high pressure and temperature gases are fed to the gas turbine through the water cooled nozzle. the gases passes through the turbine develop the useful work. The exhaust gases are fed to the steam boiler to generate steam for the steam turbine driving the compressor.

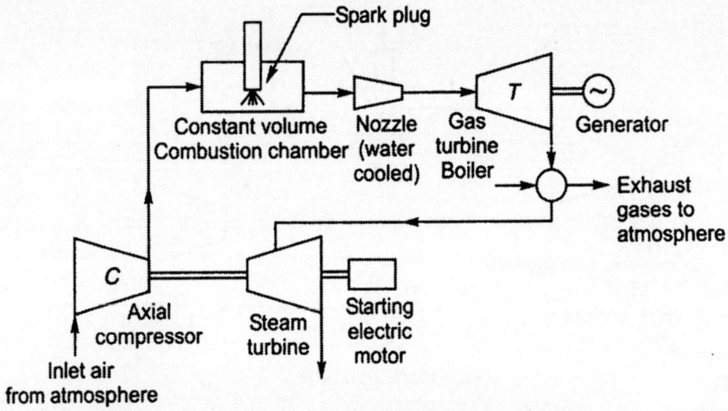

Fig. 12.2 Constant volume combustion gas turbine.

The major disadvantages of this turbine is the intermittent combustion which impairs its smooth functioning. The absolue necessity of a separate steam turbine to recover the heat from exhaust gases further discourages its use in practice.

12.7 COMPARISON OF OPEN AND CLOSED CYCLES GAS TURBINES

Particular	*Closed Cycle*	*Open Cycle*
1. Cycle pressure	Higher which results in a reduced size of plant.	Lower which results in a bigger size of plant.
2. Corrosion and erosion of blades	Avoided due to indirect heating.	More due to direct heating
3. Filtration of incoming air	Not required.	Poses a severe problem.
4. Density of working fluid	High density reduces the size of compressor, turbine and heat exchanger	Low, which increases the size of various components
5. Type of fuel used	Solid fuel can be used more economically	Fuel has to be a fluid, which is comparatively costlier.
6. Working fluid	Any thermodynamically better fuel than air can be used	Working fluid is restricted to air only
7. Maintenance costs	Low	High
8. Uses	Suitable for stationary plants and marine power as it requires abundance quantities of water	Suitable for aircraft turbines
9. Weight	High, not suitable for moving vehicles	Low, suitable for moving vehicles also
10. Heat exchange coefficient	Low	High
11. Installation cost	High	Low
12. Response to changing load	Poor	Good

12.8 POSITION OF GAS TURBINE IN POWER INDUSTRY

Constant pressure gas turbines are now used for generating electricity, driving locomotives and aeroplane propellers producing the air stream for wind tunnels and for the jet propulsion of air craft. In same types of aircraft, the exhaust gases from the main engines are used for driving a gas turbine which in turn drives the super charger. Gas turbines driven by exhaust gases are also used for supplying a pressure charge to diesel engines.

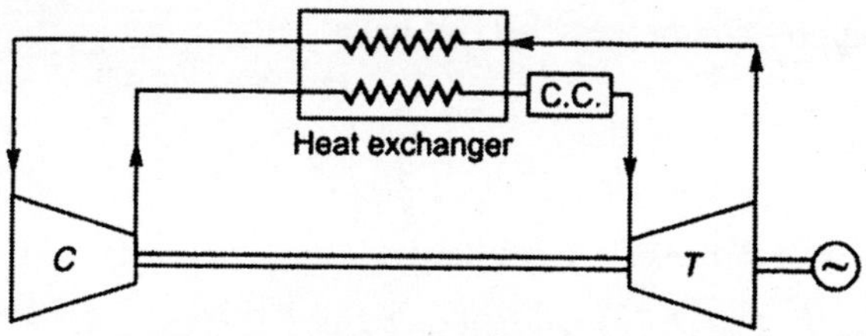

Fig. 12.3 Gas turbine utilizing exhaust heat.

Overall efficiencies of constant pressure gas turbine plants vary between 18 to 25 percent. This is based on a turbine efficiency of 78%. The efficiency of this type of turbine can be improved by using heat from exhaust gases to preheat the compressed air before it enters the combustion chamber. The hot mixture of air and burnt gases is then expanded through the tubrine from which they are exhausted into the heat exchanger. Their available heat is now given to the high pressure air from the compressor. Such a system is shown in Fig. 12.3.

12.9 THERMODYNAMICS OF CONSTANT PRESSURE GAS TURBINE : BRAYTON CYCLE

(a) Efficiency

The Brayton cycle is shown on *p*-V and T-*s* diagrams in Fig .12.4. During this cycle, air is drawn into the compressor at point 1 and then compressed isentropically along the process 1-2. Process 2-3 represents the burning of the oil at cvonstant pressure p_2. Process 3-4 is the isentropic expansion of the gases through the turbine. Process 4-1 represents the exhausting and cooling of the gases at constant pressure p_1.

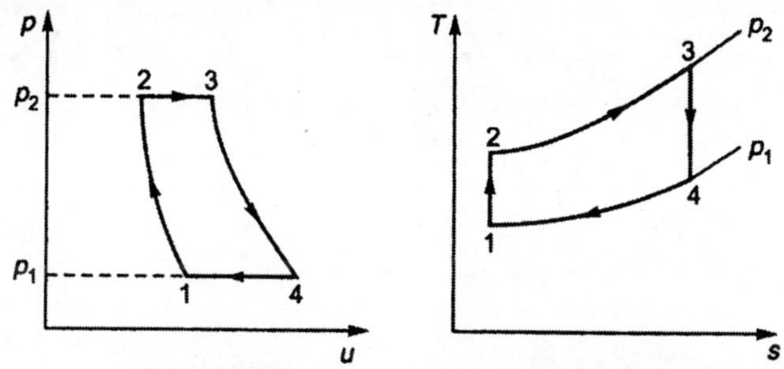

(a) *p-V* diagram (b) *T-s* diagram

Fig. 12.4 Brayton cycle for constant pressure gas turbine.

Compressor work, $w_c = w_{1-2} = h_2 - h_1 = c_p\ (T_2 - T_1)$

Heat supplied, $q = q_{2-3} = h_3 - h_2 = c_p\ (T_3 - T_2)$

Turbine work, $w_t = w_{3-4} = h_3 - h_4 = c_p\ (T_3 - T_4)$

Net work done, $$w_{net} = w_t - w_c = c_p\,[(T_3 - T_4) - (T_2 - T_1)]$$

$$= c_p T_1 \left[\frac{T_3}{T_1} - \frac{T_4}{T_1} - \frac{T_2}{T_1} + 1\right]$$

Let $r_p = \dfrac{p_2}{p_1}$ be the pressure ratio and $\alpha = \dfrac{T_3}{T_1}$

For the adiabatic compression process 1-2, we have

$$\frac{T_1}{T_2} = \frac{V_2}{V_1} = \left(\frac{p_1}{p_2}\right)^{(\gamma-1)/\gamma} \quad ...(1)$$

and for the adiabatic expansion process 3-4, we have

$$\frac{T_4}{T_3} = \frac{V_3}{V_4} = \left(\frac{p_4}{p_3}\right)^{(\gamma-1)/\gamma} \quad ...(2)$$

But $$p_1 = p_4 \text{ and } p_2 = p_3$$

$$\therefore \quad \frac{T_2}{T_1} = \frac{T_3}{T_4} = r_p^{(\gamma-1)/\gamma} = c$$

Thermal efficiency, $$\eta_{th} = \frac{W_{net}}{q}$$

$$= \frac{c_p T_1\left[\alpha\left(1 - \dfrac{1}{c}\right) - (c-1)\right]}{c_p(T_3 - T_2)}$$

$$= \frac{\alpha\left(1 - \dfrac{1}{c}\right) - (c-1)}{\dfrac{T_3}{T_1} - \dfrac{T_2}{T_1}}$$

$$= \frac{\alpha\left(1 - \dfrac{1}{c}\right) - (c-1)}{\alpha - c}$$

$$= \frac{(\alpha - c) - \left(\dfrac{\alpha - c}{c}\right)}{\alpha - c}$$

$$= 1 - \frac{1}{c} = 1 - \frac{1}{r_p^{(\gamma-1)/\gamma}} \quad ...(12.1)$$

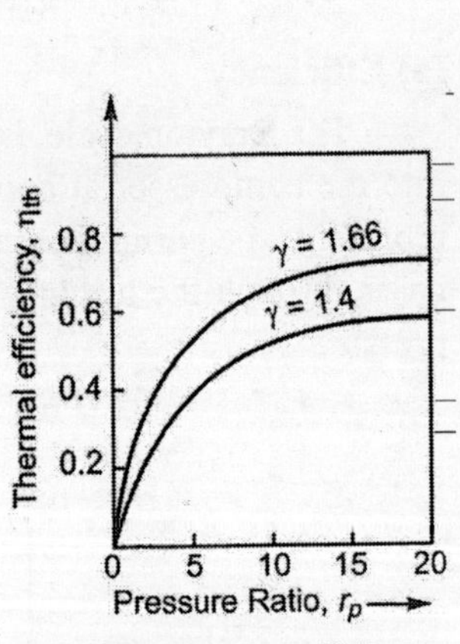

Fig. 12.5 Thermal efficiency v's pressure ratio.

The variation of thermal efficiency with r_p is shown in Fig. 12.5.

(b) Specific Output

$$\text{Specific output} = \frac{w_{net}}{c_p T_1} = \frac{T_3}{T_1} - \frac{T_4}{T_3} \times \frac{T_3}{T_1} - \frac{T_2}{T_1} + 1$$

$$= \alpha - \frac{\alpha}{c} - c + 1 = \alpha\left(1 - \frac{1}{c}\right) - (c - 1) \qquad \text{...(12.2)}$$

For maximum specific output,
$$\frac{d}{dc}\left[\frac{w_{net}}{c_p T_1}\right] = 0$$

$\therefore$
$$\alpha\left(\frac{1}{c^2}\right) - 1 = 0$$

$$\alpha = c^2 = [r_p^{(\gamma-1)/\gamma}]^2 \qquad \text{...(12.3)}$$

or
$$\left(r_p^{\frac{\gamma-1}{\gamma}}\right)_{opt} = \sqrt{\alpha} = c$$

Also
$$r_p^{\frac{\gamma-1}{\gamma}} = \frac{T_2}{T_1} = \frac{T_3}{T_4}$$

$$\alpha = \frac{T_2}{T_1} \times \frac{T_3}{T_4} = \frac{T_2}{T_4}\alpha$$

or
$$T_2 = T_4 \qquad \text{...(12.4)}$$

Hence for maximum work output the temperature after compression must be euqal to the exhaust gas temperature, which depends on the expansion ratio. The variation of specific output with r_p is shown in Fig. 12.6.

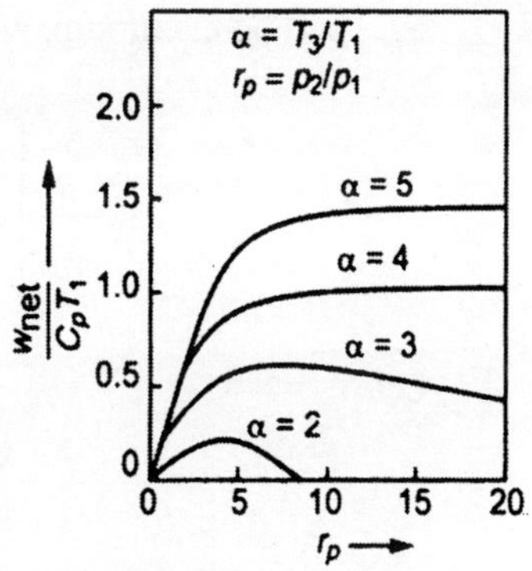

Fig. 12.6 Specific output v's pressure ratio.

(c) Maximum Work output

Maximum work output,

$$(w_{net})_{max} = c_p[(T_3 - T_4) - (T_2 - T_1)]$$
$$= c_p[T_3 + T_1 - T_4 - T_2]$$
$$= c_p[T_3 + T_1 - 2T_2] \quad (\because T_4 = T_2)$$
$$= c_p[\sqrt{T_3^2} + \sqrt{T_1^2} - 2\sqrt{T_2 T_4}]$$
$$= c_p[\sqrt{T_3^2} + \sqrt{T_1^2} - 2\sqrt{T_3 T_1}] \qquad \left(\because \frac{T_3}{T_4} = \frac{T_2}{T_1}\right)$$
$$= c_p[\sqrt{T_3} - \sqrt{T_1}]^2 = c_p[\sqrt{T_{max}} - \sqrt{T_{min}}]^2 \qquad \text{...(12.5)}$$

(d) Work Ratio

Work ratio,
$$R_w = \frac{w_{net}}{w_t} = \frac{w_t - w_c}{w_t} = 1 - \frac{w_c}{w_t}$$

$$= 1 - \frac{T_2 - T_1}{T_3 - T_4} = 1 - \frac{T_1 (r_p)^{\frac{\gamma-1}{\gamma}} - T_1}{T_3 - \dfrac{T_3}{r_p^{\frac{\gamma-1}{\gamma}}}}$$

$$= 1 - \left[\frac{T_1}{T_3}\right]\left[\frac{r_p^{\frac{\gamma-1}{\gamma}} - 1}{1 - \dfrac{1}{r_p^{(\gamma-1)/\gamma}}}\right] = 1 - \frac{T_1}{T_3}(r_p)^{\frac{\gamma-1}{\gamma}} = 1 - \frac{c}{\alpha} \qquad \text{...(12.6)}$$

(e) Optimum Pressure Ratio for Maximum Specific Work Output

$$w_{net} = w_t - w_c = c_p (T_3 - T_4) - c_p (T_2 - T_1)$$

$$= c_p T_3 \left(1 - \frac{T_4}{T_3}\right) - c_p T_1 \left(\frac{T_2}{T_1} - 1\right)$$

$$= c_p T_3 \left[1 - \frac{1}{r_p^{\frac{\gamma-1}{\gamma}}}\right] - c_p T_1 \left[r_p^{\frac{\gamma-1}{\gamma}} - 1\right]$$

w_{net} **will be maximum, when** $\dfrac{dw_{net}}{dr_p} = 0$

$$\therefore \qquad \frac{d}{dr_p}\left[c_p T_3 - c_p T_3 r_p^{-\left(\frac{\gamma-1}{\gamma}\right)} - c_p T_1 r_p^{\frac{\gamma-1}{\gamma}} + c_p T_1\right] = 0$$

Let

$$m = \frac{\gamma - 1}{\gamma}$$

$$\therefore \qquad \frac{d}{dr_p}[-T_3\, r_p^{-m} - T_1 r_p^{m}] = 0$$

$$m T_3 r_p^{-m-1} - m T_1 r_p^{m-1} = 0$$

$$\frac{T_3}{T_1} = r_p^{2m} = r_p^{\frac{2(\gamma-1)}{\gamma}}$$

$$\therefore \qquad r_p = \left[\frac{T_3}{T_1}\right]^{\frac{\gamma}{2(\gamma-1)}} \qquad \text{...(12.7)}$$

From Eq. (12.3), we have

$$(r_p)_{opt} = \left(\frac{T_3}{T_1}\right)^{\frac{\gamma}{\gamma-1}}$$

Combining with Eq. (12.7), we get

$$r_p = \sqrt{(r_p)_{opt}} \qquad \text{...(12.8)}$$

12.10 CYCLE OPERATION WITH MACHINE EFFICIENCY

(a) Maximum Pressure Ratio for Maximum Specific Work

It is not possible to achieve the isentropic compression and expansion in the actual Brayton cycle due to inevitable losses due to friction and turbulence in the compressor and turbine. Therefore, the temperature at the end of compression and expansion are higher than that in an ideal cycle for the same pressure ratio. The actual Brayton cycle is shown in Fig. 12.7.

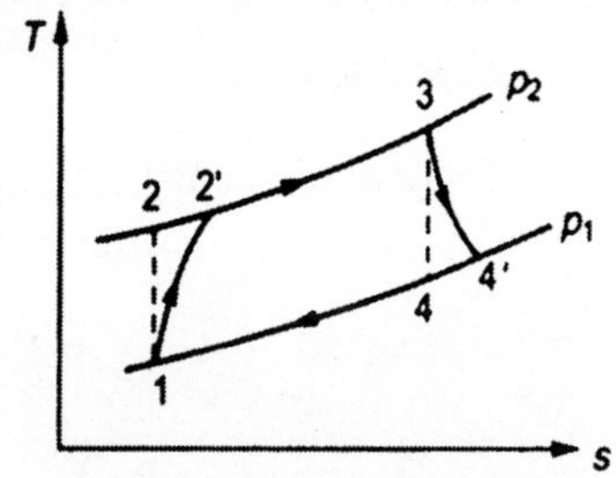

Fig. 12.7 Actual Brayton cycle with machine efficiencies.

Isentropic efficiency of compressor, $\eta_c = \dfrac{T_2 - T_1}{T_{2'} - T_1}$

Isentropic efficiency of turbine, $\eta_t = \dfrac{T_3 - T_{4'}}{T_3 - T_4}$

Net specific work available,

$$\begin{aligned} w_{net} &= w_t - w_c = c_p\,(T_3 - T_{4'}) - c_p\,(T_{2'} - T_1) \\ &= c_p \eta_t\,(T_3 - T_4) - c_p\,\frac{(T_2 - T_1)}{\eta_c} \\ &= c_p \eta_t T_3 \left(1 - \frac{T_4}{T_3}\right) - \frac{c_p T_1}{\eta_c}\left(\frac{T_2}{T_1} - 1\right) \\ &= c_p \eta_t T_3 \left[1 - \frac{1}{r_p^{\frac{\gamma-1}{\gamma}}}\right] - \frac{c_p T_1}{\eta_c}\left[r_p^{\frac{\gamma-1}{\gamma}} - 1\right] \end{aligned} \qquad \text{...(12.9)}$$

For specific work output to be maximum for given temperature limits,

$$\frac{d}{dr_p}(w_t) = 0$$

$$\eta_t T_3 \left[\left(\frac{\gamma-1}{\gamma}\right) r_p^{-\frac{\gamma-1}{\gamma}-1}\right] - \frac{T_1}{\eta_c}\left(\frac{\gamma-1}{\gamma}\right) r_p^{\frac{\gamma-1}{\gamma}-1} = 0$$

$$\eta_t T_3 r_p^{-\frac{2\gamma+1}{\gamma}-1} - \frac{T_1}{\eta_c} r_p^{\frac{-1}{\gamma}} = 0$$

$$r_p^{2\left(\frac{1-\gamma}{\gamma}\right)} = \frac{T_1}{\eta_c \eta_t T_3}$$

$$r_p = \left[\eta_c \eta_t \left(\frac{T_3}{T_1}\right)\right]^{\frac{\gamma}{2(\gamma-1)}} \quad \ldots(12.10)$$

(b) Optimum Pressure Ratio for Maximum Cycle Thermal Efficiency

Heat supplied, $q = c_p\,(T_3 - T_{2'}) = c_p(T_3 - T_2)$

Thermal efficiency, $\eta_{th} = \frac{w_{net}}{q} = \frac{c_p\,(T_3 - T_{4'}) - c_p(T_{2'} - T_1)}{c_p(T_3 - T_2)}$

$$= \frac{\eta_t\,(T_3 - T_4) - \frac{1}{\eta_c}(T_2 - T_1)}{T_3 - T_2}$$

$$= \frac{\eta_t T_3 \left[1 - \frac{T_4}{T_3}\right] - \frac{T_1}{\eta_c}\left[\frac{T_2}{T_1} - 1\right]}{T_1\left[\frac{T_3}{T_1} - \frac{T_2}{T_1}\right]}$$

Now

$$\frac{T_2}{T_1} = \frac{T_3}{T_4} = r_p^{\frac{\gamma-1}{\gamma}}$$

$$\eta_{th} = \frac{\eta_t T_3\left[1 - \frac{1}{r_p^{\frac{\gamma-1}{\gamma}}}\right] - \frac{T_1}{\eta_c}\left[r_p^{\frac{\gamma-1}{\gamma}} - 1\right]}{T_1\left[\frac{T_3}{T_1} - r_p^{\frac{\gamma-1}{\gamma}}\right]} \quad \ldots(12.11)$$

The thermal efficiency of the cycle is dependent only on r_p for the fixed values of T_1 and T_3. The condition for the maximum value of thermal efficiency for given temperature limits is

$$\frac{d\eta_{th}}{dr_p} = 0$$

Let

$$c = r_p^{\frac{\gamma-1}{\gamma}} \text{ and } \alpha = \frac{T_3}{T_1},$$

then

$$\eta_{th} = \frac{\alpha\eta_t\left(1 - \frac{1}{c}\right) - \frac{1}{\eta_c}(c-1)}{\alpha - c}$$

$$\frac{d\eta_{th}}{dc} = (\alpha - c)\left[\alpha\eta_t \times \frac{1}{c^2} - \frac{1}{\eta_c}\right] + 1\left[\alpha\eta_t\left(1 - \frac{1}{c}\right) - \frac{1}{\eta_c}(c-1)\right] = 0$$

$$\frac{\alpha^2\eta_t}{c^2} - \frac{\alpha}{\eta_c} - \frac{\alpha\eta_t}{c} + \frac{c}{\eta_c} + \alpha\eta_t - \frac{\alpha\eta_t}{c} - \frac{c}{\eta_c} + \frac{1}{\eta_c} = 0$$

$$\frac{\alpha^2\eta_t}{c^2} - \frac{\alpha}{\eta_c} - \frac{2\alpha\eta_t}{c} + \alpha\eta_t + \frac{1}{\eta_c} = 0$$

$$\alpha^2\eta_c\eta_t - \alpha c^2 - 2\alpha\eta_c\eta_t c + \alpha\eta_c\eta_t c^2 + c^2 = 0$$

$$c^2(1 + \alpha\eta_c\eta_t - \alpha) - 2\alpha\eta_c\eta_t c + \alpha^2\eta_c\eta_t = 0$$

$$c = \frac{2\alpha\eta_c\eta_t \pm [4\alpha^2\eta_c^2\eta_t^2 - 4\alpha^2\eta_c\eta_t(1 + \alpha\eta_c\eta_t - \alpha)]^{1/2}}{2(1 + \alpha\eta_c\eta_t - \alpha)}$$

$$= \frac{\alpha\eta_c\eta_t \pm \alpha[\eta_c^2\eta_t^2 - \eta_c\eta_t(1 + \alpha\eta_c\eta_t - \alpha)]^{1/2}}{(1 + \alpha\eta_c\eta_t - \alpha)}$$

$$= \frac{\alpha\left[1 \pm \left\{1 - \frac{1 + \alpha\eta_c\eta_t - \alpha}{\eta_c\eta_t}\right\}^{1/2}\right]}{\frac{1 + \alpha\eta_c\eta_t - \alpha}{\eta_c\eta_t}}$$

Taking –ve sign

$$c = \frac{\alpha\left[1 - \left\{1 - \frac{1}{\eta_c\eta_t} - \alpha + \frac{\alpha}{\eta_c\eta_t}\right\}^{1/2}\right]}{\left[1 + \left\{1 - \frac{1}{\eta_c\eta_t} - \alpha + \frac{\alpha}{\eta_c\eta_t}\right\}^{1/2}\right]\left[1 - \left\{1 - \frac{1}{\eta_c\eta_t} - \alpha + \frac{\alpha}{\eta_c\eta_t}\right\}^{1/2}\right]}$$

$$r_p^{\frac{\gamma-1}{\gamma}} = \frac{\alpha}{1 + \left\{\frac{1}{\eta_c\eta_t}(\alpha - 1) - (\alpha - 1)\right\}^{1/2}}$$

$$= \frac{\alpha}{1 + \left\{(\alpha - 1)\left(\frac{1}{\eta_c\eta_t} - 1\right)\right\}^{1/2}}$$

$$(r_p)_{opt} = \left[\frac{\frac{T_3}{T_1}}{1 + \left\{\left(\frac{T_3}{T_1} - 1\right)\left(\frac{1}{\eta_c\eta_t} - 1\right)\right\}^{1/2}}\right]^{\frac{\gamma}{\gamma-1}} \quad \ldots(12.12)$$

12.11 OPEN CYCLE CONSTANT PRESSURE GAS TURBINE

The open cycle constant pressure gas turbine plant is hsown diagrammatically in Fig. 12.8. The fuel is burned with air coming out from the compressor into the combustion chamber. The products of combustion from the combustion chamber are passed through the turbine and develop the required power. Part of the power developed is used to run the compressor. The gases leaving the turbine are exhausted to atmosphere.

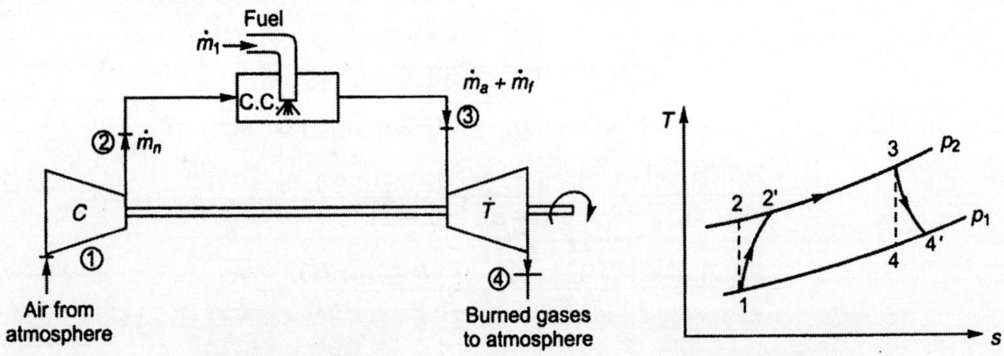

(a) Schematic diagram (b) T-s diagram

Fig. 12.8 Open cycle constant pressure gas turbine.

Let $\dot{m}_a$ = mass rate of flow of air

$\dot{m}_f$ = mass rate of flow of fuel

Compressor work, $$W_c = \dot{m}_a\, c_{pa}\, (T_2' - T_1)$$

Turbine work, $$W_t = (\dot{m}_a + \dot{m}_f)\, c_{pg}\, (T_3 - T_4')$$

where c_{pa}, c_{pg} = specific heat of air and gases respectively

Network $W_{net} = W_t - W_c = (\dot{m}_a + \dot{m}_f)\, c_{pq}\, (T_3 - T_4') - \dot{m}_a c_{pa}\, (T_2' - T_1)$

Heat supplied, $$Q = \dot{m}_f \times \text{C.V.}$$

where C.V. = calorific value of fuel

Thermal efficiency, $\eta_{th} = \dfrac{W_{net}}{Q}$

$$= \frac{(\dot{m}_a + \dot{m}_f)\, c_{pg}\, (T_3 - T_4') - \dot{m}_a c_{pa}\, (T_2' - T_1)}{\dot{m}_f \times \text{C.V.}} \quad \text{...(12.13)}$$

Now $$\eta_c = \frac{T_2 - T_1}{T_2' - T_1}$$

$\therefore$ $$T_2' - T_1 = \frac{T_2 - T_1}{\eta_c}$$

and $$\eta_t = \frac{T_3 - T_4'}{T_3 - T_4}$$

$$\therefore \qquad T_3 - T_4' = \eta_t (T_3 - T_4)$$

Hence,
$$\eta_{th} = \frac{(\dot{m}_a + \dot{m}_f) c_{pg} \eta_t (T_3 - T_4) - \dfrac{\dot{m}_a c_{pa}}{\eta_c}(T_2 - T_1)}{\dot{m}_f \times \text{C.V.}} \qquad \text{...(12.14)}$$

12.12 METHODS FOR IMPROVEMENT OF THERMAL EFFICIENCY OF OPEN CYCLE CONSTANT PRESSURE GAS TURBINE

The thermal efficiency of open cycle constant pressure gas turbine can be improved by the following methods :

1. Regeneration
2. Intercooling
3. Reheating
4. Combination of above

12.12.1 Regeneration

In this method, the heat of the exhaust gases is used to heat the air coming out from the compressor thus reducing the mass of fuel supplied in the combustion chamber. A schematic diagram of such a plant is shown in Fig. 12.9 (a).

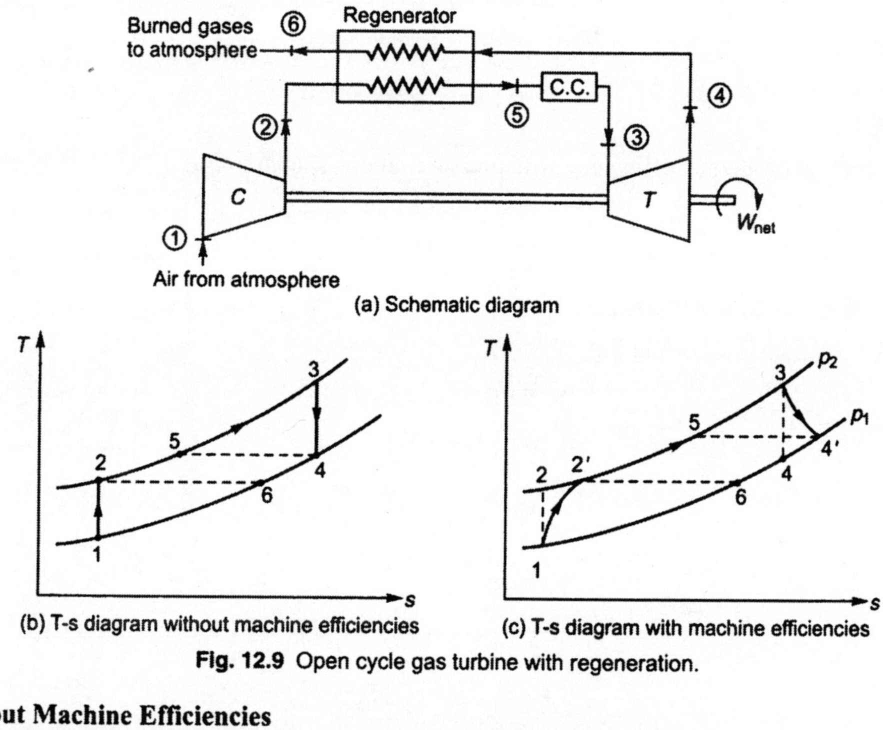

Fig. 12.9 Open cycle gas turbine with regeneration.

(a) Without Machine Efficiencies

The T-s diagram without machine efficiencies is shown in Fig. 12.9 (b).

Compressor work, $\quad w_c = w_{1-2} = c_p (T_2 - T_1)$

Heat supplied, $\quad q_s = q_{5-3} = c_p (T_3 - T_5)$

Turbine work, $\quad w_t = w_{3-4} = c_p (T_3 - T_4)$

Network, $$w_{net} = w_t - w_c = c_p\,(T_3 - T_4) - c_p\,(T_2 - T_1)$$

Thermal efficiency, $$\eta_{th} = \frac{w_{net}}{q_s} = \frac{(T_3 - T_4) - (T_2 - T_1)}{T_3 - T_5}$$

In an ideal regenerator, $$T_4 = T_5$$

$\therefore$ $$\eta_{th} = 1 - \frac{T_2 - T_1}{T_3 - T_4} = 1 - \left(\frac{T_1}{T_3}\right)\left[\frac{\dfrac{T_2}{T_1} - 1}{1 - \dfrac{T_4}{T_3}}\right]$$

$$= 1 - \frac{1}{\alpha}\left[\frac{c-1}{1-\dfrac{1}{c}}\right] = 1 - \frac{c}{\alpha} \qquad \text{...(12.15)}$$

where $$\alpha = \frac{T_3}{T_1}, \quad c = r_p^{\frac{\gamma-1}{\gamma}}$$

$$\frac{w_{net}}{c_p T_1} = \left[\frac{T_3}{T_1} - \frac{T_4}{T_1}\right] - \left[\frac{T_2}{T_1} - 1\right]$$

$$= \alpha\left(1 - \frac{1}{c}\right) - (c-1) \qquad \text{...(12.16)}$$

The variation of thermal efficiency with pressure ratio is shown in Fig. 12.10. It may be seen that thermal efficiency increases with an increase in $\alpha = \frac{T_3}{T_1}$ and decreases with increase in r_p.

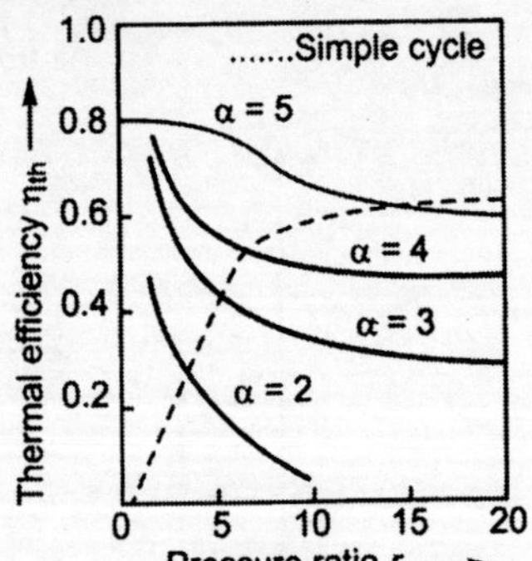

Fig. 12.10

(b) Considering Machine Efficiencies

The T-s diagram is shown in Fig. 12.9 (c).

$$w_c = w_{1-2'}$$
$$= c_p\,(T_2' - T_1)$$
$$= c_p\,\frac{(T_2 - T_1)}{\eta_c}$$

$$q_s = q_{5-3}$$
$$= c_p\,(T_3 - T_5) = c_p\,(T_3 - T_4')$$

$$w_t = w_{3-4'} = c_p\,(T_3 - T_4')$$
$$= c_p\,\eta_t\,(T_3 - T_4)$$

$$\eta_{th} = \frac{w_t - w_c}{q_s} = \frac{(T_3 - T_4') - \dfrac{T_2 - T_1}{\eta_c}}{T_3 - T_4'}$$

$$= 1 - \frac{\dfrac{T_2 - T_1}{\eta_c}}{\eta_t (T_3 - T_4)}$$

$$= 1 - \frac{T_1}{T_3} \cdot \frac{1}{\eta_c \eta_t} \left[\frac{c-1}{1 - \dfrac{1}{c}} \right] = 1 - \frac{c}{\alpha \eta_c \eta_t} \quad \text{...(12.17)}$$

$$w_{net} = w_t - w_c = c_p \eta_t (T_3 - T_4) - \frac{c_p}{\eta_c} (T_2 - T_1)$$

$$\frac{w_{net}}{c_p T_1} = \eta_t \left[\frac{T_3}{T_1} - \frac{T_4}{T_1} \right] - \frac{1}{\eta_c} \left[\frac{T_2}{T_1} - 1 \right]$$

$$= \eta_t \alpha \left[1 - \frac{1}{c} \right] - \frac{1}{\eta_c} (c - 1) \quad \text{...(12.18)}$$

(c) Effectiveness of Regenerator

In actual practice, the rise in temperature from T_2' to $T_5 = T_4'$ is not possible. The actual temperature of air will be $T_5' < T_5$. Hence, actual thermal efficiency [see Fig. (12.11)],

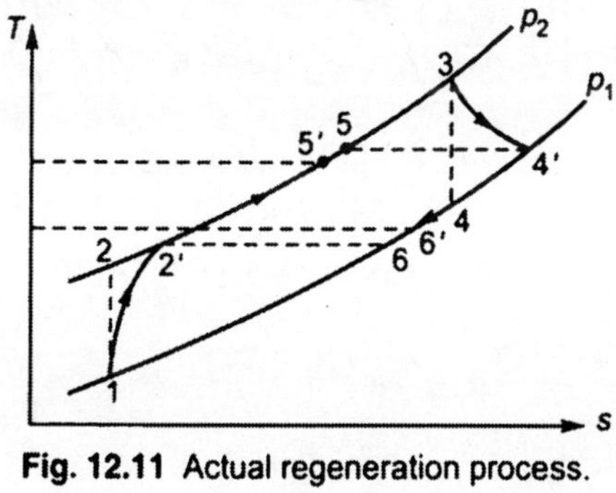

Fig. 12.11 Actual regeneration process.

$$(\eta_{th})_a = \frac{(T_3 - T_4') - (T_2' - T_1)}{(T_3 - T_5')} \quad \text{...(12.19)}$$

The effectiveness of the regeneration is given by

$$\varepsilon = \frac{\text{Actual heat transfer to air}}{\text{Maximum heat transfer theoretically possible}}$$

$$= \frac{\dot{m}_a c_{pa} (T_5' - T_2')}{\dot{m}_g c_{pg} (T_4' - T_2')} \quad \text{...(12.20)}$$

where $\dot{m}_a$, $\dot{m}_g$ = mass rate of flow of air and exhaust gases respectively

c_{pa}, c_{pg} = specific heats of air and exhaust gases respectively.

12.12.2 Intercooling

The work consumed by the compressor can be reduced by compressing the air in two stages and incorporating the intercooler in between, as shown in Fig. 12.12(a). The corresponding T-s diagram is shown in Fig. 12.12(b).

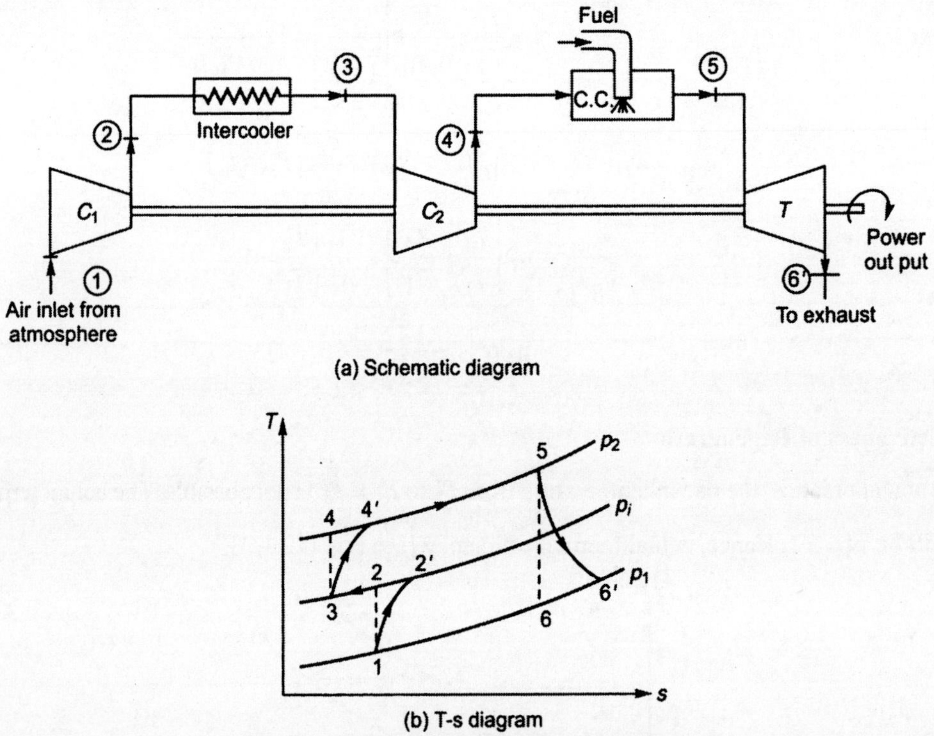

Fig. 12.12 Gas turbine plant with intercooling.

$$w_c = c_{pa}\,(T_2' - T_1) + c_{pa}\,(T_4' - T_3)$$

$$= \frac{c_{pa}}{\eta_{cl}}(T_2 - T_1) + \frac{c_{pa}}{\eta_{c2}}(T_4 - T_3)$$

$$= \frac{c_{pa}T_1}{\eta_{cl}}\left[\frac{T_2}{T_1} - 1\right] + \frac{c_{pa}T_3}{\eta_{c2}}\left[\frac{T_4}{T_3} - 1\right]$$

$$= \frac{c_{pa}T_1}{\eta_{cl}}\left[\left(\frac{p_i}{p_1}\right)^{\frac{\gamma-1}{\gamma}} - 1\right] + \frac{c_{pa}T_3}{\eta_{c2}}\left[\left(\frac{p_2}{p_i}\right)^{\frac{\gamma-1}{\gamma}} - 1\right]$$

Let $\alpha = \dfrac{T_3}{T_1}$ and $m = \dfrac{\gamma - 1}{\gamma}$

$$w_c = \frac{c_{pa}T_1}{\eta_{cl}}\left[\left(\frac{p_i}{p_1}\right)^m - 1\right] + \frac{c_{pa}\alpha T_1}{\eta_{c2}}\left[\left(\frac{p_2}{p_i}\right)^m - 1\right]$$

For minimum work, $\dfrac{dw_c}{dp_i} = 0$

$$\frac{c_{pa}T_1}{\eta_{c1}}\left[\frac{m}{p_1}\left(\frac{p_i}{p_1}\right)^{m-1}\right] + \frac{c_{pa}\alpha T_1}{\eta_{c2}}\left[m\left(\frac{p_2}{p_i}\right)^{m-1} c - \frac{p_2}{p_i^2}\right] = 0$$

$$\frac{c_{pa}T_1}{\eta_{c1}} \times p_1^{m/m} \times p_i^{m-1} + \frac{c_{pa}\alpha T_1}{\eta_{c2}}(-mp_2^m) \times \frac{1}{p_i^{m+1}} = 0$$

$$\frac{p_i^{m-1}}{\eta_{c1}\, p_1^m} = \frac{\alpha p_2^m}{\eta_{c2}\, p_i^{m+1}}$$

$$p_i^{2m} = \alpha\,\frac{\eta_{c1}}{\eta_{c2}} \times (p_1 p_2)^m$$

$$p_i = \left[\frac{\alpha\eta_{c1}}{\eta_{c2}} \times (p_1 p_2)^m\right]^{1/2m}$$

$$= \left[\frac{\alpha\ \eta_{c1}}{\eta_{c2}}\right]^{1/2m} \times (p_1 p_2)^{1/2}$$

$$= \left[\frac{T_3}{T_1} \times \frac{\eta_{c1}}{\eta_{c2}}\right]^{\frac{\gamma}{2(\gamma-1)}} \times \sqrt{p_1 p_2} \qquad \text{...(12.21)}$$

For perfect cooling, $T_3 = T_1$ and if $n_{c1} = n_{c2} = n_{c'}$ then

$$p_i = \sqrt{p_1 p_2} \qquad \text{...(12.22)}$$

$$(w_c)_{\min} = \frac{c_{pa}T_1}{\eta_c}\left[\left(\frac{p_i}{p_1}\right)^m + \left(\frac{p_2}{p_i}\right)^m - 2\right]$$

$$= \frac{c_{pa}T_1}{\eta_c}\left[\left(\frac{p_2}{p_1}\right)^{\frac{m}{2}} + \left(\frac{p_2}{p_1}\right)^{\frac{m}{2}} - 2\right]$$

$$= \frac{2c_{pa}T_1}{\eta_c}\left[\left(\frac{p_2}{p_1}\right)^{\frac{\gamma-1}{2\gamma}} - 1\right] \qquad \text{...(12.23)}$$

$$n_{th} = \frac{(T_5 - T_6') - [(T_2' - T_1) + (T_4' - T_3)]}{T_5 - T_4'}$$

If $T_3 = T_{1'}$ then $$n_{th} = \frac{(T_5 - T_6') - (T_2' + T_4' - 2T_1)}{T_5 - T_4'} \qquad \text{...(12.24)}$$

$$w_{\text{net}} = w - w_c = c_p\,(T_5 - T_6') - [c_p(T_2' - T_1) + c_p\,(T_4' - T_3)]$$

$$= c_p\eta_t(T_5 - T_6) - c_p\left[\frac{T_2 - T_1}{\eta_{c1}} + \frac{T_4 - T_3}{\eta_{c2}}\right]$$

For perfect intercooling, $T_3 = T_1$ and if $\eta_{c1} = \eta_{c2} = \eta_c$, then

$$w_{\text{net}} = c_p\eta_t\,(T_5 - T_6) - \frac{c_p}{\eta_c}(T_2 + T_4 - 2T_1)$$

$$= c_pT_1\eta_t\left[\left(\frac{T_5}{T_1} - \frac{T_6}{T_1}\right)\right] - \frac{c_pT_1}{\eta_c}\left[\frac{T_2}{T_1} + \frac{T_4}{T_1} - 2\right]$$

$$\frac{w_{\text{net}}}{c_pT_1} = \eta_t\left[\frac{T_5}{T_1} - \frac{T_6}{T_1}\right] - \frac{1}{\eta_c}\left[\frac{T_2}{T_1} + \frac{T_4}{T_1} - 2\right]$$

Let

$$\frac{T_5}{T_1} = \beta, \quad \frac{T_5}{t_6} = b, \quad \frac{T_2}{T_1} = b_1,$$

$$\frac{T_4}{T_3} = b_2 = \frac{T_4}{T_1}$$

$$\frac{w_{\text{net}}}{c_pT_1} = \eta_t\left[\beta - \frac{\beta}{b}\right] - \frac{1}{\eta_c}[b_1 + b_2 - 2]$$

Now $b = b_1b_2$ because

$$b = \left(\frac{p_2}{p_1}\right)^m, \quad b_1 = \left(\frac{p_i}{p_1}\right)^m$$

and

$$b_2 = \left(\frac{p_2}{p_i}\right)^m$$

$\therefore$

$$\frac{w_{\text{net}}}{c_pT_1} = \eta_t\beta\left(1 - \frac{1}{b}\right) - \frac{1}{\eta_c}\left(b_1 + \frac{b}{b_1} - 2\right) \quad \text{...(12.25)}$$

$$\frac{d}{db_1}\left(\frac{w_{\text{net}}}{c_pT_1}\right) = -\frac{1}{\eta_c}\left(1 - \frac{b}{b_1^2}\right) = 0$$

$$b_1 = \sqrt{b} = b_2$$

$$\frac{(w_{\text{net}})_{\max}}{c_pT_1} = \eta_t\beta\left(1 - \frac{1}{b}\right) - \frac{1}{\eta_c}\left(\sqrt{b} + \frac{b}{\sqrt{b}} - 2\right)$$

$$= \eta_t\,\beta\left(1 - \frac{1}{b}\right) - \frac{2}{\eta_c}(\sqrt{b} - 1) \quad \text{.(12.26)}$$

$$(\eta_{th})_{\max} = \frac{(w_{\text{net}})_{\max}}{q_s}$$

$$= \frac{c_pT_1\left[\eta_t\beta\left(1 - \frac{1}{b}\right) - \frac{2}{\eta_c}(\sqrt{b} - 1)\right]}{c_p(T_5 - T_4') = c_p\,(T_5 - T_4)}$$

Now $$\frac{T_5}{T_1} - \frac{T_4}{T_1} = \frac{T_5}{T_1} - \frac{T_4}{T_3} = \beta - \sqrt{b}$$

$\therefore$ $$(\eta_{th})_{max} = \frac{\eta_t \beta \left(1 - \frac{1}{b}\right) - \frac{2}{\eta_c}(\sqrt{b} - 1)}{\beta - \sqrt{b}} \quad ...(12.27)$$

$$= \frac{\beta\left(1 - \frac{1}{b}\right) - 2(\sqrt{b} - 1)}{\beta - \sqrt{b}} \text{ for } \eta_t = \eta_c = 1$$

$$= \frac{(\beta - \sqrt{b}) - \frac{\beta}{b} - \sqrt{b} + 2}{\beta - \sqrt{b}} = 1 - \frac{\frac{\beta}{b} + \sqrt{b} - 2}{\beta - \sqrt{b}} \quad ...(12.28)$$

12.12.3 Reheating

A considerable increase in power output can be achieved by expanding the gases in two stages with a reheat combustion chamber between the two as shown in Fig. 12.13 (a). The corresponding T-s diagram is shown in Fig. 12.13 (b).

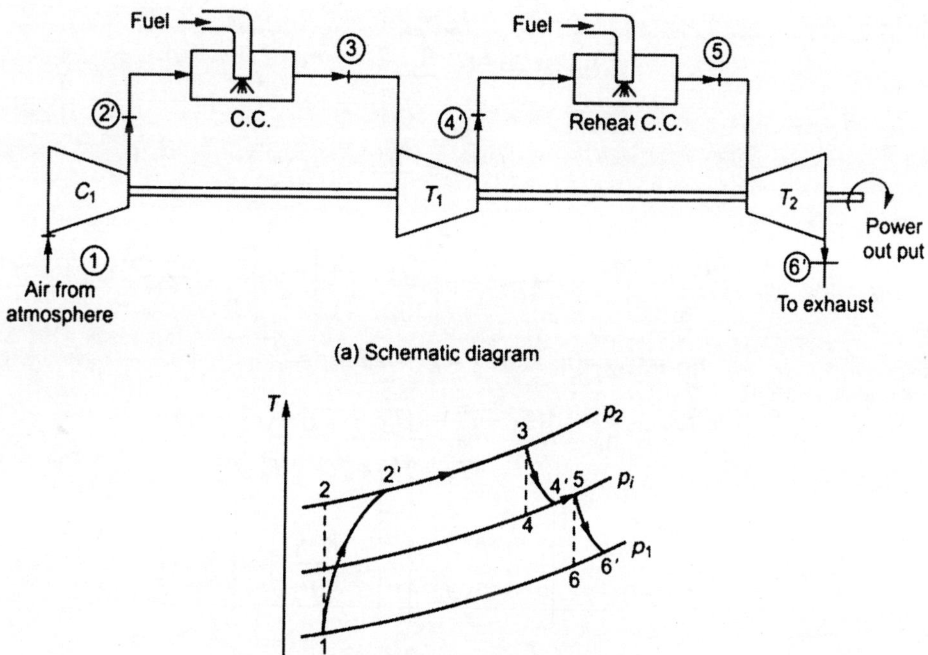

Fig. 12.13 Gas turbine plant with reheating.

$$w_t = c_{pg}(T_3 - T_4') + c_{pq}(T_5 - T_6')$$

$$= c_{pg}\,\eta_{t1}(T_3 - T_4) + c_{pq}\,\eta_{t2}(T_5 - T_6)$$

$$= c_{pg}\ \eta_{t1}\ T_3\left(1-\frac{T_4}{T_3}\right) + c_{pg}\ \eta_{t2}\ T_5\left(1-\frac{T_6}{T_5}\right)$$

$$= c_{pg}\ \eta_{t1}\ T_3\left[1-\left(\frac{p_i}{p_2}\right)^m\right] + c_{pg}\eta_{t2}T_5\left[1-\left(\frac{p_1}{p_i}\right)^m\right]$$

where $m = \frac{\gamma - 1}{\gamma}$

Let $$\alpha = \frac{T_5}{T_3},$$

then $$w_t = c_{pg}\ \eta_{t1}\ T_3\left[1-\left(\frac{p_i}{p_2}\right)^m\right] + c_{pg}\ \eta_{t2} \propto T_3\left[1-\left(\frac{p_1}{p_i}\right)^m\right] \qquad ...(12.29)$$

For w_t to be maximum, $$\frac{dw_t}{dp_i} = 0$$

$$\eta_{t1} \times (-m) \times \left(\frac{p_i}{p_2}\right)^{m-1} \times \frac{1}{p_2} + n_{t2} \propto (-m)\left(\frac{p_1}{p_i}\right)^{m-1} \times \left(\frac{-p_1}{p_i^2}\right) = 0$$

$$\frac{\eta_{t1}}{p_2^m} \times p_i^{m-1} = \frac{\eta_{t2} \propto p_1^m}{p_i^{m+1}}$$

$$p_i^{2m} = \propto \frac{\eta_{t2}}{\eta_{t1}}(p_1 p_2)^m$$

$$p_i = \left(\frac{\alpha\ \eta_{t2}}{\eta_{t1}}\right)^{\frac{1}{2m}} \times (p_1 p_2)^{\frac{1}{2}} = \left[\frac{\eta_{t2}}{\eta_{t1}} \times \frac{T_5}{T_3}\right]^{\frac{\gamma}{(2(\gamma-1)}} \times \sqrt{p_1 p_2} \qquad ...(12.30)$$

If $$\eta_{t1} = \eta_{t2} = \eta_t \text{ and } T_5 = T_3, \text{ then } p_i = \sqrt{p_1 p_2} \qquad ...(12.31)$$

$$\eta_{\text{th}} = \frac{[(T_3 - T_4') + (T_5 - T_6')] - (T_2' - T_1)}{(T_3 - T_2') + (T_5 - T_4')}$$

$$\frac{(w_t)_{\max}}{c_{pg}\eta_t T_3} = \left[1-\left(\frac{p_1}{p_2}\right)^{\frac{m}{2}}\right] + \left[1-\left(\frac{p_1}{p_2}\right)^{\frac{m}{2}}\right] \qquad ...(12.32)$$

$$= 2\left[1-\left(\frac{p_1}{p_2}\right)^{\frac{\gamma-1}{2\gamma}}\right]$$

12.12.4 Reheat and Regenerative Cycle

This cycle is shown in Fig. 12.14.

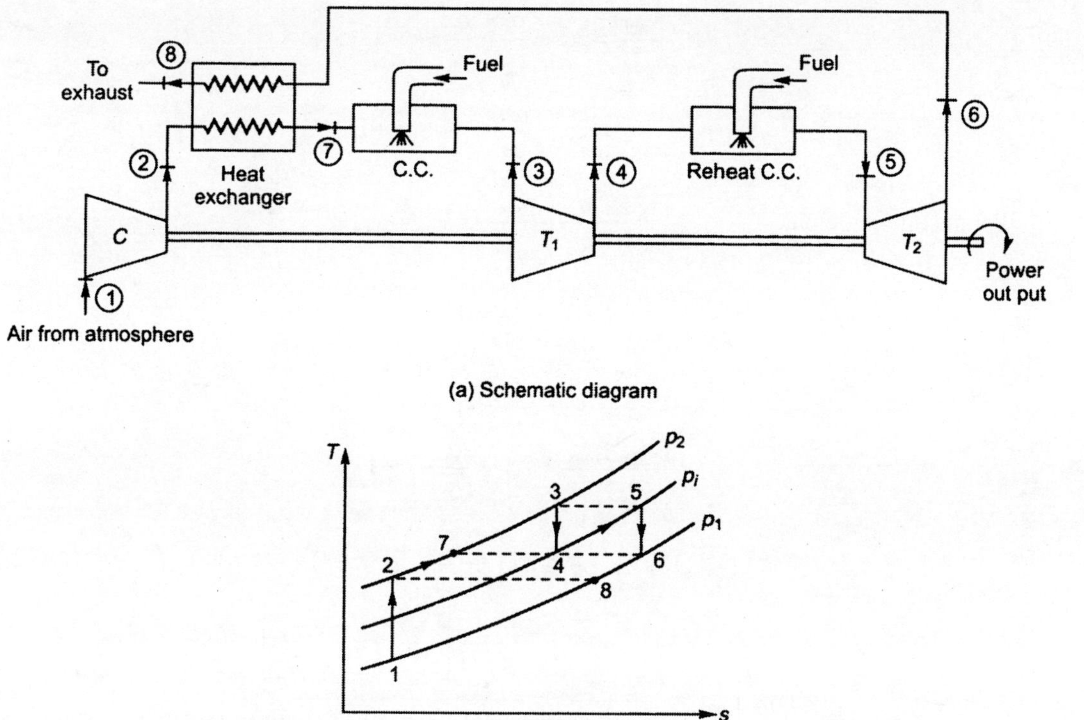

(a) Schematic diagram

(b) T-s diagram

Fig. 12.14 Gas turbine plant with reheat and regeneration.

$$w_c = c_p\,(T_2 - T_1)$$

$$q_s = q_{7-3} + q_{4-5} = c_p\,(T_3 - T_7) + c_p\,(T_5 - T_4)$$

$$w_t = w_{3-4} + w_{5-6} = c_p\,(T_3 - T_4) + c_p\,(T_5 - T_6)$$

$$w_{net} = w_t - w_c$$

$$= c_p\,[(T_3 - T_4) + (T_5 - T_6) - (T_2 - T_1)]$$

$$= c_p T_1 \left[\frac{T_3}{T_1} - \frac{T_4}{T_1} + \frac{T_5}{T_1} - \frac{T_6}{T_1} - \frac{T_2}{T_1} + 1\right]$$

Let $$r_p = \frac{p_2}{p_1},\quad r_2 = \frac{p_2}{p_i},\quad r_1 = \frac{p_i}{p_1} \text{ so that } r_p = r_1 r_2$$

$$c = r_p^{\frac{\gamma-1}{\gamma}} = \frac{T_2}{T_1},\quad c_2 = r_p^{\frac{-\gamma-1}{\gamma}} = \frac{T_3}{T_4}$$

$$c_1 = -r_1^{\frac{-\gamma-1}{\gamma}} = \frac{T_5}{T_6} \text{ so that } c = c_1 c_2$$

$$\frac{w_{net}}{c_p T_1} = \frac{T_3}{T_1} - \frac{T_4}{T_3} \times \frac{T_3}{T_1} \times \frac{T_5}{T_1} - \frac{T_6}{T_5} \times \frac{T_5}{T_1} - \frac{T_2}{T_1} + 1$$

Now, $$T_5 = T_3 \text{ and } \frac{T_3}{T_1} = \alpha$$

$$\frac{w_{\text{net}}}{c_p T_1} = 2\alpha - \frac{\alpha}{c_2} - \frac{\alpha}{c_1} - c + 1$$

$$= 2\alpha - \frac{\alpha}{2} - \frac{c_2 \alpha}{c} - c + 1$$

For maximum output,

$$\frac{\alpha}{\alpha c_2}\left[\frac{w_{\text{net}}}{c_p T_1}\right] = 0$$

$$\frac{\alpha}{c_2^2} - \frac{\alpha}{c} = 0$$

$$c_2 = \sqrt{c} = c_1$$

$$\frac{(w_{\text{net}})_{\max}}{c_p T_1} = 2\alpha - \frac{2\alpha}{\sqrt{c}} - c + 1$$

$$= 2\alpha\left(1 - \frac{1}{\sqrt{c}}\right) - (c - 1) \qquad \text{...(12.33)}$$

$$(\eta_{\text{th}})_{\max} = \frac{(w_{\text{net}})_{\max}}{q_s} = \frac{c_p T_1 \left[2\alpha - c + 1 - \frac{2\alpha}{\sqrt{c}}\right]}{c_p\,[(T_3 - T_7) + (T_5 - T_4)]}$$

$$= \frac{2\alpha - c + 1 - \frac{2\alpha}{\sqrt{c}}}{\frac{T_3}{T_1} - \frac{T_7}{T_1} + \frac{T_5}{T_1} - \frac{T_4}{T_1}}$$

Now

$$T_7 = T_4 \text{ and } T_5 = T_3$$

$$(\eta_{\text{th}})_{\max} = \frac{2\alpha - c + 1 - \frac{2\alpha}{\sqrt{c}}}{2\left(\frac{T_3}{T_1} - \frac{T_4}{T_1}\right)} = \frac{2\alpha - c + 1 - \frac{2\alpha}{\sqrt{c}}}{2\left(\alpha - \frac{\alpha}{\sqrt{c}}\right)}$$

$$= 1 - \frac{c - 1}{2\alpha\left(1 - \frac{1}{\sqrt{c}}\right)} \qquad \text{...(12.34)}$$

12.12.5 Intercooled Cycle with Regeneration

The schematic arrangement and T-s diagram of the cycle are shown in Fig. 12.15.

$$w_c = w_{1-2} + w_{3-4} = c_p\,[(T_2 - T_1) + (T_4 - T_3)]$$

$$q_s = q_{7-5} = c_p\,(T_5 - T_7)$$

$$w_t = w_{5-6} = c_p\,(T_5 - T_6)$$

$$w_{\text{net}} = w_t - w_c = c_p\,[(T_5 - T_6) - (T_2 - T_1) - (T_4 - T_3)]$$

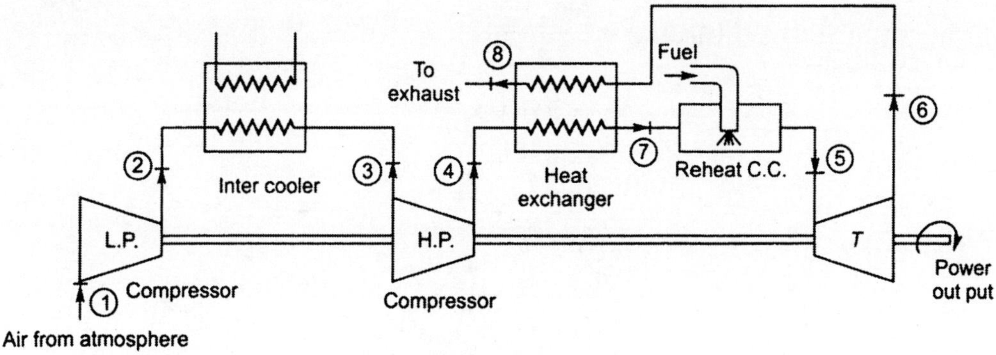

(a) Schematic diagram

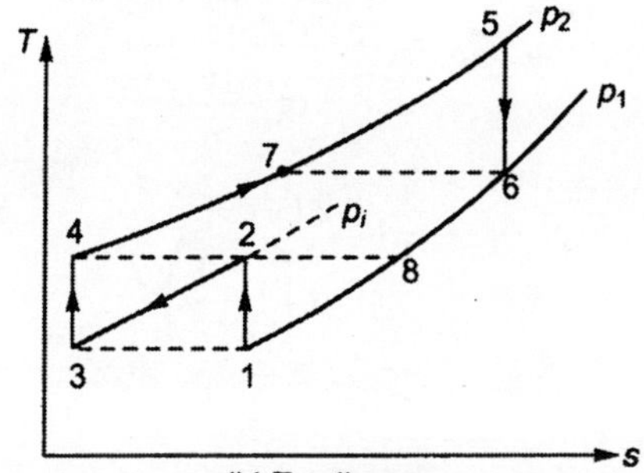

(b) T-s diagram

Fig. 12.15 Gas turbine plant with inter cooler and regeneration.

$$= c_p T_1 \left[\left(\frac{T_5}{T_1} - \frac{T_6}{T_1} \right) - \left(\frac{T_2}{T_1} - 1 \right) - \left(\frac{T_4}{T_1} - \frac{T_3}{T_1} \right) \right]$$

Let $\frac{T_5}{T_1} = \alpha, \quad \frac{T_5}{T_6} = c, \quad \frac{T_2}{T_1} = c_1, \quad \frac{T_4}{T_3} = c_2$ so that $c = c_1 c_2$ and $T_3 = T_1$

$$\frac{w_{net}}{c_p T_1} = \left(\alpha - \frac{\alpha}{c} \right) - (c_1 - 1) - (c_2 - 1)$$

$$= \alpha \left(1 - \frac{1}{c} \right) - (c_1 - 1) - \left(\frac{c}{c_1} - 1 \right)$$

$$\frac{d}{dc_1} \left[\frac{w_{net}}{c_p T_1} \right] = -1 + \frac{c}{c_1^2} = 0$$

or

$$c_1 = \sqrt{c} = c_2$$

$$\frac{(w_{net})_{max}}{c_p T_1} = \alpha \left(1 - \frac{1}{c} \right) - (\sqrt{c} - 1) - (\sqrt{c} - 1)$$

$$= \alpha \left(1 - \frac{1}{c} \right) - 2(\sqrt{c} - 1)$$

$$= \alpha - \frac{\alpha}{c} - 2\sqrt{c} + 2 \qquad \text{...(12.35)}$$

$$(\eta_{th})_{max} = \frac{(w_{net})_{max}}{q_s}$$

Now, $$q_s = c_p\,(T_5 - T_7)$$

$$\frac{q_s}{c_p T_1} = \frac{T_5}{T_1} - \frac{T_7}{T_1}$$

Also $$T_7 = T_6$$

∴ $$\frac{q_s}{c_p T_1} = \frac{T_5}{T_1} - \frac{T_6}{T_5} \times \frac{T_5}{T_1} = \alpha - \frac{\alpha}{c}$$

$$(\eta_{th})_{max} = \frac{\alpha - \dfrac{\alpha}{c} - 2\sqrt{c} + 2}{\alpha - \dfrac{\alpha}{c}}$$

$$= 1 - \frac{2(\sqrt{c} - 1)}{\alpha\left(1 - \dfrac{1}{c}\right)} \qquad \text{...(12.36)}$$

12.12.6 Intercooled Cycle with Reheating

The schematic arrangement of this cycle is shown in Fig. 12.16(a) and the corresponding T-s diagram in Fig. 12.16(b).

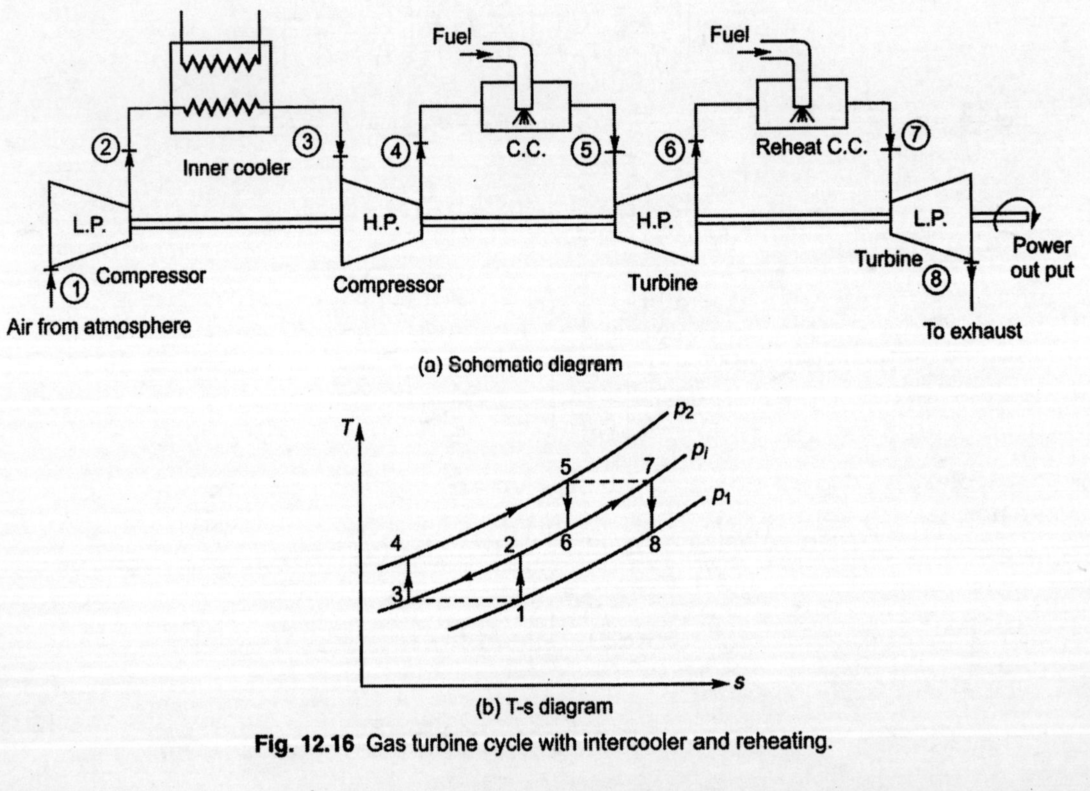

Fig. 12.16 Gas turbine cycle with intercooler and reheating.

$$w_c = c_p\,[(T_2 - T_1) + (T_4 - T_3)]$$
$$q_s = c_p\,[(T_5 - T_4) + (T_7 - T_6)]$$
$$w_t = c_p\,[(T_5 - T_6) + (T_7 - T_8)]$$
$$w_{\text{net}} = w_t - w_c = c_p\,[(T_5 - T_6) + (T_7 - T_8) - (T_2 - T_1) - (T_4 - T_3)]$$
$$= c_p T_1\left[\frac{T_5}{T_1} - \frac{T_6}{T_1} + \frac{T_7}{T_1} - \frac{T_8}{T_1} - \frac{T_2}{T_1} + 1 - \frac{T_4}{T_1} + \frac{T_3}{T_1}\right]$$

Now, $T_3 = T_1$ and $T_7 = T_5$ for perfect intercooling and reheating respectively

$$\frac{w_{\text{net}}}{c_p T_1} = \frac{T_5}{T_1} - \frac{T_6}{T_1} + \frac{T_5}{T_1} - \frac{T_8}{T_1} - \frac{T_2}{T_1} + 1 - \frac{T_4}{T_1} + 1$$
$$= \frac{T_5}{T_1} - \frac{T_6}{T_5} \times \frac{T_5}{T_1} + \frac{T_5}{T_1} - \frac{T_8}{T_7} \times \frac{T_5}{T_1} - \frac{T_2}{T_1} - \frac{T_4}{T_3} + 2$$

Let $$a = \frac{T_5}{T_1'}\ c = \frac{T_4}{T_1}, c_1 = \frac{T_2}{T_1}, c_2 = \frac{T_4}{T_3}, \text{ also } T_2 = T_3, T_5 = T_1$$

$$\frac{T_4}{T_1} = \frac{T_4}{T_3} \times \frac{T_2}{T_1}.$$

Thus, $$c = c_1 c_2$$

For maximum work output, $$p_i = \sqrt{p_1 p_2}$$

or $$\frac{p_i}{p_1} = \frac{p_2}{p_i}$$

$$\frac{T_2}{T_1} = \left(\frac{p_i}{p_1}\right)^{\frac{\gamma-1}{\gamma}} \quad \text{and} \quad \frac{T_4}{T_3} = \left(\frac{p_2}{p_i}\right)^{\frac{\gamma-1}{\gamma}}$$

∴ $$\frac{T_2}{T_1} = \frac{T_4}{T_3}$$

Also $$\frac{T_5}{T_6} = \left(\frac{p_2}{p_i}\right)^{\frac{\gamma-1}{\gamma}} = \frac{T_4}{T_3} = c_2$$

and $$\frac{T_7}{T_8} = \frac{T_5}{T_8}\left(\frac{p_i}{p}\right)^{\frac{\gamma-1}{\gamma}} = c_1$$

or $$c_1 = c_2$$

∴ $$c_1 = c_2 = \sqrt{c}\ \frac{w_{\text{net}}}{c_p T_3}$$

$$= \alpha - \frac{\alpha}{\sqrt{c_2}} + \alpha - \frac{\alpha}{c_1} - c_1 - c_2 + 2$$

∴ $$\frac{(w_{\text{net}})_{\max}}{c_p T_3} = \alpha - \frac{\alpha}{\sqrt{c}} + \alpha - \frac{\alpha}{\sqrt{c}} - \sqrt{c} - \sqrt{c} + 2$$

$$= 2\left(\alpha - \frac{\alpha}{\sqrt{c}} - \sqrt{c} + 1\right) \qquad \text{...(12.37)}$$

$$(\eta_{th})_{max} = \frac{(w_{net})_{max}}{q_s} = \frac{2c_p T_1\left(\alpha - \frac{\alpha}{\sqrt{c}} - \sqrt{c} + 1\right)}{c_p\,[(T_5 - T_4) + (T_7 - T_6)]}$$

$$= \frac{2\left(\alpha - \frac{\alpha}{\sqrt{c}} - \sqrt{c} + 1\right)}{\frac{T_5}{T_1} - \frac{T_4}{T_1} + \frac{T_7}{T_1} - \frac{T_6}{T_1}} = \frac{2\left(\alpha - \frac{\alpha}{\sqrt{c}} - \sqrt{c} + 1\right)}{2\alpha - \sqrt{c} - \frac{\alpha}{\sqrt{c}}}$$

$$= \frac{2(\alpha - \sqrt{c}) - 2\left(\frac{\alpha}{\sqrt{c}} - 1\right)}{2\alpha - \sqrt{c} - \frac{\alpha}{\sqrt{c}}}$$

$$= 1 - \frac{\frac{\alpha}{\sqrt{c}} + \sqrt{c} - 2}{2\alpha - \sqrt{c} - \frac{\alpha}{\sqrt{c}}} \qquad \text{...(12.38)}$$

12.12.7 Intercooled Cycle with Regeneration and Reheating

The schematic arrangement of the cycle and corresponding T-s diagram is shown in Fig .12.17.

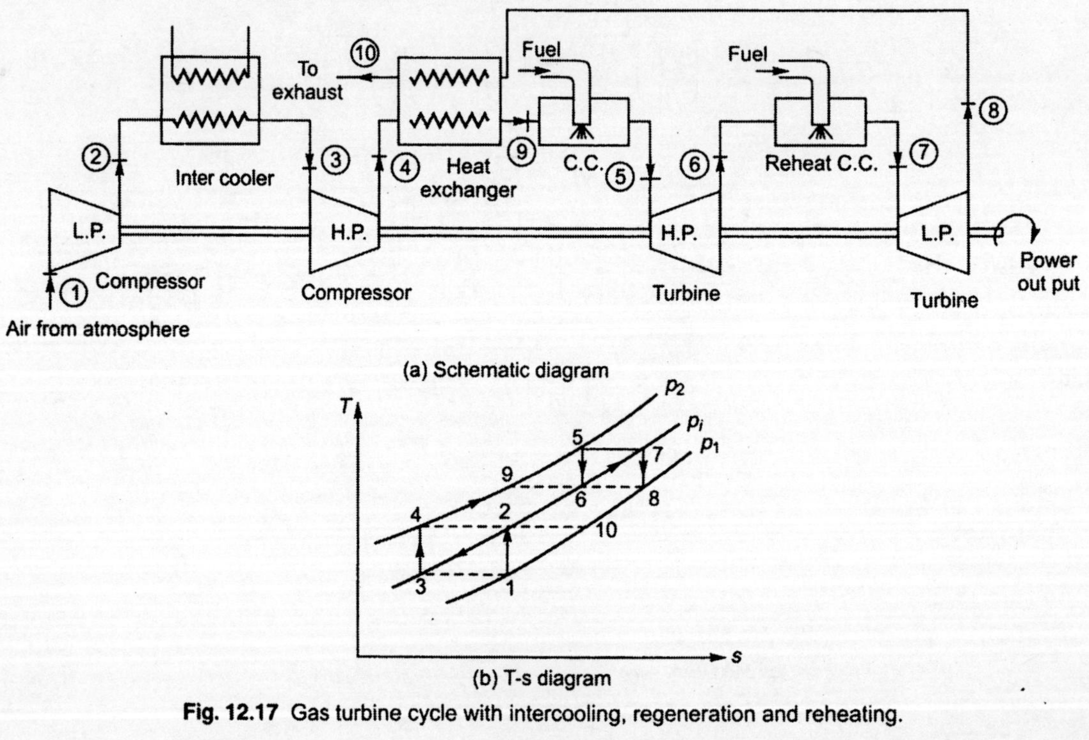

Fig. 12.17 Gas turbine cycle with intercooling, regeneration and reheating.

$$W_c = c_p\,[(T_2 - T_1) + (T_4 - T_3)]$$
$$q_s = c_p\,[(T_5 - T_9) + (T_7 - T_6)]$$
$$w_t = c_p\,[(T_5 - T_6) + (T_7 - T_8)]$$
$$w_{net} = w_t - w_c = c_p\,[(T_5 - T_6 + T_7 - T_8) - (T_2 - T_1 + T_4 - T_3)]$$
$$T_7 = T_5$$
$$\frac{w_{net}}{c_p T_1} = \frac{T_5}{T_1} - \frac{T_6}{T_5} \times \frac{T_5}{T_1} + \frac{T_5}{T_1} - \frac{T_8}{T_7} \times \frac{T_5}{T_1} - \frac{T_2}{T_1} + 1 - \frac{T_4}{T_1} + \frac{T_3}{T_1}$$

Let
$$\frac{T_4}{T_1} = c, \frac{T_1}{T_8} = \frac{T_2}{T_1} = c_1, \frac{T_5}{T_6} = \frac{T_4}{T_3} = c_2,$$
$$\frac{T_5}{T_1} = \alpha, T_6 = T_8 T_1 = T_3 \text{ and } T_5 = T_7 \text{ so that } c_1 = c_2$$

Proceeding as in case (6), we have

$$c_1 = c_2 = \sqrt{c}$$
$$\frac{w_{net}}{c_p T_1} = \alpha - \frac{\alpha}{c_2} + \alpha - \frac{\alpha}{c_1} - c_1 - c_2 + 2$$
$$\left(\frac{w_{net}}{c_p T_1}\right)_{max} = \alpha - \frac{\alpha}{\sqrt{c}} + \alpha - \frac{\alpha}{\sqrt{c}} - \sqrt{c} - \sqrt{c} + 2$$
$$\frac{(w_{net})_{max}}{c_p T_1} = 2\left(\alpha - \frac{\alpha}{\sqrt{c}} - \sqrt{c} + 1\right)$$
$$q_s = c_p\,(T_5 - T_9 + T_7 - T_6)$$
$$T_6 = T_8 = T_9$$
$$\frac{q_s}{c_p T_1} = \frac{T_5}{T_1} - \frac{T_6}{T_1} + \frac{T_5}{T_1} - \frac{T_6}{T_1}$$
$$= \alpha - \frac{\alpha}{\sqrt{c}} + \alpha - \frac{\alpha}{\sqrt{c}}$$
$$(\eta_{th})_{max} = \frac{2\left(\alpha - \frac{\alpha}{\sqrt{c}} - \sqrt{c} + 1\right)}{\alpha - \frac{\alpha}{\sqrt{c}} + \alpha - \frac{\alpha}{\sqrt{c}}}$$
$$= 1 - \frac{\sqrt{c} - 1}{\frac{\alpha}{\sqrt{c}}(\sqrt{c} - 1)} = 1 - \frac{\sqrt{c}}{\alpha} \qquad \text{...(12.40)}$$

■ **Example 12.1** Calculate the indicated mean effective pressure and efficiency of a Joule cycle if the temperature at the end of combustion is 2000 K and the temperature and pressure before compression is 350 K and 1 bar. The pressure ratio is 1:3 and assume $c_p = 1.005$ kJ/kg.K.

Solution : Refer to Fig. 12.18

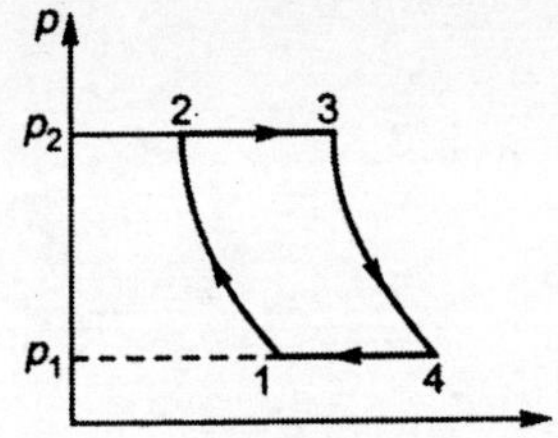

Fig. 12.18

$$p_1 = 1 \text{ bar}, T_1 = 350 \text{ K}, T_3 = 2000 \text{ K},$$

$$r_p = 1.3, c_p = 1.005 \text{ kJ/kg.K}$$

$$T_2 = T_1 r_p^{\frac{\gamma-1}{\gamma}} = 350 \times (1.3)^{\frac{0.4}{1.4}} = 377.24 \text{ K}$$

$$T_4 = T_3 \left(\frac{1}{r_p}\right)^{\frac{\gamma-1}{\gamma}} = 2000\left(\frac{1}{1.3}\right)^{\frac{0.4}{1.4}}$$

$$= 1855.56 \text{ K}$$

$$w_c = c_p\,(T_2 - T_1) = 1.005\,(377.24 - 350) = 27.376 \text{ kJ/kg}$$

$$w_t = c_p\,(T_3 - T_4) = 1.005\,(2000 - 1855.56) = 145.162 \text{ kJ/kg}$$

Network done, $$w_{net} = w_t - w_c = 145.162 - 27.376 = 117.786 \text{ kJ/kg}$$

$$q_s = c_p\,(T_3 - T_2) = 1.005\,(2000 - 377.24) = 1630.874 \text{ kJ/kg}$$

$$\eta_{th} = \frac{w_{net}}{q_s} = \frac{117.786}{1630.874} \times 100 = 7.22\%$$

$$v_4 = \frac{RT_4}{p_1} = \frac{0.287 \times 10^3 \times 1855.56}{1 \times 10^5} = 5.325 \text{ m}^3\text{/kg}$$

$$v_2 = \frac{RT_2}{p_2} = \frac{0.287 \times 10^3 \times 377.24}{1.3 \times 10^5}$$

$$= 0.833 \text{ m}^3\text{/kg}$$

$$p_m = \frac{w_{net}}{v_4 - v_2} = \frac{117.86}{5.325 - 0.833} = 26.235 \text{ kPa or } 0.26235 \text{ bar}$$

■ **Example 12.2** A gas turbine operates on a pressure ratio of 6. The inlet air temperature to the compressor is 300 K and the air entering the turbine is at a temperature of 580°C. If the volume rate of air entering the compressor is 250 m^3/s, calculate the net power output of the cycle. Also calculate the efficiency.

Solution : The T-s diagram shown in Fig. 12.19.

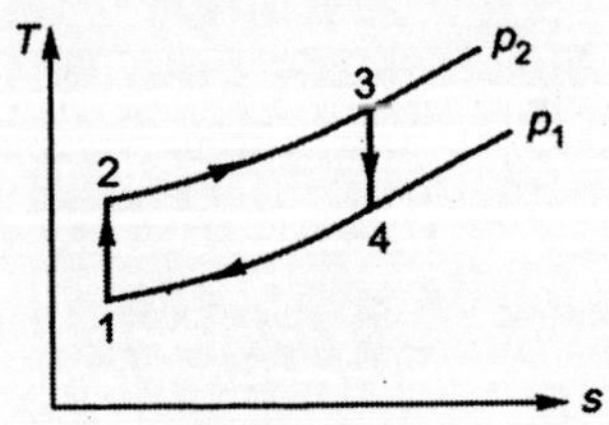

Fig. 12.19

$$r_p = 6, T_1 = 300 \text{ K}, T_3 = 273 + 580 = 853 \text{ K}$$

$$\dot{V}_1 = 250 \text{ m}^3\text{/s}$$

$$T_2 = T_1 r_p^{\frac{\gamma-1}{\gamma}} = 300 \times 6^{\frac{0.4}{1.4}} = 500.55 \text{ K}$$

$$T_4 = T_3 \left(\frac{1}{r_p}\right)^{\frac{\gamma-1}{\gamma}} = 853\left(\frac{1}{6}\right)^{\frac{0.4}{1.4}}$$

$$= 511.23 \text{ K}$$

$$w_c = c_p\,(T_2 - T_1) = 1.005\,(500.55 - 300) = 201.55 \text{ kJ/kg}$$

$$w_t = c_p\,(T_3 - T_4) = 1.005\,(853 - 511.23) = 343.48 \text{ kJ/kg}$$

$$w_{net} = w_t - w_c = 343.48 - 201.55 = 141.93 \text{ kJ/kg}$$

$$q_s = c_p\,(T_3 - T_2) = 1.005\,(853 - 500.55) = 354.21 \text{ kJ/kg}$$

$$\eta_{th} = \frac{w_{net}}{q_s} = \frac{141.93}{354.21} \times 100 = 40.07\%$$

$$p_1\dot{V}_1 = \dot{m}RT_1$$

$$\dot{m} = \frac{1 \times 10^5 \times 250}{287 \times 300} = 290.36 \text{ kg/s}$$

$$\text{Power output} = \dot{m}w_{net} = \frac{290.36 \times 141.93}{10^3} = 41.21 \text{ MW}$$

■ **Example 12.3** A gas turbine has a perfect heat exchanger. Air enters the compressor at a temperature and pressure of 300 K and 1 bar and discharges at 475 K and 5 bar. After passing through the heat exchanger the air temperature increases to 655 K. The temperature of air entering and leaving the turbine are 870°C and 450°C. Assuming no pressure drop through the heat exchanger, compute (a) the output per kg of air, (b) the efficiency of cycle, and (c) work required to drive the compressor.

Solution : The $p-V$ and T-s diagrams are shown in Fig .12.20

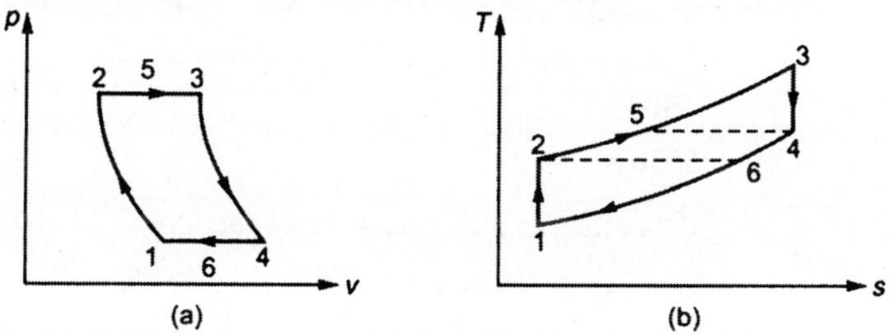

Fig. 12.20

$$T_1 = 300 \text{ K}, \quad p_1 = 1 \text{ bar},$$

$$T_2 = 475 \text{ K}, \quad p_2 = 5 \text{ bar}$$

$$T_5 = 655 \text{ K}, T_3 = 870 + 273 = 1143 \text{ K},$$

$$T_4 = 450 + 273 = 723 \text{ K}$$

$$\frac{T_2}{T_1} = r_p^{\frac{\gamma-1}{\gamma}}, \quad \frac{\gamma-1}{\gamma} = \ln\left(\frac{475}{300}\right)\Big/ \ln 5,$$

$$\gamma = 1.4, \; c_p = 1.005 \text{ kJ/kg}$$

$$w_c = c_p\,(T_2 - T_1) = 1.005\,(475 - 300) = 175.9 \text{ kJ/kg}$$

$$w_t = c_p\,(T_3 - T_4) = 1.005\,(1143 - 723) = 422.1 \text{ kJ/kg}$$

$$q_s = c_p\,(T_3 - T_5) = 1.005\,(1143 - 655) = 490.44 \text{ kJ/kg}$$

(a) $w_{net} = w_t - w_c = 422.1 - 175.9 = 246.2$ kJ/kg

(b) $\eta_{th} = \dfrac{w_{net}}{q_s} = \dfrac{246.2}{490.44} \times 100 = 50.2\%$

(c) $w_c = 175.9$ kJ/kg

■ **Example 12.4** A closed cycle regenerative gas turbine operating with air has the following data : $p_1 = 1.4$ bar, $T_1 = 310$ K, $r_p = 5$, $T_{max} = 1050$ K, effectiveness of generator = 100%, net output = 3 MW. Calculate (a) thermal efficiency, and (b) mass flow rate of air per mintue.

Solution : Refer to Fig. 12.21

$$T_2 = T_1 r_p^{\frac{\gamma-1}{\gamma}} = 310 \times 5^{\frac{0.4}{1.4}} = 491.2 \text{ K}$$

$$T_4 = \frac{T_3}{r_p^{\frac{\gamma-1}{\gamma}}} = \frac{1050}{5^{0.286}} = 662.65 \text{ K}$$

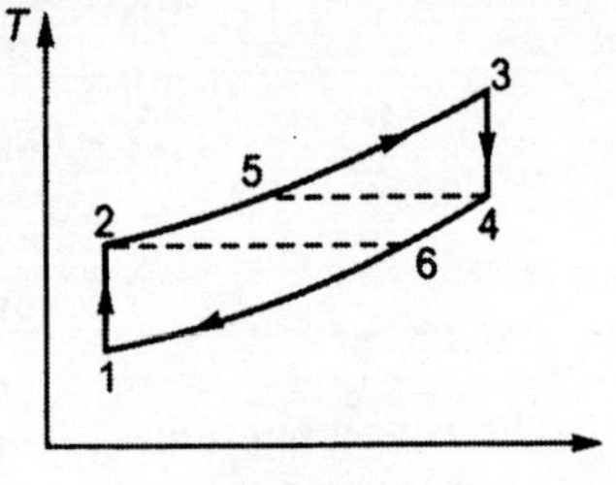

Fig. 12.21

For 100% regenerative effectiveness,

$$T_4 = T_5 = 662.65 \text{ K}$$

$$w_{net} = \dot{m}c_p (T_3 - T_4) - \dot{m}c_p (T_2 - T_1)$$

$$3000 = \dot{m} \times 1.005 [(1050 - 662.65) - (491.2 - 310)]$$

$$\dot{m} = 14.48 \text{ kg/s or } 868.8 \text{ kJ/min}$$

$$Q = \dot{m}c_p (T_3 - T_5)$$

$$= 14.48 \times 1.005 (1050 - 662.65) = 5636.87 \text{ kJ/s}$$

$$\eta_{th} = \frac{3000}{5636.87} \times 100 = 53.22\%$$

■ **Example 12.5** The following data refers to a turbine with regenerator :

$$T_1 = 290 \text{ K}, T_2 = 460 \text{ K}, T_3 = 900°\text{C}, T_4 = 467°\text{C}$$

Calculate (a) r_p, (b) specific power output, (c) overall efficiency of cycle, and (d) w_c. Assume $\eta_{mech} = 100\%$.

Solution : Refer to Fig. 12.22

(a)
$$\frac{T_2}{T_1} = r_p^{\frac{\gamma-1}{\gamma}}$$

$$\ln\left(\frac{460}{290}\right) = \left(\frac{1.4-1}{1.4}\right) \ln r_p, \; r_p = 5$$

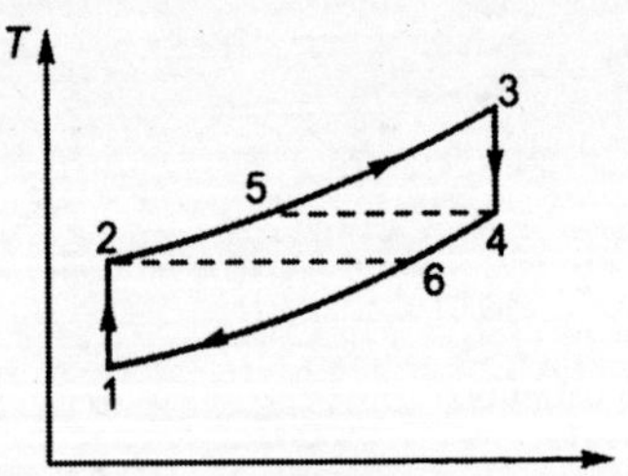

Fig. 12.22

(b)
$$w_{net} = c_p [(T_3 - T_4) - (T_2 - T_1)]$$

$$= 1.005 [(1173 - 740) - (460 - 290)]$$

$$= 264.3 \text{ kJ/kg}$$

(c)
$$q_s = c_p (T_3 - T_5) = c_p (T_3 - T_4)$$

$$= 1.005 (1173 - 740) = 435.165 \text{ kJ/kg}$$

$$\eta_{over} = \frac{w_{net}}{q_s} = \frac{264.3}{435.165} \times 100 = 60.74\%$$

(d) $w_c = c_p (T_2 - T_1) = 1.005 (460 - 290) = 170.85$ kJ/kg

■ **Example 12.6** The ratio of network to turbine work of an ideal gas turbine plant is 0.563. $T_1 = 300\,K$. Calculate the temperature drop across the turbine if thermal efficiency of the unit is 35%. Assume a mass flow rate of 10 kg/s, $c_p = 1$ kJ/kg.K and $\gamma = 1.4$.

Solution :

$$\eta_{th} = 1 - \frac{1}{r_p^{\frac{\gamma-1}{\gamma}}},\ 0.35 = 1 - \frac{1}{r_p^{0.286}},\ r_p = 4.51$$

$$T_2 = T_1\, r_p^{\frac{\gamma-1}{\gamma}} = 300 \times (4.51)^{0.286} = 461.54\ K$$

$$w_c = c_p\,(T_2 - T_1) = 1 \times (461.54 - 300) = 161.54\ kJ/kg$$

$$\frac{w_{net}}{w_t} = 0.563 = \frac{w_t - w_c}{w_t} = 1 - \frac{161.54}{w_t}$$

$$w_t = 369.66\ kJ/kg = c_p \cdot \Delta T$$

$$\Delta T = 369.66°C$$

■ **Example 12.7** Prove that the efficiency corresponding to the maximum work done in a Brayton cycle is given by, $\eta_{max} = 1 - \frac{1}{\sqrt{\alpha}}$, where $\alpha = \frac{T_3}{T_1}$.

Solution : The T-s diagram for the Brayton cycle is shown in Fig .12.23.

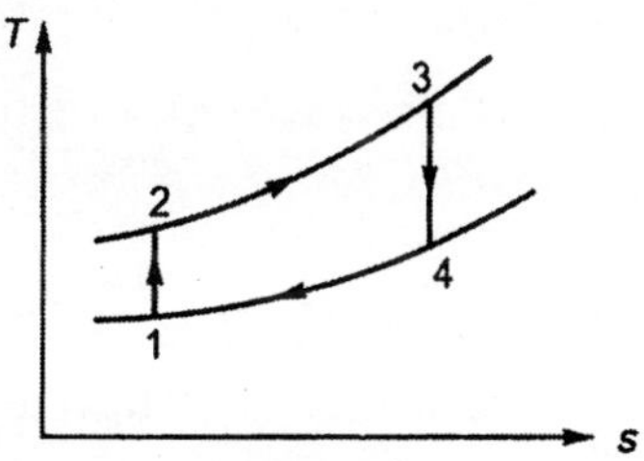

Fig. 12.23

$$w_c = c_p\,(T_2 - T_1)$$

$$w_t = c_p\,(T_3 - T_4)$$

$$w_{net} = w_t - w_c = c_p\,[(T_3 - T_4) - (T_2 - T_1)]$$

$$= c_p\left[T_3\left(1 - \frac{T_4}{T_3}\right) - T_1\left(\frac{T_2}{T_1} - 1\right)\right]$$

$$= c_p\left[T_3\left(1 - \frac{1}{r_p^{\frac{\gamma-1}{\gamma}}}\right) - T_1\left(r_p^{\frac{\gamma-1}{\gamma}} - 1\right)\right]$$

Let

$$m = \frac{\gamma - 1}{\gamma},\ \alpha = \frac{T_3}{T_1}$$

$$w_{net} = c_p T_1\left[\alpha\left(1 - \frac{1}{r_p^m}\right) - (r_p^m - 1)\right]$$

For maximum work,

$$\frac{dw_{net}}{dr_p} = 0$$

$$\alpha m r_p^{-(m+1)} - m r_p^{m-1} = 0$$

$$\alpha\, r_p^{-(m+1)} = r_p^{m-1}$$

$$r_p^{2m} = \alpha$$

$$r_p^{m} = \sqrt{\alpha}$$

$$\eta_{th} = \frac{w_{net}}{q_s}$$

$$q_s = c_p\,(T_3 - T_2)$$

$$\eta_{th} = \frac{c_p\left[\alpha\left(1 - \dfrac{1}{r_p^m}\right) - (r_p^m - 1)\right]}{c_p\left(\dfrac{T_3}{T_1} - \dfrac{T_2}{T_1}\right) = c_p\,(\alpha - r_p^m)}$$

$$= \frac{\alpha\left(1 - \dfrac{1}{\sqrt{\alpha}}\right) - (\sqrt{\alpha} - 1)}{\alpha - \sqrt{\alpha}}$$

$$= \frac{(\alpha - \sqrt{\alpha}) - (\sqrt{\alpha} - 1)}{\alpha - \sqrt{\alpha}} = 1 - \frac{\sqrt{\alpha} - 1}{\alpha - \sqrt{\alpha}}$$

$$= 1 - \frac{\sqrt{\alpha} - 1}{\sqrt{\alpha}\,(\sqrt{\alpha} - 1)}$$

$$= 1 - \frac{1}{\sqrt{\alpha}}$$

■ **Example 12.8** In a Brayton cycle the pressure ratio in the compressor is r. The minimum and maximum temperatures are T_1 and T_3. The air is expanded in two stages, each turbine having the same pressure ratio, with reheat to T_3 between the stages. Show that the specific work output will be maximum when

$$r_p = \alpha^{\frac{2}{3m}}, \text{ where } \alpha = \frac{T_3}{T_1},\ m = \frac{\gamma - 1}{\gamma},\ r_p = \frac{p_2}{p_1}$$

Solution : Refer to Fig. 12.24

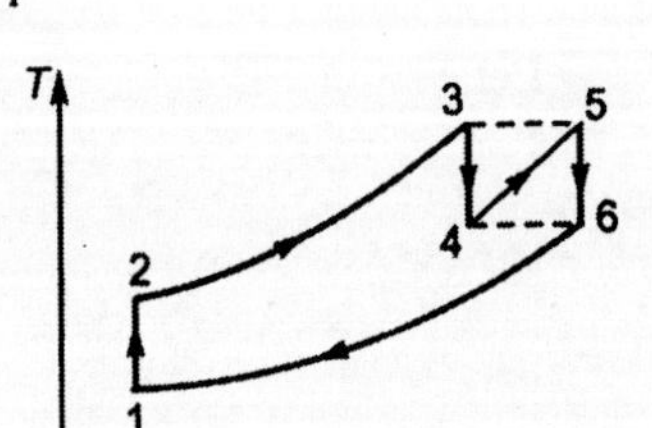

Fig. 12.24

$$\frac{T_2}{T_1} = r_p^{\frac{\gamma - 1}{\gamma}} = r_p^m = c$$

$$\frac{T_3}{T_4} = \frac{T_5}{T_6} = (\sqrt{r_p})^m = \sqrt{c},\ T_5 = T_3,\ T_4 = T_6$$

$$w_t = c_p\,(T_3 - T_4) + c_p\,(T_5 - T_6)$$

$$= c_p\,(T_3 - T_4) + c_p\,(T_3 - T_6)$$

$$= 2c_p\,(T_3 - T_4) = 2c_pT_3\left(1 - \frac{T_4}{T_3}\right)$$

$$= 2c_pT_3\left(1 - \frac{1}{\sqrt{c}}\right)$$

$$w_c = c_p\,(T_2 - T_1) = c_pT_1\left(\frac{T_2}{T_1} - 1\right)$$

$$= c_pT_1(c-1)$$

$$w_{\text{net}} = w_t - w_c = 2c_pT_3\left(1 - \frac{1}{\sqrt{c}}\right) - c_p\,T_1\,(c-1)$$

$$\frac{w_{\text{net}}}{c_pT_1} = 2\alpha\left(1 - \frac{1}{\sqrt{c}}\right) - (c-1)$$

$$\frac{d}{dc}\left[\frac{w_{\text{net}}}{c_pT_1}\right] = 2\alpha \times \frac{1}{2} \times c^{-\frac{3}{2}} - 1 = 0$$

$$\alpha c^{-\frac{3}{2}} = 1$$

$$\alpha = c^{\frac{3}{2}}$$

$$c = \alpha^{\frac{2}{3}}$$

$$r_p^m = \alpha^{\frac{2}{3}}$$

$$r_p = \alpha^{\frac{2}{3m}} = \left(\frac{T_3}{T_1}\right)^{\frac{2\gamma}{3(\gamma-1)}}$$

■ **Example 12.9** In a gas turbine, air is taken in at 1 bar pressure and 30°C. It is compressed to 6 bar with an isentropic efficiency of 87%. Heat is added by the combustion of fuel in combustion chamber to raise the temperature to 700°C. The efficiency of turbine is 85%. The calorific value of fuel used is 43.1 MJ/kg. Calculate for an air flow of 80 kg/min (a) air/fuel ratio of turbine gases, (b) final temperature of exhaust gases, (c) net power developed, and (d) overall thermal efficiency of plant.

Assume $c_{pa} = 1.005$ kJ/kg.K, $\gamma_a = 1.4$, $c_{pg} = 1.147$ kJ/kg.K, $\gamma_g = 1.33$.

Solution : The T-s diagram is shown in Fig. 12.25.

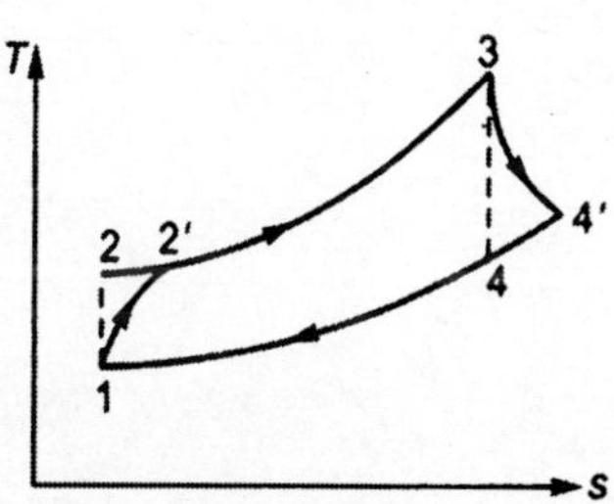

Fig. 12.25

$$p_1 = 1\text{ bar},\; T_1 = 273 + 30 = 303\text{ K}$$

$$p_2 = 6\text{ bar},\; \eta_c = 0.87,\; T_3 = 273 + 700 = 973\text{ K}$$

$$\eta_t = 0.85,\; CV = 43.1\text{ MJ/kg}$$

$$T_2 = T_1\left(\frac{p_2}{p_1}\right)^{\frac{\gamma_a - 1}{\gamma_a}} = 303\left(\frac{6}{1}\right)^{\frac{0.4}{1.4}} = 505.5\text{ K}$$

$$\eta_c = \frac{T_2 - T_1}{T_2' - T_1}, T_2' = T_1 + \frac{T_2 - T_1}{\eta_c}$$

$$= 303 + \frac{505.5 - 303}{0.87} = 535.8 \text{ K}$$

$$\dot{m}_f \times \text{C.V.} = \dot{m}_a c_{pg} (T_3 - T_2') \qquad [\because \dot{m}_a + \dot{m}_f \simeq \dot{m}_a]$$

$$\dot{m}_f = \frac{80}{60} \times \frac{1.147 \times (973 - 535.8)}{43.1 \times 10^3} = 0.0155 \text{ kg/s}$$

$$\frac{A}{F} = \frac{\dot{m}_a}{\dot{m}_f} = \frac{80}{60} \times \frac{1}{0.0155} = 86$$

$$\eta_t = \frac{T_3 - T_4'}{T_3 - T_4}, T_4 = \frac{T_3}{r_p^{\frac{\gamma_g - 1}{\gamma_g}}} = \frac{973}{6^{\frac{0.33}{1.33}}} = 623.8 \text{ K}$$

$$0.85 = \frac{973 - T_4'}{973 - 623.8'} \; T_4' = 676.2 \text{ K}$$

$$w_{\text{net}} = \dot{m}_a c_{pg} (T_3 - T_4') - \dot{m}_a c_{pa} (T_2' - T_1)$$

$$= \frac{80}{60} \times [1.147 (973 - 676.2) - 1.005 (535.8 - 303)]$$

$$\eta_{\text{th}} = \frac{w_{\text{net}}}{Q} = \frac{142 \times 100}{0.0155 \times 43100} = 21.25\%$$

■ **Example 12.10** In a gas turbine plant air enters the compressor at 1 bar and 7°C. It is compressed to 4 bar with an isentropic efficiency of 82%. The maximum temperature at the inlet to the turbine is 800°C. The isentropic efficiency of turbine is 85%. The calorific value of fuel used is 43100 kJ/kg. The heat losses are 15% of the calorific value. Calculate (a) compressor work, (b) heat supplied, (c) turbine work, (d) net work, (e) thermal efficiency, (f) $\frac{A}{F}$ ratio, (g) specific fuel consumption, and (h) ratio of compressor to turbine work.

Assume $c_{pa} = 1.005$ kJ/kg. K, $\gamma_a = 1.4$, $c_{pg} = 1.147$ kJ/kg.K, $\gamma_g = 1.33$.

Solution : $p_1 = 1$ bar, $T_1 = 280$ K, $\eta_c = 0.82$, $p_2 = 4$ bar, $T_3 = 1073$ $\eta_t = 0.85$, $\eta_{\text{comb}} = 0.85$ Refer to Fig. 12.26

$$T_2 = T_1 \left(\frac{p_2}{p_1} \right)^{\frac{\gamma_a - 1}{\gamma_a}}$$

$$= 280 (4)^{\frac{0.4}{1.4}} = 416.1 \text{ K}$$

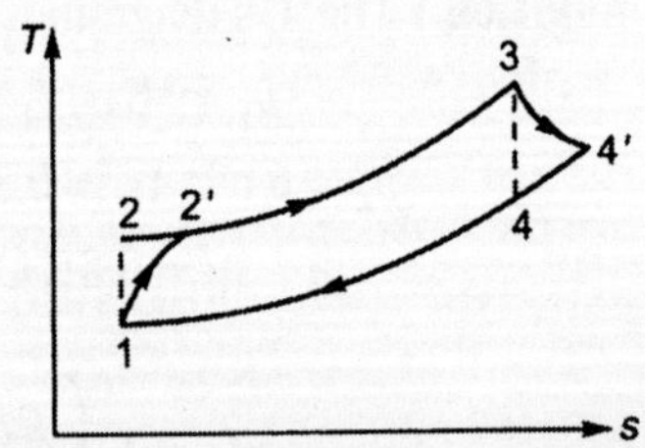

Fig. 12.26

$$T_2' = T_1 + \frac{T_2 - T_1}{\eta_c}$$

$$= 280 + \frac{416.1 - 280}{0.182} = 446 \text{ K}$$

$$w_c = c_{pa}\,(T_2' - T_1) = 1.005\,(446 - 280) = 166.82 \text{ kJ/kg}$$

$$q_s = c_{pg}\,\frac{(T_3 - T_2')}{\eta_{\text{comb}}} = \frac{1.147\,(1073 - 446)}{0.85} = 846.1 \text{ kJ/kg}$$

$$T_4 = \frac{1073}{\frac{0.33}{1.33}} = 760.7$$

$$T_4' = T_3 - \eta_t\,(T_3 - T_4) = 1073 - 0.85\,(1073 - 760.7) = 807.5 \text{ K}$$

$$w_t = c_{pg}\,(T_3 - T_4') = 1.147\,(1073 - 807.5) = 304.53 \text{ kJ/kg}$$

$$w_{\text{net}} = w_t - w_c = 304.153 - 166.83 = 137.7 \text{ kJ/kg}$$

$$\eta_{\text{th}} = \frac{w_{\text{net}}}{q_s} = \frac{137.7}{846.1} \times 100 = 16.27\%$$

$$\dot{m}_f \times \text{CV}\,(1 - 0.15) = (\dot{m}_a + \dot{m}_f)\,Q \times \eta_{\text{comb}}$$

$$\frac{\dot{m}_a}{\dot{m}_f} = \frac{\text{CV}}{q_s} - 1 = \frac{43100}{846.1} - 1 = 49.94$$

$$\text{SFC} = \frac{\text{Fuel consumed per hour}}{\text{Net power}} = \frac{\dot{m}_f \times 3600}{\dot{m}_a \times w_{\text{net}}}$$

$$= \frac{3600}{49.94 \times 137.7} = 0.524 \text{ kg/kWh}$$

$$\frac{w_c}{w_t} = \frac{166.83 \times \dot{m}_a}{304.53 \times (\dot{m}_a + \dot{m}_f)}$$

$$= \frac{166.83 \times 49.94}{304.53 \times 50.94} = 0.537$$

■ **Example 12.11** The following data refers to a gas turbine with regeneration :

Ambient temperature and pressure	= 327 K, 1 bar
Pressure ratio	= 7 : 1
Isentropic efficiency of compressor	= 82%
Maximum cycle temperature	= 1000 K
Combustion efficiency	= 97%
Mechanical efficiency	= 99%
Isentropic efficiency of turbine	= 85%
Calorific value of fuel	= 43100 kJ/kg
Air mass flow	= 20 kg/s
Regenerator effectiveness	= 75%
Regenerator pressure loss	= 0.1 bar

Calculate output, SFC and overall thermal efficiency.

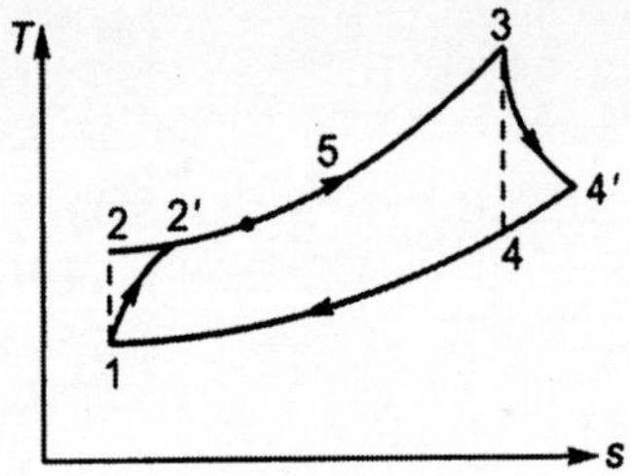

Fig. 12.27

Solution : Refer to Fig. 12.27

$$T_2 = T_1 \left(\frac{p_2}{p_1}\right)^{\frac{\gamma-1}{\gamma}} = 327\,(7)^{\frac{0.4}{1.4}} = 570.2 \text{ K}$$

$$T_2' = 327 + \frac{570.2 - 327}{0.82} = 623.6 \text{ K}$$

$$p_4' = p_1 + \Delta p = 1 + 0.1 = 1.1 \text{ bar}$$

$$T_4 = T_3 \left(\frac{p_4'}{p_3}\right)^{\frac{\gamma-1}{\gamma}} = 1000 \left(\frac{1.1}{7}\right)^{\frac{0.4}{1.4}} = 631.8 \text{ K}$$

$$T_4' = 1000 - 0.85\,(1000 - 631.8) = 687 \text{ K}$$

$$T_5 = T_2' + \varepsilon\,(T_4' - T_2')$$

$$= 623.6 + 0.75\,(687 - 623.6) = 671.2 \text{ K}$$

$$\dot{m}_f \times \text{CV} \times \eta_{\text{comb}} = \dot{m}_a\, c_{pg}\,(T_3 - T_5)$$

$$\dot{m}_f = \frac{20 \times 1.147\,(1000 - 671.2)}{43100 \times 0.97} = 0.18 \text{ kg/s}$$

$$w_{\text{net}} = w_t - w_c = (\dot{m}_a + \dot{m}_f)\, c_{pg}\,(T_3 - T_4') - \frac{\dot{m}_a c_{pa}}{\eta_{\text{mech}}}\,(T_2' - T_1)$$

$$= 20.18 \times 1.147 \times (1000 - 687) - \frac{20 \times 1.005}{0.99}\,(623.6 - 327)$$

$$= 1222.96 \text{ kW}$$

$$\text{SFC} = \frac{\dot{m}_f \times 3600}{w_{\text{net}}} = \frac{0.18 \times 3600}{1222.96} = 0.53 \text{ kg/kWh}$$

$$\eta_{\text{th}} = \frac{1222.96}{0.18 \times 43100} \times 100 = 15.8\%$$

■ **Example 12.12** The efficiencies of the compressor and turbine of a gas turbine are 70% and 72% respectively. Teh heat added in the combustion chamber per kg of air is 475 kJ/kg. Find a suitable pressure ratio such that the work ratio is 0.054. Also find the corresponding tempeature ratio. The inlet temperature of air is 300 K.

Solution :

$$\text{Work raio} = \frac{w_t - w_c}{w_t} = 1 - \frac{w_c}{w_t}$$

$$= 1 - \frac{(T_2' - T_1)}{(T_3 - T_4')} = 1 - \frac{\dfrac{T_2 - T_1}{\eta_c}}{\eta_t\,(T_3 - T_4)}$$

$$0.054 = 1 - \frac{\dfrac{T_1}{\eta_c}\left(r_p^{\frac{\gamma-1}{\gamma}} - 1\right)}{\eta_t T_3\left(1 - \dfrac{1}{r_p^{\frac{\gamma-1}{\gamma}}}\right)} = 1 - \frac{T_1}{T_2} \times \frac{1}{\eta_c \eta_t}\left[\frac{r_p^m - 1}{1 - r_p^m}\right],$$

$$m = \frac{\gamma - 1}{\gamma}, \alpha = \frac{T_3}{T_1}$$

$$0.054 = 1 - \frac{1}{\alpha} \cdot \frac{1}{\eta_c \eta_t} \cdot r_p^m = 1 - \frac{r_p^m}{\alpha} \cdot \frac{1}{0.7 \times 0.72}$$

$$\frac{r_p^m}{\alpha} = 0.4768$$

$$c_{pg}\,(T_3 - T_2') = q_s$$

$$T_2' = T_1 + \frac{T_1}{\eta_c}(r_p^m - 1)$$

$$475 = 1.147\left[T_3 - \left(T_1 + \frac{T_1}{\eta_c} r_p^m - \frac{T_1}{\eta_{cj}}\right)\right]$$

$$414.124 = T_1\left[\frac{T_3}{T_1} + \left(1 - \frac{r_p^m}{0.7} - \frac{1}{0.7}\right)\right]$$

$$= 300\left[\alpha - \left(1 + \frac{0.4768}{0.7}\alpha - \frac{1}{0.7}\right)\right]$$

$$1.3804 = \alpha - 1 - 0.6811\alpha + 1.428 = 0.3189\alpha + 0.428$$

$$\alpha = 2.9865 = \frac{T_3}{T_1}$$

$$r_p^m = 2.9865 \times 0.4768 = 1.424$$

$$m = \frac{1.4 - 1}{1.4} = 0.2857$$

$$r_p = 3.446$$

■ **Example 12.13** **In a closed circuit gas turbine plant the working fluid at 38°C is compressed with an adiabatic efficiency of 82%. It is then heated at constant pressure to 650°C. The fluid expands down to initial pressure in a turbine with an adiabatic efficiency of 80%. The pressure ratio is such that work done per kg is maximum. c_{pa} = 1.005 kJ/kg.K, c_{pg} = 1.147 kJ/kg.K. Calculate the pressure ratio and cycle efficiency.**

Solution : $$\alpha = \frac{T_3}{T_1} = \frac{923}{311} = 2.96$$

For maximum specific work,

$$(r_p)_{opt}^{\frac{\gamma-1}{\gamma}} = \sqrt{\alpha \eta_c \eta_t} = \sqrt{2.96 \times 0.82 \times 0.80} = 1.3935$$

$$(r_p)_{opt} = (1.395)^{3.5} = 3.194$$

$$T_2 = T_1 \, (r_p)_{opt}^{\frac{\gamma-1}{\gamma}} = 311 \times 1.3935 = 433.51 \text{ K}$$

$$T_2' = T_1 + \frac{T_2 - T_1}{\eta_c} = 311 + \frac{433.51 - 311}{0.82}$$

$$= 460.40 \text{ K}$$

$$T_4 = \frac{T_3}{(r_p)_{opt}^{\frac{\gamma-1}{\gamma}}} = \frac{923}{1.3935} = 692.03 \text{ K}$$

$$T_4' = 923 - 0.8\,(923 - 692.03) = 738.22 \text{ K}$$

$$\eta_{th} = \frac{1.147\,(923 - 738.22) - 1.005\,(460.40 - 311)}{1.147\,(923 - 460.40)}$$

$$= 0.1155 \text{ or } 11.55\%$$

■ **Example 12.14** In an open cycle constant pressure gas turbine, air enters the compressor at 1.0325 bar pressure and 27°C. The pressure of air after the compression is 4.13 bar. The isentropic efficiencies of compressor and turbine are 80% and 85% respectively. The A/F ratio used is 80 : 1. Find the indicated power and thermal efficiency of the cycle if the flow rate of air is 2.5 kg/s. Assume $c_p = 1$ kJ/kg. K and $\gamma = 1.4$ for air and gases. Calorific value of fuel used is 41720 kJ/kg.

Solution : Refer to Fig. 12.28

$$T_1 = 273 + 27 = 300 \text{ K}, \; p_1 = 1.0325 \text{ bar}, \; p_2 = 4.13 \text{ bar}, \; \eta_c = 0.80, \; \eta_t = 0.85$$

$$T_2 = T_1 \left(\frac{p_2}{p_1}\right)^{\frac{\gamma-1}{\gamma}} = 300 \left(\frac{4.13}{1.0325}\right)^{\frac{0.4}{1.4}} = 446 \text{ K}$$

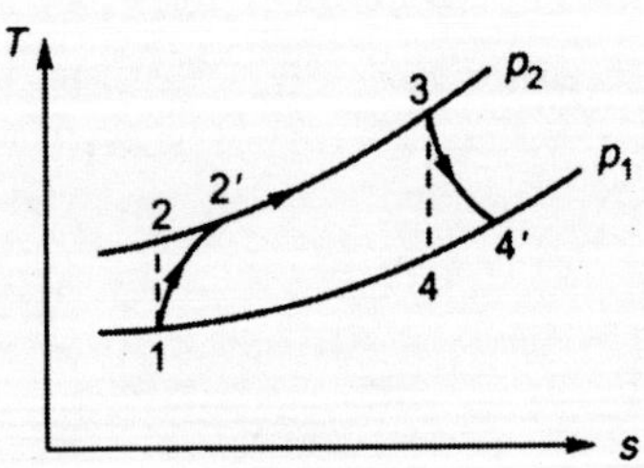

Fig. 12.28

$$\eta_c = \frac{T_2 - T_1}{T_2' - T_1'}$$

$$T_2' = T_1 + \frac{T_2 - T_1}{\eta_c}$$

$$= 300 + \frac{446 - 300}{0.80} = 4.83 \text{ K}$$

$$\dot{m}_f \times CV = (\dot{m}_a + \dot{m}_f)\, c_p \,(T_3 - T_2')$$

$$CV = \left(\frac{1 + \dot{m}_a}{\dot{m}_f}\right) c_p \, (T_3 - T_2')$$

$$41720 = (1 + 80) \times 1 \times (T_3 - 483)$$

$$T_3 = 998 \text{ K}$$

$$T_4 = \frac{T_3}{\left(\frac{p_2}{p_1}\right)^{\frac{\gamma - 1}{\gamma}}} = \frac{998}{(4)^{\frac{0.4}{1.4}}} = 672 \text{ K}$$

$$\eta_t = \frac{T_3 - T_4'}{T_3 - T_4}, \; T_4' = T_3 - \eta_t \, (T_3 - T_4)$$

$$= 998 - 0.85\,(998 - 672) = 721 \text{ K}$$

$$W_c = m_a c_{pa} \, (T_2' - T_1) = 1 \times 1 \times (483 - 300)$$

$$= 183 \text{ kJ/kg of air}$$

$$W_t = (m_a + m_f)\, c_{pg} \, (T_3' - T_4)$$

$$= \left(1 + \frac{1}{80}\right) \times 1 \times (998 - 721) = 281 \text{ kJ/kg of air}$$

$$W_{net} = W_t + W_c = 281 - 183 = 98 \text{ kJ/kg of air}$$

$$Q = \frac{1}{80} \times 41720 = 521.5 \text{ kJ/kg of air}$$

$$\eta_{th} = \frac{W_{net}}{Q} = \frac{98}{521.5} \times 100 = 18.75\%$$

$$IP = \dot{m}_a \times W_t = 2.5 \times 9.8 = 245 \text{ kW}$$

■ **Example 12.15** A constant pressure closed cycle turbine plant works between a temperature range of 800°C and 30°C. The isentropic efficiencies of compressor and turbine are 80% and 90% respectively. Find the pressure ratio of the cycle for maximum thermal efficiency and maximum specific output of the cycle.

Solution : $T_1 = 273 + 30 = 303$ K, $T_3 = 273 + 800 = 1073$ K, $\eta_c = 0.8$, $\eta_t = 0.9$

Pressure ratio r_p for maximum specific output

$$= \left[\eta_c \eta_t \frac{T_3}{T_1}\right]^{\frac{\gamma}{2(\gamma - 1)}} = \left[0.8 \times 0.9 \times \frac{1073}{303}\right]^{\frac{1.4}{2 \times 0.4}} = 5.15$$

$$r_p \text{ for maximum thermal efficiency} = \left[\frac{\dfrac{T_3}{T_1}}{1+\sqrt{\left(\dfrac{T_3}{T_1}-1\right)\left(\dfrac{1}{\eta_c\eta_t}-1\right)}}\right]^{\frac{\gamma}{\gamma-1}}$$

$$= \left[\frac{\dfrac{1073}{303}}{1+\sqrt{\left(\dfrac{1073}{303}-1\right)\left(\dfrac{1}{0.8\times0.9}-1\right)}}\right]^{\frac{1.4}{0.4}} = 25.5$$

■ **Example 12.16** Air enters at 1 bar and 15°C into the compressor of a constant pressure open cycle gas turbine plant and leaves the compressor at 6 abr. The following data is auxilable :

Temperature of gases entering the turbine = 700°C

Pressure loss in the combustion chamber = 0.1 bar

$$\eta_c = 80\%, \ \eta_t = 80\%, \ \eta_{comb} = 90\%$$

$$\gamma = 1.4, \ c_p = 1 \text{ kJ/kg.K for air and gases.}$$

Calculate (a) the quantity of air circulation in the system if the plant develops 940 kW; (b) heat supplies per kg of air circulated, and (c) thermal efficiency of the cycle.

Neglect the mass of fuel.

Solution : Refer to Fig. 12.29

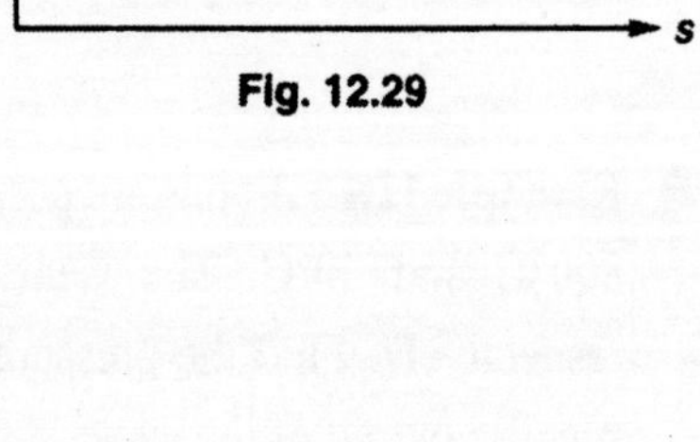

Fig. 12.29

$T_1 = 273 + 15 = 288$ K, $T_3 = 273 + 700 = 973$ K,

$p_1 = 1$ bar, $p_2 = 6$ bar, $p_i = 6 - 0.1 = 5.9$ bar

$$T_2 = T_1\left(\frac{p_2}{p_1}\right)^{\frac{\gamma-1}{\gamma}} = 286\left(\frac{6}{1}\right)^{\frac{0.4}{1.4}} = 480.78 \text{ K}$$

$$\eta_c = \frac{T_2 - T_1}{T_2' - T_1}, \ T_2' = T_1 + \frac{T_2 - T_1}{\eta_c}$$

$$= 288 + \frac{480.78 - 288}{0.8} = 528.97 \text{ K}$$

$$T_4 = \frac{T_3}{\left(\dfrac{p_i}{p_1}\right)^{\frac{\gamma-1}{\gamma}}} = \frac{973}{\left(\dfrac{5.9}{1}\right)^{\frac{0.4}{1.4}}} = 585.66 \text{ K}$$

$$T_4' = T_3 - \eta_t (T_3 - T_4) = 973 - 0.8\,(973 - 585.66) = 663.13 \text{ K}$$

$$W_c = c_p\,(T_2' - T_1) = 1 \times (528.97 - 288) = 240.97 \text{ kJ/kg}$$

$$W_t = c_p\,(T_3' - T_4') = 1 \times (973 - 663.13) = 309.87 \text{ kJ/kg}$$

$$W_{net} = W_t - W_c = 309.87 - 240.97 = 68.90 \text{ kJ/kg}$$

$$IP = \dot{m}_a \times W_{net}$$

$$\dot{m}_a = \frac{940}{68.90} = 13.5 \text{ kg/s}$$

$$q_a = \eta_{comb}\, c_p\, (T_3 - T_2')$$

$$= 0.9 \times 1 \times (973 - 528.97) = 397.5 \text{ kJ/kg}$$

$$\eta_{th} = \frac{W_{net}}{q_a} = \frac{68.9 \times 100}{397.5} = 17.5\%$$

■ **Example 12.17** The pressure ratio of an open cycle constant pressure gas turbine plant is 6. The temperaure range of the plant is 15°C and 800°C. $c_{pa} = 1$ kJ/kg.K, $c_{pg} = 1.075$ kJ/kg.K and $\gamma = 1.4$ for air and gases. CV of fuel = 43000 kJ/kg, $\eta_c = 0.85$, $\eta_t = 0.90$, $\eta_{comb} = 0.95$. Calculate (a) thermal efficiency of plant, (b) I.P. of plant if air flow rate is 5 kg/s, (c) A/F ratio, and (d) specific fuel combustion.

Solution : Refer to Fig. 12.30

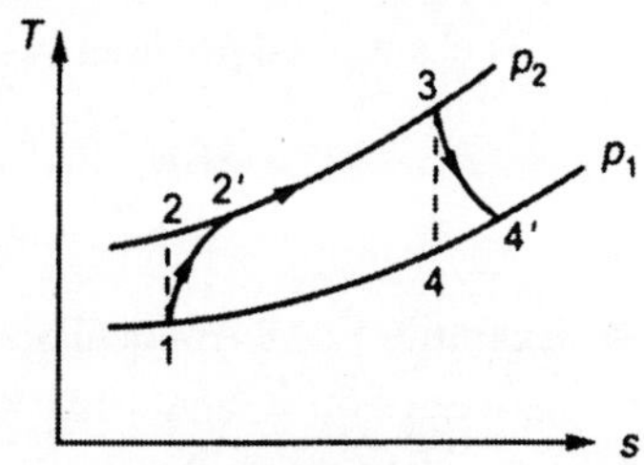

Fig. 12.30

$$T_2 = T_1\, r_p^{\frac{\gamma-1}{\gamma}} = 288(6)^{\frac{0.4}{1.4}} = 480.78 \text{ K}$$

$$\eta_c = \frac{T_2 - T_1}{T_2' - T_1},\; T_2' = T_1 + \frac{T_2 - T_1}{\eta_c}$$

$$= 288 + \frac{480.78 - 288}{0.85}$$

$$= 514.8 \text{ K}$$

$$T_4 = \frac{T_3}{r_p^{\frac{\gamma-1}{\gamma}}} = \frac{1073}{6^{\frac{0.4}{1.4}}} = 642.76 \text{ K}$$

$$T_4' = T_t - \eta_t\,(T_3 - T_4)$$

$$= 1073 - 0.90\,(1073 - 642.76) = 685.78 \text{ K}$$

$$W_c = c_{pa}\,(T_2' - 1) = 1\,(514.8 - 288) = 226.8 \text{ kJ/kg}$$

$$\dot{m}_f \times CV \times \eta_{comb} = (\dot{m}_f + \dot{m}_a)\, c_{pg}\, (T_3 - T_2')$$

$$CV \times \eta_{comb} = \left(1 + \frac{\dot{m}_a}{\dot{m}_f}\right) c_{pg}\, (T_3 - T_2')$$

$$43000 \times 0.95 = \left(1 + \frac{\dot{m}_a}{\dot{m}_f}\right) \times 1.075\,(1073 - 514.8)$$

$$\frac{\dot{m}_a}{\dot{m}_f} = \frac{67.4}{1}$$

$$W_t = \frac{(\dot{m}_a + \dot{m}_f)}{\dot{m}_a}(T_3 - T_4')$$

$$= \frac{(1+67.4)}{67.4}(1073 - 685.78)$$

$$= 393 \text{ kJ/kg of air}$$

$$W_{net} = W_t - W_c = 393 - 226.8$$

$$= 166.2 \text{ kJ/kg of air}$$

$$\text{I.P.} = \dot{m}_a \times W_{net} = 5 \times 166.2 = 831 \text{ kW}$$

$$q = \frac{1}{67.4} \times 43000 \times 0.95 = \frac{CV \times \eta_{comb}}{\dot{m}_f}$$

$$= 606 \text{ kJ/kg or air,}$$

$$\eta = \frac{166.2}{606} \times 100 = 27.4\%$$

$$\text{Fuel supplied per hour} = \frac{5}{67.4} \times 3600 = 267 \text{ kg/h}$$

$$\text{SFC} = \frac{267}{831} = 0.32 \text{ kg/kWh}$$

■ **Example 12.18** An ideal open cycle gas turbine plant has an overall pressure ratio of 4 and operates between 800 K and 1000 K. $c_p = 1$ kJ/kg.K and $C_v = 0.717$ kJ/kg.K. Calculate the specific work output and thermal efficiency for the (a) basic cycle, (b) basic cycle with heat exchanger, (c) basic cycle with two-stage inter cooled compressor, and (d) basic cycle with heat exchanger and two-stage intercooled compressor. Assume optimum stage pressure ratios, perfect intercooling and perfect regeneration.

Solution :

(a)

$$\gamma = \frac{c_p}{c_v} = \frac{1}{0.717} = 1.3947$$

$$c = r_p^{\frac{\gamma-1}{\gamma}} = 4^{\frac{0.3947}{1.3947}} = 1.4804$$

$$\alpha = \frac{T_3}{T_1} = \frac{1000}{300} = 3.333$$

$$\frac{W_{net}}{c_p T_1} = \alpha\left(1 - \frac{1}{c}\right) - (c-1)$$

$$= 3.333\left(1 - \frac{1}{1.4809}\right) - (1.4804) - 1) = 0.6012$$

$$W_{net} = 1 \times 300 \times 0.6012 = 180.36 \text{ kJ/kg}$$

$$\eta_{th} = 1 - \frac{1}{c} = 1 - \frac{1}{1.4804} = 0.3245 \text{ or } 32.45\%$$

(b)
$$W_{net} = 180.36 \text{ kJ/kg}$$
$$\eta_{th} = 1 - \frac{c}{\alpha} = 1 - \frac{1.4804}{3.333} = 0.5558 \text{ or } 55.58\%$$

(c)
$$\frac{W_{net}}{c_p T_1} = \alpha\left(1 - \frac{1}{c}\right) - 2(\sqrt{c} - 1)$$
$$= 3.333\left(1 - \frac{1}{1.4804}\right) - 2(\sqrt{1.4809} - 1) = 0.64815$$
$$W_{net} = 1 \times 300 \times 0.64815 = -194.44 \text{ kJ/kg}$$
$$\eta_{th} = 1 - \left(\frac{\frac{\alpha}{c} + \sqrt{c} - 2}{\alpha - \sqrt{c}}\right)$$
$$= \left[1 - \frac{\frac{3.333}{1.4804} + \sqrt{1.4804} - 2}{3.333 - \sqrt{1.4804}}\right]$$
$$= 0.3063 \text{ or } 30.63\%$$

(d)
$$W_{net} = 194.47 \text{ kJ/kg}$$
$$\eta_{th} = 1 - \frac{2(\sqrt{c} - 1)}{\alpha\left(1 - \frac{1}{c}\right)} = 1 - \frac{2(\sqrt{1.4804} - 1)}{3.333\left(1 - \frac{1}{1.4804}\right)}$$
$$= 0.5922 \text{ or } 59.22\%$$

■ **Example 12.19** The pressure ratio in a gas turbine is 6 and air inlet temperature is 15°C. The maximum permissible temperature is 750°C achieved by heating in a heat exchanger followed by combustion chamber. It is then expanded in two stages such that the expansion work is maximum The air is reheated to 750°C after the first stage. Determine the cycle thermal efficiency, work ratio and net shaft work per kg of air.

Solution : $T_1 = 273 + 15 = 288$ K, $T_3 = 273 + 750 = 1023$ K

$$T_2 = T_1 r_p^{\frac{\gamma - 1}{\gamma}} = 288 \times 6^{0.286} = 480.67 \text{ K}$$

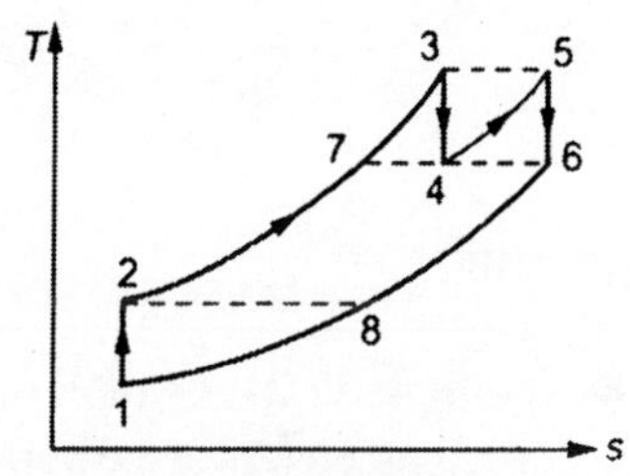

Fig. 12.31

Refer to Fig. 12.31

For maximum expansion work,

$$\frac{p_3}{p_4} = \frac{p_5}{p_6} = \sqrt{r_p} = \sqrt{6} = 2.45$$

$$T_4 = \frac{T_3}{\left(\frac{p_3}{p_r}\right)^{\frac{\gamma - 1}{\gamma}}} = \frac{1023}{2.45^{0.286}} = 791.7 \text{ K}$$

$$T_6 = T_4 = T_7 = 791.7\text{ K}$$

$$W_c = c_p(T_2 - T_1) = 1.005\,(480.67 - 288)$$
$$= 193.63\text{ kJ/kg}$$

$$W_t = c_p\,(T_3 - T_4) + c_p\,(T_5 - T_6)$$
$$= 2 \times 1.005\,(1023 - 791.7) = 464.9\text{ kJ/kg}$$

$$q_s = c_p\,(T_3 - T_7) + c_p\,(T_5 - T_4)$$
$$= 2 \times 1.005\,(1023 - 791.7) = 464.9\text{ kJ/kg}$$

$$W_{net} = W_t - W_c = 464.9 - 193.63 - 271.27\text{ kJ/kg}$$

$$\eta_{th} = \frac{W_{net}}{q_s} = \frac{271.27}{464.9} \times 100 = 58.36\%$$

$$R_W = \frac{W_{net}}{W_t} = \frac{271.27}{464.9} = 0.5836$$

■ **Example 12.20** A gas turbine plant operates between 5°C and 839°C. Find

(a) pressure ratio at which cycle efficiency equals carnot cycle efficiency,

(b) pressure ratio at which maximum work is obtained, and

(c) efficiency under conditions giving maximum work.

Solution :

$$\eta_{carnot} = 1 - \frac{T_1}{T_2} = 1 - \frac{278}{1112} = 0.75$$

For simple $6T$ cycle,

$$\eta_{th} = 1 - \frac{1}{r_p^{\frac{\gamma - 1}{\gamma}}} = 0.75$$

(i)

$$1 - \frac{1}{r_p^{0.286}} = 0.25,\ r_p = 128,\ \frac{p_2}{p_1} = 128$$

$$c = r_p^{\frac{\gamma - 1}{\gamma}} = 4,\ \alpha = \frac{T_2}{T_1} = \frac{1112}{278} = 4$$

(ii)

$$\frac{W_{net}}{c_p T_1} = \alpha\left(1 - \frac{1}{c}\right) - (c - 1),\ \frac{dW_{net}}{dc} = \frac{\alpha}{c^2} - 1 = 0,$$

$$c = \sqrt{\alpha} = \sqrt{4} = 2$$

(iii)

$$\eta = 1 - \frac{1}{c} = 1 - \frac{1}{2} = 0.5 \text{ or } 50\%$$

■ **Example 12.21** A simple open cycle gas turbine has a compressor turbine and a free power turbine. It develops electrical power output of 250 MW. The cycle takes in air at 1 bar and 288 K. Total compressor pressure ratio is 14. The turbine inlet temperature is 1500 K. The total to total isentropic efficiency of compressor and turbine are 0.86 and 0.89 , respectively. The mechanical efficiency of each shaft is 0.98. Combustion efficiency is 0.98 while combustor pressure loss is 0.03 bar. Alternator efficiency is 0.98. Take calorific value of fuel equal to 42,000 kJ/kg.

$$c_{pa} = 1.005\text{ kJ/kg.K and } c_{pg} = 1.15\text{ kJ/kg.K}$$

Calculate the following : (i) Air-fuel ratio, (ii) Specific work output, (iii) Specific fuel consumption, (iv) Mass flow rate of air, (v) Cycle thermal efficiency.

Solution : $p_1 = 1$ bar, $T_1 = 288$ K, $p_2 = 14$ bar, $p_3 = 14 - 0.03 = 13.97$ bar

$$T_3 = 1500 \text{ K}, P = 250 \text{ kW}, \eta_c = 0.86, \eta_t = 0.89, \eta_{mech} = 0.98,$$

$$\eta_{comb} = 0.98, \eta_a = 0.98, CV = 42{,}000 \text{ kJ/kg}$$

$$c_{pa} = 1.005 \text{ kJ/kg.K}, c_{pg} = 1.15 \text{ kJ/kg.K}$$

The T-s is shown in Fig. 12.32

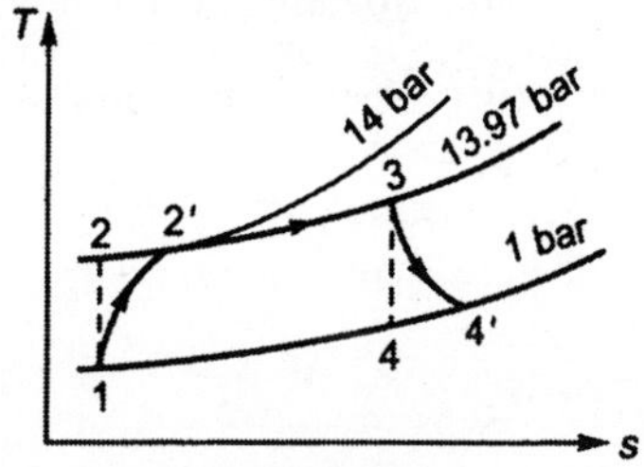

Fig. 12.32

$$\frac{T_2}{T_1} = \left(\frac{p_2}{p_1}\right)^{\frac{(\gamma-1)}{\gamma}}$$

$$T_2 = 288\,(14)^{0.4/1.4} = 612.15 \text{ K}$$

Compressor efficiency,

$$\eta_c = \frac{T_2 - T_1}{T_2' - T_1}$$

$$0.86 = \frac{612.15 - 288}{T_2' - T_1}$$

$$T_2' - T_1 = 376.92 \text{ K}, T_2' = 664.92 \text{ K}$$

$$T_4 = \frac{T_3}{\left(\frac{p_3}{p_4}\right)^{\frac{(\gamma-1)}{\gamma}}} = \frac{1500}{\left(\frac{13.97}{1}\right)^{\frac{0.4}{1.4}}} = 706.14 \text{ K}$$

Turbine efficiency,

$$\eta_t = \frac{T_3 - T_4}{T_3 - T_4'}$$

$$T_3 - T_4' = \frac{1500 - 706.14}{0.89} = 891.97 \text{ K}$$

Actual net work done, w_{net} = Turbine work, w_t − Compressor work, w_c

$$= \frac{c_{pg}\,(T_3 - T_4')\,\eta_{mech} - c_{pa}\,(T_2' - T_1)}{\eta_{mech}}$$

$$= \frac{1.15 \times 891.97 \times 0.98 - 1.005 \times 376.92}{0.98}$$

$$= 618.72 \text{ kJ/kg of air}$$

Effective heat supplied

$$q_e = c_{pg}\,(T_3 - T_2')$$

$$= 1.15\,(1500 - 664.92) = 960.34 \text{ kJ/kg}$$

Actual heat supplied,

$$q_a = \frac{q_e}{q_{comb}} = \frac{960.34}{0.98} = 979.94 \text{ kJ/kg}$$

(i) Air/fuel ratio,

$$\frac{A}{F} = \frac{\text{CV of fuel}}{q_a} = \frac{42000}{979.94} = 42.86:1$$

(ii) Specific work output = 618.72 kJ/kg of air.

(iii) Specific fuel consumption $= \dfrac{\text{Specific air consumption}}{(\text{A / F})}$

$$= \frac{\left(\dfrac{3600}{\text{Specificf work output}}\right)}{(\text{A / F})}$$

$$= \frac{\left(\dfrac{3600}{618.72}\right)}{42.86} = 0.1357 \text{ kg/kWh.}$$

(iv) Mass flow rate of air, m_a × Specific work output

= Electric power developed × Alternator efficiency

$$m_a = \frac{250 \times 10^3 \times 0.98}{618.72} = 395.98 \text{ kg/s}$$

(v) Cycle thermal efficiency

$$= \frac{\text{Actual net work done}}{\text{Actual heat suupplied}} = \frac{618.72}{979.94} = 63.14\%$$

■ **Example 12.22** An open cycle gas turbine employs a regenerative arrangement. The air enters the compressor at 1 bar and 288 K and is compressed to 10 bar with a compression efficiency of 85%. The air is heated in the regenerator and the combustion chamber till its temperature is raised to 1700 K and during the process pressure falls by 0.2 bar. The air is then expanded in the turbine and passes to regenerate which has 75% effectiveness and causes a pressure drop of 0.2 bar. The isentropic efficiency of turbine is 86%. By sketching the gas turbine system and showing the process on T-s diagram, calculate the thermal efficiency and power output if mass flow rate of air is 100 kg/s. Take mechanical and alternator efficiency as 98% each. $C_{pg} = 1.15$ kJ/kg.K and $C_{pa} = 1.005$ kJ/kg.K.

Solution : The gas turbine system and T-s diagram are shown in Fig. 12.33.

$$T_2 = T_1\left(\frac{p_2}{p_1}\right)^{\frac{(\gamma-1)}{\gamma}} = 288\left(\frac{10}{1}\right)^{\frac{0.4}{1.4}} = 556 \text{ K}$$

$$T_2' = T_1 + \frac{(T_2 - T_1)}{\eta_c} = 288 + (556 - 288)/0.85 = 603.3 \text{ K}$$

$$p_3' = 10 - 0.2 = 9.8 \text{ bar} = p_4$$

$$p_5 = 1 + 0.2 = 1.2 \text{ bar}$$

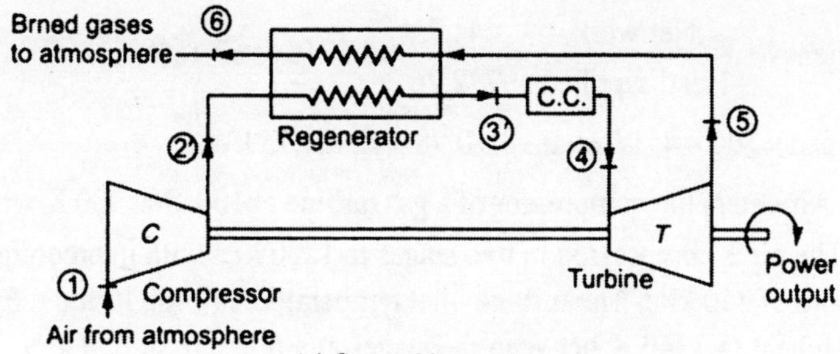

(a) Gas turbine system

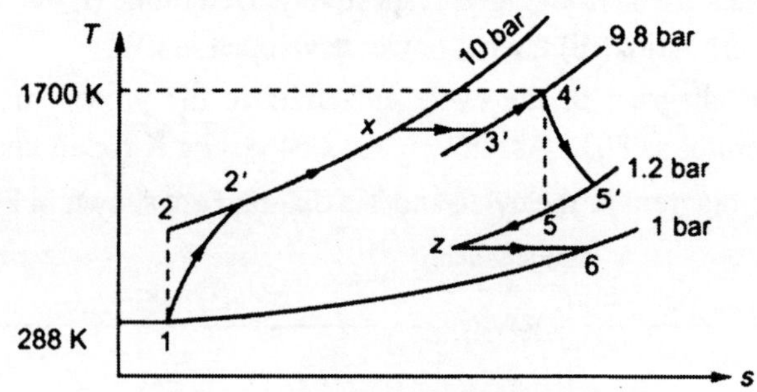

(b) T-s diagram

Fig. 12.33

$$\frac{T_4}{T_5} = \left(\frac{p_4}{p_5}\right)^{\frac{(\gamma - 1)}{\gamma}} = \left(\frac{9.8}{1.2}\right)^{\frac{0.4}{1.4}} = 1.822$$

$$T_5 = \frac{1700}{1.822} = 932.96 \text{ K}$$

$$T_5' = T_4 - (T_4 - T_5)\,\eta_t$$

$$= 1700 - (1700 - 930.96) \times 0.86 = 1040.34 \text{ K}$$

Work consumed by the compressor,

$$w_c = c_{pa}\,(T_2' - T_1)$$

$$= 1.005\,(603.3 - 288) = 316.876 \text{ kJ/kg}$$

Work done by turbine

$$w_t = c_{pg}\,(T_4 - T_5') = 1.15\,(1700 - 1040.34)$$

$$= 758.61 \text{ k/kg}$$

Net work = $w_t - w_c = 758{,}61 - 316.876 = 441.73$ kJ/kg

Regeneration effectivenes, $\varepsilon = \left(\dfrac{T_3' - T_2'}{T_5' - T_2'}\right) = 0.75$

$$T_3' = 0.75\,(1040.34 - 603.3) + 603.3 = 931.08 \text{ K}$$

Heat supplied = $c_{pa}\,(T_4 - T_3') = 1.005\,(1700 - 931.08) = 772.76$ kJ/kg

$$\text{Thermal efficiency} = \frac{\text{Net work}}{\text{Head supplied}} = \frac{441.73}{772.76} = 0.5716 \text{ or } 57.16\%$$

$$\text{Power developed} = 100 \times 441.73 \times 0.98 \times 0.98 = 42423.75 \text{ kW}.$$

■ **Example 12.23** Air enters the compressor of a gas turbine at 100 kPa, 300 K with a volumetric flow rate of 5 m^3/s. The air is compressed in two stages to 1200 kPa with intercooling to 300 K between stages at a pressure of 350 kPa. The turbine inlet temperature is 1400 K and the expansion occurs in two stages with reheat to 1340 K between the stages at a pressure of 350 kPa. The compressor and turbine stage efficiencies are 87% and 85% respectively. Determine (i) the thermal efficiency of the cycles, (ii) the back work ratio, (iii) the net power developed in kW.

Draw the schematic diagram of the cycle and indicate the process on T-s diagram. Assume effectivenes of the regenerator as 80%. Assume c_p = 1.0045 kJ/kg.K for air and gas.

Solution : The schematic diagram of the cycle and T-s diagram are shown in Fig. 12.34.

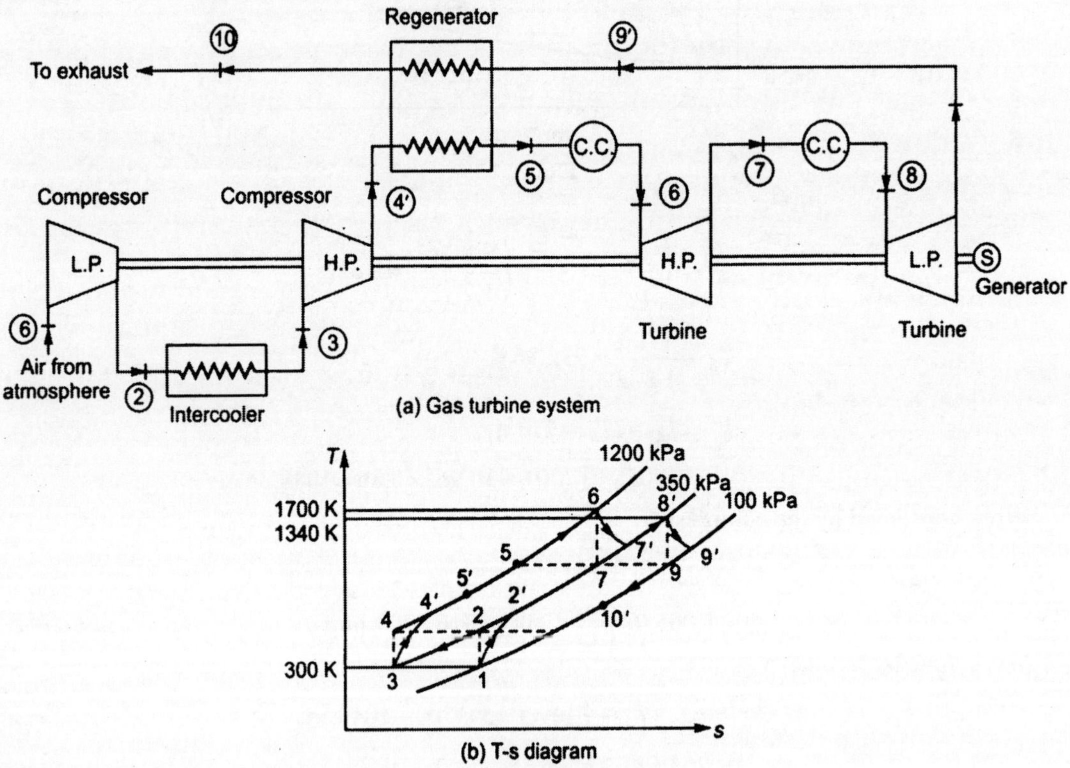

(a) Gas turbine system

(b) T-s diagram

Fig. 12.34

$$p_1 = p_a = 100 \text{ kPa}, T_1 = 300 \text{ K}, \dot{V}_1 = 5 \text{ m}^3/\text{s}$$

$$p_2 = p_3 = p_7 = p_8 = 350 \text{ kPa}, \ p_4 = p_6 \ 1200 \text{ kPa}$$

$$T_1 = T_3 = 300 \text{ K}, T_6 = 1400 \text{ K}, T_8 = 1340 \text{ K}$$

$$\eta_{c1} = \eta_{c2} = 0.87, \eta_{t1} = \eta_{t2} = 0.85, \varepsilon = 0.8$$

$$c_{pa} = c_{pg} = 1.0045 \text{ kJ/kg.K}$$

Compressor 1st stage :

$$T_2 = T_1 \left(\frac{p_2}{p_1}\right)^{\frac{\gamma-1}{\gamma}} = 300\left(\frac{350}{100}\right)^{\frac{1}{3.5}} = 429.11\,\text{K}$$

$$\eta_{c1} = \frac{T_2 - T_1}{T_2' - T_1},\ T_2' = 300 + \frac{429.11 - 300}{0.87} = 448.4\,\text{K}$$

Work done, $$w_{c1} = c_{pa}\,(T_2' - T_1) = 1.0045\,(448.4 - 300)$$

$$= 149.07\ \text{kJ/kg}$$

Compressor 2nd stage :

$$T_4 = T_3 \left(\frac{p_4}{p_3}\right)^{\frac{\gamma-1}{\gamma}} = 300\left(\frac{1250}{350}\right)^{\frac{1}{3.5}} = 426.59\,\text{K}$$

$$T_4' = 300 + \frac{426.59 - 300}{0.87} = 445.50\,\text{K}$$

Work done, $$w_{c2} = c_{pa}\,(T_4' - T_3) = 1.0045\,(445.50 - 300)$$

$$= 146.16\ \text{kJ/kg}$$

Total compressor work, $w_c = w_{c1} + w_{c2} = 295.23$ kJ/kg

Turbine Ist stage :

$$T_7 = T_6 \left(\frac{p_7}{p_6}\right)^{\frac{\gamma-1}{\gamma}} = 1400\left(\frac{350}{1200}\right)^{\frac{1}{3.5}} = 984.55\,\text{K}$$

$$\eta_{t1} = \frac{T_6 - T_7'}{T_6 - T_7},\ T_7' = 1400 - 0.85\,(1400 - 984.55)$$

$$= 1046.87\,\text{K}$$

$$w_{t1} = c_{pg}\,(T_6 - T_7') = 1.0045\,(1400 - 1046.87)$$

$$= 354.72\ \text{kJ/kg}$$

Turbine 2nd stage :

$$T_9 = T_8 \left(\frac{p_9}{p_8}\right)^{\frac{\gamma-1}{\gamma}} = 1340\left(\frac{100}{350}\right)^{\frac{1}{3.5}} = 936.82\,\text{K}$$

$$\eta_{t2} = \frac{T_8 - T_9'}{T_8 - T_9},\ T_9' = 1340 - 0.85\,(1340 - 936.82) = 997.3\,\text{K}$$

$$w_{t2} = c_{pg}\,(T_8 - T_9') = 1.0045\,(1340 - 997.3)$$

$$= 344.24\ \text{kJ/kg}$$

$$w_t = w_{t1} + w_{t2} = 698.96\ \text{kJ/kg}$$

$$w_{\text{net}} = w_t - w_c = 698.96 - 295.23 = 403.73\ \text{kJ/kg}$$

Regenerator effectiveness, $$\varepsilon = \frac{T_5' - T_4'}{T_9' - T_4'}$$

$$0.8 = \frac{T_5' - 445.5}{997.3 - 445.5}$$

$$T_5' = 886.94 \text{ K}$$

Heat supplied, $q_s = c_{pg}\,[(T_6 - T_5') + (T_8 - T_7')$

$$= 1.0045\,[(1400 - 886.94) + (1340 - 1046.87)]$$
$$= 809.82 \text{ kJ/kg}$$

(i) Thermal efficiency,

$$\eta_{th} = \frac{w_{net}}{q_s} = \frac{403.73}{809.82} = 0.4985 \text{ or } 49.85\%$$

(ii) Work ratio $= \dfrac{w_{net}}{w_t} = \dfrac{403.73}{698.96} = 0.5776$

(iii) $p_1\dot{V}_1 = \dot{m}RT_1$

$$\dot{m} = \frac{100 \times 10^5 \times 5}{287 \times 300} = 580.72 \text{ kg/s}$$

(iv) Power developed $= \dot{m} \times w_{net} = 580.72 \times 403.73 = 234454$ kW

■ **Example 12.24** In a gas turbine, the compressor takes in air at a temperature of 15°C and compresses it to 4 times the initial pressure with an isentropic efficiency of 82%. The air is then passed through a heat exchanger heated by the turbine exhaust before reaching the combustion chamber. In the heat exchanger, 78% of available heat is given to the air. The maximum temperature after constant pressure combustion is 600°C, and the efficiency of the turbine is 70%. Neglecting all losses except those mentioned, and assuming the working fluid throughout the cycle to have the characteristic of air, find the efficiency of the cycle. Assume $R = 0.287$ kJ/kg. K and $\gamma = 1.4$ for air and constant specific heats throughout.

Solution : The T-s diagram is shown in Fig .12.35

$$T_1 = 273 + 15 = 288 \text{ K}$$
$$T_3 = 273 + 600 = 873 \text{ K}$$

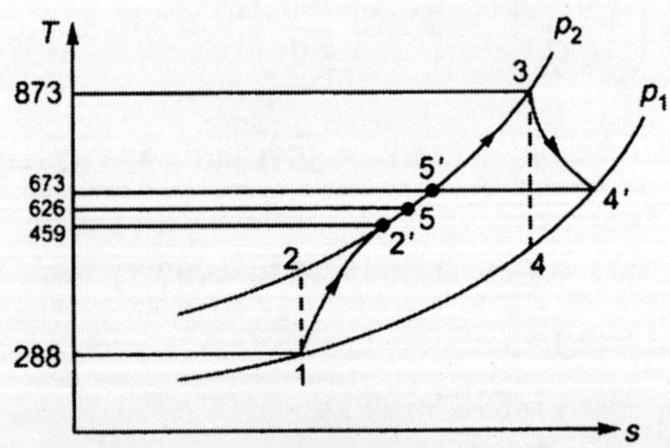

Fig. 12.35

Considering isentropic compression 1 – 2, we get

$$T_2 = T_1\left(\frac{p_2}{p_1}\right)^{\frac{\gamma-1}{\gamma}} = 288 \times 4^{0.286} = 428 \text{ K}$$

$$\eta_c = \frac{T_2 - T_1}{T_2' - T_1}$$

$$0.82 = \frac{428 - 288}{T_2' - 288}$$

$$T_2' = 459 \text{ K}$$

Considering isentropic expansion 3-4, we have

$$\frac{T_3}{T_4} = \left(\frac{p_3}{p_4}\right)^{\frac{\gamma-1}{\gamma}}$$

$$T_4 = \frac{873}{4^{0.286}} = 587.5 \text{ K}$$

$$\eta_t = \frac{T_3 - T_4'}{T_3 - T_4}$$

$$0.7 = \frac{837 - T_4'}{873 - 587.5}$$

$$T_4' = 673 \text{ K}$$

$$w_c = c_p\,(T_4' - T_1)$$

$$c_p = \frac{R\gamma}{\gamma - 1} = \frac{0.287 \times 1.4}{0.4} = 1.0045 \text{ kJ/kg.K}$$

$$w_c = 1.0045\,(459 - 288) = 171.7 \text{ kJ/kg}$$

$$w_t = c_p\,(T_3 - T_4') = 1.0045\,(873 - 673) = 200.9 \text{ kJ/kg}$$

Net work, $$w_{net} = w_t - w_c = 200.9 - 171.7 = 29.2 \text{ kJ/kg}$$

Heat exchanger effectiveness, $$\varepsilon = \frac{T_5 - T_2'}{T_4' - T_2'}$$

$$0.78 = \frac{T_5 - 459}{673 - 459}$$

$$T_5 = 626 \text{ K}$$

Heat supplied by fuel, $$q_s = c_p\,(T_3 - T_5)$$

$$= 1.0045\,(873 - 626) = 248.1 \text{ kJ/kg}$$

Cycle efficiency, $$\eta = \frac{w_{net}}{q_s} = \frac{29.2}{248.1} = 0.117 \text{ or } 11.7\%$$

■ **Example 12.25** In an open cycle constant pressure gas turbine, air enters the compressor at 1 bar and 300 K. The pressure of air after the compression is 4 bar. The isentropic efficiencies of compressor and turbine are 78% and 85% respectively. The air-fuel ratio is 80 : 1. Calculate the power developed and thermal efficiency of the cycle if the flow rate of air is 2.5 kg/s. Take $c_p = 1.005$ kJ/kg.K and $\gamma = 1.4$ for air and $c_p = 1.147$ kJ/kg.K and $\gamma = 1.3$ for gases. $R = 0.287$ kJ/kg.K. C.V. of fuel = 42000 kJ/kg.

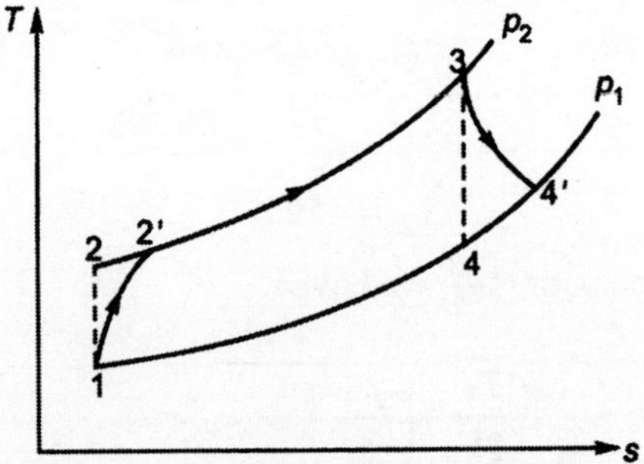

Fig. 12.36

Solution : The T-s diagram is shown in Fig. 12.36.

$$T_2 = T_1 \left(\frac{p_2}{p_1}\right)^{\frac{\gamma-1}{\gamma}}$$

$$= 300 \times 4^{0.286} = 446 \text{ K}$$

$$T_2' - T_1 = \frac{T_2 - T_1}{\eta_c}$$

$$= \frac{446 - 300}{0.78} = 187.18 \text{ K}$$

$$T_2' = 300 + 187.18 = 187.18 \text{ K}$$

Heat supplied by fuel = Heat absorbed by gases

$$m_f \times \text{CV} = (m_a + m_f)(T_3 - T_2') \times c_p$$

$$\text{CV} = \left(1 + \frac{m_a}{m_f}\right) c_p (T_3 - T_2')$$

$$42{,}000 = (1 + 80) \times 1.147 \times (T_3 - 487.18)$$

$$T_3 = 939.25 \text{ K}$$

Work required to drive the compressor

$$w_c = c_{pa}(T_2' - T_1) = 1.005 \times 187.18 = 188.12 \text{ kJ/kg of air/s.}$$

Temperature at the end of isentropic expansion,

$$T_4 = \frac{T_3}{\left(\frac{p_2}{p_1}\right)^{\frac{\gamma-1}{\gamma}}} = \frac{939.25}{4^{0.248}} = 666 \text{ K}$$

Actual temperature drop $= T_3 - T_4' = \eta_t (T_3 - T_4)$

$$= 0.85\,(939.25 - 666)$$
$$= 232.25 \text{ K}$$
$$T_4' = 939.25 - 232.25 = 707 \text{ K}$$

Total turbine work, $w_t = m_g c_{pg} (T_3 - T_4')$

$$= \frac{81}{80} \times 1.147 \times 232.25 = 269.73 \text{ kJ/kg of air/s}$$

Net output, $\quad w = w_t - w_c = 269.73 - 188.12$

$$= 81.61 \text{ kJ/kg of air/s}$$

Heat supplied per kg of air $= \frac{1}{80} \times 42{,}000 = 525$ kJ/kg of air/s

$$\eta_{th} = \frac{81.61}{525} = 0.1554 \text{ or } 15.54\%$$

Net power developed by turbine $= 81.61 \times 2.5 = 204.03$ kW.

■ **Example 12.26** A closed cycle ideal gas turbine plant operates between temperature limits of 800°C and 30°C and produces a power of 100 kW. The plant is designed such that there is no need for a regenerator. A fuel of calorific value 45000 kJ/kg is used. Calculate the mass flow rate of air through the plant and the rate of fuel consumption. (Assume $c_p = 1$ kJ/kg.K and the ratio of specific heats $= 1.4$).

Solution : $\quad T_1 = 30 + 273 = 303 \text{ K}, T_3 = 800 + 273 = 1073 \text{ K}$

without a regenerator, let $T_2 = T_4$

Now
$$\frac{p_2}{p_1} = \frac{p_3}{p_4}, \frac{p_2}{p_1} = \left(\frac{T_2}{T_1}\right)^{\frac{\gamma-1}{\gamma}}$$

and
$$\frac{p_3}{p_4} = \left(\frac{T_3}{T_4}\right)^{\frac{\gamma-1}{\gamma}}$$

$$T_2 = \sqrt{T_1 T_3} = \sqrt{303 \times 1073} = 570.2 \text{ K}$$

$$W_t - W_c = \dot{m}_f \times \text{CV} \times \left[1 - \frac{T_4 - T_1}{T_3 - T_2}\right]$$

$$100 \times 10^3 = m_f \times 45000 \times 10^3 \times \left[1 - \frac{570.2 - 303}{1073 - 570.2}\right]$$

$$\dot{m}_f = 4.74 \times 10^{-3} \text{ kg/s}$$

Also
$$(\dot{m}_a + \dot{m}_f)(T_3 - T_4) - \dot{m}_a (T_2 - T_1) = 100 \times 10^3$$

$$(\dot{m}_a + 4.74 \times 10^{-3})(1073 - 570.2) - \dot{m}_a (570.2 - 303) = 100 \times 10^3$$

$$\dot{m}_a = 0.414 \text{ kg/s.}$$

■ **Example 12.27** In a gas turbine, the compressor is driven by the high pressure turbine. The exhaust from the high pressure turbine goes to a free low pressure turbine which runs the load. The air flow rate is 20 kg/s and the minimum and maximum temperatures are respectively 300 K and 1000 K. The compressor pressure ratio is 4. Calculate the pressure ratio of the low pressure turbine and the temperature of the exhaust gases from the unit. The compressor and turbine are isentropic. C_p of air and exhaust gases = 1 kJ/kg.K and $\gamma = 1.4$.

Solution : The T-s diagram is shown in Fig. 12.37

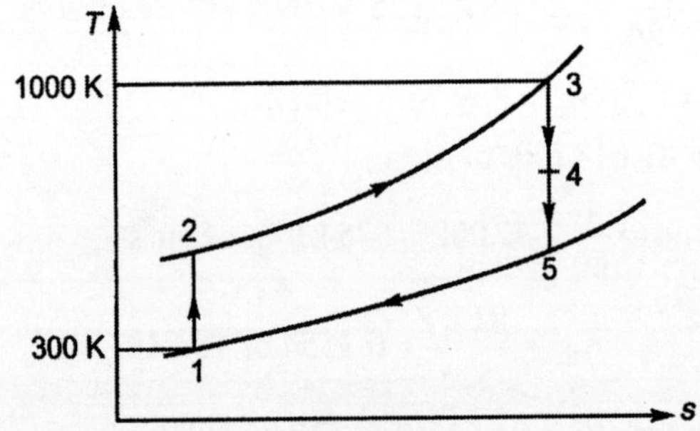

Fig. 12.37

$$T_2 = T_1\left(\frac{p_2}{p_1}\right)^{\frac{\gamma-1}{\gamma}}$$

$$= 300 \times 4^{0.286} = 445.8 \text{ K}$$

Now, $$mc_p\,(T_2 - T_1) = mc_p\,(T_3 - T_4)$$

$$T_4 = T_3 - (T_2 - T_1)$$

$$= 1000 - (445.8 - 300)$$

$$= 854.2 \text{ K}$$

$$\frac{p_3}{p_4} = \left(\frac{T_3}{T_4}\right)^{\frac{\gamma}{\gamma-1}} = \left(\frac{1000}{854.2}\right)^{3.5} = 1.736$$

Now, $$\frac{p_3}{p_4} = \frac{p_3}{p_5} \times \frac{p_5}{p_4} = 4 \times \frac{p_5}{p_4} \qquad \left(\because \frac{p_3}{p_5} = \frac{p_2}{p_1} = 4\right)$$

$$\frac{p_5}{p_4} = \frac{1}{4} \times \frac{p_3}{p_4} = \frac{1}{4} \times 1.736 = 0.434$$

Pressure ratio of low pressure turbine $= \dfrac{p_4}{p_5} = 2.304$

$$\frac{T_4}{T_5} = \left(\frac{p_4}{p_5}\right)^{\frac{\gamma-1}{\gamma}} = (2.304)^{0.286} = 1.269$$

$$T_5 = \frac{854.2}{1.269} = 672.8 \text{ K.}$$

■ **Example 12.28** Air enters the compressor of a gas turbine plant operating on Brayton cycle at 1 bar, 27°C. The pressure ratio in the cycle is 6. If $W_t = 2.5\,W_c$, where W_t and W_c are the turbine and compressor work respectively, calculate the maximum temperature and the cycle efficiency.

Solution :

$$p_1 = 1 \times 10^5 \text{ N/m}^2, T_1 = 27 + 273 = 300 \text{ K}$$

$$p_2 = 6p_1 = 6 \times 10^5 \text{ N/m}^2$$

$$\frac{T_2}{T_1} = \left(\frac{p_2}{p_1}\right)^{\frac{\gamma-1}{\gamma}}, \gamma = 1.4$$

$$T_2 = 300 \times 6^{0.286} = 500.5 \text{ K}$$

$$\frac{T_3}{T_4} = \left(\frac{p_2}{p_1}\right)^{\frac{\gamma-1}{\gamma}} = 6^{0.286} = 1.66$$

$$W_c = mc_p\,(T_2 - T_1)$$

$$W_t = mc_p\,(T_3 - T_4)$$

$$W_t = 2.5\,W_c$$

$$T_3 - T_4 = 2.5\,(T_2 - T_1)$$

$$T_3\left(1 - \frac{1}{1.66}\right) = 2.5\,(500.5 - 300)$$

$$T_3 = 1260.7 \text{ K}$$

$$\text{Cycle efficiency} = \frac{W_t - W_c}{mc_p\,(T_3 - T_2)}$$

$$= \frac{(1260.7 - 759.4) - (500.5 - 300)}{(1260.7 - 500.5)}$$

$$= 39.56\%.$$

■ **Example 12.29** In an air-standard regenerative gas turbine cycle, the pressure ratio is 5. Air enters the compressor at 1 bar, 300 K and leaves at 490 K. The maximum temperature in the cycle is 1000 K. Calculate the cycle efficiency, given that the efficiency of the generator and the adiabatic efficiency of the turbine are each 80%. Assume for air, the ratio of specific heats 1.4. Also show the cycle on T-s diagram.

Solution : The cycle on T-s diagram is shown in Fig. 12.38 assuming compressor efficiency to be 100%.

$$T_2 = T_1\left(\frac{p_2}{p_1}\right)^{\frac{\gamma-1}{\gamma}}$$

$$= 300 \times 5^{0.286} = 475.14 \text{ K}$$

$$T_4 = \frac{T_3}{\left(\frac{p_2}{p_1}\right)^{\frac{\gamma-1}{\gamma}}} = \frac{1000}{5^{0.286}}$$

$$= 631.39 \text{ K}$$

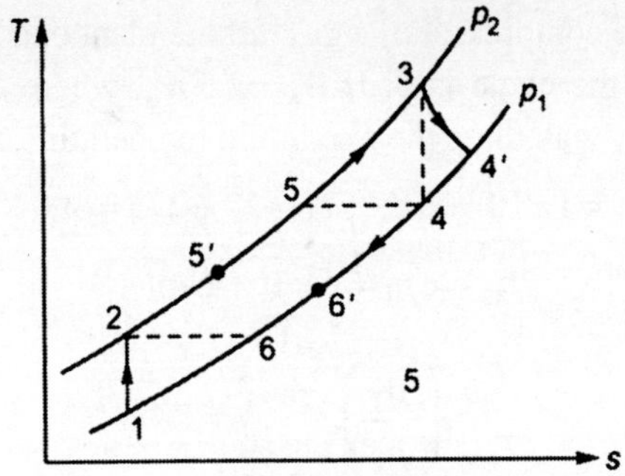

Fig. 12.38

Turbine efficiency, $$\eta_t = \frac{T_3 - T_4'}{T_3 - T_4}$$

$$0.8 = \frac{1000 - T_4}{1000 - 631.39}$$

$$T_4' = 705.11 \text{ K}$$

$$\text{Regenerative efficiency} = \frac{T_5' - T_2}{T_4' - T_2}$$

$$0.8 = \frac{T_5' - 475.14}{705.11 - 475.14}$$

$$T_5' = 659.116 \text{ K}$$

$$w_c = c_p\,(T_2 - T_1) = 1.005 \times (475.14 - 300)$$
$$= 176.01 \text{ kJ/kg}$$

$$w_t = c_p\,(T_3 - T_4') = 1.005\,(1000 - 705.11)$$
$$= 296.36 \text{ kJ/kg}$$

Net work, $$w_{net} = w_t - w_c = 296.36 - 176.01$$
$$= 120.35 \text{ kJ/kg}$$

Heat supplied, $$q_s = c_p\,(T_3 - T_5')$$
$$= 1.005 \times (1000 - 659.116) = 342.58 \text{ kJ/kg}$$

$$\text{Thermal efficiency} = \frac{w_{net}}{q_s} = \frac{120.35}{342.58} = 0.351 \text{ or } 35.1\%.$$

■ **Example 12.30** Find the required air-fuel ratio in a gas turbine whose turbine and compressor efficiencies are 85% and 80% respectively. Maximum cycle temperature is 875°C. The working fluid can be taken as air ($c_p = 1.00$ kJ/kg.K, $\gamma = 1.4$) which enters the compressor fat 1 atm and 27°C. The pressure ratio is 4. The fuel used has calorific value of 42000 kJ/kg. There is a loss of 10% of calorific value in the combustion chamber.

Solution : $T_1 = 27 + 273 = 300$ K, $T_3 = 875 + 273 = 1148$ K

$$\frac{T_2}{T_1} = \left(\frac{p_2}{p_1}\right)^{\frac{\gamma-1}{\gamma}} = 4^{\frac{0.4}{1.4}} = 0.286 = 1.485$$

$$T_2 = 300 \times 1.485 = 445.8 \text{ K}$$

$$\eta_c = \frac{T_2 - T_1}{T_2' - T_1}$$

$$0.8 = \frac{445.8 - 300}{T_2' - 300}$$

$$T_2' = 482.25 \text{ K}$$

$$Q_f = Q_g$$

$$0.9 \times \dot{m}_f \times \text{CV} = (\dot{m}_a + \dot{m}_f)\, c_p\, (T_3 - T_2')$$

$$0.9 \times 42000 \times 10^3 = \left(1 + \frac{\dot{m}_a}{\dot{m}_f}\right) \times 1 \times 10^3 \times (1148 - 482.25)$$

$$\frac{\dot{m}_a}{\dot{m}_f} = 55.78$$

$$\frac{\text{Air}}{\text{Fuel ratio}} = 55.78 : 1$$

12.13 EFFECTS OF OPERATING VARIABLES

The operating variables for the performance of open cycle gas turbine are :

1. Pressure ratio
2. Compressor inlet temperature
3. Turbine inlet temperature
4. Efficiency of the compressor
5. Efficiency of the turbine

These variables effect the thermal efficienncy and the work ratio of the gas turbine.

1. Effect of pressure ratio

For the simple Brayton cycle,

Thermal efficiency, $$\eta_{th} = 1 - \frac{1}{r_p^{\frac{\gamma-1}{\gamma}}}$$

Work ratio, $$R_W = 1 - \frac{T_1}{T_3}(r_p)^{\frac{\gamma-1}{\gamma}}$$

Table 12.1 : Effect of pressure ratio on thermal efficiency

γ	*Thermal Efficiency*				
	$r_p = 1$	$r_p = 5$	$r_p = 10$	$r_p = 15$	$r_p = 20$
1.33	0	0.329	0.435	0.489	0.524
1.40	0	0.368	0.482	0.539	0.575
1.66	0	0.472	0.600	0.659	0.696

The variation of thermal efficiency with pressure ratio for different values of adiabatic index is shown in Fig. 12.39. It may be observed that the thermal efficiency increases as pressure ratio and adiabatic index increases.

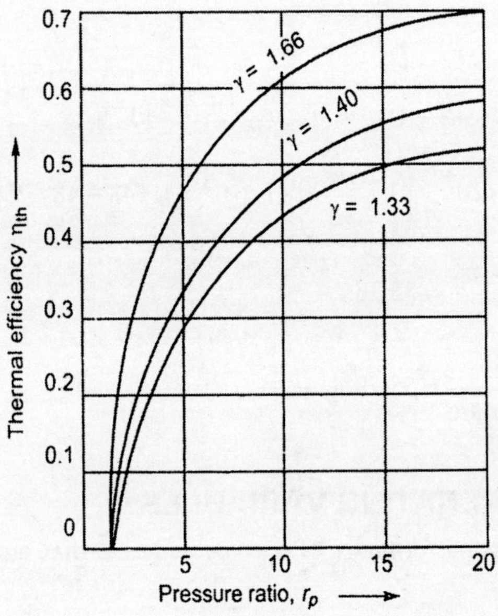

Fig. 12.39

Table 12.2 : Effect of pressure ratio and temperature ratio on work ratio (γ = 1.4).

T_3/T_1	*Work ratio*				
	$r_p = 1$	$r_p = 5$	$r_p = 10$	$r_p = 15$	$r_p = 20$
2	0.5	0.208	0.035	– 0.084	– 0.175
3	0.667	0.472	0.357	0.277	0.217
4	0.750	0.604	0.517	0.458	0.412
5	0.80	0.683	0.614	0.566	0.530

The variation of work ratio with pressure ratio for different values of temperature ratio ($\gamma = 1.4$) are shown in Fig. 12.40. It may be observed that work ratio decreases as pressure ratio increases. However the work ratio increases as temperature ratio increases.

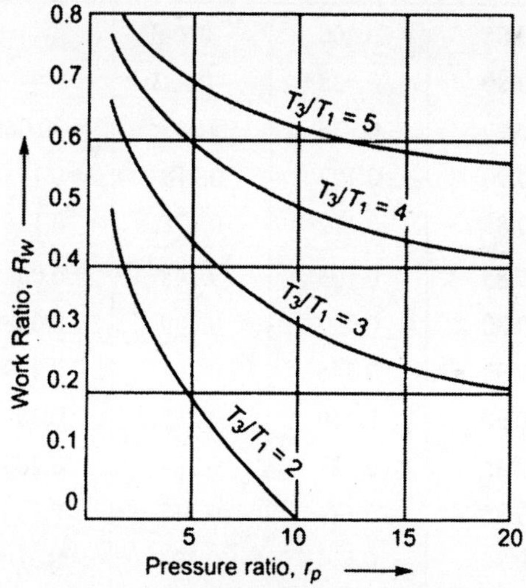

Fig. 12.40

2. Effect of efficiencies of compressor and turbine on thermal efficiency

From Eq. (12.11), we have

$$\eta_{th} = \frac{\eta_t \dfrac{T_3}{T_1}\left[1 - \dfrac{1}{r_p^m}\right] - \dfrac{1}{\eta_c}\left[r_p^m - 1\right]}{\left[\dfrac{T_3}{T_1} - r_p^m\right]}$$

where $m = \dfrac{\gamma - 1}{\gamma}$

Let $\dfrac{T_3}{T_1} = \alpha$, then

$$\eta_{th} = \frac{\alpha\eta_t\left[1 - \dfrac{1}{r_p^m}\right] - \dfrac{1}{\eta_c}\left[r_p^m - 1\right]}{\left[\alpha - r_p^m\right]}$$

For $\gamma = 1.4$, $m = \dfrac{1}{3.5}$

Table 12.3 : Effect of machine efficiencies on thermal efficiency

$\alpha = \dfrac{T_3}{T_1}$	η_c	η_t	$r_p = 2$	$r_p = 5$	$r_p = 10$	$r_p = 15$	$r_p = 20$
1	0.90	0.90	0.102	0.035	– 2.37	–	–
2	0.85	0.90	0.107	– 0.056	—	—	--
	0.85	0.85	0.061	– 0.145	—	—	--

	0.80	0.85	0.400	– 0.248	—	—	—
	0.85	0.80	0.038	– 0.233	—	—	—
	0.90	0.90	0.936	0.245	– 0.065	—	—
	0.85	0.90	0.127	0.218	0.193	0.097	– 0.057
	0.85	0.85	0.112	0.178	0.126	-3.8×10^{-4}	—
3	0.80	0.85	0.103	0.148	0.062	– 0.063	—
	0.85	0.80	0.097	0.139	0.058	– 0.090	—
	0.90	0.90	0.145	0.281	0.339	0.350	0.344
	0.85	0.90	0.140	0.265	0.309	0.308	0.291
4	0.85	0.85	0.127	0.234	0.263	0.250	0.221
	0.80	0.85	0.121	0.217	0.230	0.207	0.161
	0.85	0.80	0.114	0.203	0.216	0.191	0.151
	0.90	0.90	0.149	0.296	0.370	0.398	0.409
	0.85	0.90	0.198	0.284	0.350	0.371	0.376
5	0.85	0.85	0.134	0.257	0.310	0.323	0.322
	0.80	0.85	0.129	0.245	0.288	0.293	0.284
	0.85	0.80	0.122	0.230	0.271	0.276	0.268

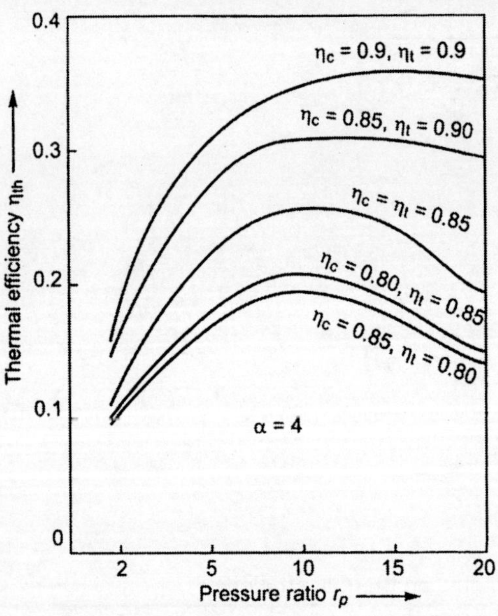

Fig. 12.41

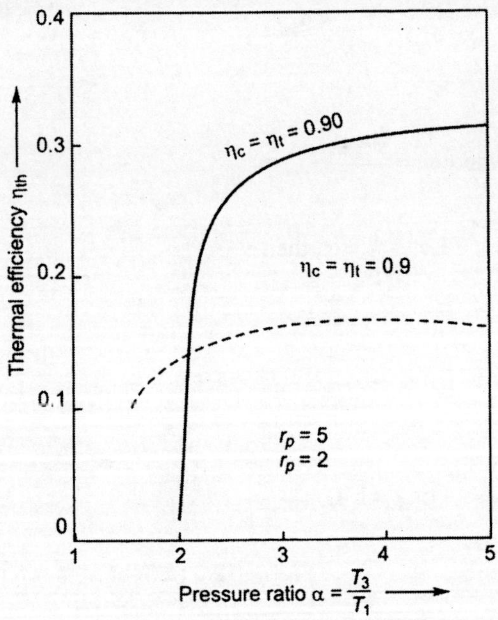

Fig. 12.42

The effect of machine efficiencies on thermal efficiency has been shown in Fig. 12.41. The thermal efficiency initially increases, becomes maximum and then starts decreasing. Fig. 12.42 shows the effect of temperature ratio on thermal efficiency. It may be observed that the thermal efficiency increases as temperature ratio increases.

12.14 MULTISHAFT SYSTEMS

It is possible to use two turbines in parallel by using a single compressor and a single combustion chamber. The arrangement for such a system is shown in Fig. 12.43. For 1 kg of air through the compressor, a certain fraction x is required to pass through turbine I in order to drive the compressor. Then $(1 - x)$ kg gpasses through turbine II to produce the power output. This has the advantage in that the speed of the compressor would not be affected by the load. A further advantage in this respect can be had by the use of two combustion chambers as shown in Fig. 12.44.

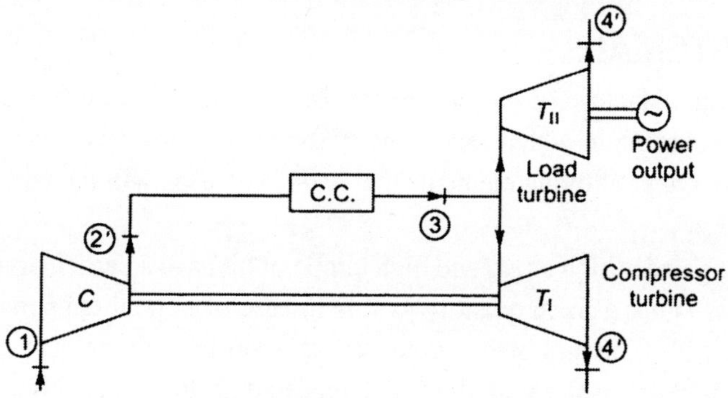

Fig. 12.43 Two turbines in parallel.

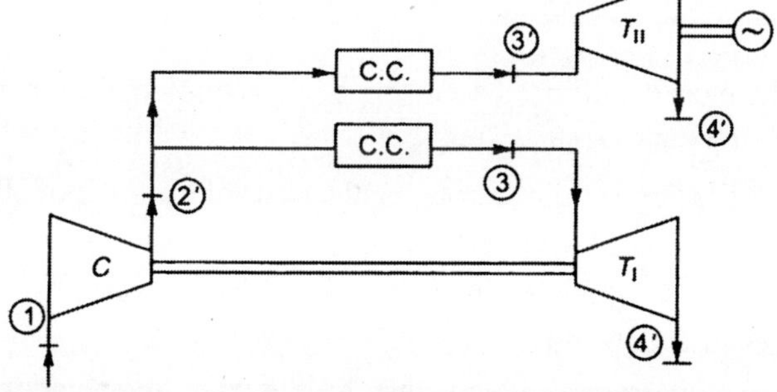

Fig. 12.44 Two turbines of parallel with two combustion chambers.

In such an arrangement, the inlet temperature of the two turbines, as well as their speeds, are separately controllable.

12.15 MULTI-SHAFT SYSTEM TURBINES IN SERIES

The arrangement shown in Fig. 12.45 has two shafts with two turbines in series. It gives advantage of independent shaft speeds.

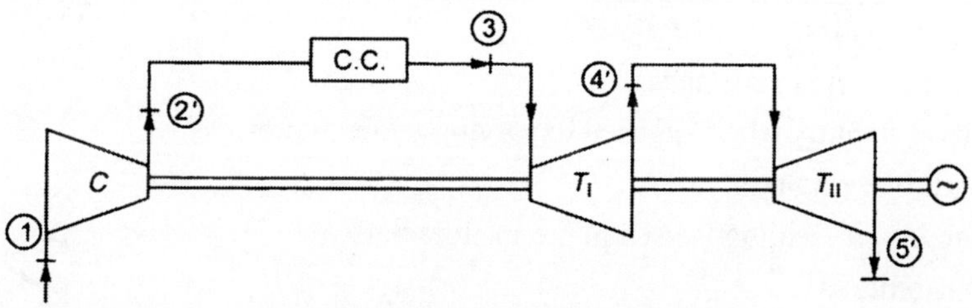

Fig. 12.45 Two turbines in series.

12.16 GAS TURBINE FUELS

The various fuels used for gas turbines are :

1. Gaseous fuels   2. Liquid fuels and   3. Solid fuels.

1. Gaseous fuels : Natural gas, blast furnace and producer gas.

2. Liquid fuels : Liquid fuels of petroleum origin such as distillate oils or residual oils.

3. Solid fuels : Pulverised coal, Crushed coal.

12.17 BLADE MATERIALS

Materials used in the manufacture of the blades consist of mainly three different elements, *i.e.*, iron, nickel and cobalt with chromium which forms one of the major alloying elements, since it gives high resistance to oxidation. Other alloying elements that have been used will include most of the metals of periodic table.

Materials working under high stress and high temperature have a particular rate of creep. It means elongation under stress is not a fixed quantity as it is in case of normal temperature. Thus elongation continues to increase with time and blades gradually take up the original gap provided at their tips. Therefore, contact with casing resulting failure. The repeated heating and cooling of the material affect the physical properties of the material.

12.17.1 Selection

The turbine blades are the most conditioned memebrs of a gas turbine, they must withstand,

(a) The high operating temperature effects.

(b) Centrifugal tensile stresses due to rotational speeds of the order of 8000 to as high as 30,000 r.p.m.

(c) Bending stress due to equivalent impulse load of the gases acting at a certain distance from fixing of the cantilever blade.

(d) Hot erosive and corrosive effects due to the high temperature combustion products *e.g.*, CO_2 and CO with O_2.

Turbine blade material must, therefore, be selected from considerations of the working conditions.

12.17.2 Requirements of Blade Material

1. Material should possess maximum strength at high temperature.
2. High creep strength.
3. High resistance of corrosion.
4. Endurance against.
5. Maximum erosion resistance.
6. Structural stability when exposed to varying temperatures.
7. Castability or forgability.
8. Weldability, if welding is used in the manufacture.
9. Machinability.
10. Absence of embrittlement in the service.

12.18 BLADE FIXING

The following are some methods of blade fixing to the wheel :

1 **Bulb root method :** (Fig. 12.46) A bulb is made in the root of the blade which is fixed to the wheel with the help of steel pins welded and inserted in the holes. This method is employed in Metro Vick gas turbine. Endwise movement of the blade is usually prevented by swaging over the metal of the rotor around the root portions of the blades.

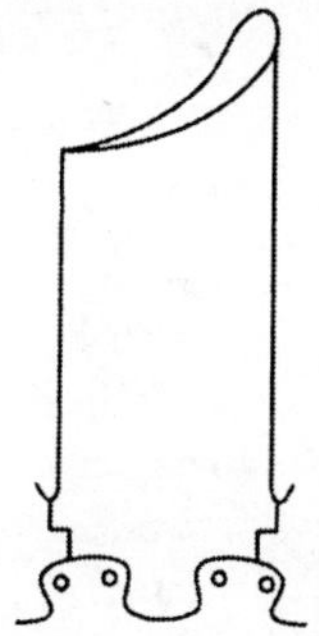

Fig. 12.46 Bulb root method of blade fixing.

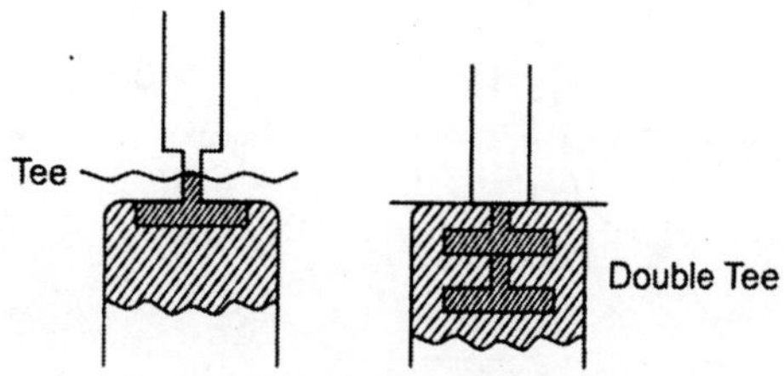

Fig. 12.47 Blade fixing by T.

2. **Tee and Double Tee :** (Fig. 12.47) A *T* slot is made in the wheel and after putting the blade in the position a wedge is placed to keep it in position.
3. **Anchor pin :** (Fig. 12.48) As shown in the figure the blades are fixed by inserting the pin through the hole in the wheel and in the blade root and screwing the nut over the threaded portion.

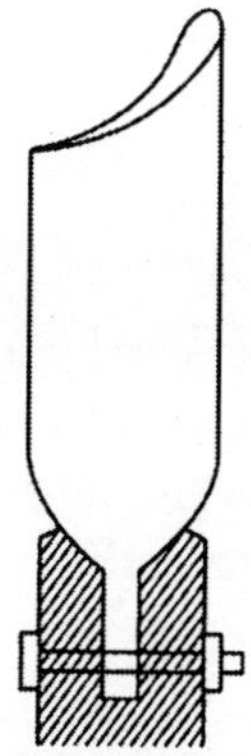

Fig. 12.48 Anchor pin method of blade fixing.

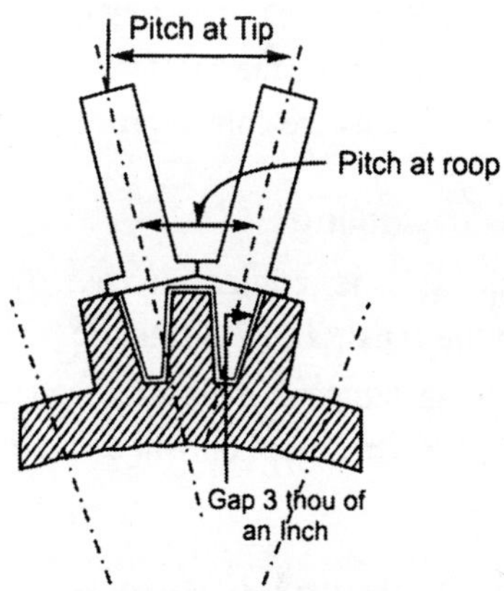

Fig. 12.49 'Fir-tree root' method of blade fixing.

4. **Fir Tree Method :** (Fig. 12.49) This method has got its own advantages.

1. Rotor is not weakened.
2. Contact is distributed.
3. Taper is provided.
4. Thermal expansion is allowed.

This type of fixing is used on the W_2B turbine and subsequently on the Rolls-Royce Derwent and other engines. The blade is driven into the rotor teeth and locked by turning over the lip at the apex of the fir cone on the blade.

5. **Grub Screw Method :** (Fig. 12.50) A screw is fitted half in the rotor drawn and half in the blade root. This method is used on German gas turbines of securing the blades in the rotor. In this case the blade base is of dovetail section and is secured by Grub-screws against end movements.

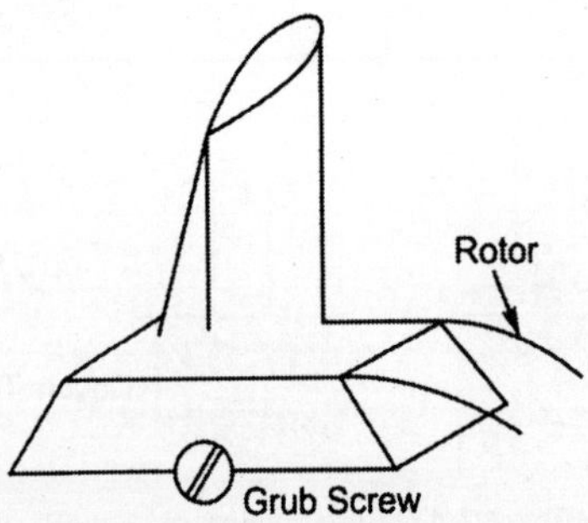

Fig. 12.50 Grub screw method of blade fixing.

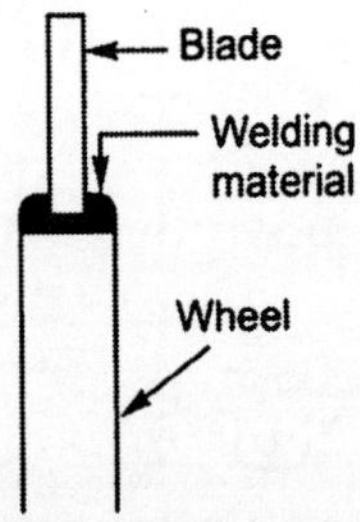

Fig. 12.51 Blade fixing by welding.

6. **Welding :** (Fig. 12.51) This blade may be fixed by welding.

12.19 COOLING OF BLADES

The thermodynamic analysis of the gas cycle shows that the efficiency of the cycle depends on the range of maximum temperature of the gas. If maximum temperature is increased, efficiency of the cycle is also increased. Any increase in the maximum temperature of gas results in an increase in the blade temperature thereby inducing more thermal stresses in the blade material. Thus this limits the capacity of in gas temperature when blades are cooled properly.

12.19.1 Advantages of Cooling

1. Specific thrust of jet engine increase by 32% for gas temperature increase from 800° to 1100°C at the expense of 7% specific fuel consumption.
2. If the maximum temperature is increased to 1600°C in place of present practice of 900°C then the specific fuel consumption will be increased by 50% and increase in specific power output by 200%.

12.19.2 Different Methods of Blade Cooling

1. **Internal air cooling :** It is done by supplying cold air through hollow blade. It employs an internal baffle or deflector to direct the flow over the hotter portions on internal surface. Refer Fig. 12.52.
2. **Film cooling :** It involves in supplying of thin film of cold compressed air through narrow slit on the blade surface to form cold boundary layer over the blades. Refer Fig. 12.53.
3. **Water cooling :** Circulation of water is maintained through hollow section of the blade from root towards tip.

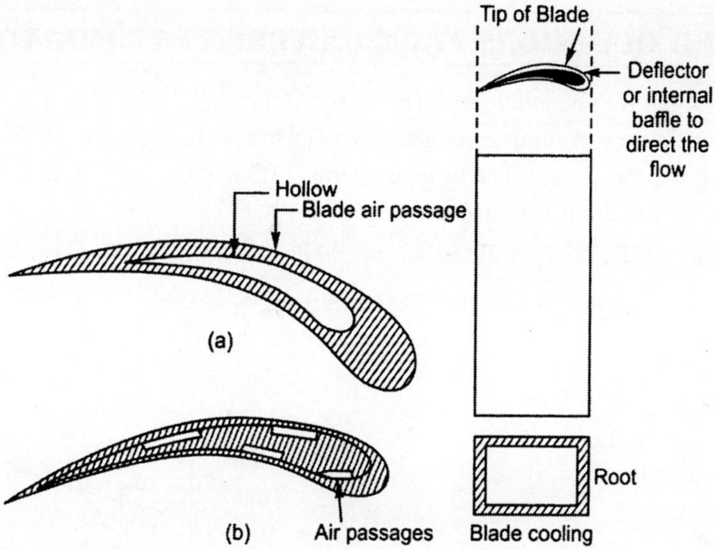

Fig. 12.52 Internal cooling by air.

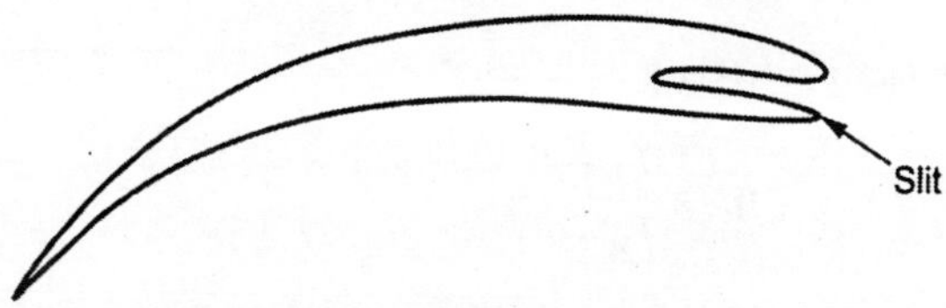

Fig. 12.53 Air film cooling of blade.

4. **Rim cooling :** Rim is cooled by circulating the cooling fluid and thus blade temperature is reduced by conduction.

   The graph Refer Fig. 12.54 shows the distribution of temperature over the blade height.

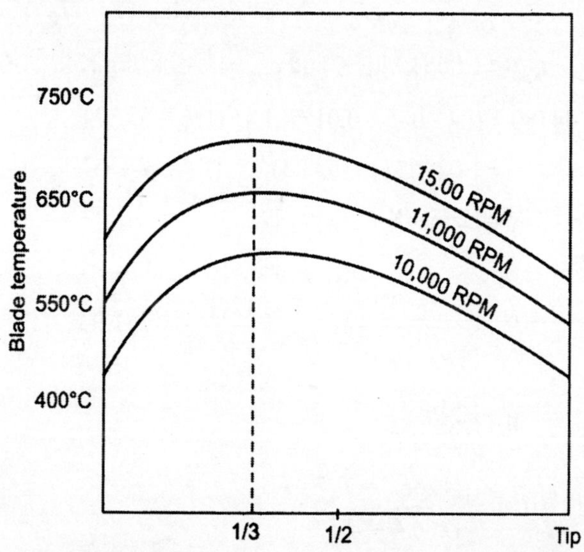

Fig. 12.54 Distribution of temperature over blade height.

SOLVED QUESTIONS FROM UNIVERSITY EXAMINATIONS

■ **Example 12.31** A gas turbine unit receives air at 100 kPa and 300 K and compresses it adiabatically to 620 kPa with efficiency of compressor 88%. The fuel has heating value of 44180 kJ/kg and fuel/air ratio is 0.017 kg fuel per kg air. The turbine internal efficiency is 90%. Calculate the compressor work, turbine work and thermal efficiency. [MDU, Dec., 2007]

Solution : Given : $p_1 = 100$ kPa, $T_1 = 300$ K, $p_2 = 620$ kPa, $\eta_c = 0.88$, $CV = 44180$ kJ/kg

$$F : A = 0.017 \text{ kj fuel/kg air}, \; \eta_t = 0.90$$

Refer to fig. 12.55.

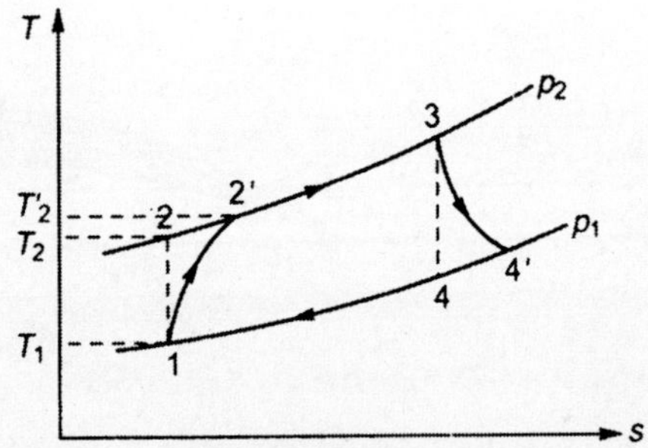

Fig. 12.55

Isentropic process 1-2 :

$$\frac{T_2}{T_1} = \left(\frac{p_2}{p_1}\right)^{\frac{\gamma - 1}{\gamma}}$$

$$= \left(\frac{620}{100}\right)^{0.4/1.4} = 1.6842$$

$$T_2 = 300 \times 1.6842 = 505.3 \text{ K}$$

Compressor efficiency, $\eta_c = \dfrac{T_2 - T_1}{T_2' - T_1}$

$$0.88 = \frac{505.3 - 300}{T_2' - 300}$$

$$T_2' = 533.3 \text{ K}$$

Combustion process 2′ – 3:

$$\dot{m}_f \times CV = \dot{m}_a \times c_{pa}(T_3 - T_2') + \dot{m}_f \times c_{pg}(T_3 - T_2')$$

Let $c_{pa} = 1.005$ kJ/kg.K and $c_{pg} = 1.15$ kJ/kg.K

$$0.017 \times 44180 = (1 \times 1.005 + 0.017 \times 1.15)(t_3 - 533.3)$$

$$= 1.0246\,(T_3 - 533.3)$$

$$T_3 = 1266.4 \text{ K}$$

Isentropic process 3-4 :

$$T_4 = \frac{T_3}{(p_3/p_4)^{(\gamma - 1)/\gamma}} = \frac{1266.4}{1.6842} = 751.9 \text{ K}$$

Turbine efficiency, $\eta_t = \dfrac{T_3 - T_4'}{T_3 - T_4}$

$$0.90 = \frac{1266.4 - T_4'}{1266.4 - 751.9}$$

$$T_4' = 803.3 \text{ K}$$

Compressor work, $w_c = c_{pa}(T_2' - T_1) = 1.005(533.3 - 300) = 234.47 \text{kJ/kg}$

Turbine work, $w_t = c_{pa}(T_3 - T_4') + \left(\dfrac{\dot{m}_f}{\dot{m}_a}\right) c_{pg}(T_3 - T_4')$

$$= (1.005 + 0.017 \times 1.15)(1266.4 - 803.3) = 474.47 \text{ kJ/kg}$$

Net work, $w_{net} = w_t - w_c = 474.47 - 234.47 = 240 \text{ kJ/kg}$

Heat supplied, $q_s = \dot{m}_f \times CV = 0.017 \times 44180 = 751.06 \text{ kJ/kg}$

Thermal efficiency, $\eta_{th} = \dfrac{w_{net}}{q_s} = \dfrac{240}{751.06} = 0.3195 \text{ or } 31.95\%$

■ **Example 12.32** A gas turbine plant with pressure ratio of 1:5 takes in air at 15°C. The maximum temperature is 600°C and develops 2200 kW. The turbine and compresor efficiencies are equal to 0.85. Take $c_p = 1$ kJ/kg.K and $c_v = 0.714$ kJ/kg.K. Determine the actual overall efficiency of the turbine and mass of air circulated by the turbine. [MDU, May, 2008]

Solution : Given $r_p = 5$, $T_1 = 273 + 15 = 288$ K, $T_3 = 273 + 600 = 873K$, $P = 2200$ kW

$$\eta_c = \eta_t = 0.85, \; c_p = 1 \text{ kJ/kg.K}, \; c_v = 0.714 \text{ kJ/kg.K}.$$

Refer to fig. 12.55. $\gamma = \dfrac{c_p}{c_v} = \dfrac{1}{0.714} = 1.4$

Isentropic process 1-2 :

$$T_2 = T_1 (r_p)^{(\gamma - 1)/\gamma} = 288(5)^{1/3.5} = 456.14 \text{ K}$$

$$\eta_c = \frac{T_2 - T_1}{T_2' - T_1}$$

$$0.85 = \frac{456.14 - 288}{T_2' - 288}$$

$$T_2' = 485.81 \text{ K}$$

Isentropic process 3-4 :

$$\frac{T_3}{T_4} = (r_p)^{(\gamma - 1)/\gamma} = 1.5838$$

$$T_4 = 551.2 \text{ K}$$

$$\eta_t = \frac{T_3 - T_4'}{T_3 - T_4}$$

$$0.85 = \frac{873 - T_4'}{873 - 551.2}$$

$$T_4' = 599.47 \text{ K}$$

Compressor work per kg of air, $w_c = c_p(T_2' - T_1) = 1 \times (485.81 - 288) = 197.81$ kJ/kg.

Turbine work per kg of air, $w_t = c_p(T_3 - T_4') = 1 \times (873 - 599.47) = 273.53$ kJ/kg

Net work, $w_{net} = w_t - w_c = 273.53 - 197.81 = 75.72$ kJ/kg

Heat supplied, $q_s = c_p(T_3 - T_2') = 1 \times (873 - 485.81) = 387.19$ kJ/kg

Overall efficiency, $\eta_{overall} = \frac{w_{net}}{q_s} = \frac{75.72}{387.19} = 0.1950$ or 19.56%

Let $\dot{m}_a$ = mass of air circulated, kg/s

$$\dot{m}_a \times w_{net} = P$$

$$\dot{m}_a = \frac{2200}{75.72} = 29.05 \text{ kg/s}$$

■ **Example 12.33** The pressure ratio of an open cycle constant pressure gas turbine plant is 6. The temperature range of the plant is 20°C and 850°C. Using the following data; $c_{pa} = 1$ kJ/kg.K, $c_{pg} = 1.05$ kJ/kg.K, $\gamma_a = \gamma_g = 1.4$, C.V. of fuel = 44,000 kJ/kg, $\eta_c = 0.85$, $\eta_t = 0.90$, $\eta_{comb} = 0.95$.

Find : (a) Thermal efficiency of the plant, (b) Net power developed if circulation of air is 5 kg/s, (c) Air-fuel ratio, (d) Specific fuel consumption. [MDU, Dec. 2008]

Solution : Given : $r_p = 6$, $T_1 = 273 + 20 = 293$ K, $T_3 = 273 + 850 = 1223$ K, $c_{pa} = 1$ kJ/kg.K

$$c_{pg} = 1.05 \text{ kJ/kg.K}, \gamma_a = \gamma_g = 1.4, CV = 44{,}000 \text{ kJ/kg},$$

$$\eta_c = 0.85, \eta_t = 0.90, \eta_{comb} = 0.95, \dot{m}_a = 5 \text{ kg/s}$$

Refer to fig. 12.55

Isentropic process 1-2 :

$$T_2 = T_1(r_p)^{(\gamma_a - 1)/\gamma_a} = 293(6)^{0.4/1.4} = 488.87 \text{ K}$$

$$\eta_c = \frac{T_2 - T_1}{T_2' - T_1}$$

$$0.85 = \frac{488.87 - 293}{T_2' - 293}, T_2' = 523.43 \text{ K}$$

Combustion process 2′-3 :

$$\dot{m}_f \times CV \times \eta_{comb} = (\dot{m}_a \times c_{pa} + \dot{m}_f \times c_{pg})(T_3 - T_2')$$

or

$$CV \times \eta_{comb} = \left(\frac{\dot{m}_a}{\dot{m}_f} \times c_{pa} + c_{pg}\right)(T_3 - T_2')$$

$$44{,}000 \times 0.95 = [(A/F) \times 1 + 1.05](1123 - 523.43)$$

$$\frac{A}{F} = 68.67$$

Isentropic process 3-4 :

$$T_4 = \frac{T_3}{r_p^{(\gamma_g - 1)/\gamma_g}} = \frac{1123}{(6)^{0.4/1.4}} = 673 \text{ K}$$

$$\eta_t = \frac{T_3 - T_4'}{T_3 - T_4}$$

$$0.90 = \frac{1123 - T_4'}{1123 - 673}, T_4' = 718 \text{ K}$$

Compressor work, $w_c = c_{pa}(T_2' - T_1) = 1 \times (523.43 - 293) = 230.43$ kJ/kg

Turbine work, $\quad w_t = 1 \times c_{pa}(T_3 - T_4') + \frac{1}{(A/F} \times c_{pg}(T_3 - T_4')$

$$= 1 \times 1.0(1123 - 718) + \frac{1}{68.67} \times 1.05(1123 - 718)$$

$$= 411.19 \text{ kJ/kg}$$

Net work per kg of air, $w_{net} = w_t - w_c = 411.19 - 230.43 = 180.76$ kJ/kg

Heat supplied, $q_s = \dot{m}_f \times CV \times \eta_{comb} = \frac{1}{68.67} \times 44{,}000 \times 0.95 = 608.7$ kJ/kg

(a) Thermal efficiency of plant,

$$\eta_{th} = \frac{w_{net}}{q_s} = \frac{180.76}{608.7} = 0.2969 \text{ or } 29.69\%$$

(b) Net power developed,

$$P = \dot{m}_a \times w_{net} = 5 \times 180.76 = 903.8 \text{ kW}$$

(c) Specific fuel consumption,

$$\text{SFC} = \frac{\dot{m}_f \times 360}{P} = \frac{\dot{m}_a}{(A/F)} \times \frac{3600}{P}$$

$$= \frac{5}{68.69} \times \frac{3600}{903.8} = 0.29 \text{ kg/kW.h}$$

■ **Example 12.34** A closed cycle gas turbine consists of a 2 stage compressor and a 2 stage turbine. All the components are mounted on the same shaft. The pressure and temperature at the inlet of the first stage compressor are 2 bar and 250°C, the maximum cycle temperature and pressure are limited to 900°C and 8 bar. A perfect intercooler is used between the two compressors and a reheater is used between the two turbines. Gases are heated in the reheater to 900°C before entering into the L.P. turbine.

Assuming the compressor and turbine efficiencies as 0.85, find (a) the cycle efficiency (i) without regenerator (ii) with regenerator whose effectiveness is 0.65 (b) If intermediate pressure may be taken as 4 bar and I.P. developed by the plant is 310 kW, find the mass of fluid circulated. Air is used in the cycle as working fluid, $\gamma = 1.4$ and $c_p = 1$ kJ/kg.K. [MDU, May, 2009]

Solution : Given : $p_1 = 2$ bar, $T_1 = 273 + 250 = 523$ K,

$$p_3 = 8 \text{ bar}, T_6 = 273 + 900 = 1173 \text{ K},$$

$$T_3 = T_1, T_8 = T_6, \eta_c = \eta_t = 0.85, \varepsilon = 0.65, p_2 = 4 \text{ bar},$$

$$IP = 310 \text{ kW}, \gamma = 1.4, c_p = 1 \text{ kJ/kg.K}$$

The T-s diagram for the cycle of plant is shown in fig. 12.56.

$$p_2 = p_3 = p_7 = p_8 = 4 \text{ bar}$$

$$p_4 = p_6 = p_3 = 8 \text{ bar}$$

$$p_1 = p_9 = 2 \text{ bar}$$

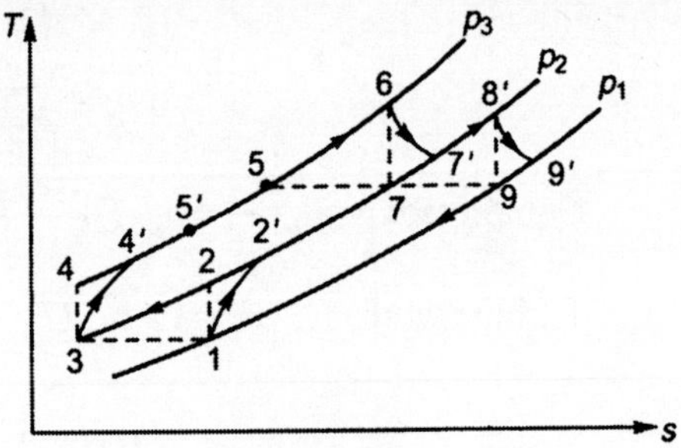

Fig. 12.56

LP compressor :

$$T_2 = T_1\left(\frac{p_2}{p_1}\right)^{(\gamma-1)/\gamma} = 523\left(\frac{4}{2}\right)^{0.4/1.4} = 637.5 \text{ K}$$

$$\eta_c = \frac{T_2 - T_1}{T_2' - T_1}$$

$$0.85 = \frac{637.5 - 523}{T_2' - 523}$$

$$T_2' = 657.7 \text{ K}$$

H.P. compressor :

$$T_4 = T_3\left(\frac{p_4}{p_3}\right)^{(\gamma-1)/\gamma} = 523\left(\frac{8}{4}\right)^{0.4/1.4} = 637.5 \text{ K}$$

$$\eta_c = \frac{T_4 - T_3}{T_4' - T_3}$$

$$0.85 = \frac{637.5 - 523}{T_4' - 523}$$

$$T_4' = 657.5 \text{ K}$$

H.P. turbine :

$$\frac{T_6}{T_7} = \left(\frac{p_6}{p_7}\right)^{(\gamma-1)/\gamma}\left(\frac{p_3}{p_2}\right)^{(\gamma-1)/\gamma} = \left(\frac{8}{4}\right)^{0.4/1.4} = 1.219$$

$$T_7 = \frac{1173}{1.219} = 962.2 \text{ K}$$

$$\eta_t = \frac{T_6 - T_7'}{T_6 - T_7}$$

$$0.85 = \frac{1173 - T_7'}{1173 - 962.2}$$

$$T_7' = 993.8 \text{ K}$$

L.P. turbine :

$$\frac{T_8}{T_9} = \left(\frac{p_8}{p_9}\right)^{(\gamma-1)/\gamma} = \left(\frac{p_2}{p_1}\right)^{(\gamma-1)/\gamma} = \left(\frac{4}{2}\right)^{0.4/1.4} = 1.219$$

$$T_9 = \frac{1173}{1.219} = 962.2 \text{ K}$$

$$\eta_t = \frac{T_8 - T_9'}{T_8 - T_9}$$

$$0.85 = \frac{1173 - T_9'}{1173 - 962.2}$$

$$T_9' = 993.8 \text{ K}$$

(a) (i) Cycle efficiency without regenerator :

Total compressor work,

$$w_c = (w_c)_{LP} + (w_c)_{HP} = c_p[(T_2' - T_1) + (T_4' - T_3)]$$

$$= 2c_p(T_2' - T_1) \qquad [\because T_2' = T_4' \text{ and } T_1 = T_3]$$

$$= 2 \times 1 \times (657.7 - 523) = 269.4 \text{ kJ/kg}$$

Similarly total turbine work, $w_t = 2c_p(T_6 - T_7') = 2 \times 1 \times (1173 - 993.8)$

$$= 358.4 \text{ kJ/kg}$$

Net work, $w_{net} = w_t - w_c = 358.4 - 269.4 = 89$ kJ/kg

Heat supplied,

$$q_s = c_{pg}[(T_6 - T_4') + (T_8 - T_7')] = 1 \times [(1173 - 657.5) + (1173 - 993.8] = 694.7 \text{ kJ/kg}$$

$$\text{Cycle efficiency} = \frac{w_{net}}{q_s} = \frac{89}{694.7} = 0.1281 \text{ or } 12.81\%$$

(ii) Cycle efficiency with regenerator

Regenerator effectiveness, $\varepsilon = \dfrac{T_5' - T_4'}{T_9' - T_4'}$

$$0.65 = \frac{T_5' - 657.5}{993.8 - 657.5}$$

$$T_5' = 876 \text{ K}$$

Heat supplied, $q_s = c_{pg}[(T_6 - T_5') + (T_8 - T_7')]$

$$= 9 \times [(1173 - 876) + (1173 - 993.8)]$$

$$= 476.2 \text{ kJ/kg}$$

$$\text{Cycle efficiency} = \frac{w_{net}}{q_s} = \frac{89}{476.2} = 0.1869 \text{ or } 18.69\%$$

(b)

$$\dot{m}_a \times w_{net} = P$$

$$\dot{m}_a = \frac{310}{89} = 3.483 \text{ kg/s}$$

■ **Example 12.35** A closed cycle gas turbine consists of a 2 stage compressor and a 2 stage turbine. All the components are mounted on the same shaft. The pressure and temperature at the inlet of the first stage compressor are 2 bar and 250°C. The maximum cycle temperature and pressure are limited to

850°C and 8 bar. A perfect intercooler is used between the two compressors and a reheater is used between the two turbines. Gases are heated in the reheater to 850°C before enetering into I.P. turbine.

Assuming the compressor and turbine efficiencies as 0.83, find :

(i) The cycle efficiency without regenerator (ii) with regenerator whose effectiveness is 0.65

(iii) If the I.P. developed by the plant is 310 kW, find the mass of fluid circulated. Air is used in the cycle as working fluid. $\gamma = 1.4$ and $c_p = 1$ kJ/kg.K. [MDU, Dec., 2009]

Solution : Refer to example 12.33.

■ **Example 12.36** Determine the efficiency of a gas turbine plant fitted with a heatexchanger of 75% effectiveness. The presure ratio is 4 : 1 and the compression is carried out in two stages of equal pressure ratio with entercooling back to initial temperature of 290 K. The maximum temperature is 925 K. The turbine isentropic efficiency is 88% and each compressor isentropic efficiency is 85%. For air $\gamma = 1.4$, $C_p = 1.005$ kJ/kg.K. [KU, Jan., 2006]

Solution : Given : $\varepsilon = 0.75$, $r_p = 4$, $T_1 = 290\text{ K} = T_3$, $\frac{p_2}{p_1} = \frac{p_3}{p_2} = \sqrt{4} = 2$,

$$T_6 = 925\text{ K}, \eta_t = 0.88, \eta_c = 0.85, \gamma = 1.4, c_p = 1.005\text{ kJ/kg.K.}$$

Refer to fig. 12.57.

L.P. compressor :

$$T_2 = T_1\left(\frac{p_2}{p_1}\right)^{(\gamma-1)/\gamma} = 290(2)^{0.4/1.4} = 353.5\text{ K}$$

$$\eta_c = \frac{T_2 - T_1}{T_2' - T_1}$$

$$0.85 = \frac{353.5 - 290}{T_2' - 290}$$

$$T_2' = 364.7\text{ K}$$

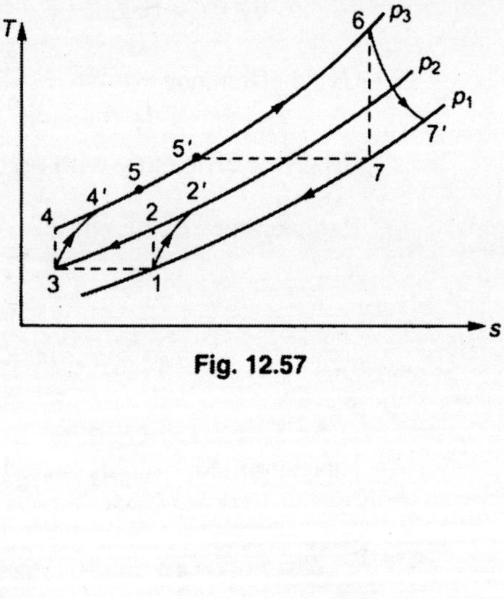

Fig. 12.57

Since $\quad T_3 = T_1$ and $\frac{p_3}{p_2} = \frac{p_2}{p_1} = 2$

$\therefore T_4' = 364.7$ K for the H.P. compressor.

H.P. turbine :

$$\frac{T_6}{T_7} = \left(\frac{p_3}{p_1}\right)^{(\gamma-1)/\gamma} = (8)^{0.4/1.4} = 1.811$$

$$T_7 = \frac{925}{1.811} = 510.6\text{ K}$$

$$\eta_t = \frac{T_6 - T_7'}{T_6 - T_7}$$

$$0.88 = \frac{925 - T_7'}{925 - 510.6}$$

$$T_7' = 560.3\text{ K}$$

Heat exchanger effectiveness, $\varepsilon = \dfrac{T_5' - T_4'}{T_7' - T_4'}$

$$0.75 = \frac{T_5' - 364.7}{5603 - 364.7}$$

$$T_5' = 511.4 \text{ K}$$

Heat supplied, $q_s = c_p(T_6 - T_5') = 1.005(925 - 511.4) = 415.67 \text{ kJ/kg}$

Compressor work, $w_c = 2c_p(T_2' - T_1) = 2 \times 1.005(364.7 - 290) = 150.147 \text{ kJ/kg}$

Turbine work, $w_t = c_p(T_6 - T_7') = 1.005(925 - 5003) = 36652 \text{ kJ/kg}$

Net work, $w_{net} = w_t - w_c = 36652 - 150.147 = 216376 \text{ kJ/kg}$

Efficiency of gas turbine, $\eta = \dfrac{w_{net}}{q_s} = \dfrac{216376}{415.67}$

$$= 05205 \text{ or } 52.05\%$$

■ **Example 12.37**

(a) For a two stage gas turbine, the minimum work required, condition is $p_2 = \sqrt{p_1 p_3}$ with perfect intercooling.

(b) A closed cycle gas turbine consists of 2-stage compressor with perfect intercooling and a two stage turbine, with a reheater. All the components are mounted on the same shaft. The pressure and temperatures at the inlet of the low compressor are 2 bar and 300 K. The maximum pressure and temperature are limited to 8 bar and 1000 K. The gases are reheated in the heater to 1000 K. Calculate the mass of the fluid circulated in the turbine if the net power developed is 370 kW. Also find the amount of heat supplied per sec from the external source. [KU, Dec., 2006]

Solution : (a) Refer to art. 12.12.2.

(b) Given : $T_1 = T_3 = 300$ K, $p_1 = p_2 = 2$ bar,

$$p_3 = 8 \text{ bar}, T_5 = 1000 \text{ K}, T_7, P = 370 \text{ kW}$$

With perfect intercooling, $p_2 = \sqrt{p_1 p_3} = \sqrt{2 \times 8} = 4$ bar

Refer to fig. 12.58

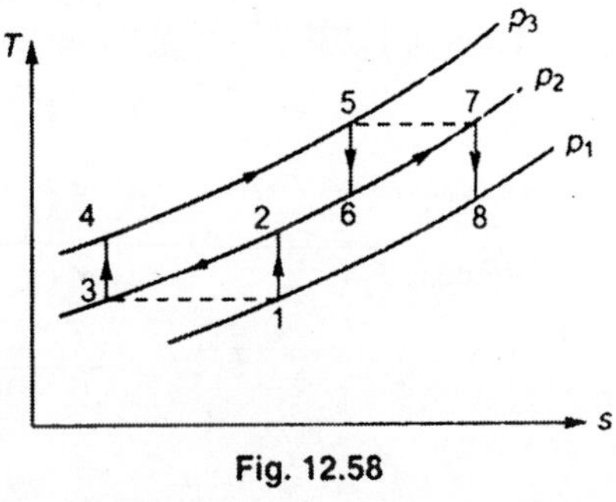

Fig. 12.58

$$r_p = \frac{p_2}{p_1} = \frac{p_3}{p_2} = \frac{4}{2} = 2$$

L.P. compressor :

$$T_2 = T_1 \left(\frac{p_2}{p_1}\right)^{(\gamma - 1)/\gamma} = 300(2)^{0.4/1.4} = 365.7 \text{ K}$$

H.P. compressor :

Since $T_3 = T_1$ and $\dfrac{p_3}{p_2} = \dfrac{p_2}{p_1}$, therefore

$$T_4 = T_2 = 365.7 \text{ K}$$

H.P. turbine :

$$\frac{T_5}{T_6} = \left(\frac{p_3}{p_2}\right)^{(\gamma-1)/\gamma} = (2)^{0.4/1.4} = 1219$$

$$T_6 = \frac{1000}{1219} = 8203 \text{ K}$$

L.P. turbine :

Since $T_7 = T_5$ and $\frac{p_2}{p_1} = \frac{p_3}{p_2} = 2$, therefore,

$$T_8 = T_6 = 8203 \text{ K}$$

Compressor work, $w_c = 2c_p(T_2 - T_1) = 2 \times 1.005(365.7 - 300) = 132.057$ kJ/kg

Turbine work, $w_t = 2c_p(T_5 - T_6) = 2 \times 1.005(100 - 8203) = 361.197$ kJ/kg

Net work, $w_{net} = w_t - w_c = 361.197 - 132.057 = 229.14$ kJ/kg

Let $\dot{m}_a$ = mass of fluid circulated, kg/s

$$\dot{m}_a \times w_{net} = P$$

$$\dot{m}_a = \frac{370}{229.14} = 1.614 \text{ kg/s}$$

■ **Example 12.38**

(a) Explain with neat sketch and T-s diagram of the closed and open gas turbine cycle.

(b) Air enters the compressor of a gas turbine plant operting on Brayton cycle at 1 bar and 27°C. The pressure ratio in the cycle is 6. Calculate the maximum temperature in the cycle and cycle efficiency. Assume the turbine work 2.5 times the compressor work. Take $\gamma = 1.4$

[KU, Dec., 2007]

Solution : (a) Refer to art. 12.6.

(b) Given : $p_1 = 1$ bar, $T_1 = 273 + 27 = 300$ K, $r_p = 6$, $T_3 = ?$

$$w_t = 25\, w_c$$

Refer to Fig. 12.4(b)

Process 1-2 :

$$T_2 = T_1(r_p)^{(\gamma-1)/\gamma} = 300\,(6)^{0.4/1.4} = 50055 \text{ K}$$

$$w_c = c_p(T_2 - T_1) = 1.005(50055 - 300) = 20155 \text{ kJ/kg}$$

Process 3-4 :

$$T_4 = \frac{T_3}{(r_p)^{(\gamma-1)/\gamma}} = \frac{T_3}{(6)^{0.4/1.4}} = \frac{T_3}{1.6685} \text{ K}$$

$$w_t = c_p(T_3 - T_4) = 1.005\left[T_3 - \frac{T_3}{1.6685}\right] = 0.40266 T_3 \text{ kJ/kg}$$

Now $\quad 0.40266 T_3 = 25 \times 20155$

Maximum temperature :

$$T_3 = 125136 \text{ K}$$

$$w_t = 2.5 \times 20155 = 503.87 \text{ kJ/kg}$$

Net work, $w_{net} = w_t - w_c = 503.87 - 20155 = 30232 \text{ kJ/kg}$

Heat supplied, $q_s = c_p(T_3 - T_2) = 1.005(125136 - 500.55)$

$$= 754564 \text{ kJ/kg}$$

Cycle efficiency, $\eta = \dfrac{w_{net}}{q_s} = \dfrac{30232}{754564} = 0.40 \text{ or } 40\%$

■ **Example 12.39**

(a) With the help of neat sketch and T-s diagram explain the working of closed gas turbine cycle.

(b) Determine the efficiency of a gas turbine plant fitted with heat exchanger of 75% effectiveness. The pressure ratio is 4 : 1 and compression is carried out two stages of equal pressure ratio with interlocking back to initial temperature of 290 K. The maximum temperature is 925 K. The turbine isentropic efficiency is 88% and each compressor isentropic efficiency 85%. For air $\gamma = 1.4$, $c_p = 1.005$ kJ/kg K. [KU, Jan. 2008]

Solution : (a) Refer to art. 12.9

(b) Refer to example 12.35.

■ **Example 12.40** A gas turbine cycle takes in air at 25°C and atmospheric pressure. The compression pressure ratio is 4. The compressor efficiency is 75%. The inlet temperature to turbine is limited to 750°C. What turbine efficiency would give overall cycle efficiency zero percent? [KU, Dec., 2008]

Solution : Given : $T_1 = 273 + 25 = 298$ K, $p_1 = 1.013$ bar,

$$r_p = 4, \eta_c = 0.75, T_3 = 273 + 750 = 1023 \text{ K}, \eta_t = ?$$

Process 1-2 :

$$T_2 = T_1 (r_p)^{(\gamma - 1)/\gamma} = 298(4)^{0.4/1.4} = 442.8 \text{ K}$$

$$\eta_c = \frac{T_2 - T_1}{T_2' - T_1}$$

$$0.75 = \frac{442.8 - 298}{T_2' - 298}$$

$$T_2' = 491 \text{ K}$$

Process 3-4 :

$$\frac{T_4}{T_3} = \frac{1}{(r_p)^{(\gamma - 1)/\gamma}} = \frac{1}{(4)^{0.4/1.4}} = 0.673$$

$$T_4 = 1023 \times 0.673 = 688.4 \text{ K}$$

$$\eta_t = \frac{T_3 - T_4'}{T_3 - T_4} = \frac{1023 - T_4'}{1023 - 688.4} = \frac{1023 - T_4'}{334.6}$$

$$T_4' = 1023 - 334.6\ \eta_t$$

Compressor work, $w_c = c_p(T_2' - T_1) = 1.005(491 - 298) = 193.96$ kJ/kg

Turbine work, $w_t = c_p(T_3 - T_4') = 1.005(1023 - 1023 + 334.6\eta_t)$

$$= 336.27\ \eta_t$$

Net work, $w_{net} = w_t - w_c = 336.27\eta_t - 193.96$

Heat supplied, $q_s = c_p(T_3 - T_2') = 1.005(1023 - 491) = 534.66$ kJ/kg

Cycle efficiency, $\eta = \dfrac{w_{net}}{q_s} = 0$

or $336.27\eta_t - 193.96 = 0$

$$\eta_t = 0.5768 \text{ or } 54.68\%$$

■ **Example 12.41** In a simple gas turbine plant air enters the compressor at 1 bar and 27°C and leaves at 6 bar. It is then heated in the combustion chamber to 700°C and then enters the turbine to expand to 1 bar. The isentropic efficiency of compressor is 0.80 and turbine 0.85 and the combustion efficiency is 0.98. The fall in pressure through combustion chamber is 0.1 bar. Find (a) thermal efficiency, (b) work ratio, (c) air rate in kg/kW, (d) specific fuel consumption, and (e) air-fuel ratio.

[KU, Dec., 2010]

Solution : Given : $p_1 = 1$ bar, $T_1 = 273 + 27 = 300$ K, $p_2 = 6$ bar, $T_3 = 273 + 700 = 973$ K,

$\eta_c = 0.80$, $\eta_t = 0.85$, $\eta_{comb} = 0.98$, $\Delta p = 0.1$ bar , $p_3 = p_2 - \Delta p = 5.9$ bar

Refer to fig. 12.59.

Compressor process 1-2 :

$$T_2 = T_1\left(\frac{p_2}{p_1}\right)^{(\gamma-1)/\gamma} = 300\left(\frac{6}{1}\right)^{0.4/1.4} = 500.55 \text{ K}$$

$$\eta_c = \frac{T_2 - T_1}{T_2' - T_1}$$

$$0.80 = \frac{500.55 - 300}{T_2' - 300}$$

$$T_2' = 550.69 \text{ K}$$

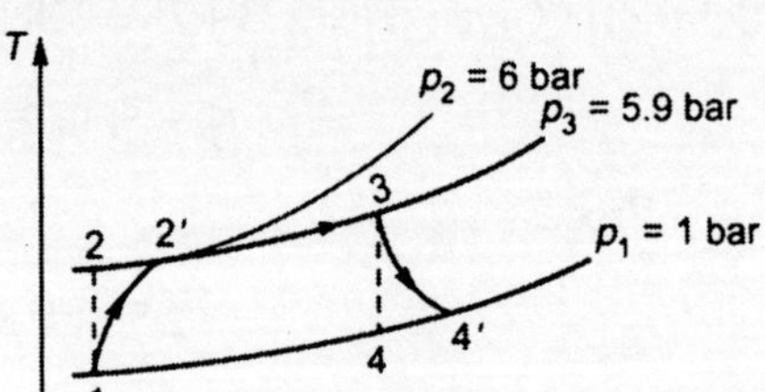

Fig. 12.59

Turbine process 3-4 :

$$\frac{T_3}{T_4} = \left(\frac{p_3}{p_4}\right)^{(\gamma-1)/\gamma} = \left(\frac{5.9}{1}\right)^{0.4/1.4} = 1.6605$$

$$T_4 = \frac{973}{1.6605} = 585.96 \text{ K}$$

$$\eta_t = \frac{T_3 - T_4'}{T_3 - T_4}$$

$$0.85 = \frac{973 - T_4'}{973 - 585.96}$$

$$T_4' = 644 \text{ K}$$

Compressor work, $w_c = c_p\ (T_2' - T_1) = 1.005\ (550.69 - 300) = 251.94$ kJ/kg

Turbine work, $w_t = c_p\ (T_3 - T_{4'}) = 1.005\ (973 - 644) = 330.64$ kJ/kg

Network, $w_{net} = w_t - w_c = 330.64 - 251.94 = 78.70$ kJ/kg

Effective heat supplied, $q_s = c_{pg}\ (T_3 - T_2')$

Assume $c_{pg} = c_{pa} = 1.005$ kJ/kg.K

$$q_s = 1.005\ (973 - 550.69) = 424.42 \text{ kJ/kg}$$

Actual heat supplied, $q_a = \dfrac{q_s}{\eta_{comb}} = \dfrac{424.42}{0.98} = 433.08$ kJ/kg

Let CV of fuel = 42,000 kJ/kg

$$\text{A : F ratio} = \frac{CV}{q_a} = \frac{42{,}000}{433.08} = 97:1$$

(a) Thermal efficiency, $\eta_{th} = \dfrac{w_{net}}{q_a} = \dfrac{78.70}{433.08} = 0.1817$ or 18.17%

(b) Work ratio $= \dfrac{w_{net}}{w_t} = \dfrac{78.70}{330.64} = 0.238$

(c) Let

$$P = 250 \text{ kW}$$

$$\text{Air rate} = \dot{m}_a$$

$$\dot{m}_a \times w_{net} = P$$

$$\dot{m}_a = \frac{250}{78.7} = 3.17 \text{ kg/s}$$

(d) Specific fuel consumption, SFC $= \dfrac{\left(\dfrac{360}{w_{net}}\right)}{(A/F)} = \dfrac{\left(\dfrac{3600}{78.70}\right)}{97} = 0.47$ kg/kWh

(e) A : F = 97.

■ **Example 12.42** In a gas turbine plant, operating on Joule cycle, maximum and minimum temperatures are 825°C and 27°C respectively. The pressure ratio is 4.5. Calculate the specific work output, cycle efficiency and work ratio. Assuming isentropic efficiency of compressor as 85% and turbine is 90%. Find out the heat rate in kJ/kW-hour. If the rating of the turbine is 1300 kW, what is the mass flow in kg/sec? Neglect the mass flow of fuel. Given $c_p = 1.005$ kJ/kg.K. [KU, Dec., 2009]

Solution : Given : $T_3 = 273 + 825 = 1098$ K, $T_1 = 273 + 27 = 300$ K, $r_p = 4.5$, $\eta_c = 0.85$, $\eta_t = 0.90$, $P = 1300$ kW, $c_p = 1.005$ kJ/kg.K.

Refer to Fig. 12.60.

Process 1-2 :

$$T_2 = T_1\ (r_p)^{(\gamma - 1)/\gamma} = 300\ (4.5)^{0.4/1.4} = 461 \text{ K}$$

$$\eta_c = \frac{T_2 - T_1}{T_2' - T_1}$$

$$0.85 = \frac{461 - 300}{T_2' - 300}$$

$$T_2' = 489.4 \text{ K}$$

$$w_c = c_p (T_2' - T_1) = 1.005 (489.4 - 300)$$
$$= 190.347 \text{ kJ/kg}$$

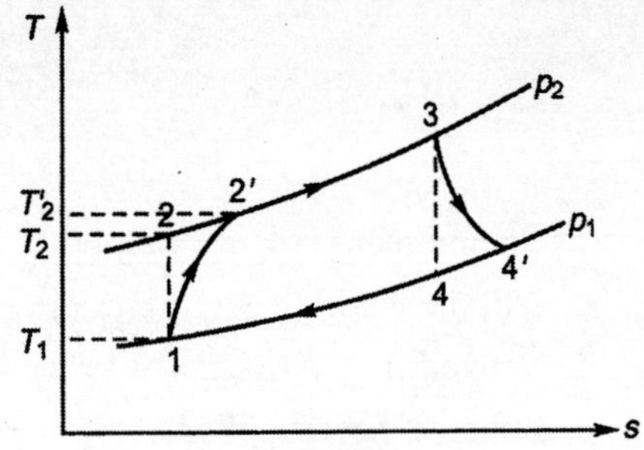

Fig. 12.60

Heat supplied, $q_s = q (T_3 - T_2') = 1.005 (1098 - 489.4)$
$$= 602.598 \text{ kJ/kg}$$

Process 3-4 :

$$\frac{T_3}{T_4} = (r_p)^{(\gamma - 1)/\gamma} = (4.5)^{0.4/1.4} = 1.5368$$

$$T_4 = \frac{1098}{1.5368} = 714.45 \text{ K}$$

$$\eta_t = \frac{T_3 - T_4'}{T_3 - T_4}$$

$$0.90 = \frac{1098 - T_4'}{1098 - 714.45}$$

$$T_4' = 752.8 \text{ K}$$

$$w_t = c_p (T_3 - T_4') = 1.005 (1098 - 752.8) = 346.926 \text{ kJ/kg}$$

Network, $w_{net} = w_t - w_c = 346.926 - 190.347 = 156.579 \text{ kJ/kg}$

(i) Heat rate $= \dfrac{3600 \times q_s}{P} = \dfrac{3600 \times 602.598}{1300} = 1668.7 \text{ kJ/kW.h}$

(ii) $\dot{m}_a \times w_{net} = P$

$$m_a = \frac{1300}{156.579} = 8.3 \text{ kg/s}$$

(iii) Specific work output $= w_{net} = 156.579 \text{ kJ/kg}$

(iv) Cycle efficiency, $\eta_{th} = \dfrac{w_{net}}{q_s} = \dfrac{156.579}{602.598} = 0.2598$ or 25.98%

(v) Work ratio $= \dfrac{w_{net}}{w_t} = \dfrac{156.579}{346.926} = 0.451$

■ **Example 12.43** A gas turbine plant with a pressure ratio of 1:5 takes in air at 15°C. The maximum temperature in 600°C and develops 2200 kW. The turbine and compressor efficiencies are 0.85. Taking $c_p = 1$ kJ/kg.K and $c_v = 0.714$ kJ/kg.K, calculate (a) actual overall efficiency of the turbine (b) mass of air circulated by the turbine. [KU, Sept., 2009]

Solution : Given : $r_p = 5, T_1 = 273 + 15 = 288 \text{ K}, T_3 = 273 + 600 = 873 \text{ K}, P = 2200 \text{ kW}, \eta_c = \eta_t = 0.85,$

$c_p = 1 \text{ kJ/kg.K}, c_v = 0.714 \text{ kJ/kg.K}.$

Refer to Fig. 12.60, $\gamma = \frac{c_p}{c_v} = \frac{1}{0.714} = 1.4$

$$T_2 = T_1 (r_p)^{(\gamma - 1)/\gamma} = 288 (5)^{0.4/1.4} = 456.14 \text{ K}$$

$$\eta_c = \frac{T_2 - T_1}{T_2' - T_1}$$

$$0.85 = \frac{456.14 - 288}{T_2' - 288}, \quad T_2' = 485.81 \text{ K}$$

$$w_c = c_p (T_2' - T_1) = 1 (485.89 - 288) = 197.81 \text{ kJ/kg}$$

$$q_s = c_p (T_3 - T_2') = 1 (873 - 485.81) = 387.19 \text{ kJ/kg}$$

$$T_4 = \frac{T_3}{(r_p)^{(\gamma - 1)/\gamma}} = \frac{873}{(5)^{0.4/1.4}} = 551.2 \text{ K}$$

$$\eta_t = \frac{T_3 - T_4'}{T_3 - T_4}$$

$$0.85 = \frac{873 - T_4'}{873 - 551.2}, \quad T_4' = 599.47 \text{ K}$$

$$w_t = c_p (T_3 - T_4') = 1 (873 - 599.47) = 273.53 \text{ kJ/kg}$$

$$w_{net} = w_t - w_c = 273.53 - 197.81 = 75.72 \text{ kJ/kg}$$

(a) Actual overall efficiency $= \frac{w_{net}}{q_s} = \frac{75.72}{387.19} = 0.1955$ or 19.55%

(b) $\dot{m}_a \times w_{net} = P$

$$\dot{m}_a = \frac{220}{75.72} = 29.05 \text{ kg/s}$$

■ **Example 12.44** The pressure-ratio of an open-cycle constant pressure gas turbine plant is 6. The temperature range of the plant is 15°C and 800°C.

Using the following data :

$c_{pa} = 1$ kJ/kg.K, $c_{pg} = 1.075$ kJ/kg.K and $\gamma = 1.4$ for air and gases.

C.V. of fuel = 43,000 kJ/kg

$\eta_c = 0.85$

$\eta_t = 0.90$, $\eta_{combustion} = 0.95$

Find :

(i) Thermal efficiency of the plant.

(ii) I.P. of the plant, if the circulation of air is 5 kg/s.

(iii) A : F ratio.

(iv) Specific fuel-consumption.

Neglect the losses inthe system. [MDU, Dec., 2010]

Solution : Given : $r_p = 6$, $T_1 = 273 + 15 = 288$ K, $T_3 = 273 + 800 = 1073$ K, $c_{pa} = 1$ kJ/kg.K, $c_{pg} = 1.075$ kJ/kg.K, $\gamma_a = \gamma_g = 1.4$, CV = 43,000 kJ/kg, $\eta_c = 0.85$, $\eta_t = 0.90$, $\eta_{comb} = 0.95$, $\dot{m}_a = 5$ kg/s.

Isentropic compression process 1-2 :

$$T_2 = T_1\,(r_p)^{(\gamma-1)/\gamma} = 288\,(6)^{0.4/1.4} = 480.5\text{ K}$$

$$\eta_c = \frac{T_2 - T_1}{T_2' - T_1}$$

$$0.85 = \frac{480.5 - 288}{T_2' - 288}, \quad T_2' = 514.5\text{ K}$$

$$w_c = c_{pa}\,(T_2' - T_1) = 1 \times (514.5 - 288) = 266.5\text{ kJ/kg}$$

Isentropic expansion process in turbine 3-4 :

$$T_4 = \frac{T_3}{(r_p)^{(\gamma-1)/\gamma}} = \frac{1073}{(6)^{1/3.5}} = 643.1\text{ K}$$

$$\eta_t = \frac{T_3 - T_4'}{T_3 - T_4}$$

$$0.90 = \frac{1073 - T_4'}{1073 - 643.1}$$

$$T_4' = 686.1\text{ K}$$

$$w_t = c_{pg}\,(T_2 - T_4') = 1.075\,(1073 - 686.1) = 415.93\text{ kJ/kg}$$

Combustion process 2′-3 :

$$\dot{m}_f \times CV \times \eta_{\text{comb}} = \dot{m}_a \times c_{pa}\,(T_3 - T_2') + \dot{m}_f \times c_{pg}\,(T_3 - T_2')$$

$$CV = \frac{\dot{m}_a}{\dot{m}_f} \times c_{pa}\,(T_3 - T_2') + c_{pg}\,(T_3 - T_2')$$

$$43{,}000 \times 0.95 = (\text{A}:\text{F}) \times 1.0\,(1073 - 514.5) + 1.075\,(1073 - 514.5)$$

$$40850 = (\text{A}:\text{F}) \times 558.5 + 600.4$$

$$\text{A}:\text{F} = 72:1$$

(i) Heat supplied, $q_s = \dfrac{\dot{m}_f}{\dot{m}_a} \times CV \times \eta_{\text{comb}} = \dfrac{1}{72} \times 43{,}000 \times 0.95 = 567.36\text{ kJ/kg}$

Network, $w_{\text{net}} = w_t - w_c = 415.93 - 226.5 = 189.43\text{ kJ/kg}$

Thermal efficiency of plant, $\eta_{\text{th}} = \dfrac{\eta_{\text{net}}}{q_s} = \dfrac{189.43}{567.36} = 0.3339$ or 33.39%

(ii) $\text{IP} = \dot{m}_a = w_{\text{net}} = 5 \times 189.43 = 947.15\text{ kW}$

(iii) $\text{SFC} = \dfrac{3600 \times \dot{m}_f}{\text{IP}} = \dfrac{3600 \times \dfrac{5}{72}}{947.15} = 0.264\text{ kg/kW.h.}$

SHORT ANSWER TYPE QUESTIONS

1. List the fields of application of gas turbines.
2. What are the limitations of a gas turbine?
3. List four advantages of gas turbine over I.C. engine.
4. List four disadvantages of gas turbine over I.C. engine.
5. Enumerate five advantages of gas turbine over steam turbine.
6. How do you classify gas turbines?
7. Made a compression of open and closed cycles gas turbines.
8. Explain the position of gas turbine in the power industry.
9. Write the formula for the thermal efficiency of the Brayton cycle.
10. Define specific output and work ratio.
11. Write the expressions for maximum work out and work ratio of the Brayton cycle.
12. Draw the schematic arrangement of open cycle constant pressure combustion gas turbine.
13. What do you understand by regeneration?
14. Define effectiveness of a regenerator.
15. What is the intermediate pressure for intercooling to have minimum compressor work?

EXERCISES

1. A Brayton cycle operates with ideal air between 1 bar, 300 K and 5 bar, 1000 K. The air is compressed in two stages with perfect intercooling. Similarly in the turbine, expansion occurs in two stages with perfect reheating. Calculate the optimum pressure, net work output and the fraction of turbine output that has to be put back to the compressor. **[Ans.** 2.236 bar, 257.3 kJ/kg, 0.378]
2. A gas turbine unit operates at a mass flow of 30 kg/s. Air enters the compressor at a pressure of 1 bar, 288 K and is discharged from the compressor at a pressure of 10.5 bar. Combustion occurs at constant pressure and resuslts in a temperature rise of 420 K. If the flow leaves the turbine at a pressure of 1.2 bar, determine the net power output from the unit and also the thermal efficiency. **[Ans.** 5388.74 kW, 42.55%]
3. In a regenerative gas turbine cycle, air enters the compressor at a temperature and pressure of 30°C and 1.5 bar and discharges at 220°C and 5.2 bar. After passing through the regenerator the air temperature is 395°C. The temperature of air entering and leaving the gas turbine are 900°C and 510°C respectively. Assuming no pressure drop through the regenerator, determine (a) the output per kg of air, (b) the efficiency of cycle, and (c) the work required to drive the compressor. Take c_p = 1.005 kJ/kg.K. **[Ans.** 201 kJ/kg, 39.6%, 190.95 kJ/kg]
4. A Brayton cycle works between 1 atm, 27°C and 5 atm, 977°C. There are two stages of compression with perfect intercooling and two stages of expansion. The work output of first expansion stage being used to drive the two compressors, where the interstage pressure is optimised for the compressor. The air from the first stage turbine is again heated to 977°C and expanded. Calculate the power output of free power turbine and cycle efficiency without and with a perfect heat exchanger and compate them. Also calculate the percentage improvement in the efficiency because of the addition of heat exchangers. **[Ans.** 350.8 kJ/kg, 33.96%, 69.2%, 50.9%]

5. The following data apply to a gas turbine set employing a separate power turbine, regenerator, and intercooler between two-stage compression :

Isentropic efficiency of compression in each stage	= 80%
Isentropic efficiency of compressor turbine	= 88%
Isentropic efficiency of power turbine	= 88%
Turbine to compressor transmission efficiency	= 98%
Pressure ratio in each stage of compression	= 3:1
Temperature after intercooler	= 297 K
Air mass flow rate	= 15 kg/s
Regenerator effectiveness	= 80%
Regenerator gas-side pressure loss	= 0.1 bar
Maximum turbine temperature	= 1000 K
Ambient temperature	= 327 K
Ambient pressure	= 1 bar
Calorific value of fuel	= 43.1 MJ/kg

Calculate (a) the net power output, (b) specific fuel consumption, and (c) over all thermal efficiency. Assume that the pressure losses in the air-side of the regenerator and combustion chamber are accounted for in the compressor efficiency. Take c_{pa} = 1.005 kJ/kg.K, c_{pg} = 1.147 kJ/kg.K, γ_a = 1.4 and γ_g = 1.33.

[Ans. 1811.67 kW, 0.32 kg/kWh, 26.1%**]**

6. In a gas turbine plant, air is compressed from 1 bar, 15°C through a pressure ratio of 4 : 1. It is then heated to 650°C in a combustion chamber and expanded back to atmospheric pressure of 1 bar in a turbine. Calculate the cycle efficiency and the work ratio if a perfect heat exchanger is used. The isentropic efficiencies of turbine and compressor are 85% and 80% respectively. Take c_{p_a} = 1.00 kJ/kg.K. **[Ans.** 31.6%, 0.316**]**

7. In a gas turbine plant, the pressure ratio through which air at 15°C is compressed is 6. The same air is then heated to a maximum permissible temperature of 750°C, first in a heat exchanger which is 75% effecient, and then in the combustion chamber. The same air at 750°C is expanded in two stages such that the expansion work is maximum. The air is reheated to 750°C after the first stage. Determine (a) the cycle thermal efficiency, (b) the work ratio, and (c) the net shaft work per kg of air. The machine efficiencies may be assumed to be 80% and 85% for the compressor and turbine respectively. Take c_{p_a} = 1.00 kJ/kg.K.

[Ans. 33.6%, 0.398, 160 kJ/kg**]**

8. In a gas turbine plant, the air at 283 K and 1 bar is compressed to 4 bar with compression efficiency of 80%. The air is heated in the regenerator and in the combustion chamber till its temperature is raised to 973 K, and during the process the pressure falls by 0.1 bar. The air is then expanded in the turbine and passes to the regenerator which has 75% effectiveness and causes a pressure drop of 0.14 bar. If the isentropic efficiency of the turbine is 85%, determine the thermal efficiency of the plant. **[Ans.** 23.1%**]**

9. A gas turbine works on Brayton cycle between 27°C and 827°C. Determine the maximum net work per kg and the cycle efficiency. Take c_{p_a} = 1.005 kJ/kg.K. **[Ans.** 252.34 kJ/kg, 47.77%**]**

10. Atmospheric air enters the compressor of an open cycle constant pressure gas turbine at a pressure of 1 bar, 293 K. The pressure of air after compression is 4 bar. The isentropic efficiencies of compressor and turbine are 80% and 85% respectively. The air-fuel ratio used is 90 : 1. The mass flow rate of air is 3 kg/s, calculate (a) the power developed, and (b) thermal efficiency of the cycle.

Assume $c_{pa} = c_{pg}$ = 1.0 kJ/kg.K and $\gamma_a = \gamma_g$ = 1.4. Calorific value of fuel = 41800 kJ/kg.

[Ans. 252.84 kW, 18.14%**]**

11. In a constant pressure open cycle gas turbine, air enters at 1 bar, 20°C and leaves the compressor at 5 bar. Temperature of gases entering the turbine is 680°C, pressure loss in the combustion chamber is 0.1 bar, $\eta_c = 0.85, \eta_t = 0.80, \eta_{combustion} = 0.85, \gamma_a = 1.4$ and $c_{pa} = c_{pg} = 1.024$ kJ/kg.K. Calculate (a) the quantity of air circulated if power developed is 1065 kW, (b) the heat supplied per kg of air circulated, and (c) the thermal efficiency of the cycle. Neglect the mass of fuel as compared to that of air.

[**Ans.** 13.43 kg, 552.9 kJ/kg, 14.34%]

12. An open cycle constant pressure gas turbine draws air at 1.01 bar, 15°C and pressure ratio of 7:1. The compressor is driven by the H.P. turbine and L.P. turbine drives a separate power shaft. The isentropic effciencies of compressor, H.P. and L.P. turbines are 0.82, 0.85 and 0.85 respectively. If the maximum cycle temperature is 610°C. Calculate (a) the pressure and temperature of gases entering the power turbine, (b) the net power developed, (c) the work ratio, and (d) the thermal efficiency of the plant.
Assume $c_{pa} = 1.005$ kJ/kg.K, $\gamma_a = 1.4$; $c_{pg} = 1.15$ kJ/kg.K, $\gamma_g = 1.333$. Neglect mass of fuel.

[**Ans.** 1.636 bar, 654.4 K; 72.1 kW; 0.25; 18.8%]

13. A gas turbine employs a heat exchanger with a thermal ratio of 0:72. The turbine operates between 1.01 bar and 4.04 bar and ambient temperature is 20°C. $\eta_a = 0.8$ and $\eta_t = 0.85$. The pressure drop on each side of the heat exchanger is 0.05 bar and in the combustion chamber 0.14 bar. Assume $\eta_{combustion} = 1.0$ and C.V. of fuel = 41800 kJ/kg.

Calculate the increase in thermal efficiency due to heat exchanger.

14. A gas turbine has a pressure ratio of 6:1 and a maximum cycle temperature of 600°C. The isentropic efficiencies of the compressor and turbine are 0.82 and 0.85 respectively. Calculate the power output when the air enters the compressor at 15°C at the rate of 15 kg/s. $c_{pa} = 1.005$ kJ/kg.K, $\gamma_a = 1.4$; $c_{pg} = 1.11$ kJ/kg.K, $\gamma_g = 1.333$.

[**Ans.** 920 kW]

15. A gas turbine plant operates between the temperatures of 1200 K and 320 K and corresponding pressures of 6 bar and 1.2 bar. The ratio of turbine and compressor efficiencies is 0.95. Determine the two efficiencies if the plant operates at optimum pressure ratio for maximum specific output. Assume ratio of specific heats to be 1.4.

[**Ans.** $\eta_t = 79.72\%, \eta_c = 83.90\%$]

16. The maximum and minimum temperatures occurring in a closed cycle gas turbine plant are 927°C and 37°C. The pressure at the outlet and the inlet of the compressor are 5 bar and 1 bar respectively per kg of air flow in the cycle, determine the compressor work, the turbine work, the heat supplied to the cycle and the net work done in the cycle. Also determine the thermal efficiency of the cycle and the optimum pressure ratio for maximum power output for the given temperatures. Assume γ for air to be 1.4 and $c_p = 1.005$ kJ/kg. K.

17. An open cycle constant pressure gas turbine plant operates with a pressure ratio of 6:5. The temperatures at the inlet to the compressor and turbine are 17°C and 820°C respectively. Assume : specific heat at constant pressure of air and gas to be 1.005 kJ/kg.K and 1.079 kJ/kg.K respectively; ratio of specific heats for air and gas to be 1:4; calorific value of fuel 44350 kJ/kg; efficiency of compressor and turbine to be 0.86 and 0.92 respectively. Calculate : (i) Power of the plant for air circulation of 6 kg/s, (ii) Thermal efficiency of plant, (iii) Air/fuel ratio, and (iv) Specific fuel consumption.
Take the mass of the fuel into account.

18. A gas turbine power plant works between pressures of 1 bar and 5 bar and temperatures of 285 K and 1100 K. The intercooler cools the air at 2.3 bar to 285 K before the air is sent to the second stege compressure. The compressor air from the second-stage compressor passes through a regenerator, whose effectiveness is 0.72 and then through the combustion chamber. The heated air is then expanded in a high pressure turbine to 2.3 bar and is then reheated to 1100 K. The air is finally expanded in the low-pressure turbine to 1 bar. Assuming the compressor and turbine efficiencies to be 85%, determine : (i) the ratio of compression work to the turbine work, (ii) power developed for an air flow of 3 kg/s, (iii) the thermal efficiency of the cycle, (iv) heat rejected per second to the cooling water in the intercooler, and (v) heat rejected per second to the atmosphere. Assume that all the components are mounted on the same shaft. Sketch the flow diagram of the turbine and represent the process on T-s plane. Also assume c_p = 1.005 kJ/kg.K and γ = 1.4.

19. A gas turbine set takes in air at 27°C and 1 atm. The pressure ratio is 4 and the maximum temperature is 560°C. The compressor and turbine efficiencies are 0.83 and 0.85 respectively. Determine the overall efficiency if the regenerator effectiveness is 0.75.

❑❑❑

2210

B.E. Vth Semester (Mech. Engg.)
Examination December, 2009
INTERNAL COMBUSTION ENGINES AND GAS TURBINES
Paper-ME-307-E

Time : Three Hours Maximum Marks : 100

Before answering the question, candidates should ensure that they have been supplied the correct and complete question paper. No complaint in this regard, will be entertained after examination.

Note : Attempt any five questions. Assume missing data suitably.

1. For an engine working on the ideal Dual cycle, the compression ratio is 10 and the max-pressure is limited to 70 bar. If the heat-supplied is 1680 kJ/kg, find the pressures and temperatures at the various salient points of the cycle and the cycle efficiency. The pressure and temp. of air at the commencement of the compression are 1 bar and 100°C resp. Assume, $c_p = 1.004$ kJ/kg.K and $c_v = 0.717$ kJ/kg.K for air. 20

2. (a) State the essential requirements of an ignition system. 5

   (b) Derive an expression for the calculation of air-fuel ratio for the carburettor. 15

3. (a) Briefly explain the stages of combustion in SI engine, elaborating the flame-front propagation. 10

   (b) Explain the phenomenon of knock in CI-engine and compare it with SI engine knock. 10

4. (a) Explain the various properties of an ideal lubricant. 10

   (b) Explain the following with neat sketches : 10

   (i) Evaporative cooling system.

   (ii) Pressure cooling system.

5. The power output of a 6-cylinder 4-stroke engine is absorbed by a water brake for which the law is WN/20000. Where the brake load, *W* is in Newton and the speed *N*, is in rpm. The air-consumption is measured by an air-box with sharp edged orifice system. The following readings are obtained : 20

Orifice diameter	30 mm
Bore	100 mm
Stroke	120 mm
Brake load	560 N
C/H ratio by mass	83/17
Co-efficient of discharge	0.6
Ambient-pressure	1 bar
Pressure-drop across orifice	14.5 cm of Hg
Time taken for 100 cc of fuel consumption	20 s
Ambient temperature	27°C
Fuel density	831 kg/m^3

Calculate,

Brake power, the torque, the brake specific fuel consumption, the percentage of excess air and volumetric efficiency.

6. (a) Name 3 categories of S.I. emissions. Explain various factors affecting exhaust emission. 10

(b) Explain briefly various alternative fuels for IC engines. 10

7. (a) Explain various types of rotary compressor with diagrams. 12

(b) Define slip factor and derive an expression for same. 8

8. A closed cycle gas turbine consists of a 2-stage compressor and a 2-stage turbine. All the components are mounted on the same shaft. The pressure and temp. at the inlet of the first stage compressor are 2 bar and 25°C. The max. cycle temp. and pressure are limited to 850°C and 8 bar. A perfect inter-cooler is used between the two compressors and a reheater is used between the two turbines. Gases are heated in the reheater to 850°C before entering into the L.P. turbine. 20

Assuming the compressor and turbine efficiencies are 0.83. Finds :

(i) Cycle-efficiency without regenerator.

(ii) With regenerator whose effectiveness is 0.65.

(iii) If the I.P. developed by the plant is 310 kW. Find the mass of the fluid circulated. Air is used as the working fluid.

$$\gamma = 1.4$$

$$c_p = 1 \text{ kJ/kg.K}$$

❑

2210

B.E. Vth Semester (Mech. Engg.)
Examination December, 2010
INTERNAL COMBUSTION ENGINES AND GAS TURBINES
Paper-ME-307-E

Time : Three Hours — Maximum Marks : 100

Note : (i) Attempt any five questions.
(ii) Assume any missing data.

1. In an engine working on Diesel cycle inlet pressure and temperature are 1 bar and 17°C resp. Pressure at the end of adiabatic compression is 35 bar. The ratio of expansion, *i.e.*, after constant pressure heat-addition is 5. Calculate the heat-addition, heat-rejection and efficiency of the cycle.

Assume, $\gamma = 1.4, c_p = 1.004$ kJ/kg.K
$c_v = 0.717$ kJ/kg.K — 20

2. (a) State the essential requirements of a diesel injection system. — 5

(b) What is air-fuel mixture?
Explain, Rich mixture
Stoichiometric mixture
Lean mixture — 15

3. (a) Explain the effect of various engine variables on SI engine knock. — 10

(b) Name various theories of detonation. Explain the pre-ignition with neat sketch. — 10

4. (a) Explain the two types of cooling systems and compare them. — 10

(b) Explain briefly the essential properties of an ideal lubricants and explain wet sump lubrication briefly. — 10

5. The following observations were recorded during a trial of a 4-stroke, single cylinder oil engine. Duration of trial is 30 minute. Oil-consumption is 4 litres. Calorific value of oil is 43 MJ/kg, specific-gravity of fuel is 0.8, average area of the indicator diagram = 8.5 cm^2, length of indicator diagram is 8.5 cm, spring-constant = 5.5 bar/cm, brake load = 150 kg, spring balance reading is 20 kg, effective brake-wheel diameter is 1.5 m, speed 200 rpm, cylinder-diameter is 30 cm, stroke = 45 cm, jacket cooling-water = 10 kg/min temperature rise is 36°C.

Calculate :
(i) Indicated power
(ii) Brake power
(iii) Mechanical efficiency
(iv) Brake specific fuel consumption in kg/kWh
(v) Indicated thermal efficiency. — 20

6. (a) Explain the various alternatives fuels for an IC engine. 10
   (b) What are the major constituents which contribute to air-pollution in SI engines? 5
   (c) State the current pollution scenario. 5

7. (a) With a neat sketch, explain the inlet and exit velocity triangles for various types of blades.
   (b) Explain the following :
   (i) Degree of reaction (ii) Surging
   (iii) Choking (iv) Polytropic efficiency. 20

8. The pressure-ratio of an open-cycle constant pressure gas turbine plant is 6. The temperature range of the plant is 15°C and 800°C.

   Using the following data :

$$c_{pa} = 1 \text{ kJ/kg.K}, \; c_{pg} = 1.075 \text{ kJ/kg.K}$$

and

$$\gamma = 1.4 \text{ for air and gases.}$$

$$\text{C.V. of fuel} = 43{,}000 \text{ kJ/kg},$$

$$\eta_c = 0.85$$

$$\eta_t = 0.90,$$

$$\eta_{combustion} = 0.95$$

   Find :
   (i) Thermal efficiency of the plant.
   (ii) I.P. of the plant, if the circulation of air is 5 kg/s.
   (iii) A : F ratio.
   (iv) Specific fuel-consumption.

   Neglect the losses in the system. 20

❑❑❑

INDEX

❑❑❑

Win Prizes

ATTENTION READERS

We request for your frank assessment regarding various aspects of the book as given below:

INTERNAL COMBUSTION ENGINES AND GAS TURBINES

By :

Dr. Sadhu Singh

Please fill up the given space in neat capital letters.

(i) What topics of your syllabus that are important from you examination point of view are not covered in the book ?

....

....

....

....

(ii) What are the chapters wherein the treatment of the subject matter is not systematic or organised or updated ?

....

....

....

....

(iii) Have you come across misprints/mistakes/factual inaccuracies in the book? Please specify the chapters and the page numbers.

....

....

....

....

(iv) What other books on the same subject have you found/heard better than the present book? Please specify in terms of price and quality.

....

....

....

....

(v) Further suggestions and comments for the improvement of the book:

....

....

....

....

(PS: If need be please attach a separate sheet to write your views.)

Other Details

(i) Who recommended you the book ? (Please tick in the box near the option relevant to you.)

☐ Teachers ☐ Friends ☐ Bookseller

(ii) Name of the recommending teacher, his designation and address.

....

....

....

....

(iii) Name and address of the bookseller you purchased the book from:

....

....

....

....

(iv) Name and address of your institution (Please mention the University or Board, as the case may be)

....

....

....

....

(v) Your name and complete postal address:

....

....

....

....

(vi) Course you are enrolled in:

....

....

....

....

The best assessment will be awarded each month. The award will be in the form of our publications as decided by the Editorial Board.

Please mail the filled up coupon at your earliest to:

S.K. KATARIA & SONS®

Publisher of Engineering Books

4885/109, Prakash Mahal, Dr. Subhash Bhargav Lane,
Opposite Delhi Medical Association, Daryaganj, New Delhi–110002 (INDIA)
Phone: +91-11-43551243, *Mobile:* +91-9871775858, 7678425209
e-mail: katariabook@gmail.com, katariabooks@yahoo.com
Website: www.skkatariaandsons.com